# A Shock-Fitting Primer

CHAPMAN & HALL/CRC APPLIED MATHEMATICS AND NONLINEAR SCIENCE SERIES

Series Editors *Goong Chen and Thomas J. Bridges*

**Published Titles**

*Advanced Differential Quadrature Methods,* Zhi Zong and Yingyan Zhang
*Computing with hp-ADAPTIVE FINITE ELEMENTS, Volume 1, One and Two Dimensional Elliptic and Maxwell Problems,* Leszek Demkowicz
*Computing with hp-ADAPTIVE FINITE ELEMENTS, Volume 2, Frontiers: Three Dimensional Elliptic and Maxwell Problems with Applications,* Leszek Demkowicz, Jason Kurtz, David Pardo, Maciej Paszyński, Waldemar Rachowicz, and Adam Zdunek
*CRC Standard Curves and Surfaces with* Mathematica®*: Second Edition,* David H. von Seggern
*Exact Solutions and Invariant Subspaces of Nonlinear Partial Differential Equations in Mechanics and Physics,* Victor A. Galaktionov and Sergey R. Svirshchevskii
*Geometric Sturmian Theory of Nonlinear Parabolic Equations and Applications,* Victor A. Galaktionov
*Introduction to Fuzzy Systems,* Guanrong Chen and Trung Tat Pham
*Introduction to non-Kerr Law Optical Solitons,* Anjan Biswas and Swapan Konar
*Introduction to Partial Differential Equations with MATLAB®,* Matthew P. Coleman
*Introduction to Quantum Control and Dynamics,* Domenico D'Alessandro
*Mathematical Methods in Physics and Engineering with Mathematica,* Ferdinand F. Cap
*Mathematical Theory of Quantum Computation,* Goong Chen and Zijian Diao
*Mathematics of Quantum Computation and Quantum Technology,* Goong Chen, Louis Kauffman, and Samuel J. Lomonaco
*Mixed Boundary Value Problems,* Dean G. Duffy
*Multi-Resolution Methods for Modeling and Control of Dynamical Systems,* Puneet Singla and John L. Junkins
*Optimal Estimation of Dynamic Systems,* John L. Crassidis and John L. Junkins
*Quantum Computing Devices: Principles, Designs, and Analysis,* Goong Chen, David A. Church, Berthold-Georg Englert, Carsten Henkel, Bernd Rohwedder, Marlan O. Scully, and M. Suhail Zubairy
*A Shock-Fitting Primer,* Manuel D. Salas
*Stochastic Partial Differential Equations,* Pao-Liu Chow

CHAPMAN & HALL/CRC APPLIED MATHEMATICS
AND NONLINEAR SCIENCE SERIES

# A Shock-Fitting Primer

**Manuel D. Salas**
NASA Langley Research Center
Hampton, Virginia, U.S.A.

CRC Press
Taylor & Francis Group
Boca Raton London New York

CRC Press is an imprint of the
Taylor & Francis Group, an **informa** business
A CHAPMAN & HALL BOOK

CRC Press
Taylor & Francis Group
6000 Broken Sound Parkway NW, Suite 300
Boca Raton, FL 33487-2742

First issued in paperback 2017



ISBN 13: 978-1-138-11663-4 (pbk)
ISBN 13: 978-1-4398-0758-3 (hbk)



**Library of Congress Cataloging-in-Publication Data**

Salas, M. D.
A shock-fitting primer / Manuel D. Salas.
p. cm. -- (Chapman & Hall/CRC applied mathematics and nonlinear science series)
Includes bibliographical references and index.
ISBN 978-1-4398-0758-3 (hardcover : alk. paper)
1. Lagrange equations--Numerical solutions. 2. Differential equations, Partial--Numerical solutions. 3. Shock waves--Mathematics. 4. Fluid dynamics--Mathematics. I. Title. II. Series.

QA911.S25 2010
518'.6--dc22 2009040940

**Visit the Taylor & Francis Web site at**
**http://www.taylorandfrancis.com**

**and the CRC Press Web site at**
**http://www.crcpress.com**

*I dedicate this book to my teachers,*
*Gino Moretti and Leonard Peikoff,*
*for sharing their love of physics and philosophy with me;*
*and to the memory of my friend,*
*David Gottlieb.*

# Contents

**Supplementary Resources Disclaimer**

Additional resources were previously made available for this title on CD. However, as CD has become a less accessible format, all resources have been moved to a more convenient online download option.

You can find these resources available here: www.routledge.com/9781138116634

Please note: Where this title mentions the associated disc, please use the downloadable resources instead.

# *Preface*

"This is the place I told you to expect.
Here you shall pass among the fallen people,
souls who have lost the good of intellect."
So saying, he put forth his hand to me,
and with a gentle and encouraging smile
he led me through the gate of mystery.

Canto 3, Inferno, *Divine Comedy*

Dante Alighieri (1265–1321)

Half a century ago, Sir James Lighthill lamented about the volume of publications: "I think that a lot of us are publishing too much. We turn things out as soon as we have finished them, instead of putting them in a drawer and just thinking about them from time to time. If, when you have done one of these brilliant, involved bits of mathematics, you would just put it in a drawer, there would be so much the less for people to have to read!"[127]. The culture of *publish* or *perish** has not changed. Today, we drown in an avalanche of books, journals, technical papers, conference proceedings, and electronic media. It was not, therefore, without trepidation that I ventured to write this book. I kept asking myself: Do we need one more textbook on computational fluid dynamics?

This book has it origins in a letter from Gino Moretti to me, dated January 1, 1990. Moretti had started writing a book on one-dimensional problems that he planned as the first volume of a two-volume set. Two-dimensional problems would be covered in the second volume. Moretti started the letter with "I rejoice in the idea that, with your help, this book may finally be completed." We collaborated on the manuscript for a couple of years, but unfortunately other work commitments pulled me away and the manuscript was put in a drawer. In 1993, David Gottlieb invited me to spend a sabbatical in the Department of Applied Mathematics at Brown University, Providence, Rhode Island. I took that opportunity to test some portions of the manuscript on a small group of graduate students and develop, with the help of Angelo Iollo, the notes on generalized functions. However, with my return to NASA and assuming responsibility for the Institute of Computer Applications in Science and Engineering (ICASE) in 1996, the manuscript had to be put aside again. The senseless closure of ICASE in 2002, done with good intentions, but

* The phrase is said to have originated in the early 1950s.

poor judgment, had the unintended consequence of giving me time to rethink the scope of the manuscript and translate from FORTRAN to the MATLAB® environment many of the programs in the accompanying CD. I also convinced myself that I had met Lighthill's recommended incubation period and that the subject matter was worth risking being part of the avalanche.

This book is intended for graduate students and researchers interested in the proper numerical treatment of shock waves. I assume that the reader has a fair knowledge of fluid dynamics. Soon after I started working with Moretti, it became clear to me that to fully understand a computational concept you had to be able to program it. Programming is very unforgiving; it allows little room for error. Thus, the codes in the CD are intended as concrete examples of how the ideas discussed in the book are to be implemented.

Chapter 1 is a recount of the events leading to our understanding of the theory of shock waves and of the early developments related to their computation. As Jouguet has justly observed, the discovery of shock waves was an intellectual achievement only later confirmed by experiments. However, the path to that achievement was not an easy one. The story provides a good lesson on how important it is to have sound mathematical and physical foundations. It also shows how idiosyncrasies about how nature is perceived can impact scientific progress.

Chapter 2 presents the key shock-fitting ideas in the context of a simple scalar equation. All of the important concepts are developed here, and, if time is limited, this chapter and Chapter 4 are the chapters to concentrate on. Chapter 2 also provides an introduction to generalized function theory. Critical to the application of shock-fitting techniques is the knowledge of the shock jump conditions associated with a system of differential equations. Generalized function theory provides a means of obtaining the jump conditions for systems that cannot be expressed in divergence form.

Chapter 3 is a recapitulation of the Euler equations and of the definitions and assumptions used throughout the rest of the book. In this chapter, the theory of generalized functions is applied to the Euler equations to demonstrate that it recovers the well-known results. In addition, the analysis provides a new, more in-depth understanding into the nature of the jump conditions.

Chapter 4 applies the ideas developed in Chapter 2 to the one-dimensional and quasi-one-dimensional Euler equations. The treatment of shocks of different families and their interactions, which was only briefly discussed at the end of Chapter 2, is discussed in detail here. Many examples of flows resulting from the motion of pistons and from area variations, both in space and time, are discussed in this chapter.

In Chapter 5, shock-fitting techniques are applied to two-dimensional flows. Three problems are investigated: (1) the classical blunt body problem, (2) the conical flow over an external corner, and (3) steady supersonic flow. A new, simplified shock-fitting method is introduced in this chapter and results are presented demonstrating that the method performs as well as the standard technique. The study of conical flow over an external corner leads to a review of conical cross-flow singularities, and the results presented in this chapter focus on the numerical simulation of flow bifurcations. Finally, the study of steady supersonic flows looks into the calculation of simple cones, ogives, and elliptic wings at angles of attack.

Chapter 6 looks into the current and future developments in shock-fitting methods within the context of unstructured grid methods. Work in this area has the potential of making shock-fitting a practical tool for the study of very complex flows. Although much progress has already been made, there is still room for improvement. Some new ideas are advanced in this chapter.

MATLAB® is a registered trademark of The MathWorks, Inc. For product information, please contact:

The MathWorks, Inc.
3 Apple Hill Drive
Natick, MA 01760-2098 USA
Tel: 508 647 7000
Fax: 508-647-7001
E-mail: info@mathworks.com
Web: www.mathworks.com

# Chapter 1

# Introduction

> ... read Euler, because in his writings all is clear,
> well calculated, because they teem with beautiful examples,
> and because one must always study the sources.*
>
> Joseph L. Lagrange (in his deathbed, 1813).

## 1.1 Prelude

We begin the account of the events leading to the development of the theory of shock waves (reviewed in Section 1.2) around the time of the creation of the École Polytechnique of Paris, just prior to the end of the eighteenth century. A better understanding of those events requires a review of the changes in mathematical thinking that took place during the preceding half century. Of particular importance is the emergence of the *concept of a function* and the introduction of partial differential equations as models of physical phenomena. The latter was a paradigm shift in which equations replaced axioms for the description of physical phenomena. The key players in these events were the mathematical savants Jean d'Alembert, Leonhard Euler, Luigi de la Grange (Lagrange), and the experimentalist Daniel Bernoulli.

Today, when we talk about a function we think of some relation connecting some objects, namely:

> Let $X$ and $Y$ be two given sets, which may or may not be distinct. The function $f$ is defined on the set $X$ and has values in the set $Y$ if there is assigned to each element $x$ of the set $X$ one and only one element $f(x)$ of the set $Y$.

The modern concept of a function is based on the theory of sets. However, the concept has changed over the years. The concept first evolved, through the d'Alembert–Euler debates over the solution to the wave equation, to an acceptance

* Grattan-Guinness, I. 1985. A Paris curiosity, 1814: Delambre's obituary of Lagrange, and its "supplement," in [76].

of discontinuous functions. This period lasted until around 1840. Between 1840 and the beginning of the twentieth century, the period critical to the development of the theory of shock waves, the concept became more restrictive, and nondifferentiable functions were not acceptable as solutions of differential equations. After the beginning of the twentieth century, the problem received rigorous discourse leading by 1950 to our current way of thinking [131].

In 1747, d'Alembert published a seminal paper on research on the curve formed by a stretched string set into vibration [50], in which he wrote: It is plain by Lemma XI, Section I, Book I of [Newton's] Principia that

$$\frac{1}{c^2}\frac{\partial^2 y}{\partial t^2} = \frac{\partial^2 y}{\partial x^2}, \tag{1.1}$$

and with those few lines the first partial differential equation took shape in the form of the wave equation, and we can herald the beginnings of modern mathematical physics.* In Equation 1.1, $y$ is the displacement of the string from the horizontal, $x$ is the abscissa, $t$ is time, and $c^2$ is the ratio of the string linear mass to the string tension. With a change of variables, d'Alembert rewrites the equation as

$$\frac{\partial^2 y}{\partial t^2} = \frac{\partial^2 y}{\partial s^2}.$$

Applying the boundary conditions that the string is fixed at both ends, $y(0, t) = 0$ and $y(l, t) = 0$, d'Alembert obtains the solution as the sum of a left-running and a right-running wave:

$$y = \Psi(x + t) + \Psi(x - t).$$

d'Alembert asserts that the *generating curves* (initial conditions) for $y(x)$ and $v(x)$ (the velocity $= \partial y/\partial t$) "may not be given at will," and he restricts $y$ and $v$ to be odd functions (functions where only odd powers of $x$ occur) of period $2l$ which must be *given by an equation* (meaning that it must be analytic). Truesdell observes [224, p. 244] that d'Alembert considered it obvious that *mechanical* (meaning discontinuous) functions had no place in mathematical physics. This way of thinking was a result of his interpretation of Leibniz's *law of continuity*†: *only continuous functions occur in the solution of physical problems*. It is this restriction on the initial conditions that sparks the controversy between d'Alembert and Euler.

It is obvious from Euler's book *Introduction to Analysis of the Infinite* (*Introductio in Analysin Infinitorum*) [71], written in 1745 and published in 1748, that Euler had been carefully thinking about the concept of a function. In the opening chapter, Euler provides the following definition for a function:

> A function of a variable quantity is an analytic expression composed in any way whatsoever of the variable quantity and numbers or constant quantities.

---

* The notation of the period has little resemblance to today's notation and is not used here.

† See the footnote "*" on page 10.

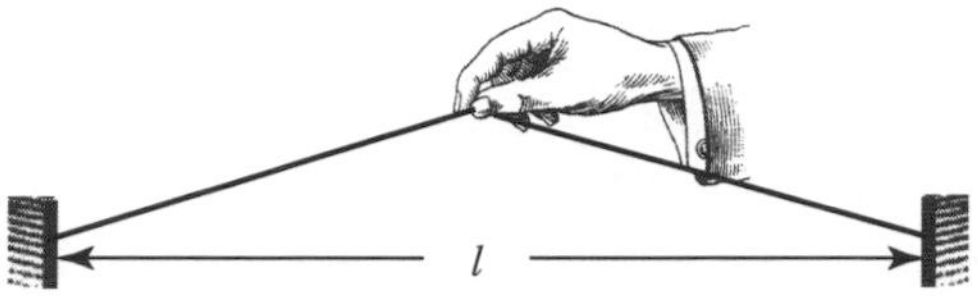


**FIGURE 1.1:** Illustration of the plucked string with an initial slope discontinuity.

Thus, around the time of d'Alembert's paper on the vibrating string, Euler's notion of a function was not unlike d'Alembert's. However, the restriction imposed by d'Alembert on the initial conditions must have made Euler realize that something was amiss. d'Alembert's mathematics did not reflect all the possible physical conditions, for example, it does not allow the simple case illustrated in Figure 1.1 where the string is plucked from its resting position, and this was contrary to Euler's belief that the physics was captured by the mathematics, i.e., by Equation 1.1.* This is a very radical thought. Therefore, with the publication in 1755 of his book *Foundations of Differential Calculus* (*Institutiones Calculi Differentialis*) [72, p. vi] Euler redefines a function, as follows:

> Those quantities that depend on others in this way, namely, those that undergo a change, are called functions of these quantities. This definition applies rather widely and includes all ways in which one quantity can be determined by other.

This is now a much weaker definition, allowing the kind of nonanalytic (discontinuous) functions Euler wanted to consider as initial conditions for the string. To be more specific, Euler's concept of a discontinuous function corresponds to a set of arbitrary piecewise continuous functions with piecewise continuous first and second derivatives [240].

Euler's response to d'Alembert's 1747 paper came quickly by way of a memoir read to the Berlin Academy on May 16, 1748 [67]. In this paper, Euler derives again the wave equation and says: *In order that the initial shape of the string may be adjusted arbitrarily, the solution must be as inclusive as possible.*

d'Alembert's strongest argument, from the perspective of classical solutions, was that the right-hand and left-hand second derivatives of $y$ must be equal at every point in order to satisfy Equation 1.1 [51]. Obviously, this is not the case for the plucked string, Figure 1.1. Euler's counter argument was that "such an error committed in one or several elements is always infinitely small and will not disturb the total result of the calculation. . . ."† The controversy persisted for years until both, d'Alembert and Euler, died in 1783.

* This way of thinking is evident in Euler's conclusion to [69, p. 315]: "Nevertheless all that the theory of fluids includes is contained in the two equations presented above, so that it is not the principles of Mechanics which we need for the pursuit of these researches, but analysis alone, which is not yet sufficiently cultivated for this end."

† Letter to Lagrange dated February 16, 1765, [224, p. 285].

It is not our intention to cover the d'Alembert–Euler debate in detail, for this the reader should consult [112,131,224,240], it is worthwhile, however, to mention some interesting aspects of Lagrange's contributions to the debate. Lagrange's paper of 1759 [117], written when he was an unknown 23 year old, catapulted him into the debate. In this paper, Lagrange claims an "entirely new" derivation of Euler's results by looking at a semidiscrete representation of the string consisting of $n$ equally spaced point masses:

$$\frac{\mathrm{d}y_k}{\mathrm{d}t} = v_k,$$
$$\frac{\mathrm{d}v_k}{\mathrm{d}t} = c^2(y_{k+1} - 2y_k + y_{k-1}).$$

After solving the problem for a finite number of masses, he then obtains the solution for the continuous case by letting the number of masses go to infinity and the spacing between masses go to zero. While deriving Euler's formulas Lagrange almost discovered *Fourier's theorem* for the representation of a periodic function by a series consisting of sine and cosine terms. But more interesting is Lagrange's paper of 1760 [118]. In this paper, Lagrange multiplies both sides of the wave equation (Equation 1.1) by a function $M(x)$, and integrating partially over the interval $[0, l]$ obtains

$$\int_0^l \frac{\partial^2 y}{\partial t^2} M \mathrm{d}x = c^2 \left( \frac{\partial y}{\partial x} M - y \frac{\partial M}{\partial x} \right)_0^l + c^2 \int_0^l y \frac{\partial^2 M}{\partial x^2} \mathrm{d}x.$$

Since $y$ is zero at the end points and since $M$ was defined to be zero at the end points he obtained

$$\int_0^l \frac{\partial^2 y}{\partial t^2} M \mathrm{d}x = c^2 \int_0^l y \frac{\partial^2 M}{\partial x^2} \mathrm{d}x.$$

He then proceeded to solve the above equation. Aside from the fact that the expression still contains differentiation of $y$ with respect to $t$, Lagrange discovered (without realizing it) the idea of using a test function to remove the differential operator from the unknown function by transferring it onto a test function with compact support. By this method, he wrote, "the differentiation of $y$ with respect to $x$ vanishes." The significance of this went unnoticed until 1926 when it was rediscovered by Wiener [237].

Euler's concept of a discontinuous function continued to be championed by Gaspard Monge. As we will see in Section 1.2, Monge was responsible for the creation of the École Polytechnique of Paris. In his memoir of 1771,* he studies the solution of certain first-order equations in terms of geometrical surfaces that satisfy specific curves and concludes that these curves need not be continuous or *expressible analytically*.

* The memoir was read by Monge to the *Académie Royale des Science de Paris* on November 27, 1771. The memoir was not published at the time. The first part of the memoir is lost; the second part of the memoir can be found in [217].

Four years after Euler's death, the Academy of St. Petersburg offered a prize for determining: "Whether the arbitrary functions which are obtained from the integration of equations with three or more variables represent any curves or surfaces whatsoever, either algebraic or transcendental, either mechanical, discontinuous, or produced by an arbitrary movement of the hand."* The competition was won by L.F. Arbogast [5] and represents a vindication of Euler. In the winning entry, Arbogast defines two classes of noncontinuous functions†:

This continuity may be destroyed in two ways:

1. The function may change its form, that is to say, the law by which the function depends on the variable may change all at once. A curve formed by the assemblage of many portions of different curves is of this kind... it may be entirely irregular and not follow any law for any interval however small. Such would be a curve traced at hazard by the free movement of the hand.

2. The law of continuity is again broken when different parts of a curve do not join with each other.

The first is what Euler had in mind; the second is what today we call a $C^0$-discontinuous function.

---

## 1.2 The Curious Events Leading to the Theory of Shock Waves

> The shock wave represents a phenomenon of rare peculiarity such that it has been uncovered by the pen of mathematicians, first by Riemann, then by Hugoniot. The experiments followed not until later.
>
> J.C. Emile Jouguet (1871–1943) [111].

The period between Poisson's 1808 paper on the theory of sound and Hugoniot's fairly complete 1887 exposition of the theory of shock waves is characterized by much insecurity brought about by weak foundations in mathematics and thermodynamics.‡ In the early 1800s few British scientists had read the works of Johann and Daniel Bernoulli, d'Alembert, and Euler [46]. Truesdell summarizes the prevailing current in England thus:

> The mathematics taught in Cambridge in the early nineteenth century was so antiquated that experiment and mathematical theory had turned their backs upon each other. In order to set up a mathematical framework

---

* For original statement of the problem see [112, p. 675].

† English translation in [112].

‡ An earlier version of this section was published in [194].

> general enough to cover the phenomena of tides and waves and resistance and deformation and heat flow and attraction and magnetism, the young British mathematicians had to turn, finally, straight to what had been until then the enemy camp: the French Academy, where the mantle of the Basel school, inherited from Euler by Lagrange, had been passed on to Laplace, Legendre, Fourier, Poisson and Cauchy [226].

However, to be fair the truth is that the mathematical apparatus needed to effectively deal with discontinuous functions* did not exist anywhere and the long established attitude among British scientists to ignore the work of scientists in the Continent was equally reciprocated by their peers in the Continent. In addition there was a generally held belief that nature would not tolerate a discontinuity and this, compounded by a lack of appreciation of the singular nature of the inviscid equations, cast a fog over the problem.

The main events that follow unfold as a tale of two cities, in Cambridge and in Paris. In order to appreciate the events in Paris leading to Hugoniot's paper of 1887, we have to start by considering the contributions of Gaspard Monge (1746–1818). Monge's work is unique in that he approached the solution of partial differential equations by means of geometrical constructions. In 1773, he presented his approach to the French academy [142], in addition to other works on the calculus of variation, infinitesimal geometry, the theory of partial differential equations, and combinatorics. His work on solutions to first-order partial differential equations established the foundations for the method of characteristics that would be later expanded by Earnshaw, Riemann, and Hadamard. But beyond his technical contributions, Monge is also important to this story for his role in the creation in 1794 of what would become one of the most prestigious education centers in the world, the École Polytechnique of Paris. The political turmoil of the French revolution from 1790 to 1793, culminating in the Reign of Terror, resulted in the closing of all institutions of higher learning and left the republic without a much needed supply of civilian and military engineers. The École was created to meet this demand. As Dickens observed about Paris in 1794: "it was the best of times, it was the worst of times, it was the age of wisdom, it was the age of foolishness. . . . " On September 1793 a law was passed for the arrest of all foreigners born in enemy countries. Antoine Lavoisier, the founder of modern chemistry, intervened on behalf of Lagrange saving him from arrest. But on May 1794, with a thud, Antoine Lavoisier's head rolled under the blade of *la machine*. The prosecutor proclaimed that the Republic had no need for savants. In the contradictions of the times, 400 students, the best the country had to offer, were

* Our understanding of the meaning of a function has its roots in the acrimonious debate between d'Alembert and Euler over the solution of a vibrating string [132]. Euler's view employed the notion of "improper" functions which allowed for the representation of discontinuities consistent with the physical observations of D. Bernoulli [133]. The dispute declined with the passing of both protagonists in 1783, see Section 1.1. The issue of the regularity of solution did not resurface until it was forced on mathematicians by Riemann and other physicists dealing with discontinuous waves in the latter part of the nineteenth century. The first steps toward a theory of generalized solutions to hyperbolic partial differential equations were taken only at the beginning of the twentieth century [131]. The theory reached maturity in the works of Sobolev [206], 1936 and Schwartz [199], 1950.

enrolled that first year for a 3 year curriculum in "revolutionary courses" in mathematics and chemistry [90]. Over the years, the École would count among its faculty and students Lagrange, Poisson, Fourier, Laplace, Cauchy, Carnot, Biot, Fresnel, Hugoniot, Navier, Saint-Venant, Sturm, Liouville, Hadamard, and Poincaré, to list but a few. The school's great success, particularly in mathematics, must have been in part due to Lagrange. The greatest mathematician of the times is reputed as the greatest teacher of mathematics.

Siméon-Denis Poisson (1781–1840) entered the École at age 17. There he was trained by Laplace* and Lagrange who quickly recognized his mathematical talents. As a student Poisson had troubles with Monge's descriptive geometry which was a requirement for students going into public service, but because he was interested in a career in pure science he was able to avoid taking the course. Soon after completing his studies, Poisson was appointed repetiteur† at the École. Appointment to full professor was a difficult proposition, but in a lucky break a vacancy was created in 1802 by Napoleon when he sent Fourier to Grenoble in south-east of France to be the prefect of Isére. With the support of Laplace, Poisson took the position in 1806. A year later, Poisson delivered his lecture "On the theory of sound," which appeared the following year (1808) in the École's journal [167]. The opening paragraph begins by giving credit to Lagrange and continues with:

> However, at the time of their publication, very little was known about the use of partial differential equations on which the solution for these types of problems depend. There was disagreement on the use of discontinuous functions which are nevertheless fundamental for representing the status of the air at the origin of the motion: thankfully, these difficulties have been removed with the progress made in the analysis, whilst those which persist relate to the nature of the problem.

Without further reference to discontinuous functions, the paper proceeds to prove several general theorems for the solution of partial differential equations governing the propagation of sound waves. In a section dealing with disturbances of finite amplitude he introduces the governing equation for the velocity potential $\varphi$ and particle velocity $\mathrm{d}\varphi/\mathrm{d}x$:

$$\frac{\mathrm{d}\varphi}{\mathrm{d}t} + a\frac{\mathrm{d}\varphi}{\mathrm{d}x} + \frac{1}{2}\frac{\mathrm{d}^2\varphi}{\mathrm{d}x^2} = 0,$$

where $a$ is the speed of sound which is assumed constant. Poisson finds the exact solution for a traveling wave in one direction in the form,

$$\mathrm{d}\varphi/\mathrm{d}x = f(x - (a + \mathrm{d}\varphi/\mathrm{d}t)t), \tag{1.2}$$

---

* Laplace considered second only to Lagrange as a mathematical savant was not a professor at the École Polytechnique, but an examiner. As an examiner he traveled to cities throughout France administering public, individual, oral exams to aspiring students. Laplace's influence on the school was considerable.

† A repetiteur is a professor's aid who explains the lectures to the students.

where $f$ is an arbitrary function. Poisson's other major contribution to the theory of sound was his derivation in 1823 of the gas law for sound waves with infinitesimal amplitudes, $p \propto \rho^{\gamma}$, later called *Poisson isentrope* [168].*

Twenty years younger than Poisson, George Airy (1801–1892) graduated top of his class at Trinity College, Cambridge in 1823. That year, Airy was awarded the first Smith's prize, given for proficiency in mathematics and natural philosophy. Among Airy's examiners for the Smith's prize were Robert Woodhouse and Thomas Torton, both former Lucasian Chair holders. Only 3 years after graduating from Trinity College, Airy was elected to the Lucasian Chair. The Chair paid Airy very little† and thus in 1828, less than 2 years after being appointed Lucasian Chair, Airy took a much higher paying job by replacing Woodhouse as Plumian Professor of astronomy at Cambridge and director of the Cambridge Observatory. In 1835, Airy moved to Greenwich to become the Astronomer Royal of the Royal Observatory at Greenwich. In his position as Astronomer Royal, Airy publishes a long article on tides and waves [2] in volume 3 of *Encyclopaedia Metropolitana*. In the article he makes the following reference to Poisson and Cauchy:

> ... [He] does not comprehend those special cases which have been treated at so great length by Poisson ... and Cauchy ... With respect to these we may express here an opinion, borrowed from other writers, but in which we join, that as regards their physical results these elaborate treatises are entirely uninteresting; although they rank among the leading works of the present century in regard to the improvement of pure mathematics.

Airy's remark captures the views of his colleagues toward the works of Poisson, Cauchy, and other leading scientists in the Continent. In the *Metropolitana* article Airy studies waves of finite amplitudes in water canals and makes the observation that the crests tend to gain upon the hollows so that the fore slopes become steeper and steeper. The significance of Airy's observation is recognized by James Challis (1803–1882), a colleague of Airy, who is best known for his role in the British failure to discover Neptune [202]. In 1845 Challis was director of the Cambridge Observatory, the same position previously held by Airy. That year, John Adams, a young mathematical prodigy, approached both Challis and Airy with calculations he had made based on irregularities on the orbit of Uranus predicting the position of a new planet. His calculations were ignored until late in June of 1846 when calculations by Joseph Le Verrier, a French mathematician, became known in England. After seeing Le Verrier's prediction, Airy suggested that Challis should conduct a search for the planet, but by this time it was too late. The discovery of Neptune was snatched from the British by the Berlin Observatory on September 23, 1846. Challis'

---

* Although the isentropic relation is credited to Poisson's 1823 paper, Lagrange discussed it in 1760 [77], and Laplace developed it in a short note [120] published in 1816, in the same journal as Poisson's 1823 paper. Laplace wrote: "The real speed of sound equals the product of the speed according to the Newtonian formula by the square root of the ratio of the specific heat[s] ...," i.e., $c^2 = \gamma \partial p / \partial \rho$ at constant entropy.

† The Lucasian Chair paid Airy £99 per year compared to £500 per year as Plumian Professor.

record in fluid dynamics was not impressive either [46]; however, in 1848 Challis published an article in the *Philosophical Magazine* [31] based on Airy's observation about the behavior of waves of finite amplitude in the *Metropolitana*. In this article Challis writes that if we consider a sinusoidal motion,

$$w = m\sin\left[\frac{2\pi}{\lambda}(z-(a+w)t)\right],$$

an apparent contradiction occurs. The contradiction comes about because when the velocity $w$ vanishes at some time $t=t_1$ we have the wave located at $z=at_1+n\lambda/2$ and since at the same time, $t=t_1$, $w$ reaches its maximum value ($\pm m$) when $z=at_1+n\lambda/2+mt_1-\lambda/4$, and since $m$, $t_1$, and $\lambda$ are arbitrary, we may have $mt_1-\lambda/4=0$, in which case the maximum and the points of no velocity occur at the same point, i.e., $w$ becomes a multivalued function.

George Stokes (1819–1903) graduated from Pembroke College, Cambridge, in 1841. Like Airy, he was first of his class, was a recipient of the Smith's prize, and was appointed to the Lucasian Chair. While attending Pembroke College, Stokes had already crossed path with Challis and had quarreled with his views on fluid mechanics on several occasions [47]. Thus it appears that Stokes seized the opportunity to embarrass Challis with a comment on the next issue of the *Philosophical Magazine* entitled "On a difficulty in the theory of sound" [212]. Unlike Challis and many others of his British contemporaries, Stokes had both read and understood the importance of the works of Fourier, Cauchy, and Poisson. Stokes begins his article by writing down Poisson's exact solution to the traveling wave problem, Equation 1.2. He follows with an illustration that shows how a sinusoidal solution satisfying Equation 1.2 would change in time. He then describes the motion as follows: "It is evident that in the neighborhood of the points $a$, $c$ [the compression side] the curve becomes more and more steep as $t$ increases, while in the neighborhood of the points $o$, $b$, $z$ [the expansion side] its inclination becomes more and more gentle." He continues by finding the rate of change of the tangent to the velocity curve,

$$\frac{\mathrm{d}w/\mathrm{d}z}{1+t\,\mathrm{d}w/\mathrm{d}z}, \tag{A}$$

and remarks that: "At those points of the original curve at which the tangent is horizontal, $\mathrm{d}w/\mathrm{d}z=0$, and therefore the tangent will constantly remain horizontal at the corresponding points of the altered curve. For the points for which $\mathrm{d}w/\mathrm{d}z$ is positive, the denominator of the expression (A) increases with $t$, and therefore the inclination of the curve continually decreases. But when $\mathrm{d}w/\mathrm{d}z$ is negative, the denominator of (A) decreases as $t$ increases, so that the curve becomes steeper and steeper. At last, for sufficiently large values of $t$, the denominator of (A) becomes infinite* for some value of $z$. Now the very formation of the differential equations of motion with which we start, tacitly supposes that we have to deal with finite and continuous functions; and therefore in the case under consideration we must not,

* Stokes should have said that the denominator of (A) becomes zero. Equation A represents the new tangent of inclination, when $1+t\,\mathrm{d}w/\mathrm{d}z\to 0$ the tangent $\to\infty$.

without limitation, push our results beyond the least value of $t$ which renders (A) infinite. This value is evidently the reciprocal, taken positively, of the greatest negative value of $dw/dz$; where, as in the whole of this paragraph, denoting the velocity when $t=0$." After finding the breakdown time for Challis' problem ($t=\lambda/2\pi m$), Stokes explains that Challis' paradox occurs because Challis considered times larger than the breakdown time. Stokes continues with:

> Of course, after the instant at which the expression (A) becomes infinite, some motion or other will go on, and we might wish to know what the nature of that motion was. Perhaps the most natural supposition to make for trial is that a surface of discontinuity is formed, in passing across which there is an abrupt change of density and velocity. The existence of such a surface will presently be shown to be possible, on the two suppositions that the pressure is equal in all directions about the same point, and that it varies as the density... even on the supposition of the existence of a surface of discontinuity, it is not possible to satisfy all the conditions of the problem by means of a single function of the form $f\{z-(a+w)t\}$.

Stokes then proceeds to find the jump conditions for mass and momentum:

$$\rho w - \rho' w' = (\rho - \rho')\gamma, \tag{1.3}$$

$$(\rho w - \rho' w')\gamma - (\rho w^2 - \rho' w'^2) = a^2(\rho - \rho'), \tag{1.4}$$

where $\gamma$ is the speed of the discontinuity. We immediately recognize Equation 1.3 as the conservation of mass across a shock wave. However, Equation 1.4 does not look familiar. Conceptually, Equation 1.4 expresses a correct balance of momentum across a shock wave. The reason it does not look familiar is that Stokes has replaced the right-hand side term which corresponds to the pressure difference across the shock with the density difference using the prevalent Newtonian theory of sound: Boyle's law and constant speed of sound (isothermal flow). Stokes then writes: "The equations (1.3), (1.4) being satisfied, it appears that the discontinuous motion is dynamically possible. This result, however, is so strange, that it may be well to consider more in detail the simplest possible case of such a motion." After considering in detail the motion of a shock moving with a constant velocity $\gamma$ and finding no contradictions, Stokes writes:

> The strange results at which I have arrived appear to be fairly deducible from the two hypotheses already mentioned. It does not follow that the discontinuous motion considered can ever take place in nature,* for we have all along been reasoning on an ideal elastic fluid which does not exist in nature.

* Stokes is echoing the popular adage attributed to the Latin philosopher Lucretius Caro: *Natura non facit saltus* (Nature makes no jumps). Leibniz expressed the same thought in his *Nouveaux essays sur l'entendement humain* (1705): "C'est une de mes grandes maximes et des plus vérifiées, que la nature ne fait jamais des sauts (It is one of my great maxims and the most verified, that nature never makes jumps)," and Darwin quotes it seven times in his *Origin of Species* (1859).

He then discusses the effects of heat addition and friction, concluding that:

> It appears, then, almost certain that the internal friction would effectively prevent the formation of a surface of discontinuity, and even render the motion continuous again if it were for an instant discontinuous.

The following year, Stokes was appointed Lucasian Professor of Mathematics at Cambridge. He retained this chair until his death in 1903. Stokes' note on Challis' paradox goes unnoticed for many years.

Lord Rayleigh (1842–1919) entered Cambridge as a student in 1861. There he was a student of the mathematician E.J. Routh.* As an undergraduate, Rayleigh attended the lectures of Stokes and was inspired by Stokes' approach which combined experimental and theoretical methods. Rayleigh's first volume of *The Theory of Sound* was published the same year he wrote the following letter to Stokes. The letter is in response to a conversation Rayleigh had with Stokes concerning Challis' paradox [226]:

*4 Carlton Gardens, S.W.*
*June 2/77*

*Dear Prof. Stokes,*
*In consequence of our conversation the other evening I have been looking at your paper "On a difficulty in the theory of sound", Phil. Mag. Nov. 1848. The latter half of the paper appears to me be liable to an objection, as to which (if you have time to look at the matter) I should be glad to hear your opinion.*

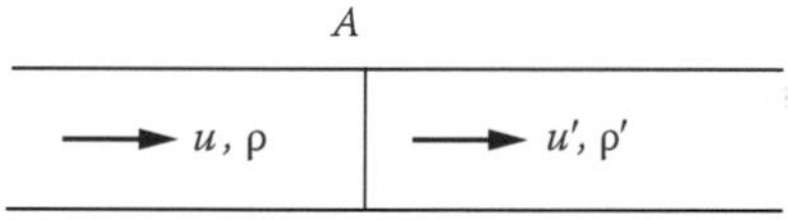


*By impressing a suitable velocity on all the fluid the surface of separation at A may be reduced to rest. When this is done, let the velocities and densities on the two sides be* $u, \rho, u', \rho'$. *Then by continuity*

$$u\rho = u'\rho'.$$

*The momentum leaving a slice including A in unit time* $= \rho u \cdot u'$, *momentum entering* $= \rho u^2$.

*Thus*† $p - p' = a^2\,(\rho - \rho') = \rho u(u' - u)$.

*From these two equations*

$$u = a\sqrt{\rho'/\rho}, \quad u' = a\sqrt{\rho/\rho'}.$$

---

* Routh is best known for the Routh–Hurwitz theorem which can be used to establish if a polynomial is stable. Here stable is in the sense that the roots lie to the left of the imaginary axis.

† The second equal sign appears as a minus sign in [226].

*This, I think, is your argument, and you infer that the motion is possible. But the energy condition imposes on u and u′ a different relation, viz.*

$$u'^2 - u^2 = 2a^2 \log\frac{\rho}{\rho'},$$

*so that energy is lost or gained at the surface of separation A.*

*It would appear therefore that on the hypotheses made, no discontinuous change is possible.*

*I have put the matter very shortly, but I dare say what I have said will be intelligible to you.*

In order to follow Rayleigh's argument, consider the energy balance across the shock:

$$\rho'\left(\tfrac{1}{2}u'^2 + e'\right)\tilde{u}' - \rho\left(\tfrac{1}{2}u^2 + e\right)\tilde{u} = pu - p'u',$$

where

$\tilde{u}$ is the velocity relative to the shock wave

$e$ is the internal energy

A change in internal energy is given by

$$\mathrm{d}e = T\mathrm{d}S + \frac{p}{\rho}\frac{\mathrm{d}\rho}{\rho},$$

where

$T$ is the temperature

$S$ is the entropy

If, as assumed by Rayleigh, the shock is stationary and the flow is reversible and it obeys Boyle's law, then*

$$u'^2 - u^2 = 2(e - e') = 2\int_{\rho'}^{\rho} a^2\frac{\mathrm{d}\rho}{\rho} = 2a^2 \log\frac{\rho}{\rho'}.$$

Stokes replies to Rayleigh [226]:

*Cambridge,*
*5th June, 1877.*

*Dear Lord Rayleigh,*

*Thank you for pointing out the objections to the queer kind of motion I contemplated in the paper you refer to Sir W. Thomson*† *pointed the same out to me many years*

---

* $pu - p'u' = 0$, since Rayleigh assumes that $p = a^2\rho$ and $p' = a^2\rho'$.

† W. Thomson (1824–1907) was named Lord Kelvin in 1866.

*ago, and I should have mentioned it if I had had occasion to write anything bearing on the subject, or if, without that, my paper had attracted attention. It seemed, however, hardly worthwhile to write a criticism on a passage in a paper which was buried among other scientific antiquities.*

*P.S. You will observe I wrote somewhat doubtfully about the possibility of the queer motion.*

It is apparent that Stokes doesn't have the determination or confidence in his position to defend the convincing case he had presented in 1848.

Years later, in a letter to W. Thomson, dated October 15, 1880 [238], Stokes tells Thomson that he is reviewing his paper "On a difficulty in the theory of sound" for inclusion in his collected works. Stokes reminds Thomson that both he and Rayleigh had pointed out years earlier that his analysis violated the principle of conservation of energy. Then Stokes argues: "The conservation of energy gived [sic] another relation, which can be satisfied, so that it appears that such a motion is possible." Two weeks later, on November 1 [238], Stokes changes his mind and writes to Thomson: "On futher [sic] reflection I see that I was wrong, and that a surface of discontinuity in crossing which the density of the gas changes abruptly is impossible." In this letter, Stokes gives no details for this conclusion. The details come 2 days later [238], he writes:

> I mentioned to you I think in a letter that I had found that my surface of discontinuity was bosh. In fact, the equation of energy applied to a slice infinitely near the surface of discontinuity leads to
>
> $$\int_{\rho'}^{\rho} \frac{p}{\rho^2}\,d\rho = \frac{1}{2}\left(w'^2 - w^2\right),$$
>
> where
>
> $\rho, \rho'$ are the densities
>
> $w, w'$ the velocities on the two sides of the surface of discontinuity
>
> But this equation is absurd, as violating the second law of motion. In this way the existence of a surface of discontinuity is proved to be impossible.

On December 4, Stokes again writes to Thomson [238]: "I have cut out the part of the paper which related to the formation of a surface of discontinuity... The equation I sent you was wrong, as I omitted the considerations of the work of the pressures at the two ends of the elementary portion...." And thus, in his collected works of 1883 [214], Stokes adds the following footnote to his 1848 paper at the point where he said that such a surface was possible, "Not so: see substituted paragraph at the end," and removes the entire section dealing with the description of the discontinuity. After claiming that he had made a mistake by considering only conservation of mass and momentum, he says: "It was however pointed out to me by Sir William Thomson, and afterward by Lord Rayleigh, that the discontinuous motion supposed above involves a violation of the principle of the conservation of energy."*

The difficulties that these prominent scientists were having originated from, as Truesdell has succinctly put it, "the insufficiency of thermodynamics as it was then (and often still now is) understood" [226]. Through much of the first half of the nineteenth century, particularly in Britain, the Newtonian theory of sound, based on Boyle's law, $p \propto \rho$, constant speed of sound, and isothermal conditions was accepted, even though it clearly contradicted experimental observations [77]. For a reenactment of the painful birth of thermodynamics in the nineteenth century read Truesdell's play in five acts [225].

The first step in improving thermodynamics was taken by William Rankine (1820–1872). Rankine attended Edinburgh University for 2 years and left without a degree to practice engineering. Around 1848, Rankine started developing theories on the behavior of matter, particularly a theory of heat. In 1851, before taking the chair of civil engineering and mechanics at the University of Glasgow, Rankine writes: "Now the velocity with which a disturbance of density is propagated is proportional to the square root, not of the total pressure divided by the total density, but of the variation of pressure divided by the variation of density ... " [172].†

Samuel Earnshaw (1805–1888) studied at St. John's College, Cambridge and later became a cleric and tutor of mathematics and physics. In 1860 Earnshaw submitted for publication to the *Philosophical Transactions of the Royal Society* a paper on the theory of sound of finite amplitudes [62]. Stokes, in his capacity of secretary of the Royal Society, asked Thomson, in a letter dated April 28, 1859 [238], to review the paper. Thomson's review stretches through seven letters to Stokes, from May 11, 1859 to June 20, 1860 [238]. Thomson is decidedly against publication. In the paper, Earnshaw develops a simple wave solution in one dimension for gases satisfying an arbitrary relation between pressure and density. Earnshaw works with the Lagrangean formulation of the equations to find the relation between the velocity and the density for $p \propto \rho^{\gamma}$. Like others before him, he observes that the differential equations might not have a unique solution. He remarks:

> I have defined a bore to be a tendency to discontinuity of pressure; and it has been shown that as a wave progresses such a tendency necessarily arises. As, however, discontinuity of pressure is a physical impossibility, it is certain Nature has a way of avoiding its actual occurrence.

Thomson has trouble with Earnshaw's Lagrangean formulation and feels that his "aerial bore" is a rehash of Stokes' paper on "A difficulty in the theory of sound." In his last letter to Stokes on this subject Thomson writes [238]:

> On the whole I think if called on to vote, it would be against the publication ... On speaking to Rankine I found *the* idea he had taken from

* What is so odd here is that both Thomson and Stokes were familiar with Rankine's paper of 1870, yet failed to understand its significance for Stokes' 1848 paper.

† One step forward, two steps back, see the footnote "†" on page 8.

> Earnshaw's paper, ... was superposition of transm$^n$ vel. On wind vel.: & he thought it good. This however is of course fully expressed in Poisson's solution.

Bernhard Riemann (1826–1866) received his PhD from the University of Göttingen in 1851. After Dirichlet died in 1857, vacating the chair previously held by Gauss, Riemann became a full professor. Most of Riemann's papers were in pure mathematics and differential geometry, and they have been extremely important to theoretical physics. Unique among his contributions is his more applied paper on the propagation of sound waves of finite amplitude published in 1860 [176]. The paper is very easy to read with notation very similar to that used today. Early in the paper, Riemann introduces what we know today as Riemann variables which he denotes as $r$ and $s$. For an isentropic gas he writes the governing equations as

$$\sqrt{\varphi'(\rho)} + u = \frac{k+1}{2}r + \frac{k-3}{2}s,$$
$$\sqrt{\varphi'(\rho)} - u = \frac{k-3}{2}r + \frac{k+1}{2}s,$$

where

$k$ is the ratio of specific heats
$\varphi' = \mathrm{d}p/\mathrm{d}\rho = a^2$

Shortly after introducing $r$ and $s$ Riemann describes how a compression wave would necessarily steepen leading to multiple vales of $\rho$ at one point. Then he says:

> Now since in reality this cannot occur, then a circumstance would have to occur where this law will be invalid..., and from this moment on a discontinuity occurs...so that a larger value of $\rho$ will directly follow a smaller one....The compression wave [Verdichtungswellen], that is, the portions of the wave where the density decreases in the direction of propagation, will accordingly become increasingly more narrow as it progresses, and finally goes over into compression shocks [Verdichtungsstösse].

He derives the jumps in mass and momentum for an isentropic (reversible) flow and establishes that the speed of the shock wave, $\mathrm{d}\xi/\mathrm{d}t$, is bounded by

$$u_1 + \sqrt{\varphi'(\rho_1)} > \frac{\mathrm{d}\xi}{\mathrm{d}t} > u_2 + \sqrt{\varphi'(\rho_2)}.$$

He then discusses the Riemann problem, i.e., the wave patterns corresponding to various initial conditions with jumps in $u$ and $\rho$ at $x=0$. Riemann, like Stokes before him, failed to understand the true nature of the shock layer. The problem is one of physics, not mathematics, and its solution must wait for a better understanding of thermodynamics.

Rankine, Figure 1.2, makes his main contribution in his 1870 paper on the thermodynamic theory of waves [173] published in the *Philosophical Transactions*

**FIGURE 1.2:** William Rankine. (Photo courtesy of Glasgow University Archive Services, University of Glasgow, Glasgow, U.K.)

*of the Royal Society of London*. Previous papers by Earnshaw and Riemann shared some similarities; not so with Rankine's paper which was more focused on thermodynamics.* He begins with:

> The object of the present investigation is to determine the relations which must exist between the laws of the elasticity of any substance, whether gaseous, liquid, or solid, and those of the wave-like propagation of a finite longitudinal disturbance in that substance.

Here "elasticity" is the eighteenth-century term for what we now call pressure. Later he writes:

> It is to be observed, in the first place, that no substance yet known fulfills the condition expressed by the equation $\mathrm{d}p/\mathrm{d}S = -m^2 = \text{constant}$,† between finite limits of disturbance, at a constant temperature, nor in a

* In papers written in 1850 and 1851, Rankine developed a theory of thermodynamics which included an entropy function [225], but it will take another 15 years for R. Clausius to coin the term and fully develop the concept.

† Rankine denotes by $S$ the "bulkiness" $= 1/\rho$, and by $m$ the "mass velocity" $= \rho\tilde{u}$, where $\tilde{u}$ is the velocity relative to the shock wave. In its discrete form, we call this expression Prandtl's relation: $[[p]]/[[S]] = -m^2$.

> permanency of type may be possible in a wave of longitudinal disturbance, there must be both change of temperature and conduction of heat during the disturbance.

Therefore, Rankine by explaining that the shock transition is an adiabatic process, where the particles exchange heat with each other, but no heat is received from the outside, resolved the objections that had been raised by Rayleigh and others concerning the conservation of energy. He goes on to find, for a perfect gas, the jump conditions for a shock wave moving with speed $a$ into an undisturbed medium with pressure and specific volume defined respectively by $P$ and $S$. He writes:

$$m^2 = \frac{1}{S}\left\{\frac{1}{2}(\gamma+1)p + \frac{1}{2}(\gamma-1)P\right\},$$

$$a^2 = S\left\{\frac{1}{2}(\gamma+1)p + \frac{1}{2}(\gamma-1)P\right\},$$

$$u = (p-P)\sqrt{\frac{S}{\frac{1}{2}(\gamma+1)p + \frac{1}{2}(\gamma-1)P}}.$$

Thomson in a letter to Stokes dated March 7, 1870 [238] writes:

> I have read Rankine's paper with great interest. The simple elementary method by which he investigates the condition for sustained uniformity of type is in my opinion very valuable. It ought as soon as it is published to be introduced into every elementary book henceforth written on the subject.

Pierre-Henri Hugoniot (1851–1887) entered the École Polytechnique of Paris in 1870. That summer, France declared war on Germany* and patriotic feelings ran high among the students, see Figure 1.3. In 1872 Hugoniot entered the marine artillery service. The artillery service turned Hugoniot's attention to research on the flight characteristics of projectiles. In 1879 he was appointed professor of mechanics and ballistics at the Lorient Artillery School, and 3 years later he became assistant director of the Central Laboratory of Marine Artillery. He returned to the École as an auxiliary assistant in mechanics in 1884. Here, in the course of a few months, he completed his memoir entitled "On the propagation of motion in bodies" which he submitted for publication on October of 1885. The publication is delayed because as the editor explains: "... The author, carried off before his time, was unable to make the necessary changes and additions to his original text...." The memoir appears in two parts. The first is published in 1887 [102]; it consists of three chapters. Chapter 1 begins with an exposition of the theory of characteristic curves for partial differential equations of which Hugoniot says: "The theories set out herein

* The Franco-Prussian war lasted from July 19, 1870 to May 10, 1871.

**FIGURE 1.3:** Hugoniot with classmates from the École Polytechnique, 1870. Hugoniot is second from left, front row, see insert. (Photo courtesy of Professor Jean-Francois Gouyet of the École Polytechnique, Paris, France, and the Collections Archives de École Polytechnique, Paris, France.)

are not entirely new; however, they are currently being expounded in the works of Monge and Ampère and have not, to my knowledge, been brought together to form a body of policy." In Chapter 2 he sets down the equations of motion for a perfect gas, and in Chapter 3 he discusses the motion in gases in the absence of discontinuities. It is the second memoir, published in 1889 [103], that is most interesting. Chapter 4 covers the motion of a nonconducting fluid in the absence of external forces, friction, and viscosity.

Hugoniot analysis is very similar to that of Earnshaw. Finally, Chapter 5 "examine[s] the phenomena which occur when discontinuities are introduced into the motion." It is in this last chapter that Hugoniot writes the famous *Hugoniot equation* relating the internal energy to the kinetic energy. It appear in Section 150, not in the usually quoted form,

$$e' - e = \tfrac{1}{2}(p + p')(v' - v), \tag{1.5}$$

but as

$$\frac{p + p_1}{2} = \frac{p_1 - p}{m - 1}\,\frac{1}{z_1 - z} + \frac{p_1 z_1 - p z}{m - 1}\,\frac{1}{z_1 - z}, \tag{1.6}$$

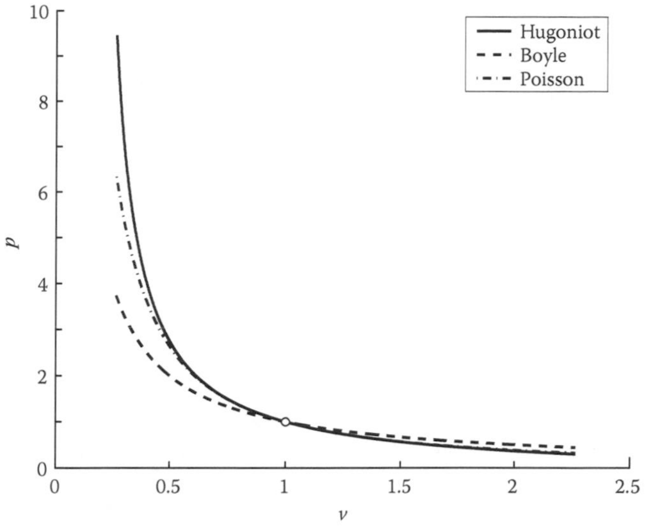


**FIGURE 1.4:** The three $p$–$v$ relations.

where $m = \gamma$, the ratio of specific heats, $v = z + 1$, and $e = pv/(\gamma - 1)$ is the internal energy. Of course, Equation 1.5 follows easily from Equation 1.6, but the elegant way in which Equation 1.5 connects the three thermodynamic variables, $p$, $v$, and $e$ is lost in Equation 1.6. Equation 1.5 states that the increase in internal energy across a shock is due to the work done by the mean pressure in compressing the flow by an amount $v' - v$. Figure 1.4 shows the three $p$–$v$ relations used by various authors for $\gamma = 1.4$. Boyle's law and Poisson's isentrope are constitutive relations, while the Hugoniot curve establishes what states are possible across a shock wave. Later, in Section 155, Hugoniot explains that in the absence of viscosity and heat conduction, the conservation of energy implies that $p/\rho^m = \text{constant}$, but that across a shock this relation is no longer valid and is replaced by

$$p_1 = p\frac{(m+1)\rho_1/\rho - (m-1)}{(m+1) - (m-1)\rho_1/\rho}.$$

It is Hadamard's *Lectures on the Propagation of Waves* [93] that brings Hugoniot's work to the attention of the Cambridge community. In the preface to his *Lectures*, Hadamard explains that Chapters I through IV were prepared during the years 1898 to 1899, but the publication was delayed. He also acknowledges his friend Pierre M. Duhem* for pointing out the theory of Hugoniot:

* When Jacques Hadamard (1865–1963) entered the École Normale Supérieure in 1885, Pierre M. Duhem (1861–1916) was a third year student there, and the two became close friends. Duhem is well known for his work on thermodynamics and history and philosophy of science.

> Dans le cas des gaz, on est, au contraire, conduit à la théorie d'Hugoniot, sur laquelle l'attention a été attiré depuis quelques années, grace aux *leçons d'Hydrodynamique, Elasticitè et Acoustique* de M. Duhem.*

However, Duhem's lectures [61] deal primarily with Hugoniot's treatment of waves of small amplitude (see Figure 1.5), not discontinuities. Hadamard writes the *Hugoniot equation* in Chapter IV, Section 209, of his *Lectures*:

$$\frac{(p_1 + p_2)(\omega_2 - \omega_1)}{2} = \frac{1}{m-1}(p_1\omega_1 - p_2\omega_2),$$

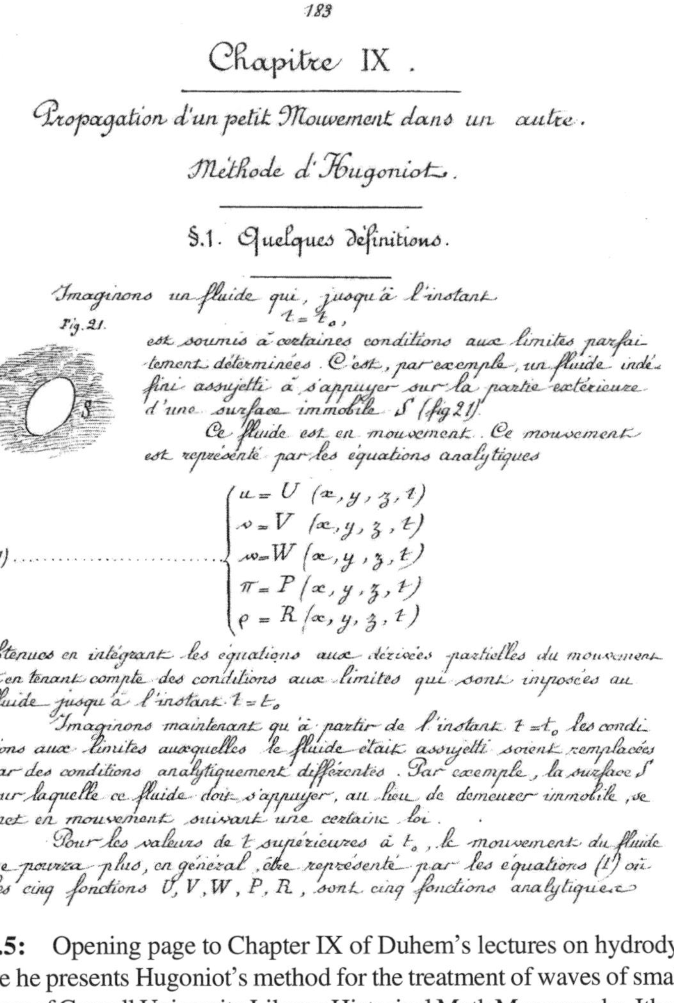

183

Chapitre IX.

Propagation d'un petit Mouvement dans un autre.

Méthode d'Hugoniot.

§.1. Quelques définitions.

Imaginons un fluide qui, jusqu'à l'instant $t = t_0$,

Fig. 21.

est soumis à certaines conditions aux limites parfaitement déterminées. C'est, par exemple, un fluide indéfini assujetti à s'appuyer sur la partie extérieure d'une surface immobile $S$ (fig 21).

Ce fluide est en mouvement. Ce mouvement est représenté par les équations analytiques

$$(1)\quad \begin{cases} u = U(x, y, z, t) \\ v = V(x, y, z, t) \\ w = W(x, y, z, t) \\ \pi = P(x, y, z, t) \\ \rho = R(x, y, z, t) \end{cases}$$

obtenues en intégrant les équations aux dérivées partielles du mouvement et en tenant compte des conditions aux limites qui sont imposées au fluide jusqu'à l'instant $t = t_0$

Imaginons maintenant qu'à partir de l'instant $t = t_0$ les conditions aux limites auxquelles le fluide était assujetti soient remplacées par des conditions analytiquement différentes. Par exemple, la surface $S$ sur laquelle ce fluide doit s'appuyer, au lieu de demeurer immobile, se met en mouvement suivant une certaine loi.

Pour les valeurs de $t$ supérieures à $t_0$, le mouvement du fluide ne pourra plus, en général, être représenté par les équations (1) où les cinq fonctions $U, V, W, P, R$, sont cinq fonctions analytiques

**FIGURE 1.5:** Opening page to Chapter IX of Duhem's lectures on hydrodynamics. In this lecture he presents Hugoniot's method for the treatment of waves of small amplitude. (Courtesy of Cornell University Library Historical Math Monographs, Ithaca, NY.)

* "In the case of the gases, one is, on the contrary, led to the theory of Hugoniot, toward which the attention has turned for a few years, thanks to the lessons of Hydrodynamics, Elasticitè, and Accoustics of Mr. Duhem."

here $\omega = v$, and he attributes it to Hugoniot:

> Telle est la relation qu'Hugoniot a substituée à (66) pour exprimer que la condensation ou dilatation brusque se fait sans absorption ni dégagement de chaleur. On lui donne actuellement le nom de *loi adiabatique dynamique*, la relation (66), qui convient aux changements lents, étant désignée sous le nom de *loi adiabatique statique*.*

The *adiabatique statique*, equation (66), that Hadamard mentions is, of course, Poisson's isentrope.

By 1910, all the principal players, Stokes, Earnshaw, Riemann, Rankine, and Hugoniot, had passed away. Thus the review article by Rayleigh, "Aerial plane waves of finite amplitude" [174], is intended for a new generation of scientists. Rayleigh divides the review into two main parts: "Waves of finite amplitude without dissipation" and "Permanent regime under the influence of dissipative forces." The first part, aside from a review of the work of Earnshaw and Riemann, is a rehash of his letter to Stokes of 1877. Once again he states:

> . . . , I fail to understand how a loss of energy can be admitted in a motion which is supposed to be subject to the isothermal or adiabatic laws, in which no dissipative action is contemplated.

In the second part of the paper, he reviews Rankine's 1870 paper calling it "very remarkable . . . although there are one or two serious deficiencies, not to say errors. . . . " He also reviews Hugoniot's 1887–1889 memoirs, thus:

> The most original part of Hugoniot's work has been supposed to be his treatment of discontinuous waves involving a sudden change of pressure, with respect to which he formulated a law often called after his name by French writers. But a little examination reveals that this law is *precisely the same* as that given 15 years earlier by Rankine, a fact which is the more surprising in as much as the two authors start from quite different points of view.

Of the Hugoniot curve he says:

> . . . however valid [it] may be, its fulfillment does not secure that the wave so defined is possible. As a matter of fact, a whole class of such waves is certainly impossible, and I would maintain, further, that a wave of the kind is never possible under the conditions, laid down by Hugoniot, of no viscosity or heat-conduction.

* "Such is the relation that Hugoniot substituted for (66) to express that condensation or abrupt dilation is made without absorption or release of heat. One currently gives it the name of adiabatic dynamic law, the relation (66), which is appropriate for the slow changes, being designee under the name of static adiabatic law."

Rayleigh makes two small contributions in the article. He shows that the increase in $\int dQ/\theta$ (the entropy) across the shock, for weak shocks, is of the order of the third power of the pressure jump, and he estimates that the shock wave thickness for air under ordinary conditions is of the order of $\frac{1}{3} \times 10^{-5}$ cm.

We can conclude that the early treatment of shock waves was hampered by three factors: first, a lack of understanding of what is an admissible solution to a partial differential equation; second, the incomplete knowledge of thermodynamics at the times; and third, as is evident in Rayleigh's paper, the lack of understanding that the shock wave manifested itself in the inviscid equations as a singular limit* of the viscous, heat conducting, Navier–Stokes equations. As a postscript, consider how Lamb perpetuated the folly.

Horace Lamb (1849–1934) was a student of Stokes and Maxwell at Cambridge. Lamb was a prolific writer who authored many books in fluid mechanics, mathematics, and classical physics. His texts were used in British universities for many years. In his Presidential address to the British Association in 1904, he provided the following insight into his writings:

> It is . . . essential that from time to time someone should come forward to sort out and arrange the accumulated material, rejecting what has proved unimportant, and welding the rest into a connected system.

His acclaimed† book *Hydrodynamics* [119], based on his brief 250 page "Treatise on the mathematical theory of the motion of fluids" of 1879, was first published in 1895 and was then revised and expanded until the current 700-page 6th edition of 1932. In it Lamb discusses the conditions for a discontinuous wave in Sect. 284. Lamb mentions the works of Rankine [173] and Hugoniot, as described by Hadamard [93], but he sides with the Stokes–Rayleigh way of thinking:

> These results are [the jump conditions for mass and momentum], however, open to the criticism that in actual fluids the equation of energy cannot be satisfied consistently with (1) [mass conservation] and (2) [momentum conservation].

Of Hugoniot's result he says in a footnote:

> . . . the argument given in the text [referring to Hadamard's book [93]] is inverted. The possibility of a wave of discontinuity being *assumed*, it is

* The conceptual leap needed is to see the inviscid shock jumps as the outer limits of the viscous shock layer as viscosity vanishes. Curiously, Stokes made a significant contribution to asymptotic theory with what we call today Stokes phenomenon [213], see [78] for an overview.

† "The leading treatise on classical hydrodynamics," *Mathematical Gazette*; "Difficult to find a writer on any mathematical topic with equal clearness and lucidity," *Philosophical Magazine*; ". . . it has become the foundation on which nearly all subsequent workers in hydrodynamics have built. The long-continued supremacy of this book in a field where much development has been taking place is very remarkable, and is evidence of the complete mastery which its author retained over his subject throughout his life." From *G. I. Taylor's eulogy, Nature*, 1934.

> pointed out that the equation of energy will be satisfied if we equate expression (10) $\left[\frac{1}{2}(p_1+p_0)(u_0-u_1)\right]$ to the increment of intrinsic energy. On this ground the formula
>
> $$\frac{1}{2}(p_1+p_0)(v_0-v_1)=\frac{1}{\gamma-1}(p_1v_1-p_0v_0)$$
>
> is propounded, as governing the transition from one state to the other... But no physical evidence is adduced in support of the proposed law.

The Cambridge legacy is still nourished by Hawking, current Lucasian Chair holder, who writes:

> It seems to be a good principle that the prediction of a singularity by a physical theory indicates that the theory has broken down, i.e., it no longer provides a correct description of observations [97].

---

## 1.3 Early Attempts at Computing Flows with Shocks

Mathematical physics changed forever with the onset of the Second World War. Prior to the war, mathematicians' and physicists' main business was first to conceive the partial differential equations governing the motions of interest. In fluid mechanics, this was accomplished by men like Newton, Euler, Navier, and Stokes in the eighteenth and nineteenth centuries. Next the physics and thermodynamics of gases had to be understood. The foundation for this was laid in the later part of the nineteenth century. Problems then were solved by finding solutions *in-the-large*, i.e., looking for functions that satisfied the partial differential equations governing some problem or class of problems. Thus, as we have discussed in Section 1.2, Poisson in 1808 expressed the formal solution for the propagation of sound by Equation 1.2. Knowledge of these formal solutions provided a *global* understanding of the *structure* of the solution. The dirty business of finding solutions to specific problems was usually left to engineers. However, around the time of the Second World War, the problems of interest, for example an airplane flying at transonic speeds or the effect of a blast wave on a building, did not yield to the classical approach. The need for solutions to specific problems brought on by the war forced mathematicians, scientists, and engineers to look for numerical solutions obtained *by computers*. Of course, until the mid-to-late 1950s the word "computers" referred to people (usually women) who performed computations, not to machines. The following recollection by Godunov [87] is particularly apropos:

> [The calculations] were performed by a large group of exclusively female operators on Mercedes-brand electronic adding machines using the method of characteristics... The technique of these calculations was very well

> developed. The operators played on the keys of the calculators like pianists while actively discussing household matters and other problems that usually excite female interest.... They often acidly remarked that our pseudo-scientific explanations for the various numerical troubles which frequently hindered the calculations usually turned to be useless. One should keep in mind that they were paid for the volume of calculations, as determined by the number of filled-up lines in the tables. Any calculations with errors were not taken into account.

Before the decade ended, the term "digital computer" gained usage to refer to the "supercomputers" of the day, such as the IBM 650, and distinguish them from their human counterpart.

A remarkable example of the shift to numerical solutions is provided by Emmons' work on transonic flows. From 1944 to 1948, under support from NACA,* Emmons [64,65] produced a series of transonic solutions that were at least 30 years ahead of their time. He solved the Euler equations in terms of a velocity potential and a stream function using the relaxation method of Southwell in the subsonic region and a finite difference method of his own combined with a large doze of computational steering based on personal insight and good judgment in the supersonic region. Emmons describes his *code* for the subsonic region as follow (paraphrased):

1. Draw the airfoil and flow region to a scale such that the distance between net points is about 1.5 in. Do not use too many points at the start.

2. With the boundary conditions in mind, guess values of the stream function at the net points, and compute the residuals. To aid the accuracy of guessing, a freehand sketch of the streamlines and potential lines is somewhat useful. Use whole values of the stream function ranging from say 0 to 1000.

3. The residuals are relaxed, each time recording at each point the change in stream function and the resultant residual. In this way the points at which the residual is largest can be spotted at a glance and relaxed next.

4. After all the residuals have values between ±2, add changes to the stream function to get final values at each point.

5. Recompute the residual to locate any computation error.

6. If the solution is not accurate enough, additional points are added where needed.

The *code* for fitting shocks is described as follows (paraphrased):

1. Solve problem as previously described in the absence of shocks.

2. A shock is arbitrarily placed in some location in the supersonic region.

---

* National Advisory Committee for Aeronautics, created in 1915.

3. With this shock fixed the flow in the region following the shock is determined by the shock boundary conditions of stream function and entropy distribution.

4. On completing this solution it will be found that the streamline directions following the shock do not agree with the assumed shock inclination. Change the shock inclination to get agreement and repeat Step 3.

5. A few repetitions suffice to get an accurate solution.

In the concluding section to [64] Emmons adds (italics mine): "All of the methods described have one enormous advantage over analytical methods of solution of these problems. They permit the *computer* to use all of the facts *he* knows about the phenomena throughout the computations."

Emmons' calculation of a compressible flow in a hyperbolic channel was, to my knowledge, the first transonic calculation with a fitted shock wave, see Figure 1.6. His other landmark calculation of the transonic flow over a NACA 0012 airfoil in free air and inside a tunnel, see Figure 1.7, was not only the first transonic airfoil calculation with a fitted shock wave (see [178] for subsequent studies), but it also discovered a pressure singularity in the subsonic region at the foot of the shock. Emmons correctly explained the observed pressure behavior as follows:

> Since the stream is required to follow the airfoil surface, the curvature of the streamlines adjacent to the surface is specified by that surface. Since

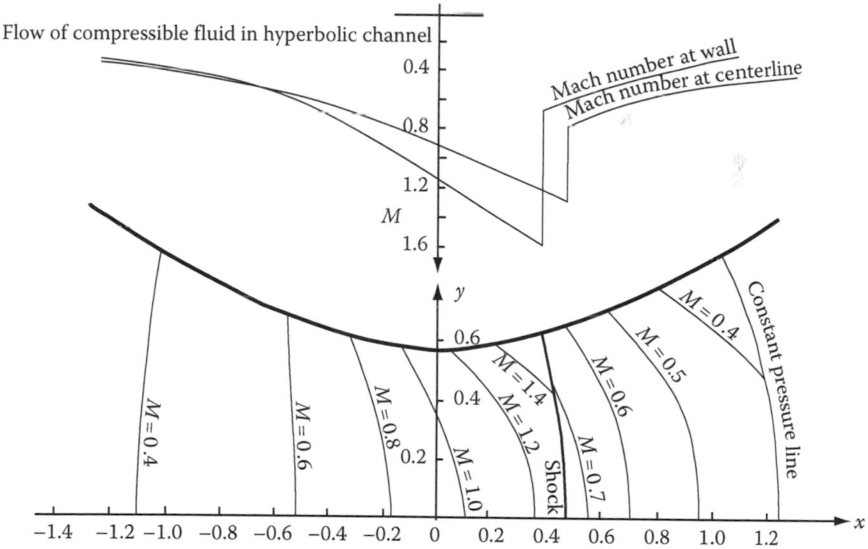


**FIGURE 1.6:** Computed iso-Mach lines for flow through a hyperbolic channel. The calculated flow includes a fitted shock spanning the high of the channel near $x = 0.5$. This is the first known shock-fitting result for a transonic problem. (Adapted from Emmons, H.W., The numerical solution of compressible fluid flow problems, NACA Tech. Note No. 932, 1944, Figure 10.)

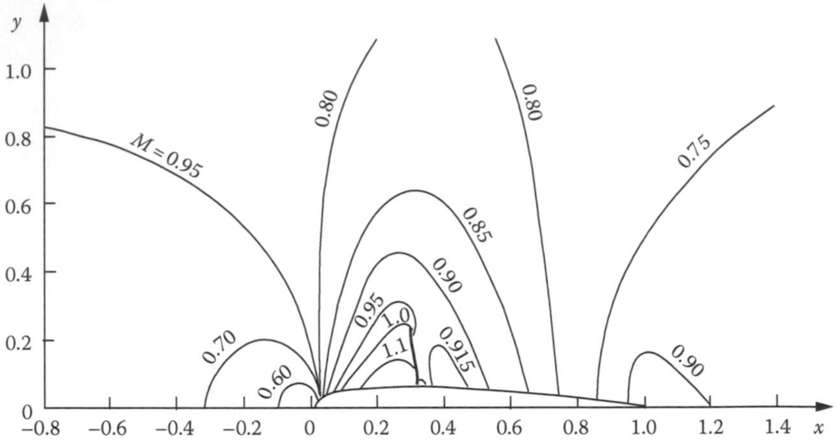


**FIGURE 1.7:** The first transonic airfoil calculation with a fitted shock. NACA airfoil inside a wind tunnel, wind tunnel height is 1.8 chords, $M_\infty = 0.75$ and zero angle of attack. (Adapted from Emmons, H.W., Flow of a compressible fluid past a symmetrical airfoil in a wind tunnel and in free air, NACA Tech. Note No. 1746, 1948, Figure 5.)

> the surface is everywhere convex, there must be a pressure increase normal to the airfoil surface in order to cause the velocity vector to turn as the fluid flows along the surface. Following the shock, however, the shock conditions have produced a pressure variation normal to the surface dependent on the pressure variation just prior to the shock, and in general for Mach numbers near 1 the pressure variation normal to the surface will be reversed by the shock. Hence the fluid must readjust itself very rapidly if it is to follow the airfoil surface. This curvature condition is responsible for the rapid pressure rise [65].

Many years later, Gadd [80] and then Oswatitsch and Zierep [160] worked out the mathematical nature of the singularity showing that, immediately downstream of the shock, a logarithmically infinite acceleration develops along the surface.

The findings of Emmons raise the question: Can we gain insight into the structure of a flow from numerical calculations? The answer is yes, we can gain insight, but not *definite* knowledge. While Emmons was able to identify the singularity in his calculations and correctly explain its cause, full, precise understanding required mathematical analysis. That is why the singularity is known as Zierep's singularity, not Emmon's singularity. The danger of using numerical results to establish structural behavior is illustrated by the calculations of Ziff, Merajver, and Stell [246]. Through extensive numerical calculations, they tried to verify the conjecture that all derivatives with respect to time of the entropy function approach their equilibrium value of zero monotonically. Their numerical results showed the conjecture to be true for the first 30 derivatives. The authors concluded their work with the hope that their results "will stimulate further work that will yield a rigorous proof." Only

1 month after the publication of [246], Lieb [125] proved, in half a page of analysis, that the conjecture was false.

In 1950, von Neumann and Richtmyer [235] introduced the idea of capturing shocks using an artificial dissipative term "to give shocks a thickness comparable to (but preferably somewhat larger than) the spacing of the points of the network."* The motivation for their idea was to simplify the numerical solution of the equations of fluid motion which is "usually severely complicated by the presence of shocks . . . , [because] the motion of the [shock] surfaces is not known in advance but is governed by the differential equations and boundary conditions themselves." The great appeal of this approach is that all points are treated the same way, hence a local analysis of the scheme is globally valid (except at boundaries): in Moretti's words "one code that can describe any flow" [152]. The paper consisted of an analysis of the time-dependent, one-dimensional Euler equations written in terms of the specific volume, the fluid velocity, and the internal energy per unit mass with an additional artificial dissipation term of the form $\sim(\Delta x)^2\ u_x|u_x|$ added to the momentum equation. By doing what we today call a von Neumann stability analysis, they showed that both the modified differential equations and a proposed second-order difference scheme were stable. They asserted that "the method [had] been applied, so far, only to one-dimensional flows, but [appeared] to be equally suited to study more complicated flows." However, no results were presented. At the time of their report the need to express the equations in conservation form in order to conserve mass across the shock and capture the right shock jumps and speed had not been established. Thus, the equations they used were not written in conservation form, and we suspect that the results they had did not show the right shock speeds. In 1964, Burstein [23] presented results for an oblique shock reflection which included the method suggested by von Neumann and Richtmyer, but with the equations written in conservation form. The result for this case is shown in Figure 1.8. It is not particularly impressive.

One of the first problems involving shocks to be solved by the electronic computer was the supersonic blunt body problem. The problem is very difficult to treat, other than numerically, because the shock is an unknown boundary and the flow in the layer between the shock and the body changes type, being subsonic near the nose of the body and supersonic away from the nose, see Figure 1.9. Three methods were popular in the late 1950s to early 1960s. All three treated the shock wave as a boundary of the flow. One was the inverse method of Van Dyke [229], another was the direct integral relations method of Dorodnitsyn [60] implemented by Belotserkovskii [14], and finally the direct time-asymptotic method of Rusanov [186,187] and Moretti [154]. Both the inverse and integral relations methods start from a steady state formulation. In Van Dyke's formulation of the inverse method the problem is reduced to the numerical integration of two differential equations for the density and stream function. These are integrated proceeding downstream from an assumed shock wave shape toward the body, an approach that completely defies the

* See Godunov's *Reminiscences* [87] for parallel developments in Russia.

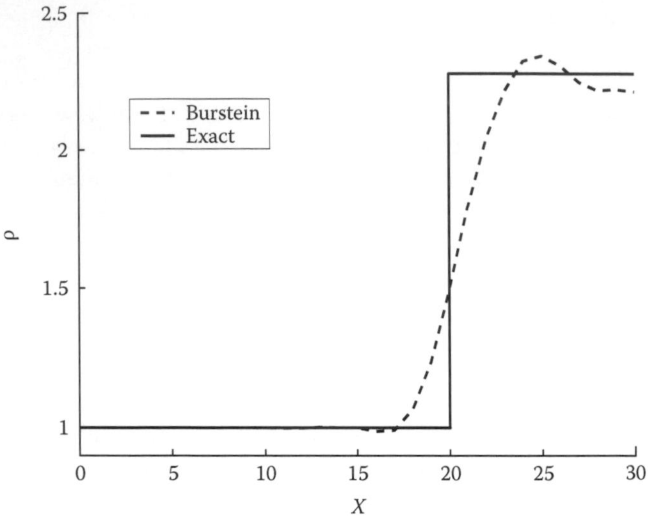


**FIGURE 1.8:** Density for shock reflection calculated by Burstein. (Reproduced from Burstein, S.Z., *AIAA J.*, 2, 2111, 1964, Table 2.)

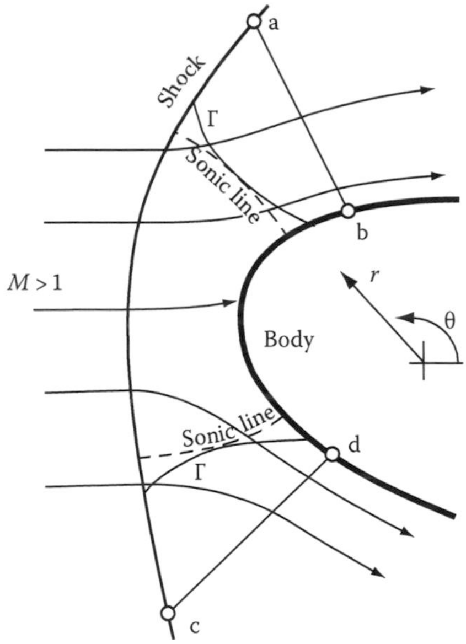


**FIGURE 1.9:** Sketch of blunt body shock layer structure.

elliptic character of the equations in the subsonic region. The equations are integrated until the stream function becomes negative. The body corresponding to the assumed shock shape is found by locating the zeroes of the stream function. The inverse method is ill-posed, since the shock shape is not very sensitive to changes in the body shape, and the approach is frail due to instabilities. Nevertheless, Van Dyke was able to obtain solutions for many body shapes of interest on one of the first general purpose electronic computers, the IBM 650.*

The method of integral relations begins with the steady Euler equations written in divergence form in a suitable $\xi, \eta$ coordinate system:

$$\frac{\partial F_i}{\partial \eta} + \frac{\partial G_i}{\partial \xi} = H_i, \quad i = 1, 2, \ldots, n, \tag{1.7}$$

where

$F_i$ and $G_i$ are the components of the flux vectors
$n$ is the number of equations

To find a solution we divide the shock layer into strips as shown in Figure 1.10. The direction of the strips, whether parallel to $\xi$ or $\eta$, is decided by prior knowledge about the dependence of the flow variables on $\xi$ and $\eta$. Let us say that we use $N$ strips parallel to $\eta=$constant lines. Each strip is then bounded by lines $\eta = \eta_k(\xi)$, with $k=0, 1, \ldots, N$. If we integrate Equation 1.7 along an $\xi=$constant line from the body

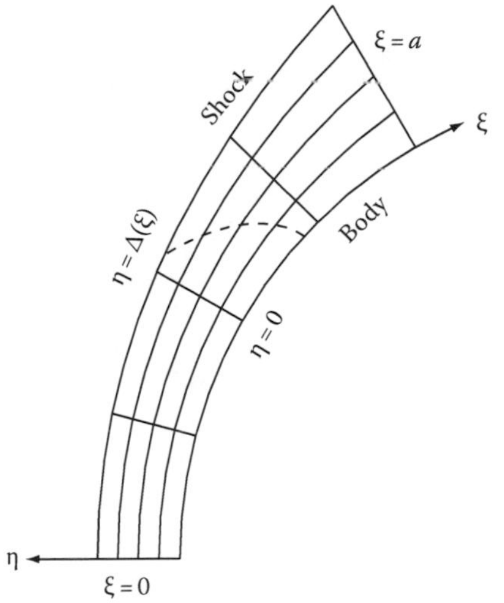


**FIGURE 1.10:** Domain partitioned into strips for method of integral relations; dashed line is sonic line.

* The 650 had only 2000 words of memory. Initially it was programmed in machine language, then in SOAP (Symbolic Optimal Assembly Program). By 1957 a FORTRAN compiler was available which compiled FORTRAN into SOAP.

($\eta = 0$) to the shock ($\eta = \Delta(\xi)$), we end up with a system of $n \times N$ integral relations of the form

$$\frac{d}{d\xi}\int_0^{\eta_k(\xi)} F_i d\eta - \frac{d\eta_k}{d\xi} F_{i,k} + G_{i,k} - G_{i,0} = \int_0^{\eta_k(\xi)} H_i d\eta. \tag{1.8}$$

Each of the components $F_i$, $G_i$, and $H_i$ is approximated by an interpolation function, for example

$$F_i \approx f_{i,0}(\xi) + \sum_{j=1}^{N} f_{i,j}(\xi)\eta^j. \tag{1.9}$$

The coefficients $f_{i,j}$ are linearly dependent on the values of $F_i$ on the strip boundaries. The choice of the interpolation function, a polynomial in $\eta$ in this example, would benefit from prior knowledge of the function behavior. Introducing the interpolation functions (Equation 1.9) into Equation 1.8 we obtain a system of ordinary differential equations which, with the appropriate boundary conditions, is integrated from $\xi = 0$ to $\xi = a$. The unknown shock position, $\Delta(\xi)$, and the shock slope, $d\Delta/d\xi$, are found as part of the solution. The transonic nature of the solution in the shock layer manifests itself in the integration of the ordinary differential equation along the strip as a singularity where the strip crosses the sonic line. To overcome this problem, additional regularizing conditions must be introduced at these points. For more details on the calculation of flow past a sphere and a cylinder using the integral relation method see [99].

The blunt body problem is ideally suited for the time-asymptotic method. With this method, the problem is framed as an initial–boundary value problem which is integrated in time until a steady state is reached. Using this approach the steady state problem which is a mixed elliptic–hyperbolic problem becomes strictly hyperbolic. Once the problem is formulated this way, the definition of initial conditions is perhaps the only challenge. The initial conditions require guessing at the shape of the bow shock and defining a shock layer flow field that is reasonably consistent with the governing equations, and not too far from the steady state solution. If the initial conditions are too far from the steady state, then the path to the steady state might go through an intermediate flow structure different from that depicted in Figure 1.9. Fortunately, the bow shock shape is not very sensitive to the body shape and there is a large knowledge base on what the shock shape and flow shock layer should look like. The boundary conditions are relatively simple. The free stream is uniform and constant and, since the flow is supersonic, no signals propagate upstream. Therefore, only the values of the free stream immediately upstream of the bow shock are needed. These, together with the shock shape and Rankine–Hugoniot jumps, define the inflow boundary at the shock. Downstream of the sonic lines, the region of integration is terminated by outflow lines, see lines $\overline{ab}$ and $\overline{cd}$ in Figure 1.9. The only criteria for selecting these lines are (a) that they are sufficiently downstream for the flow to be fully supersonic, and (b) that they are sufficiently downstream from the limiting characteristic touching the sonic line, lines labeled Γ in Figure 1.9. Since these outflow boundaries lie in the supersonic region, extrapolation from inside the layer is valid. The boundary condition on the blunt body surface is of course the vanishing of the velocity component normal to the surface. In Moretti's approach to this problem, the problem is formulated in polar or spherical coordinates depending on the problem being two dimensional or three dimensional. Considering only the two-dimensional case, the physical plane delimited by the lines $\overline{ab}$, $\overline{bd}$, $\overline{dc}$, and $\overline{ca}$ is transformed to a rectangular computational plane by the coordinate transformation

$$\zeta = \frac{r - r_b(\theta)}{r_s(t, \theta) - r_b(\theta)},$$
$$Y = \pi - \theta,$$
$$T = t,$$

where

$r_b$ is the radial coordinate of the blunt body
$r_s$ is the radial coordinate of the bow shock
$r,\theta$ are the polar coordinates

Grid points on the shock layer are computed using a modified Lax–Wendroff scheme, while body and shock points are computed by a modified method of characteristics. Using this method, Moretti conducted an extensive study of two-dimensional and axisymmetric blunt bodies [145]. Since this problem will be covered in Chapter 5, we will not go into any more details here.

The blunt body problem has been the focus of recent interest because when the bow shock wave is captured with some upwind schemes that correctly treat contact surfaces a numerical instability now known as the *carbuncle phenomenon* is found to occur [165]. The instability originates in the subsonic region where the bow shock is normal to the free stream. The instability leads to very large scale effects in the bow shock shape. There has been some speculation as to the possible physical nature of this instability, but this has been shown not to be the case [38]. The instability does not occur with shock-fitting methods.

# Chapter 2

## *Shock-Fitting Principles*

Look ma, no wiggles!*

Gino Moretti [148].

When air is compressed by a pressure wave its temperature raises, thus increasing the speed of sound of the air left behind. Subsequent waves traveling through this compressed region travel faster than the front wave and begin to catch up with it. As waves overtake each other, the front of the wave becomes steeper, eventually developing very high gradients. Within this narrow region of high gradients the steepening effect of convection reaches a balance with the smearing effect of diffusion and a shock wave is formed. The thickness of the shock wave is typically just a few mean free paths. This is illustrated in Figure 2.1 where the shock wave thickness in multiples of the mean free path, $\lambda$, is plotted as a function of free stream Mach number following a definition of shock wave thickness by Prandtl [170]. When this flow phenomenon is modeled with the Euler equations, the thin shock layer is replaced by a discontinuity. These shock discontinuities are one of the most important features of the Euler equations and present a great challenge to their numerical representation. Over the last 40 years, much work has been devoted to the numerical treatment of these discontinuities with the purpose of developing schemes that represent them accurately. The prevalent approach today is *shock capturing*, which has its roots in the work of von Neumann and Richtmyer [235] as we discussed in Section 1.3. In shock capturing, by means of a local modification to the governing equations, the discontinuity is replaced by a thin *viscous-like* layer which is computed using the same discretization used elsewhere in the flow. The other approach called *shock-fitting* is the subject of this book. In shock-fitting the discontinuities are treated as true discontinuities governed by their own set of partial differential equations. Shock-fitting was used by many investigators from the late 1940s through the 1960s, and unlike shock capturing there is no one original

* The first part of the title of this paper was Moretti's best effort to concoct a pompous, nonsensical title. The second part, *Look Ma, no wiggles!*, captured perfectly the essence of the paper. The journal to which it was submitted for publication demanded to have the second part removed. The paper was never published; nevertheless it is a classic in computational fluid dynamics.

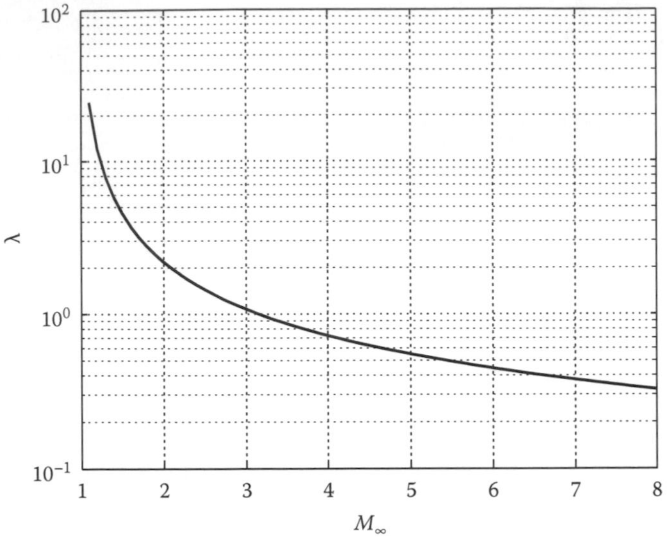


**FIGURE 2.1:** Shock wave thickness, λ, as a function of the free stream Mach number.

manifesto that we can single out for the origin of shock-fitting. In the United States, Gino Moretti is the undisputed virtuoso of shock-fitting and, although occasionally we might present our own variations, it is Moretti's theme that we try to follow.

The purpose of this chapter is to introduce the basic concepts, ideas, and techniques used in shock-fitting. For simplicity, in this introductory chapter we have chosen to work with the inviscid form of Burgers' equation. By working with this equation, we can go more deeply into mathematical details and take advantage of exact solutions. The viscous form of the equation was originally proposed by Burgers [22] in 1948 as a model equation for the study of turbulence. It is given by,

$$u_t + uu_x = \varepsilon u_{xx}, \tag{2.1}$$

where $\varepsilon$ is a small positive parameter. The exact solution to (2.1) for a wave traveling at speed $w$ is given by

$$u(x,t) = 2\varepsilon\Delta \tanh\left(\Delta(x - wt)\right) + w,$$

where

$$w = (u_\infty + u_{-\infty})/2$$
$$\Delta = (u_\infty - u_{-\infty})/4\varepsilon$$

The equation models the effects of convection and dissipation. The equation has proven useful in what is called the *viscosity method* which establishes the existence and stability of solutions of the inviscid equation by considering their behavior in the limit $\varepsilon \rightarrow 0$. A related equation, modeling the behavior of solitons, is the Korteweg and de Vries (KdV) equation [115],

$$u_t + uu_x = -\delta u_{xxx}, \tag{2.2}$$

which models convection and dispersion. The combination of diffusion and dispersion,

$$u_t + uu_x = \varepsilon u_{xx} - \delta u_{xxx},$$

leads to significantly different solutions depending on the relative values of the coefficients $\varepsilon$ and $|\delta|$. An example of the high-frequency waves that can be generated is shown in Figure 2.2 for $\varepsilon = 0.001$ and $\delta = 0.00009$.

The inviscid form of Burgers' equation,

$$u_t + uu_x = 0, \tag{2.3}$$

only considers convection. Burgers' equation (2.1) is more than an abstract model. It is a proper asymptotic limit of the Navier–Stokes equations [244] and its inviscid form is a correct limit representing weakly nonlinear waves [128]. The equation and its properties are well known to most students of numerical analysis and computational fluid dynamics. Like the compressible Euler equations, the inviscid Burgers' equation is a quasi-linear hyperbolic equation, but unlike the Euler equations it is a scalar equation. We will take advantage of its simplicity to work several examples in detail.

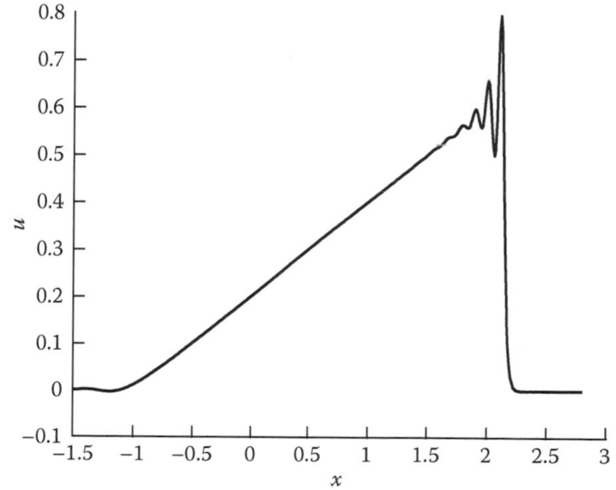


**FIGURE 2.2:** Example of high-frequency waves resulting from diffusion–dispersion equation.

## 2.1 The Inviscid Burgers' Equation

### 2.1.1 Continuum Hypothesis

Liquids and gases are made up of molecules, many molecules. A drop of water the size of the period at the end of this sentence would contain about 10 trillion water molecules at ordinary temperature and pressure. In 1 s a single molecule of water undergoes $10^{14}$ collisions with its neighbors. However, fluid dynamics is not concerned with the behavior of individual molecules, but with the behavior of fluids at large. Therefore, a fundamental assumption of fluid dynamics is the existence of a continuum. Physical quantities such as mass and temperature are treated as uniformly spread over small elemental volumes of the fluid. When we take measurements over short periods of time and small volumes, we effectively measure average quantities over those intervals. A natural representation of this process would involve distribution functions, but our classical analysis is based on point functions. Thus, we take a limit in which our elemental volumes and time intervals are reduced to a point. A continuum is defined as the limit in which the number of molecules in a unit volume tends to infinity, while the mean time and distance between successive collisions for any individual molecule tends to zero when compared to relevant units of time and length. The continuum hypothesis allows us to introduce the notion of material volumes, surfaces, and lines which consist always of the same fluid particles and move with them, and of the fluid velocity vector $\boldsymbol{u}(\boldsymbol{x},t)$ as a function of the position vector $\boldsymbol{x}$ and time $t$. It also allows us to talk of the value of a fluid property at a *point* and in general to represent these properties by continuous functions of $(\boldsymbol{x},t)$.

### 2.1.2 Material Derivative

Another important concept is the acceleration of a material element. Consider an element of fluid at position $\boldsymbol{x}$ at time $t$. After a brief instant of time its position changes to $\boldsymbol{x}+\boldsymbol{u}\delta t$, and the change in velocity over this interval is

$$\boldsymbol{u}(\boldsymbol{x}+\boldsymbol{u}\delta t, t+\delta t)-\boldsymbol{u}(\boldsymbol{x},t)=\delta t\left(\frac{\partial \boldsymbol{u}}{\partial t}+(\boldsymbol{u}\cdot\nabla)\boldsymbol{u}\right)+O(\delta t^2),$$

where $\nabla$ is the gradient operator (vector differential operator). Dividing by $\delta t$ and taking the limit as $\delta t \rightarrow 0$ we find the acceleration of an element of fluid at $(\boldsymbol{x},t)$,

$$\frac{\partial \boldsymbol{u}}{\partial t}+(\boldsymbol{u}\cdot\nabla)\boldsymbol{u}.$$

where

$\partial \boldsymbol{u}/\partial t$ is the local rate of change due to temporal changes

$(\boldsymbol{u}\cdot\nabla)\boldsymbol{u}$ is the convective rate of change due to the transport of the element to a different position

The inviscid Burgers' equation (2.3) says that the acceleration of a fluid element in one space dimension is zero. This will be the case in the absence of both intermolecular interactions and external forces. It is convenient to introduce the notation

$$\frac{D}{Dt} = \frac{\partial}{\partial t} + \boldsymbol{u} \cdot \nabla$$

which we call the material or convective derivative. Thus the inviscid Burgers' equation may be also written as

$$\frac{Du}{Dt} = 0.$$

### 2.1.3 Divergence Form

The inviscid Burgers' equation (2.3) may also be written in the form

$$u_t + f(u)_x = 0, \quad f(u) = \tfrac{1}{2}u^2. \tag{2.4}$$

We call this *conservative* or *divergence* form. Both (2.3) and (2.4) are equivalent from a mathematical point of view, but not necessarily so when expressed discretely. If we integrate (2.4) from some $x = x_a$ to some $x = x_b$ we get

$$\int_{x_a}^{x_b} u_t \mathrm{d}x = f_a - f_b,$$

that is, the integral of the rate of change of $u$ over the interval $[x_a, x_b]$ has to equal the difference of the flux entering at $x_a$ and exiting at $x_b$. If we perform this operation numerically, by subdividing the interval $[x_a, x_b]$ into $m$ cells of size $h = (x_b - x_a)/m$, we want all the fluxes $f_i$, $1 < i < m$, at the cell faces $x_i$, $x_a < x_i < x_b$, to cancel out identically. When this happens, we say that the discretization is conservative, if not we say it is nonconservative. Although, it is possible (and not difficult) to write a conservative discretization starting from (2.3), this is not the case for more general, multidimensional problems and, therefore, from the point of view of conservation of fluxes, (2.4) is preferred.

If $u$ is a continuous function, it makes little difference whether we use (2.3) or (2.4) for the discretization. Certainly, in the limit $h \to 0$ the discretization of (2.3) and (2.4) should be the same. However, if the discretization is done across a shock (shock capturing), the nonconservative discretization can result in incorrect shock jump conditions. This adversely affects the shock velocity and, consequently, the shock location.

### 2.1.4 Characteristics

Characteristics are curves or surfaces along which information is propagated. Characteristics have an intrinsic, absolute, relation to the governing equations that is independent of how the equations are expressed. If we were given data on a characteristic to solve a hyperbolic problem we would not be able to find a solution, because we would not be able to propagate information perpendicular to the characteristic. This property can be used to define a characteristic as a curve or surface such that, if data is given on that curve or surface, the differential equation does not determine the solution at any point away from it.

Consider again Equation 2.3 with initial conditions

$$u(x,0) = u_0(x), \qquad -\infty < x < \infty \tag{2.5}$$

and let us look for characteristic curves defined by $x(\xi)$, $t(\xi)$ in $t \geq 0$. That is, given the initial data $u_0(x)$ along the $x$-axis, we look for curves along which we can propagate the initial data into the half plane $t \geq 0$. Applying the chain rule to $u(x(\xi), t(\xi))$, we find

$$\frac{\mathrm{d}}{\mathrm{d}\xi} u(x(\xi), t(\xi)) = u_x \frac{\mathrm{d}x}{\mathrm{d}\xi} + u_t \frac{\mathrm{d}t}{\mathrm{d}\xi}.$$

Using (2.4) we can write

$$\frac{\mathrm{d}u}{\mathrm{d}\xi} = \{-f_u t_\xi + x_\xi\} u_x.$$

Now, if the curve defined by $x(\xi)$, $t(\xi)$ is a characteristic we would not be able to determine $u_x$ from $u(\xi)$, therefore we must require that

$$-f_u t_\xi + x_\xi = 0.$$

This will be the case if we choose

$$t_\xi = 1, \quad x_\xi = f_u. \tag{2.6}$$

Therefore along the characteristic curve

$$\frac{\mathrm{d}u}{\mathrm{d}\xi} = u_t + f_u u_x,$$

but from (2.4) it follows generally that

$$\frac{\mathrm{d}u}{\mathrm{d}\xi} = u_t + f_u u_x = 0, \tag{2.7}$$

and specifically for Burgers' equation that

$$\frac{du}{d\xi} = u_t + uu_x = 0. \tag{2.8}$$

We conclude that for Burgers' equation $u$ is constant along the characteristic. From (2.6) we know that in the $(x,t)$ plane the slope of the characteristic is $1/u$, hence the characteristic curves are straight lines.

Associated with the characteristic surfaces or curves are the important notions of *domain of dependence* and *range of influence*. For example, these concepts play an important role on ensuring the stability of a numerical scheme by requiring that the Courant–Friedrichs–Lewy rule [43] be satisfied. The domain of dependence of a point $P$ somewhere in the half plane $t > 0$ is the set of points on the initial value line that contribute to the solution at $P$. The range of influence of a point $Q$ on the initial value line is the set of points on the half plane $t > 0$ which are influenced by the initial data at point $Q$. For initial value problems typical of the one-dimensional Euler equations, we will find that in the $(x,t)$ plane the domain of dependence of a point $P$ is a segment of the $x$-axis and the range of influence of a point $Q$ on the $x$-axis is a wedge with apex at $Q$ and opening into the $t > 0$ plane. For the inviscid Burgers' equation, with only one characteristic curve, the domain of dependence of a point $P$ is only the point corresponding to the origin, on the $x$-axis, of the characteristic reaching the point $P$, and the range of influence of a point $Q$ is the set of points on thc characteristic curve originating at $Q$.

## 2.1.5 Characteristics for a System of Equations

In Section 2.10 we will treat a system of two equations not too unlike Burger's equation. Thus, let us consider how to find the characteristics and compatibility relations for a system of equations. Consider a system of equations written in the form

$$\boldsymbol{u}_t + \mathbf{A}(\boldsymbol{u})\boldsymbol{u}_x = 0, \tag{2.9}$$

where

$\boldsymbol{u}$ is a column vector
$\mathbf{A}$ is a matrix

If the governing equations are written in divergence form, then

$$\mathbf{A}(\boldsymbol{u}) = \frac{d\mathbf{f}}{d\boldsymbol{u}},$$

where $d\mathbf{f}/d\boldsymbol{u}$ is the Jacobian* of the flux vector with respect to $\boldsymbol{u}$.

* The Jacobian is shorthand for the Jacobian matrix—the matrix of all first-order partial derivatives of a vector-valued function. Likewise, the Hessian is shorthand for the Hessian matrix—the square matrix of second-order partial derivatives of a function.

We want to know whether a linear combination of the equations of motion can be obtained, in each of which the derivatives of anyone of the components of the original vector $\boldsymbol{u}$ appear only in the form

$$u_t + \Lambda u_x, \tag{2.10}$$

where $\Lambda$ is a scalar. A generic linear combination of the equations of motion can be written as

$$\vec{\mu}(\boldsymbol{u}_t + \mathbf{A}\boldsymbol{u}_x) = 0, \tag{2.11}$$

where $\vec{\mu}$ is a row vector, since $\boldsymbol{u}$ is a column vector. We require the terms in parenthesis in (2.11) to have the form (2.10). Therefore, the following condition must be satisfied:

$$\vec{\mu}\mathbf{A} = \Lambda\vec{\mu}, \tag{2.12}$$

since $\boldsymbol{u}_t$ and $\boldsymbol{u}_x$ are both generally different from zero. A number of scalar equations equal to the number of equations in the original system (2.9) are synthesized by (2.12), for the same number of unknowns in $\vec{\mu}$. For the solution $\vec{\mu}$ not to be trivial, the condition

$$\det(\mathbf{A} - \Lambda\mathbf{I}) = 0, \tag{2.13}$$

must be satisfied, where $\mathbf{I}$ is the identity or unit matrix. In matrix algebra, the members of $\Lambda$ satisfying (2.13) are called *eigenvalues* of $\mathbf{A}$. Whatever the choice of dependent variables is for system (2.9), the slope of the characteristics, $\Lambda$, in the $(x,t)$-plane are the eigenvalues of the matrix $\mathbf{A}$.

*Eigenvectors* are commonly associated with *eigenvalues*. There are two different sets of eigenvectors associated with the matrix $\mathbf{A}$. One is the set of row vectors, $\vec{\mu}$, solutions of (2.12), and called *left eigenvectors* because they appear to the left of $\mathbf{A}$. The other is the set of *column* vectors, $\vec{v}$, solutions of

$$\mathbf{A}\vec{v} = \vec{v}\Lambda, \tag{2.14}$$

which are called *right eigenvectors* because they appear to the right of $\mathbf{A}$. The eigenvalues for (2.14) are the same as for (2.12) because (2.13) still holds. The left eigenvectors are proportional to the cofactors of elements of $\mathbf{A} - \Lambda\mathbf{I}$, orderly chosen from any column. Occasionally, for a certain eigenvalue, the eigenvector is null; in such a case, its only significant element can be found by direct inspection. We will not use right eigenvectors in this section. Let us instead focus our attention on the left eigenvectors. Since $\vec{\mu}$ is a left eigenvector of $\mathbf{A}$, (2.11) is now

$$\vec{\mu}(\boldsymbol{u}_t + \lambda\boldsymbol{u}_x) = 0, \tag{2.15}$$

where

$\lambda$ is one of the eigenvalues of $\Lambda$

$\vec{\mu}$ the corresponding eigenvector

Therefore, (2.15) synthesizes a set of scalar equations, as many as in (2.9), which are the *compatibility* equations valid along the slopes $\lambda_i$. See Section 2.10 for a simple example.

### 2.1.6 Breakdown of the Classical Solution

If thc initial data is smooth, the Cauchy–Kowalewski theorem [42] asserts a unique smooth solution in the small* for the initial value problem of (2.3). As is well known, the solution to (2.3) with initial data (2.5) is

$$u(x,t) = u_0(x - ut). \tag{2.16}$$

This is easy to show by substitution of (2.16) into (2.3). However, we note that if $u_0(x_a) > u_0(x_b)$ and $x_a < x_b$, in other words if $u_0' < 0$,† the characteristics originating from $x_a$ and $x_b$ would necessarily intercept at some point in $t > 0$. At the point of interception the solution would have two values, $u_0(x_a)$ and $u_0(x_b)$, therefore, the solution would no longer be smooth. This result is purely a consequence of the nonlinear character of (2.3). Let us take a closer look. From (2.16) we find

$$u_t = -\frac{uu_0'}{1 + u_0' t}$$

$$u_x = \frac{u_0'}{1 + u_0' t}$$

from which we can conclude that if $u_0' \geq 0$ for all $x$, then the gradients of $u$ remain bounded for all $t > 0$, but if $u_0' \leq 0$ at any point, then both gradients become singular. If we let $t = \hat{t}$ be the time at which the singularity occurs, i.e., the time when the classical solution breaks down, then $\hat{t}$ follows from the solution of

$$1 + u_0' \hat{t} = 0.$$

### 2.1.7 Weak Solutions

Consider then what takes place with our functions $u$ and $f$ near $\hat{t}$. Just before $\hat{t}$, the functions are smooth and well behaved, but their gradients are becoming unbounded. At $t = \hat{t}$, the functions are torn apart. However, this seemingly cataclysmic event is happening under the control of the conservation law (2.4) following the script embodied in the initial conditions (2.5). Depending on the initial conditions,

---

* Meaning: for a short time.

† The prime denotes differentiation with respect to the argument.

the functions $u$ and $f$ immediately after $\hat{t}$ could have large jumps or the jumps could start at zero and grow smoothly. In any case, after $\hat{t}$ we need to be able to take gradients of step functions, and we don't know how to do this under the classical theory. Thus we look for more *general* functions.

When we accepted the continuum hypothesis, we also accepted, perhaps for its simplicity, pointwise functions, $u = u(\boldsymbol{x}, t)$, to describe the physics. We could have instead used functions defined as averages over some volume, $\tau$, of space and some interval of time, $\bar{u} = \iint u \mathrm{d}\tau \mathrm{d}t$,* or better yet, functions defined as weighted averages, $\tilde{u} = \iint u\phi \mathrm{d}\tau \mathrm{d}t$, in which different values of $u$ are weighted according to their frequency $\phi$ of occurrence. The latter approach leads to the distribution theory and it is in the sense of the distribution theory that we want to generalize the concept of a function.

The creation of the distribution theory is traceable to the work of Sobolev [206], who defined distributions rigorously as functionals in 1935 and to L. Schwartz† [199] who, unaware of the work of Sobolev, invented the theory of distributions in 1945 to resolve issues arising in the theory of generalized solutions to partial differential equations. The starting point for Schwartz was to provide a better definition of a generalized function. He was focused on the fact that the generalized solution to a partial differential equation acted as a convolution operator taking a compact infinitely differentiable function into an infinitely differentiable function. This work naturally led to defining the derivative of the convolution operator. He resolved this and a number of other issues in the course of one night in October of 1944. However, he ran into difficulties with some unwieldy operators. Six months later, he realized that if he defined his generalized functions not as operators but as functionals he could overcome the difficulties. He called these functionals distributions. For a detailed history of this branch of mathematics see [131].

#### 2.1.7.1 Functions and Distributions

The function $\phi$, which we will call a *test function*, will play a crucial role in resolving operational issues with our functions $u$ and $f$. Our plan will be to transfer the operations that we cannot handle with the functions $u$ and $f$, particularly differentiation, to the test function. For this reason, we use test functions that are very well behaved; functions that have continuous derivatives of all orders, $C^\infty$ functions, and that have compact support. By compact support we mean functions that vanish outside some bounded set. In one space dimension, a typical test function, in the interval [0, 1] could be defined by

$$\phi(x) = h(x)h(1 - x),$$

* To simplify the notation we omit the scaling by $\Delta\tau\Delta t$.

† It is interesting to note that one of Schwartz professors at the École Normale Supérieure was J. Hadamard who, see Section 1.2, lectured on Hugoniot's work.

where

$$h(x) = \begin{cases} e^{-\frac{1}{x}} & x > 0, \\ 0 & x \leq 0. \end{cases}$$

This function is shown in Figure 2.3. The extension to several dimensions is done by multiplication. To build a test function with support on $x$ [0, 1] and $t$ [0, 1], we could use $\varphi(x, t) = \phi(x)\phi(t)$. To shift and contract the support from the interval [0, 1] to the interval [a, b], we use the transformation

$$\phi_{a,b}(x) = \phi\left(\frac{x - a}{b - a}\right).$$

A distribution

$$U = \int_{-\infty}^{\infty} \int_{-\infty}^{\infty} u(x, t)\varphi(x, t)\mathrm{d}x\mathrm{d}t$$

is a mapping from test functions $\phi$'s to numbers, not unlike the function $u$ is a mapping of $(x, t)$ to numbers. The inner product notation is the standard notation used to describe the action of a distribution $U$ on a test function $\phi$, that is we write $\langle U, \phi \rangle$ for $\int U\phi \mathrm{d}x$. Distributions satisfy the properties of *linearity*: $\langle U, a\phi + b\varphi \rangle = a \cdot \langle U, \phi \rangle + b \cdot \langle U, \varphi \rangle$ for all test functions $\phi$, $\varphi$, and all constants $a$, $b$; and of *continuity*: if a sequence of test functions $\phi_1, \phi_2, \ldots,$ and its derivatives $\phi_1^{(k)}, \phi_2^{(k)}, \ldots,$ converges to zero uniformly, then $\langle U, \phi_n \rangle \to 0$. This is a very strong convergence requirement. *Pointwise convergence* of a sequence of functions $\phi_n(x)$,

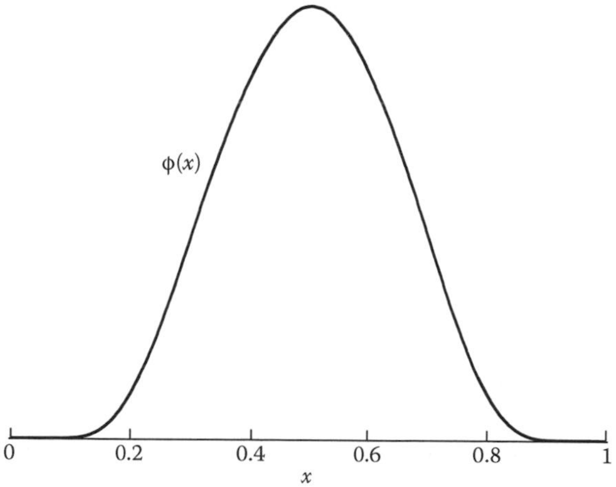


**FIGURE 2.3:** A pulse-like test function with support on [0,1].

defined on some set $\mathbb{S}$, to a function $\phi(x)$ as $n \to \infty$ occurs if for each $x$ in $\mathbb{S}$ there is an $\varepsilon > 0$ such that,

$$|\phi_n(x) - \phi(x)| < \varepsilon,$$

whenever $n$ is greater than or equal to some integer $N$. $N$ here can have different values depending both on $\varepsilon$ and $x$. However, for *uniform convergence* we require the existence of a single $N$ for all $x$'s in $\mathbb{S}$. Uniform convergence implies pointwise converge, but the converse is not necessarily true. As an example consider the series

$$\phi_n(x) = 1 - \frac{1}{(1+x^2)^{n-1}}, \quad 0 \le x \le 1,$$

shown in Figure 2.4. The sequence converges pointwise to the function

$$\phi = \begin{cases} 1 & 0 < x \le 1, \\ 0 & x = 0, \end{cases}$$

but fails to converge uniformly, since we need to find one integer $N$ such that $|\phi_n(x) - \phi(x)| < \varepsilon$ for all $x$ which requires that $(1+x^2)^{1-N} < \varepsilon$. But this implies that

$$(1-N)\ln(1+x^2) < \ln \varepsilon$$

or that

$$N > 1 + \ln \varepsilon / \ln(1+x^2)$$

and, therefore, $N$ is a function of $x$ and grows as $x \to 0$.

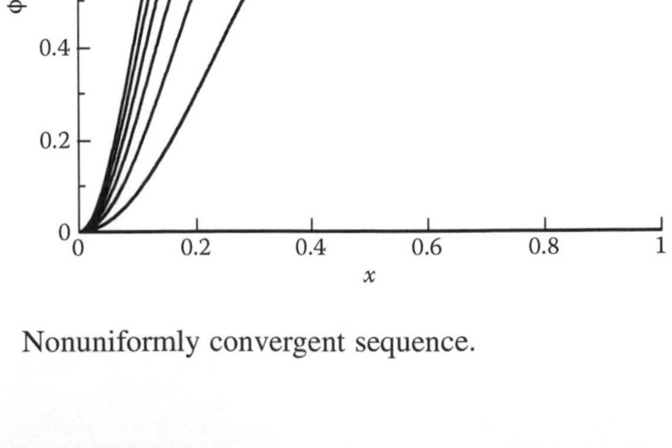

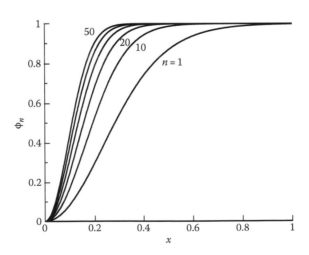

**FIGURE 2.4:** Nonuniformly convergent sequence.

We call our distribution functions, generalized functions. If $\Omega$ is an open subset of $\mathbb{R}^n$ (Euclidean $n$-dimensional vector space), we denote by $\mathcal{D}'(\Omega)$ the space of generalized functions and by $\mathcal{D}(\Omega)$ the space of $C^\infty$ test functions with compact support. $\mathcal{D}'(\Omega)$ is a very large vector space. It includes, as a subset, all our familiar classical functions, i.e., sin($x$), $x$, $e^{-x}$, etc., as well as functions that do not behave as classical functions, i.e., the Heaviside function and its derivative, the Dirac delta function. We refer to the former as regular generalized functions and to the latter as singular generalized functions.

### 2.1.7.2 Generalized Derivative

Let us see how we can use the distribution theory to evaluate the derivative of a continuous function $u(x)$. We want to evaluate

$$\int_{-\infty}^{\infty} u'(x)\phi(x)dx.$$

Using integration by parts we transfer the operation on $u$ to $\phi$:

$$\int_{-\infty}^{\infty} u'(x)\phi(x)dx = u(x)\phi(x)|_{-\infty}^{\infty} - \int_{-\infty}^{\infty} u(x)\phi'(x)dx,$$

but since $\phi$ vanishes as $x \to \pm\infty$, we get

$$\int_{-\infty}^{\infty} u'(x)\phi(x)dx = -\int_{-\infty}^{\infty} u(x)\phi'(x)dx.$$

The same methodology is used if $u$ were not differentiable or when dealing with distributions, thus

$$\langle U', \phi \rangle = -\langle U, \phi' \rangle.$$

Therefore by transferring the differential operation to the test function, the concept of distribution provides a setting for the differentiation of functions which are not differentiable in the classical sense.

Distributions have derivatives of any order, defined by

$$\langle U^{(n)}, \phi \rangle = (-1)^n \langle U, \phi^{(n)} \rangle, \tag{2.17}$$

and partial derivatives given by,

$$\left\langle \frac{\partial U}{\partial x_i}, \phi \right\rangle = -\left\langle U, \frac{\partial \phi}{\partial x_i} \right\rangle.$$

As an example, consider the *step* or Heaviside distribution function $H(x)$, defined by*

$$H(x) = \begin{cases} 1 & x > 0 \\ 0 & x < 0, \end{cases}$$
$$= \frac{1}{2}\left(1 + \frac{x}{|x|}\right), \tag{2.18}$$

which could represent the behavior of a state variable near a shock wave. To evaluate $H'$ we proceed as follows

$$\begin{aligned} \langle H', \phi\rangle &= -\langle H, \phi'\rangle, \\ &= -\int_{-\infty}^{\infty} H\phi' \mathrm{d}x, \\ &= -\int_{0}^{\infty} \phi' \mathrm{d}x, \\ &= \phi(0). \end{aligned}$$

The third step above uses the definition of $H$, (2.18), while the last step follows since $\phi$ has compact support for $x \to \infty$. $H'$ is the transformation that maps every test function $\phi$ into its value at the origin. It is known as the unit impulse function, the Dirac delta function, or just the delta function, and is symbolically represented by $\delta(x)$. The above result is referred to as the sampling property of the delta function. The derivatives of the delta function follow directly from (2.17):

$$\left\langle \delta^{(k)}, \phi \right\rangle = (-1)^k \left\langle \delta, \phi^{(k)} \right\rangle = (-1)^k \phi^{(k)}(0).$$

This is the sampling property of the $k$th derivative of the delta function acting on a function with continuous derivatives at least up to the $k$th order in some neighborhood of the origin.

Let $U(x)$ be a piecewise smooth function with a jump at $x_0$. We can write $U(x)$ as

$$U(x) = u_{\mathrm{l}}(x)H(x_0 - x) + u_{\mathrm{r}}(x)H(x - x_0),$$

where $u_{\mathrm{l}}$ and $u_{\mathrm{r}}$ are smooth functions. The generalized derivative of $U$ is

$$U'(x) = u_{\mathrm{l}}'(x)H(x_0 - x) + u_{\mathrm{r}}'(x)H(x - x_0) + [\![u]\!]\delta(x - x_0),$$

* This definition goes back to Cauchy, 1849. He called the Heaviside function "coefficient limitateur" and defined it as $u(t) = \frac{1}{2}\left(1 + t\Big/\sqrt{t^2}\right)$ [28]. Cauchy also used the impulse function in his derivation of the Fourier-integral theorem [27].

where $[\![u]\!] = u_r(x_0) - u_l(x_0)$.* If $U$ has $n$ jumps, say $[\![u]\!]_i$ at $x_i$, $i = 1, 2, \ldots n$, it is easy to show that the generalized derivative of $U$ is

$$U'(x) = u_l'(x)H(x_1 - x) + \sum_{i=1}^{n-1} u'_{i+\frac{1}{2}}(x)M(x_i, x_{i+1}) + u_r'(x)H(x - x_n) + \sum_{i=1}^{n} [\![u]\!]_i \delta(x - x_i),$$

where $M(x_i, x_{i+1})$ is a *tabletop* function:

$$M(x_i, x_{i+1}) = H(x - x_i) - H(x - x_{i+1}).$$

We can extend these results to functions of more than one variable as follows. Let $U(\boldsymbol{x})$ have a discontinuity across a surface $\Sigma$ defined by $\mathbb{F}(\boldsymbol{x})$. Let $\boldsymbol{n}$ be the unit normal to $\mathbb{F}$ at some point $A$ on $\Sigma$. Let $\eta$ be a coordinate in the direction of $\boldsymbol{n}$ and let $(\xi, \zeta)$ be the other two orthogonal coordinates on the surface of $\Sigma$, see Figure 2.5. Let the jump at $A$ be $[\![u]\!] = u(\mathbb{F}_+) - u(\mathbb{F}_-)$, where $\mathbb{F}_+$ is on the side of $\Sigma$ to which $\boldsymbol{n}$ points. The function $U$ is given by

$$U(\boldsymbol{x}) = u_-(\boldsymbol{x})H(-\eta) + u_+(\boldsymbol{x})H(\eta).$$

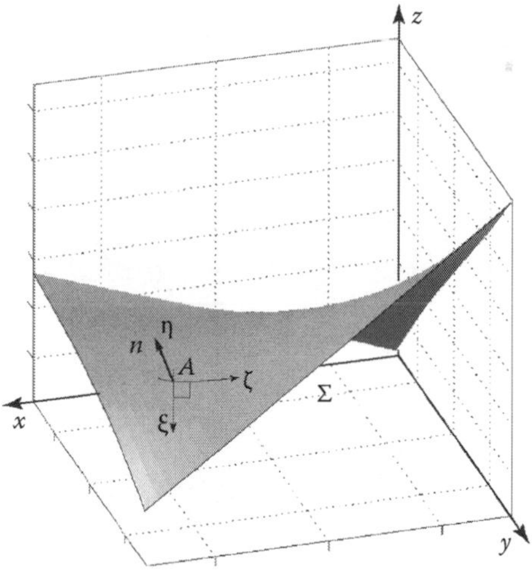


**FIGURE 2.5:** Coordinate system on a surface discontinuity Σ.

* The notation [[ ]] used to denote a jump was introduced by Christoffel in 1877 [35].

On $\Sigma$, $u$ is continuous, therefore

$$\frac{\partial U}{\partial \xi} = \frac{\partial u}{\partial \xi}, \quad \frac{\partial U}{\partial \zeta} = \frac{\partial u}{\partial \zeta}.$$

Across $\Sigma$ we have

$$\frac{\partial U}{\partial \eta} = \frac{\partial u_-}{\partial \eta} H(-\eta) + \frac{\partial u_+}{\partial \eta} H(\eta) + [\![u]\!]\delta(\eta).$$

To simplify the notation we will write the pointwise smooth part as $\partial u/\partial \eta$, meaning

$$\frac{\partial u}{\partial \eta} = \frac{\partial u_-}{\partial \eta} H(-\eta) + \frac{\partial u_+}{\partial \eta} H(\eta).$$

Therefore, by the chain rule we have

$$\frac{\partial U}{\partial x} = \frac{\partial u}{\partial \xi}\frac{\partial \xi}{\partial x} + \frac{\partial u}{\partial \zeta}\frac{\partial \zeta}{\partial x} + \left(\frac{\partial u}{\partial \eta} + [\![u]\!]\delta(\eta)\right)\frac{\partial \eta}{\partial x},$$

with similar expressions for $\partial U/\partial y$ and $\partial U/\partial z$. It follows then that the generalized gradient operation is given by

$$\nabla U = \nabla u + [\![u]\!]\nabla \mathbb{F}\delta(\mathbb{F}),$$

since $\eta = 0$ corresponds to $\mathbb{F}(\boldsymbol{x}) = 0$. Similarly, if $\boldsymbol{U}$ is a vector, the generalized divergence and curl operators are defined by

$$\nabla \cdot \boldsymbol{U} = \nabla \cdot \boldsymbol{u} + \nabla \mathbb{F} \cdot [\![\boldsymbol{u}]\!]\delta(\mathbb{F}),$$
$$\nabla \times \boldsymbol{U} = \nabla \times \boldsymbol{u} + \nabla \mathbb{F} \times [\![\boldsymbol{u}]\!]\delta(\mathbb{F}).$$

#### 2.1.7.3 Other Generalized Operations

The scaling of $U(x) \to U(ax)$ is defined as follows. Let $u(x)$ be a regular generalized function; then

$$\int_{-\infty}^{\infty} u(ax)\phi(x)\mathrm{d}x = \frac{1}{a}\int_{-\infty}^{\infty} u(z)\phi(z/a)\mathrm{d}z.$$

Since the limits on the right-hand side depend on the sign of $a$, we have the following two results:

$$\frac{1}{a}\int_{-\infty}^{\infty} u(z)\phi(z/a)\mathrm{d}z = \frac{1}{a}\langle u(x), \phi(z/a)\rangle, \quad a > 0,$$

$$\frac{1}{a}\int_{\infty}^{-\infty} u(z)\phi(z/a)\mathrm{d}z = -\frac{1}{a}\langle u(x), \phi(z/a)\rangle, \quad a < 0.$$

For any generalized function, scaling is defined by

$$\langle U(ax), \phi(x)\rangle = \frac{1}{|a|}\langle U(x), \phi(x/a)\rangle, \quad a \neq 0.$$

Applying this result to the delta function, we get

$$\langle \delta(-x), \phi(x)\rangle = \langle \delta(x), \phi(x/(-1))\rangle = \phi(0) = \langle \delta(x), \phi(x)\rangle,$$

hence the delta function is even.

The shift of $U(x) \rightarrow U(x-a)$ is defined as follows. Let $u(x)$ be a regular generalized function; then

$$\begin{aligned}\langle u(x-a), \phi(x)\rangle &= \int_{-\infty}^{\infty} u(x-a)\phi(x)\mathrm{d}x,\\ &= \int_{-\infty}^{\infty} u(z)\phi(z+a)\mathrm{d}z = \langle u(x), \phi(x+a)\rangle,\end{aligned}$$

hence for any generalized function we define the shift operation by

$$\langle U(x-a), \phi(x)\rangle = \langle U(x), \phi(x+a)\rangle.$$

Applying a shift to the delta function we get

$$\langle \delta(x-a), \phi(x)\rangle = \langle \delta(x), \phi(x+a)\rangle = \phi(a).$$

Let $\mathcal{L}_i$ stand for one of the following operations $\mathrm{d}/\mathrm{d}x_i$, $\sum$, $\int$, $\lim\limits_{n\rightarrow\infty}$, then the following exchange of order is permissible with generalized functions: $\mathcal{L}_i\mathcal{L}_j(\,) = \mathcal{L}_j\mathcal{L}_i(\,)$.

### 2.1.8 Problem with Products: Colombeau's Generalized Functions

In general, two generalized functions cannot be multiplied. If $u$ is a scalar or a smooth function, then the product $uU = \langle U, u\phi \rangle$ is perfectly valid, but the product of singular distributions is not permitted. In 1954, Schwartz [200] published a theorem on the "impossibility of the multiplication of distributions" that says that "there does not exist a differential algebra containing the vector space $\mathcal{D}$' of distributions and having the classical properties of differentiation and the algebraic operations of addition and multiplications." This is a very severe problem, since we are interested in the solution of nonlinear equations which more often than not involve products of discontinuous functions. To deal with this problem, J. F. Colombeau and collaborators [16,37,157] have developed a differential algebra $\mathcal{G}(\Omega)$ of generalized functions, where $\Omega$ is an open set of $\mathbb{R}^n$. The algebra of all continuous functions on $\Omega$ is not a subalgebra of $\mathcal{G}(\Omega)$, but the algebra of all $C^\infty$ functions with compact support is a subalgebra. $\mathcal{D}'(\Omega)$ is imbedded as a vector subspace of $\mathcal{G}(\Omega)$. The product of two continuous functions $f$ and $g$ in $\mathcal{G}(\Omega)$, denoted by $f \odot g$, reduces to the classical product $fg$ in the sense of the distribution theory. In the parlance of Colombeau's theory, the product $f \odot g$ in $\mathcal{G}(\Omega)$ has a *shadow* on $\mathcal{D}'(\Omega)$ which is the classical product $fg$. This *shadow* is an important concept in Colombeau's algebra and is called *association. Association* is a generalization of the classical equality of functions and distributions. The *association process* involves a limit as some parameter $\varepsilon \to 0$. It is represented by the symbol $\approx$, which denotes a weak equality. Two generalized functions $G_1$ and $G_2$ in $\mathcal{G}(\Omega)$ are associated with each other if and only if for any $\phi$ in $\mathcal{D}(\Omega)$ we have

$$\int (G_1(x) - G_2(x))\phi(x)\mathrm{d}x \approx 0,$$

meaning that the left-hand side in $\mathcal{G}(\Omega)$ is weakly equal to the distribution 0 in $\mathcal{D}'(\Omega)$. That is, the difference between $G_1$ and $G_2$, while not zero, is so small that it can be ignored as long as it is not multiplied by an infinite quantity. This property is denoted by $G_1 \approx G_2$. As long as only classical operations on continuous functions are involved, the classical computations and the computations in $\mathcal{G}(\Omega)$ always give the same results. However this might not be true if singular distributions are involved. The need for a concept of weak equality is a result of Schwartz's impossibility theorem.

Consider the two functions shown in Figure 2.6. $R_\varepsilon$ is defined by

$$R_\varepsilon = .5(1 + \tanh(x/\varepsilon)),$$

and $R_\varepsilon^n$ is $R_\varepsilon$ raised to the $n$th power, $n > 1$. In the limit $\varepsilon \to 0$ each becomes a generalized Heaviside function, although $R_\varepsilon$ and $R_\varepsilon^n$ represent different generalized functions, for example

$$\lim_{\varepsilon \to 0} \left(R_\varepsilon - R_\varepsilon^n\right)_{x=0} = \tfrac{1}{2} - \left(\tfrac{1}{2}\right)^n \neq 0.$$

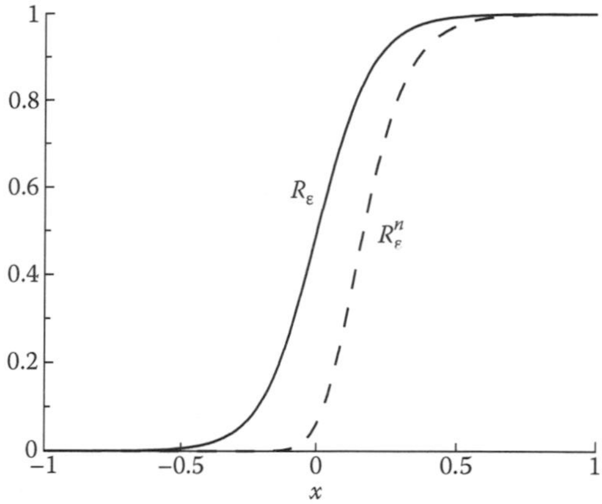


**FIGURE 2.6:** Two generalized functions that in the limit of vanishing ε become Heaviside functions.

If $L$ and $K$ are two Heaviside generalized functions, represented by $R_L$, $R_K$ in $\mathcal{G}(\Omega)$, then we write $L \approx K$ and

$$\left.\begin{array}{l}\int R_L(\varepsilon, x)\phi(x)\mathrm{d}x \\ \int R_K(\varepsilon, x)\phi(x)\mathrm{d}x\end{array}\right\} \rightarrow \int_0^\infty \phi(x)\mathrm{d}x = \langle H, \phi\rangle \text{ as } \varepsilon \rightarrow 0,$$

where $H$ is the classical Heaviside function in $\mathcal{D}'(\Omega)$, and we say that $H$ is the macroscopic aspect of any Heaviside function. It follows that if

$$H^n \approx H, \tag{2.19}$$

then differentiating,

$$H^{n-1}\delta \approx \frac{1}{n}\delta, \quad n \geq 1. \tag{2.20}$$

Note that $H^n \approx H$, but $H^n\delta$ is not associated with $H\delta$, i.e., multiplication of (2.19) by $\delta$ is not valid. Also note that if the association symbol in (2.19) is replaced by the equal sign we get meaningless results, for example if $HH' = \frac{1}{2}\delta$, then $H^2\delta = \frac{1}{2}H\delta$, and using (2.20) with an equal sign results in the absurdity $\frac{1}{3}\delta = \frac{1}{4}\delta$.

We can understand the sampling property of the delta function by studying its behavior as $\varepsilon \rightarrow 0$. First take $\partial/\partial x$ of $R_\varepsilon$ and consider the limit as $\varepsilon \rightarrow 0$, as shown in Figure 2.7. Now, the integral

$$\int_{-\infty}^{\infty} f(x)\delta(x-a)\mathrm{d}x = \lim_{\varepsilon\rightarrow 0}\int_{-\infty}^{\infty} f(x)\delta_\varepsilon(x-a)\mathrm{d}x,$$

$$= f(a)\lim_{\varepsilon\rightarrow 0}\int_{-\infty}^{\infty} \delta_\varepsilon(x-a)\mathrm{d}x,$$

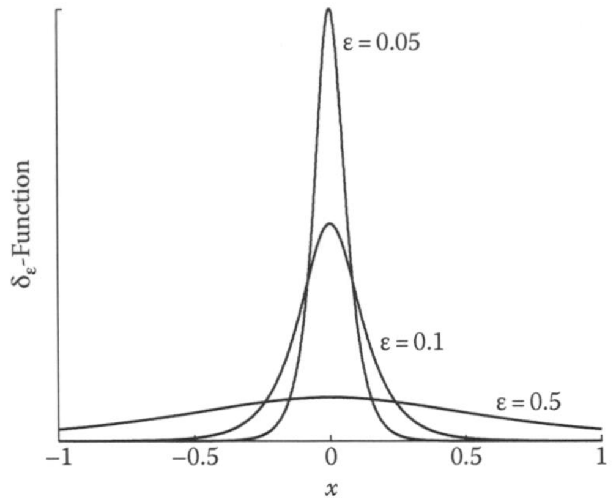


**FIGURE 2.7:** Behavior of delta function as $\varepsilon \to 0$.

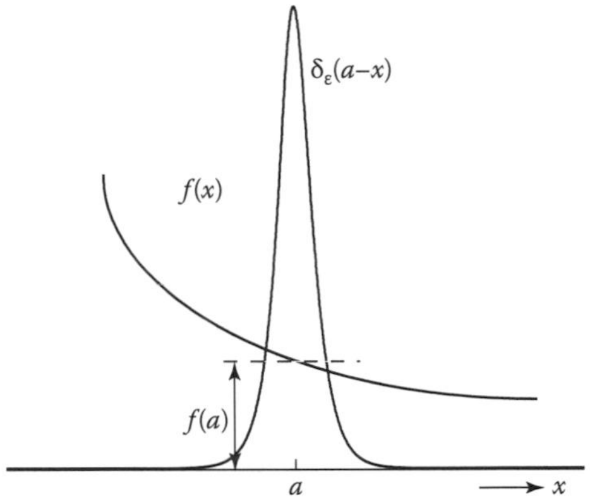


**FIGURE 2.8:** Illustration of sampling property of the delta function.

is illustrated by Figure 2.8. It shows that the product of $f(x)$ with the delta function vanishes outside the neighborhood of $a$ and as $\varepsilon \to 0, f(x) \to f(a)$. Thus the integral equals $f(a)$ times the area under the delta function. Since this area equals to one in the limit of $\varepsilon \to 0$, the delta function acts as a sieve picking out the value of $f(a)$.

#### 2.1.8.1 Application of Colombeau's Generalized Functions

In Section 2.1.10 we will show how to find the shock jump condition for the conservation form of Burgers' equation using the standard distribution theory. Here we show how the same result can be obtained using Colombeau's generalized functions. The significance of this work is that Colombeau's theory is applicable also to the nonconservative form of the governing equations, where the standard distribution theory fails, and thus it opens the way to finding jump conditions for equations that cannot be written in conservative form [29].

If we introduce the traveling wave solution $u = u_r + [\![u]\!]H(-\xi)$, where $\xi = x - wt$, $w$ is the wave speed, and $[\![u]\!] = u_l - u_r$, into the inviscid Burgers' equation (2.4) we obtain the following:

$$w[\![u]\!]\delta + \tfrac{1}{2}\left(u_r^2 + 2u_r[\![u]\!]H + [\![u]\!]^2H^2\right)_x = 0.$$

But by (2.19) $H^2 \approx H$, thus after differentiation we have

$$\left(w - u_r - \tfrac{1}{2}[\![u]\!]\right)\delta = 0,$$

and therefore $w = \frac{1}{2}(u_l + u_r)$; this result, as we will show in Section 2.1.10 using standard procedures is the correct jump condition. If we try this with the nonconservation form, $u_t + uu_x = 0$, we get

$$(w - u_r)\delta - [\![u]\!]H\delta = 0. \tag{2.21}$$

It would appear that we can factor $\delta$ to obtain

$$(w - u_r - [\![u]\!]H)\delta = 0.$$

The result $w = u_r + [\![u]\!]H$ makes no sense. The problem is that the product $H\delta$ is not meaningful under the standard distribution theory and should not be factored. Returning to (2.21), first we write it with the association sign, because of the occurrence of $H\delta$:

$$(w - u_r)\delta - [\![u]\!]H\delta \approx 0,$$

then using (2.20), we get

$$(w - u_r)\delta - \tfrac{1}{2}[\![u]\!]\delta \approx 0.$$

It follows that $w = \frac{1}{2}(u_l + u_r)$. To obtain this result, we have made use of a chain of events in Colombeau's theory that resulted in Equations 2.19 and 2.20. It is therefore worthwhile to explain what it is we have done. Equation 2.21 does not have a solution within the standard theory of distributions, because of the ambiguous product $H\delta$. There is however a solution in $\mathcal{G}(\Omega)$ in terms of generalized functions $(U_\varepsilon, H_\varepsilon)$ whose macroscopic aspect are respectively $(u,H)$ in $\mathcal{D}'(\Omega)$. In passing from $(U_\varepsilon, H_\varepsilon)$ to $(u,H)$ there is loss of information which results in the ambiguity of (2.21). This ambiguity is resolved by using the proper limits represented by Equations 2.19 and 2.20. Technically, $(u,H)$ are not solutions of $u_t + uu_x = 0$, but they have exactly the same mean value properties as the genuine solutions $(U_\varepsilon, H_\varepsilon)$.

### 2.1.9 Returning to the Divergence Equation

With this introduction to the distribution theory we can return to (2.4) and (2.5), for further reading consult [74,129,199]. Let us begin by assuming that $u$ is a classical solution. Following the previous discussion, we multiply (2.4) by a test function $\phi$ with compact support inside the rectangle $0 \leq t < \tau$, $a < x < b$ (we give $\phi$ support along $t = 0$), and integrate by parts, to find

$$\iint_{t>0} (u_t + f_x)\phi \mathrm{d}x\mathrm{d}t = \int_0^\tau \int_a^b (u_t + f_x)\phi \mathrm{d}x\mathrm{d}t = 0,$$

$$= -\int_a^b u_0(x)\phi(x,0)\mathrm{d}x - \int_0^\tau \int_a^b u\phi_t \mathrm{d}x\mathrm{d}t - \int_0^\tau \int_a^b f\phi_x \mathrm{d}x\mathrm{d}t = 0.$$

Hence,

$$\iint_{t\geq 0} (u\phi_t + f\phi_x)\mathrm{d}x\mathrm{d}t + \int u_0(x)\phi(x,0)\mathrm{d}x = 0. \tag{2.22}$$

If $u$ is a classical solution of (2.4) and (2.5), then (2.22) is valid for all test functions $\phi$. As a minor point, we defined our test functions as $C^\infty$ functions, but to obtain (2.22) we only require the test functions to be $C^1$. Equation 2.22 would be valid as long as $u$ and $u_0$ are bounded and are well behaved, in the sense that we can probe them with test functions (i.e., they are measurable). Therefore, we generalize the class of solutions to (2.4) and (2.5) to bounded measurable functions and refer to them as *weak* solutions.

### 2.1.10 Shock Waves: Standard Analysis

We want to find the conditions that a discontinuity must satisfy to be compatible with (2.22). Let the curve $\Sigma: x = x(t)$ in Figure 2.9 around the point $p$. We said that

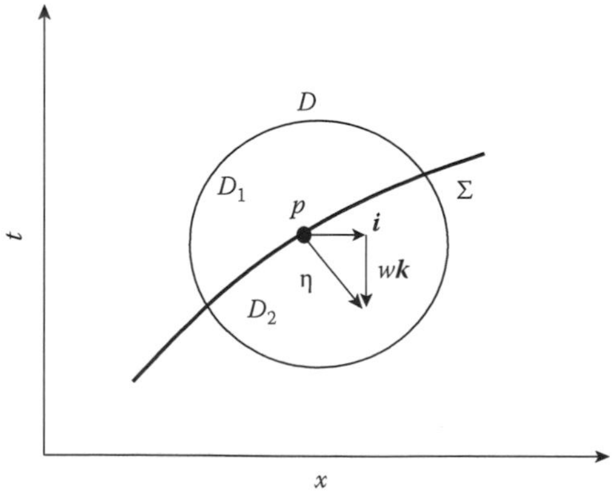


**FIGURE 2.9:** Determination of shock jump conditions.

$\Sigma$ is an isolated discontinuity to ensure that $u$ is differentiable on either side of $\Sigma$ within $D$. Let $D_1$ and $D_2$ be the components of $D$ ($D = D_1 \cup D_2$) on either side of $\Sigma$. Define the time–space vector $z$ by $z = u\boldsymbol{k} + f\boldsymbol{i}$, where $\boldsymbol{i}$ and $\boldsymbol{k}$ are the unit vectors in the direction of the $(x,t)$ coordinates, respectively. Let $\phi$ have support on the $x$-axis, then we write (2.22) as

$$\iint_D \nabla\phi \cdot z\mathrm{d}x\mathrm{d}t = \iint_{D_1} \nabla\phi \cdot z\mathrm{d}x\mathrm{d}t + \iint_{D_2} \nabla\phi \cdot z\mathrm{d}x\mathrm{d}t = 0,$$

where we take $\phi$ to be zero everywhere on the boundary of $D$, but not along $\Sigma$. Since we assume that $u$ is a classical solution of (2.4), we can rewrite the two integrals on the right-hand side as

$$\iint_{D_i} \nabla \cdot (\phi z)\mathrm{d}x\mathrm{d}t, \quad i = 1, 2.$$

Since $u$ is differentiable in $D_i$, we can use the divergence theorem to transform the volume integrals to contour integrals over the surface enclosing $D_i$ ($\partial D_i + \Sigma$). Since $\phi$ is zero on $\partial D_i$, the contour integrals are zero everywhere except along $\Sigma$. Therefore, we are left with

$$\int_\Sigma \phi z_1 \cdot \boldsymbol{\eta}\mathrm{d}s - \int_\Sigma \phi z_r \cdot \boldsymbol{\eta}\mathrm{d}s = 0, \tag{2.23}$$

where

$z_1$ denotes the limit of $z$ as $x \to x(t)-$
similarly $z_r$ denotes the limit as $x \to x(t)+$
$\boldsymbol{\eta}$ is the unit normal to $\Sigma$
$s$ is the length along $\Sigma$, see Figure 2.9

The $\pm$ sign comes about because we integrate counterclockwise, thus it is positive on $D_1$ and negative on $D_2$. Equation 2.23 may be written as

$$\int_\Sigma \phi(z_1 - z_r) \cdot \boldsymbol{\eta}\mathrm{d}s = 0.$$

Since the integrand is valid for all $\phi$'s over any interval of $\Sigma$, we must have

$$(z_1 - z_r) \cdot \boldsymbol{\eta} = 0, \tag{2.24}$$

at every point along $\Sigma$. Let $\mathbb{F} = x - x_x(t) = 0$ define the space–time shock wave surface. Its unit normal is given by

$$\boldsymbol{\eta} = \frac{\nabla\mathbb{F}}{|\nabla\mathbb{F}|} = \frac{\boldsymbol{i} - w\boldsymbol{k}}{\sqrt{1 + w^2}}, \tag{2.25}$$

where $w = \mathrm{d}x_s/\mathrm{d}t$ is the speed of the discontinuity. With (2.25) we can write the jump condition (2.24) as

$$-w[\![u]\!] + [\![f]\!] = 0. \tag{2.26}$$

This relation is known as the *Rankine–Hugoniot* jump.

For Burgers' equation the Rankine–Hugoniot jump condition is

$$-w[\![u]\!] + \left[\!\left[\tfrac{1}{2}u^2\right]\!\right] = 0,$$

and therefore

$$w = \langle u \rangle = \tfrac{1}{2}(u_\mathrm{l} + u_\mathrm{r}). \tag{2.27}$$

For infinitesimally weak discontinuities, we can write $[\![u]\!] \approx \mathrm{d}u$ and $[\![f]\!] \approx f_u \mathrm{d}u$ and therefore the speed of the discontinuity becomes $\mathrm{d}x/\mathrm{d}t = [\![f]\!]/[\![u]\!] = f_u$. But from (2.6) we know that this is the "speed" of a characteristic. Therefore, we conclude that weak discontinuities, i.e., discontinuities in first or higher derivatives of $u$, are supported *across* characteristics. This is explored further in Section 3.5.

### 2.1.11 Entropy Conditions

The Rankine–Hugoniot jump limits the weak solutions that are valid solutions of (2.4), but not enough. In the Euler limit, sending $\varepsilon \rightarrow 0$ in (2.1), information about the irreversible processes that take place within the shock layer is lost. A consequence of this is that the weak solutions that satisfy (2.4) and (2.24) are not unique. To remedy this, we impose an additional condition on the discontinuities which is known as the *entropy condition*. The name is in reference to the fact that across a real shock wave the entropy (which is a measure of irreversible processes) must increase. Thus, the condition is a condition on the structural nature of the discontinuity. The steady-state solution to the inviscid Burgers' equation for a shock wave traveling at constant speed $w$ (traveling wave solution) can be written in the form,

$$u_i(\xi) = u_\mathrm{r} + [\![u]\!](1 - H(\xi)), \tag{2.28}$$

where

$\xi = x - wt$

$H$ is the Heaviside function

$u_i$ is a weak solution to (2.4) that satisfies the jump condition (2.27). Trivially, for $\xi \rightarrow +$, we get $u_i = u_r$ and for $\xi \rightarrow -$, we get $u_i = u_\mathrm{l}$, independent of the actual values of $u_\mathrm{l}$ and $u_\mathrm{r}$. Consider the following two cases,

Case 1

$$\begin{cases} u_\mathrm{l} = 1 \\ u_\mathrm{r} = 0, \end{cases}$$

and Case 2

$$\begin{cases} u_l = 0 \\ u_r = 1. \end{cases}$$

Both cases correspond to a shock speed $w = \frac{1}{2}$. We want to find if one or both cases are valid solutions.

It can be shown that the solutions of Burgers' equation (2.1) are unique, smooth, and exist for all $t \geq 0$ [114]. This suggest investigating a limiting process, the so-called viscous regularization, that validates the solution of the inviscid Burgers' equation by considering it as the limit $u_i = \lim_{\varepsilon \to 0} u_\varepsilon$, where $u_\varepsilon$ is the solution of (2.1). Let us then consider a traveling wave of the form $u_\varepsilon(x,t) = v(\xi_\varepsilon)$ where $\xi_\varepsilon = (x - wt)/\varepsilon$ and such that $v(\xi_\varepsilon) \to u_l$ as $\xi_\varepsilon \to -\infty$, $v(\xi_\varepsilon) \to u_r$ as $\xi_\varepsilon \to \infty$, and $v' \to 0$ as $\xi_\varepsilon \to \pm\infty$. We expect that the outer limit, $\xi_\varepsilon \to \pm\infty$, of this problem corresponds to the inviscid traveling wave problem. We find the viscous solution by substituting $v$ into (2.1), with the following result,

$$v'(v - w) = v''.$$

Integrating twice we find,

$$u_\varepsilon(x,t) = u_r + \tfrac{1}{2}[\![u]\!]\left\{1 - \tanh\left(\tfrac{1}{4}[\![u]\!]\xi_\varepsilon\right)\right\}. \tag{2.29}$$

Evaluating $u_\varepsilon$ and $du_\varepsilon/d\xi_\varepsilon$ at $\xi_\varepsilon = 0$, we find

$$\begin{aligned} u_\varepsilon(0) &= w, \\ \left.\frac{du_\varepsilon}{d\xi_\varepsilon}\right|_{\xi_\varepsilon = 0} &= -\frac{1}{8}[\![u]\!]^2, \\ &< 0. \end{aligned}$$

Hence, $u_l > w > u_r$. This result rules out case 2 as a limit $u_i = \lim_{\varepsilon \to 0} u_\varepsilon$. Inspection of (2.29) shows that it is only consistent with case 1, i.e., $u_l > u_r$. Since

$$\lim_{\varepsilon \to 0} \tanh\left([\![u]\!]\xi/\varepsilon\right) \to 2H([\![u]\!]\xi) - 1, \tag{2.30}$$

we can make (2.28) a consistent limit of (2.29) by rewriting it in the form,

$$u_i(\xi) = u_r + [\![u]\!](1 - H([\![u]\!]\xi)), \quad t > 0. \tag{2.31}$$

Equation 2.31 is remarkable. First, it cannot be evaluated until we know the sign of $[\![u]\!]$; second, it only allows for $u_i(\xi \to -) > u_i(\xi \to +)$.

The convergence of $u_i = \lim_{\varepsilon \to 0} u_\varepsilon$ evaluated by means of,

$$\Delta(\varepsilon_n) = \int_{-\xi_a}^{0} (u_i(\xi) - u_{\varepsilon_n}(\xi/\varepsilon_n))d\xi, \quad \text{with } \varepsilon_n \to 0,$$

is linear as shown in the left panel of Figure 2.10. Similarly, the shock wave thickness defined by

$$\tau = \frac{-[\![u]\!]}{\left.\dfrac{\partial u_\varepsilon}{\partial \xi}\right|_{\xi=0}} = \frac{8\varepsilon}{[\![u]\!]},$$

vanishes linearly with ε.

For the parabolic equation

$$u_t = \varepsilon u_{xx} + a u_x + b u, \quad \varepsilon > 0,$$

Matano [138] has shown that the number of $u$-sign changes does not increase with time. We can use this result to show that for Burgers' equation $u_x$ does not change sign. Consider again (2.1) with initial data $u(x,0) = u_0(x)$. Let $v = u_x$ and introduce $v$ in (2.1) after taking $\partial/\partial x$ of (2.1), thus

$$v_t + u v_x + v^2 = \varepsilon v_{xx}, \quad v(x,0) = u_{0x}(x).$$

Therefore, if $u_{0x}$ is of one sign, either a compression or an expansion, it remains of one sign for all times.

A study of the viscous regularization problem can be formalized to obtain an entropy condition (actually, several entropy conditions). We will not follow that path. The interested reader is referred to [123] for further details. Instead we stipulate the entropy condition as follows. We begin by looking at the flow of information.

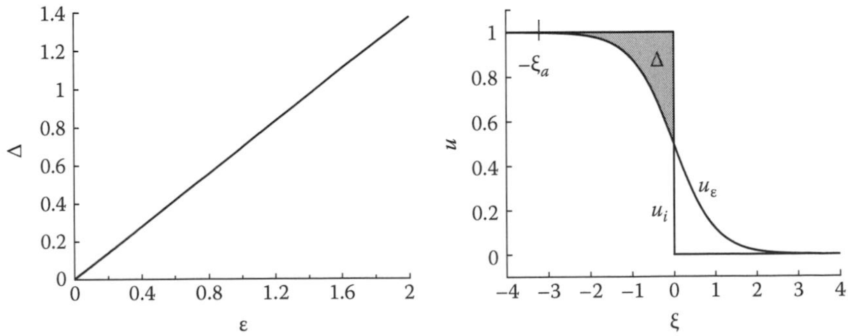


**FIGURE 2.10:** Convergence of viscous solution to inviscid solution for vanishing ε.

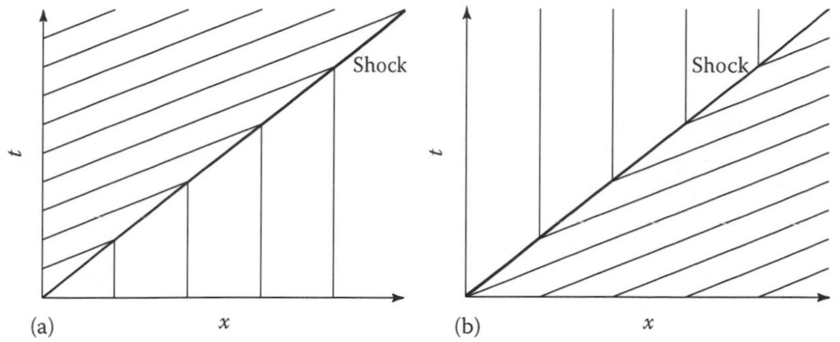


**FIGURE 2.11:** Characteristic patterns associated with (a) case 1 and (b) case 2.

The characteristics patterns associated with cases 1 and 2 are shown in Figure 2.11. The characteristic pattern for case 1 shows the characteristics on either side of the shock converging on the shock with increasing time. This pattern tells us that the shock depends on the initial data. Case 2 shows characteristics leaving the shock with increasing time, i.e., the shock does not depend on the initial data. For this reason, we conclude that case 2 is not plausible. Let us consider case 1, $u_r < u_l$, in more detail. Returning to (2.4), since $f$ is convex up, i.e., $f_{uu} > 0$, $f_u$ is increasing, and since

$$w = \frac{f(u_l) - f(u_r)}{u_l - u_r},$$

see Figure 2.12, it follows from the mean value theorem that $w = f_u(\tilde{u})$ for some $\tilde{u}$ between $u_l$ and $u_r$. Therefore, recalling that the characteristic direction satisfies Equation 2.6,

$$\frac{dx}{dt} = f_u(u),$$

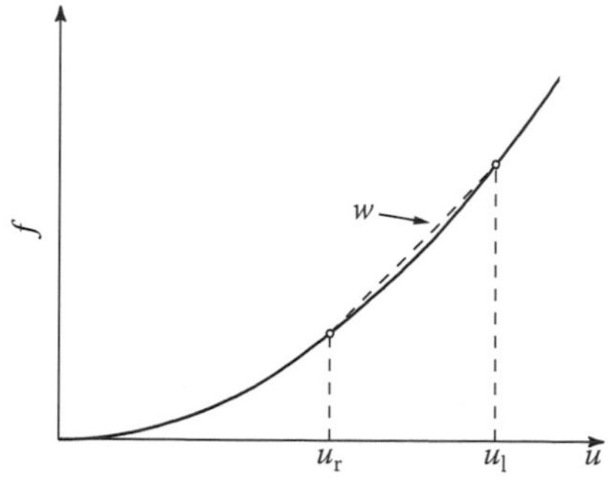


**FIGURE 2.12:** Flux function for the inviscid Burgers' equation.

it follows that $f_u(u_r) < w < f_u(u_l)$, i.e., the shock wave speed lies between the characteristic to the left of the shock wave and the characteristic to the right of the shock wave. This is known as Lax's entropy condition [121]. If the discontinuity becomes infinitesimally weak, see remarks at the end of Section 2.1.10, the discontinuity becomes a characteristic (a contact surface). To include this case, we modify the entropy condition to read

$$f_u(u_r) \leq w \leq f_u(u_l). \tag{2.32}$$

---

## 2.2 The One-Saw tooth Problem

> One must learn by doing the thing;
> for though you think you know it,
> you have no certainty until you try.
>
> Sophocles (496–406 BC)

Consider the inviscid Burgers' equation for the following Cauchy problem. For the function $u_0(x)$ defined by

$$u_0(x) = \begin{cases} 0 & x \leq -1, \\ 1+x & -1 < x \leq 0, \\ 1-x & 0 < x < 1, \\ 0 & x \geq 1, \end{cases} \tag{2.33}$$

find, for $t \geq 0$, the function $u(x,t)$ such that $u(x,t)$ satisfies the equation

$$\begin{aligned} &u_t + f_x = 0, \quad f = \tfrac{1}{2}u^2, \\ &\text{and} \\ &u(x,0) = u_0(x). \end{aligned} \tag{2.34}$$

Because of the shape of the initial profile, we will refer to this problem as the "*one-saw-tooth*" problem. The solution to this problem follows from (2.16). It is given by

$$u_1(x,t) = \begin{cases} 0 & x \leq -1, \\ \dfrac{1+x}{1+t} & -1 < x \leq t, \\ \dfrac{1-x}{1-t} & t < x < 1, \\ 0 & x \geq 1, \end{cases} \qquad t < \hat{t} \tag{2.35}$$

where $\hat{t}$ is the time at which the assumption of continuity breaks down. Equation 2.35 can also be written in the form

$$u_1(x,t) = \frac{1+x}{1+t}(H(x+1) - H(x-t)) + \frac{1-x}{1-t}(H(x-t) - H(x-1)),$$
$$t < \hat{t}.$$

The origin of the discontinuity can be found by evaluating the time when $u_x$ becomes infinite. Since

$$\frac{\partial u}{\partial x} = \frac{-1}{1-t}, \tag{2.36}$$

we find $\hat{t} = 1$. The behavior of $u_x$ leading to the breakdown of the classical solution and the formation of a shock wave is shown in Figure 2.13.

For $t > \hat{t}$ a shock wave becomes part of the solution. The path of the shock, $x_s(t)$, is defined by

$$\frac{dx_s}{dt} = w = \tfrac{1}{2}(u_l + u_r).. \tag{2.37}$$

Here $u_l$ and $u_r$ are the values of $u$ on the left- and right-hand side of the discontinuity, respectively.

Integrating Equation 2.37, with $u_r = 0$ and $u_l = \dfrac{1+x_s}{1+t}$, we find

$$\int_1^{x_s} \frac{dx_s}{1+x_s} = \frac{1}{2}\int_1^{t} \frac{dt}{1+t},$$

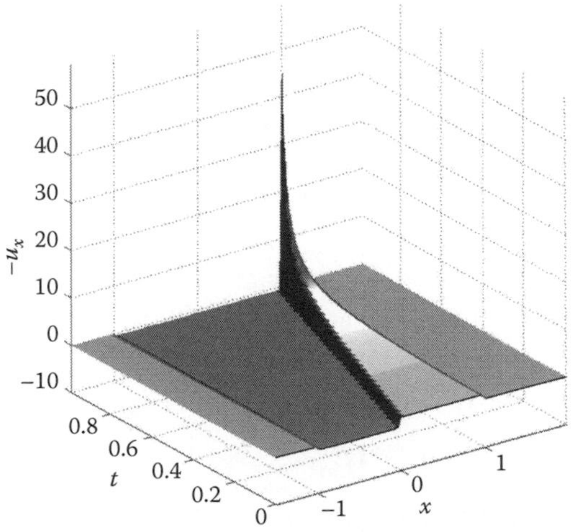


**FIGURE 2.13:** Behavior of the gradient of $u$ leading to break down of classical solution.

hence,

$$x_s(t) = \sqrt{2(1+t)} - 1, \quad t \geq 1. \tag{2.38}$$

For $t > \hat{t}$ the solution is given by

$$u_2(x,t) = \begin{cases} 0 & x \leq -1, \\ \dfrac{1+x}{1+t} & -1 < x < x_s(t), \\ 0 & x > x_s(t), \end{cases} \tag{2.39}$$

which can be written in a more compact form using the step function $H$, (2.18), as

$$u_2(x,t) = \frac{1+x}{1+t}(H(x_s(t) - x) - H(-x-1)).$$

To build the full solution, it will be useful to make the Heaviside function a function of two variables,

$$\hat{H}(x,t) = \frac{1}{2}\left(1 + \frac{xt}{|xt|}\right).$$

This function is one in the quarter-plane $(0 < x < \infty, 0 < t < \infty)$ and zero elsewhere. We can now write the full solution as

$$\begin{aligned} u(x,t) = u_1(x,t)(\hat{H}(1,t) - \hat{H}(1,t-\hat{t})) + \\ + \frac{1+x}{1+t}(\hat{H}(x_s(t) - x, t - \hat{t}) - \hat{H}(-x-1, t-\hat{t})). \end{aligned} \tag{2.40}$$

The first Heaviside-term of (2.40), multiplying the classical solution, is one for all $x$ within $(0 < t < \hat{t})$ and zero elsewhere. The second Heaviside-term is one within $(-1 < x < x_s(t), \hat{t} < t < \infty)$ and zero elsewhere. Figure 2.14a shows the characteristics and shock wave path and Figure 2.14b shows a carpet plot of $u$. The shock wave forms as a result of the focusing of characteristics emanating from $0 \leq x \leq 1$ at $t = 0$. The shock is strongest at its origin with a jump $[\![u]\!] = 1$. The shock strength then decays at a rate proportional to $1/\sqrt{t}$.

### 2.2.1 Area Conservation Rule

For this problem, the simplicity of the initial conditions allowed us to find the exact shock path as a function of time. In general this is not an easy task. However, sometimes a geometrical technique based on the "conservation of area rule" can be used.

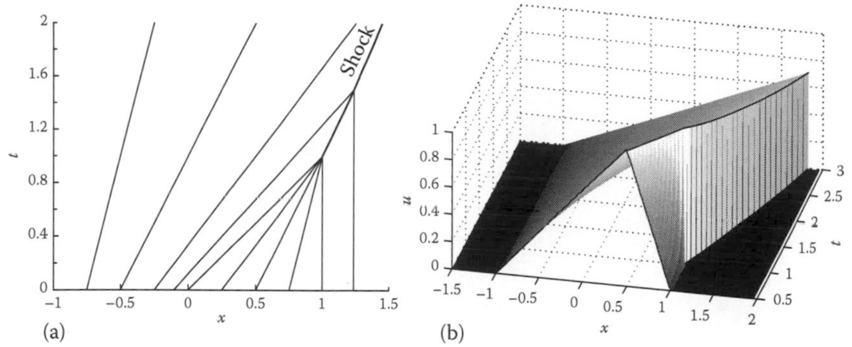


**FIGURE 2.14:** (a) Characteristics and shock wave path for *one-saw-tooth* problem; (b) carpet plot of $u$.

For Equation 2.34, if the flux $f$ vanishes as $x \to \pm\infty$ it is easy to show that the integral of $u$ from $-\infty$ to $+\infty$ is conserved in time. The "area" under the curve $u(x)$ at any time $t$ is

$$A = \int_{-\infty}^{+\infty} u(x, t)\mathrm{d}x, \tag{2.41}$$

and from Equation 2.34 it follows that

$$A_t = \int_{-\infty}^{+\infty} u_t \mathrm{d}x = -\int_{-\infty}^{+\infty} f_x \mathrm{d}x = -(f(\infty, t) - f(-\infty, t)), \tag{2.42}$$

if the fluxes are equal or vanish at $\pm\infty$. The presence of shocks does not affect this conservation property. For example, if a jump from $f(x_{0-}, t)$ to $f(x_{0+}, t) = f(x_{0-}, t) - \Delta$ occurs at some point $x_0$ at any time $t$, the integral above can be evaluated using the Dirac delta function as follows:

$$\begin{aligned} A_t &= -\left\{ \int_{-\infty}^{x_0} f_x \mathrm{d}x - \Delta \int_{-\infty}^{+\infty} \delta(x - x_0)\mathrm{d}x + \int_{x_0}^{+\infty} f_x \mathrm{d}x \right\} \\ &= -\{f(x_{0-}, t) - f(-\infty, t) - \Delta + f(\infty, t) - f(x_{0+}, t)\} = 0. \end{aligned} \tag{2.43}$$

In the absence of a shock, the initial conditions (2.33) would have resulted in the multivalued profile shown in Figure 2.15 at $t = 2$. According to the area conservation rule, the shock should be located such that the area indicated by the shaded region 1 is equal to the area in the shaded region 2. Referring to Figure 2.15, the area in region 1 is

$$A_1 = \frac{1}{2}\left(\frac{1 + x_s}{1 + t} - \frac{x_s - 1}{t - 1}\right)(t - x_s), \tag{2.44}$$

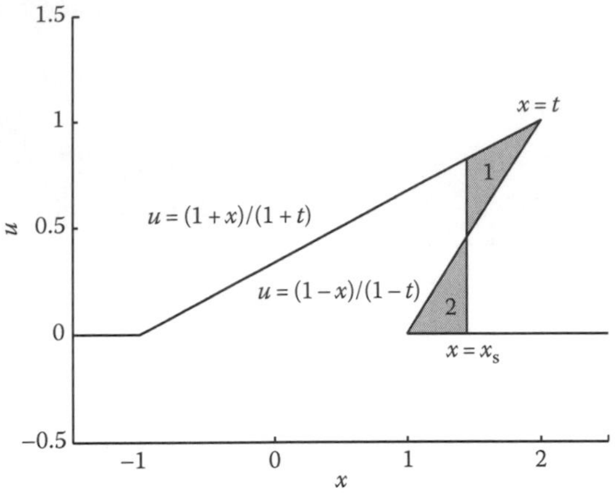


**FIGURE 2.15:** $u$-profile at $t = 2$ in the absence of a shock wave.

and the area in region 2 is

$$A_2 = \frac{1}{2}(x_s - 1)\frac{x_s - 1}{t - 1}. \tag{2.45}$$

Equating (2.44) and (2.45), we find the expression for the shock path

$$x_s(t) = \frac{t + \sqrt{\frac{1}{2}(t+1)}}{1 + \sqrt{\frac{1}{2}(t+1)}} = \sqrt{2(1+t)} - 1 \tag{2.46}$$

in agreement with Equation 2.38.

A word of caution, conservation of area is not a sufficient condition for uniqueness. For example, Figure 2.16 depicts two shock-expansion fan–shock combinations that satisfy the area conservation rule and the entropy inequality, but are nevertheless inadmissible. The pattern depicted in Figure 2.16a is ruled out because

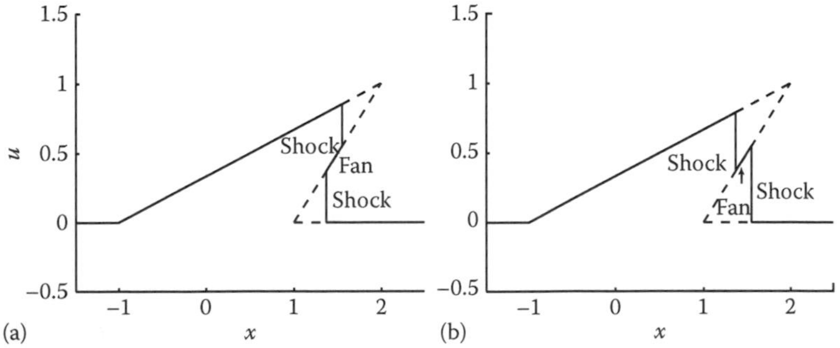


**FIGURE 2.16:** Area conservation is not sufficient to establish a unique solution.

it does not resolve the original multiple solution issue (i.e., three values of $u$ at one $x$ location). For the pattern depicted in Figure 2.16b, the shock to the left of the expansion fan has a speed greater than the speed of the shock to the right of the fan. Therefore, the shock to the left will overtake the shock to the right. This argument has to be carried all the way to the origin of the shock, with the conclusion that only one shock can connect the upper and lower branches of $u$.

---

## 2.3 Background Numerical Schemes

The purpose of computing is insight, not numbers.

Richard Hamming (1915–1998)

### 2.3.1 Preliminaries

Our treatment of discontinuities will be imbedded in some numerical scheme that solves the overall problem: initial conditions, boundary conditions, and the smooth regions of the field. Numerous numerical schemes are available for this purpose; see for example [92,116,220]. Although our purpose is not to write another volume on numerical integration schemes, we need to outline some criteria for the selection and implementation of the numerical scheme that would work with our treatment of discontinuities.

1. *Semi-discretization.* We prefer to decouple the space–time discretization, because it opens up the design space for the numerical scheme and allows developing and debugging codes incrementally. The procedure consists of first discretizing the space derivatives and then introducing a time integration method, typically a Runge–Kutta method.

2. *Physics-based discretization.* By this we mean, a method that takes advantage of the wave propagation nature of the problem being modeled and respects the domain of dependence—range of influence rules.

3. *Conservative differencing.* Although the techniques that we will be developing are specifically designed to limit discretization to smooth regions, thus eliminating the need for conservative differencing, occasionally we will find problems where shock waves develop under such adverse conditions that it is best to delay their fitting. A conservative difference scheme with limiters would allow postponing the fitting without compromising (too much) the quality of the solution. An alternative approach, that has the same goal, is used in Section 4.2.

4. *At least second-order accurate methods.* First-order methods are too diffusive and require too many mesh points to achieve good accuracy.

5. *Methods that have compact stencils*. The main reason for this is to simplify the implementation of boundary conditions, particularly in the presence of shocks.

6. *Methods where the work increases only linearly with the numbers of mesh points*. This is primarily for very large problems and is beyond the scope of this monograph.

7. *Simplicity with accuracy and efficiency trumps everything else.*

Keep in mind that these are only goals and in some cases are competing against each other. For example, the goal of having a physics-based discretization and second-order accuracy might require a larger stencil than some other second-order scheme. The code developer needs to balance these requirements.

Throughout this chapter, we will develop full codes to illustrate the implementation of the ideas and techniques that are developed. The codes are written for the MATLAB® computing environment and are provided in the accompanying CD.

### 2.3.2 Truncation Error and Consistency

Let a partial differential equation, or a system of partial differential equations, be symbolically written as

$$E = 0. \tag{2.47}$$

Let each time derivative be replaced by differences of values at certain points, divided by $\Delta t$, and each space derivative replaced by similar differences divided by $\Delta x$. The catalog of possible replacements is very extensive, but the final outcome should be a *discretized* equation represented by

$$F = 0. \tag{2.48}$$

Suppose we know the solution of (2.47) and we replace its values at a given point in the $(x,t)$-plane into $F$. The result is generally different from zero; (2.48) is not satisfied by the exact solution of the problem. The value of $F$ is called the *truncation error*, $\varepsilon$. For a given discretizing scheme, at each point $\varepsilon$ depends on $\Delta x$ and $\Delta t$. We expect $\varepsilon$ to decrease and tend to zero as $\Delta x$ and $\Delta t$ tend to zero. If this happens, we say that the scheme is *consistent* with the original partial differential equation. The concept of consistency is a *local* one. The replacement of the exact solution into $F$ is intended to be made at every point, and the truncation error evaluated via a local analysis, *disregarding any influence of other points* on the point in question. Clearly, consistency is thus a necessary condition for the scheme represented by (2.48) to yield a set of results over an extended region of the $(x,t)$-plane which approximates the global exact solution of (2.47), but it is by no means a sufficient condition. In practice, the analysis of consistency is performed using Taylor expansions about the point in question. If we use a subscript $n$ to label points on the $x$-axis, and a superscript $k$ to label different time levels, we can write for any function, $f$

$$
\begin{aligned}
&\vdots \\
f_{n-1}^{k} &= f_{n}^{k} - f_x \Delta x + \tfrac{1}{2} f_{xx} \Delta x^2 + O(\Delta x^3), \\
f_{n}^{k+1} &= f_{n}^{k} + f_t \Delta t + \tfrac{1}{2} f_{tt} \Delta t^2 + O(\Delta t^3), \\
f_{n+1}^{k} &= f_{n}^{k} + f_x \Delta x + \tfrac{1}{2} f_{xx} \Delta x^2 + O(\Delta x^3), \\
&\vdots
\end{aligned}
\qquad (2.49)
$$

all derivatives being calculated at some point $(x_0, t_0)$. Similar expressions can be written for $f_n^{k-1}$, etc, as needed. Such expressions are replaced into (2.48). For consistency, the zero-order terms in (2.49) should cancel out and the first-order terms (after dividing by $\Delta x$ and $\Delta t$) should reproduce (2.47).

### 2.3.3 Order of Accuracy

If, after replacing the Taylor expansions (2.49) into (2.48), dividing by $\Delta x$ and $\Delta t$ as required and subtracting (2.47), the lower order terms contain a factor $\Delta t^p$, we commonly say that (2.48) is $p$th-order accurate in $t$; similarly, if such terms contain a factor $\Delta x^p$, we say that (2.48) is $p$th-order accurate in $x$. The general wisdom is that the higher the order of accuracy the better the results, as $\Delta x$ and $\Delta t$ are made smaller and smaller. For a number of problems, it is not necessarily so.

One of the major shortcomings of the analysis of consistency, as outlined in Section 2.3.2, is that it relies on Taylor expansions. An evaluation of the order of accuracy of a given scheme makes sense only if the Taylor series converges monotonically, so that each additional term contributes to the sum by less than the preceding one. This is not necessarily the case in most of the problems of practical interest, when singularities such as gradient discontinuities and shock waves are present or are about to appear within a flow which is still continuous and differentiable. An example of this problem related to Burger's equation is discussed in Section 2.6.1.

### 2.3.4 Stability

A second necessary condition to be satisfied by a numerical scheme is that of stability, in the sense that the results of the finite-difference calculation must remain bounded as the calculation proceeds. Little exposure to finite-difference work is needed to learn that, when instability sets in, the values explode, frequently in not more than three computational steps. A well-known analysis, due to von Neumann, shows indeed that the error growth is exponential and, for a given scheme, shows under what limitations, if any, instability can be avoided. The analysis is, as in the one of Section 2.3.2, local, this time in the sense that all coefficients in (2.47) are frozen at their values at a point, to make the problem *linear*. A system of equations, after linearization, can be written in the form

$$\boldsymbol{u}_t + \mathbf{A}\boldsymbol{u}_x = 0, \tag{2.50}$$

where $\mathbf{A}$ is a matrix with constant elements. Using the method of separation of variables, a simple solution of (2.50) is obtained in the form

$$\boldsymbol{u} = \boldsymbol{w}e^{i(\alpha x+\beta t)}, \tag{2.51}$$

with $\boldsymbol{w}$, a constant vector and $\alpha$ and $\beta$ also constant. Replacing (2.51) into (2.50) and dropping the common factor, $\exp(i(\alpha x + \beta t))$, we obtain the condition

$$\beta\boldsymbol{w} = \mathbf{A}\alpha\boldsymbol{w},$$

which takes on the form, if we let $\Lambda = -\beta/\alpha$,

$$\Lambda\boldsymbol{w} = \mathbf{A}\boldsymbol{w}.$$

For (2.51) to be a nontrivial solution of (2.50), $\Lambda$ has to be any eigenvalue of $\mathbf{A}$; consequently (2.51) is written as

$$\boldsymbol{u} = \boldsymbol{w}e^{i\alpha(x-\Lambda t)}.$$

If the number of scalar equations and unknowns in (2.50) is $n$, $\mathbf{A}$ has $n$ eigenvalues. Each eigenvalue, $\Lambda$, provides a simple solution, which in turns depends on an arbitrary choice of the parameter $\alpha$. Each simple solution can be considered as a wave, traveling at a speed $\Lambda$. Its propagation is not affected by other waves (depending on different wave number, $\alpha$, or wave speed $\Lambda$), because of the linearity of the system. Moreover, its speed of propagation, $\Lambda$, is the same, regardless of the wave number, $\alpha$. We can easily relate this linear problem to the theory of Fourier series and integrals. Any set of initial conditions, at time $t = 0$, can be written as a Fourier series

$$\boldsymbol{u}(x,0) = \sum_{-\infty}^{+\infty} \boldsymbol{w}_m e^{imx},$$

if $\boldsymbol{u}(x,0)$ is a periodic function of $x$ (for any value of $x$ between $-\infty$ and $+\infty$), or as a Fourier integral

$$\boldsymbol{u}(x,0) = \int_{-\infty}^{+\infty} \boldsymbol{w}(\alpha)e^{i\alpha x},$$

if $\boldsymbol{u}(x, 0)$ is not periodic in $x$. Consistently, an exact solution of (2.50) at time $t = \Delta t$ is

$$\boldsymbol{u}(x, \Delta t) = \sum_{-\infty}^{+\infty} \boldsymbol{w}_m e^{im(x-\Lambda t)}$$

or

$$\boldsymbol{u}(x, \Delta t) = \int_{-\infty}^{+\infty} \boldsymbol{w}(\alpha) e^{i\alpha(x-\Lambda t)}, \tag{2.52}$$

respectively. Limiting ourselves to the case of a nonperiodic $\boldsymbol{u}$, we see that, at any point $x = n\Delta x$, the original elementary component of $\boldsymbol{u}$ at a time level $k$

$$\boldsymbol{w}(\alpha) e^{i\alpha n \Delta x} = \boldsymbol{w}^k(\alpha) e^{i\alpha n \Delta x},$$

corresponding to an arbitrarily chosen time $t = 0$, becomes

$$\boldsymbol{w}(\alpha) e^{i\alpha n \Delta x} e^{-i\alpha \Lambda \Delta t} = \boldsymbol{w}^{k+1}(\alpha) e^{i\alpha n \Delta x}$$

at level $k + 1$, corresponding to $t = \Delta t$, so that

$$\boldsymbol{w}^{k+1}(\alpha) = e^{-i\alpha\Lambda\Delta t} \boldsymbol{w}^k(\alpha), \tag{2.53}$$

shows how the *Fourier spectrum*, $\mathbf{w}(\alpha)$, of the exact solution changes from level $k$ to $k + 1$. Note that, regardless of the wave number, $\alpha$, the modifying factor, $\exp(i\alpha\Lambda\Delta t)$ has modulus 1; therefore, *it does not alter the modulus of the spectral component*. Note also that, as we said above, all elements in the integral (2.52) travel in space and time at the same speed, $\Lambda$, regardless of the value of $\alpha$. The linearized stability analysis, according to von Neumann, parallels the above considerations. Let us resume studying (2.48), and expand a set of initial data (at step $k$) into a Fourier integral (or, for problems periodic in $x$, a Fourier series). We will consider a vector $\mathbf{f}(x)$ as such a set, for generality, and call $\mathbf{F}(\alpha)$ its Fourier spectrum

$$\mathbf{f}^k(x) = \int_{-\infty}^{+\infty} \mathbf{F}^k(\alpha) e^{i\alpha x} d\alpha$$

so that, at a point $x = n\Delta x$,

$$\mathbf{f}_n^k = \int_{-\infty}^{+\infty} \mathbf{F}^k(\alpha) e^{i\alpha n \Delta x} d\alpha. \tag{2.54}$$

Since (2.48) is now assumed linear, we can treat each term in the integral of (2.54) and each component of $\mathbf{f}_n^k$ separately. Therefore, from the general relations among $f_n^{k+1}$ and $f_{n-1}^k, f_n^k, f_{n+1}^k$, as expressed by (2.49), single relations can be obtained between $\mathbf{F}^{k+1}(\alpha)e^{i\alpha n\Delta x}$ and $\mathbf{F}^k(\alpha)e^{i\alpha n\Delta x}$, $\mathbf{F}^k(\alpha)e^{i\alpha(n-1)\Delta x}$, $\mathbf{F}^k(\alpha)e^{i\alpha(n+1)\Delta x}$,.... All these terms contain the factor $\exp(i\alpha n\Delta x)$ which can, thus, be dropped. Note that neither $\mathbf{F}^{k+1}(\alpha)$ nor $\mathbf{F}^k(\alpha)$ depends on $\Delta x$. Some algebraic manipulations allow the relation between $\mathbf{F}^{k+1}(\alpha)$ and $\mathbf{F}^k(\alpha)$ to be written in the form

$$\mathbf{F}^{k+1}(\alpha) = \mathbf{g}\mathbf{F}^k(\alpha), \tag{2.55}$$

where $\mathbf{g}$ (the *amplification matrix*) is a (generally complex) matrix depending on $\Delta x$, as well as on $\alpha$ and on the coefficients of the finite-difference equation. Again, we can reduce the analysis of (2.55) to a discussion of eigenvalues. Indeed, we can write

$$\mathbf{g} = \mathbf{M}^{-1}\mathbf{D}\mathbf{M} \tag{2.56}$$

where $\mathbf{D}$ is a diagonal matrix, and $\mathbf{M}$ is a matrix that satisfies the equation

$$\mathbf{M}\mathbf{g} = \mathbf{D}\mathbf{M}.$$

The latter is the same as (2.12) with all the left eigenvectors, $\vec{\mu}$, arranged in a matrix form, and the scalars, $\Lambda$, arranged on the diagonal of $\mathbf{D}$. Therefore, the nonzero elements of $\mathbf{D}$ are the eigenvalues of $\mathbf{g}$. Using (2.56), (2.55) becomes:

$$\mathbf{F}^{k+1}(\alpha) = \mathbf{M}^{-1}\mathbf{D}\mathbf{M}\mathbf{F}^k(\alpha).$$

Consequently,

$$\mathbf{M}\mathbf{F}^{k+1}(\alpha) = \mathbf{D}\mathbf{M}\mathbf{F}^k(\alpha),$$

that is,

$$\Phi^{k+1}(\alpha) = \mathbf{D}\Phi^k(\alpha), \tag{2.57}$$

if we let $\Phi = \mathbf{MF}$. Let $\Omega$ be any eigenvalue of $\mathbf{D}$. We can interpret (2.57) by saying that, for any $\Omega$, there is a certain combination of the numerical solutions which moves at the speed $\Omega$ as a numerical wave. The local growth of the spectral component of such a wave is $|\Omega|$. Obviously, if any one of the eigenvalues has a modulus larger than 1, the related spectral component grows exponentially in successive steps. A necessary condition for von Neumann stability, is thus

$$\max|\Omega| \leq 1. \tag{2.58}$$

It is convenient to plot $\Omega$ in a complex plane and see whether it falls within the circle defined by (2.58). Moreover, since not only the modulus but also the argument of

$\Omega$ is important for a preliminary analysis of a numerical scheme, it is convenient to write $\Omega$ in the form

$$\Omega = Ge^{i\omega}, \tag{2.59}$$

and study plots of $G(\alpha\Delta x)$ and $\omega(\alpha\Delta x)$ separately.

### 2.3.5 Dissipation and Dispersion

If we compare (2.59) to (2.53), we see that for a perfect numerical solution, **g** should have the same eigenvalues as the matrix **A** of the original equation, and for each eigenvalue the following two conditions should apply:

$$G = 1, \quad \omega = -\alpha\Lambda\Delta t.$$

If proper discretizing schemes are chosen, forcing $G$ to remain less than 1, the calculation is certainly stable. When $G < 1$, however, the computed values tend to decrease in magnitude from step to step. This phenomenon is called *dissipation*. A strong dissipation can do a lot of damage to a calculation, particularly in a long calculation; it should be feared as a treacherous enemy because, contrary to instability which brings thc computation to a catastrophic stop, dissipation lets the computation continues indefinitely, with deceivingly smooth results. The other bad feature of finite-difference techniques is that $\omega$ is generally different from $-\alpha\Lambda\Delta t$. The consequence is what is commonly called a phase shift in $f$, as it is updated from step to step. To make things worse, the phase shift is a function of $\alpha$. In a long calculation, the distribution of $\mathbf{f}(x)$ may get strongly distorted.

### 2.3.6 Lax's Equivalence Theorem

In 1957 [121], Lax proved that, if (1) a scheme is consistent, in the sense of Section 2.3.2; and (2) it is stable, according to von Neumann analysis of Section 2.3.4 (or in mathematical terms, stable in the $L_2$ norm); (3) the original problem is linear; and (4) the initial value problem is well posed, in the sense of Hadamard [94] (that is, the solution of the partial differential equation depends continuously on the initial conditions), then convergence is assured, in the sense that results computed from (2.48) tend to get closer and closer to the exact solution of (2.47) on the entire region of the $(x,t)$-plane covered by the calculation. When all conditions are met, one can safely infer *global* convergence from local analysis of consistency and stability. Lax's equivalence theorem is an important tool of numerical analysis, particularly in the area of code verification. In practice, however, the equivalence theorem has been often misused, if not abused, not so much because of the nonlinearity of the problems in gas dynamics, as for other reasons, which are mostly interlocked: (1) A formal analysis of consistency, such as in Section 2.3.2, does not reveal its failure where the Taylor expansion argument is invalid, see Section 2.6.1. (2) Such failure is generally associated with the formation of discontinuities; when discontinuities exists, two separate Taylor expansions can be defined on each side of the singularity, but a single, bilateral Taylor expansion is out of the question. (3) In the presence of discontinuities, the initial value problem is no longer well posed (which does not mean that the solution does not exist!). (4) Proof of consistency does not imply that the scheme is physically correct. (5) A numerical calculation may converge to the exact solution if the size of the intervals is made extremely small, but such an ideal situation is seldom reached in practice, particularly in multidimensional calculations. (6) Finally, neither the theorem nor the previous analyses of consistency and stability deal with the numerical effects of the treatment of boundary conditions on interior points. Instabilities and other departures from the correct solution are, very often, due to the choice of a physically unacceptable scheme and to the consequent mishandling of boundary conditions.

### 2.3.7 Lax–Wendroff and MacCormack Schemes

In 1960, Lax and Wendroff [122] introduced a scheme based on a Taylor expansion in time designed to achieve second-order accuracy. Let us linearize Burger's equation by freezing $u$ at some value $c$, namely

$$u_t + cu_x = 0, \tag{2.60}$$

and expand in time about $u_n^k$:

$$u_n^{k+1} = u_n^k + u_t\Delta t + u_{tt}\Delta t^2/2$$

Now, with the aid of (2.60) replace $u_t$ and $u_{tt}$ by $-cu_x$ and $c^2u_{xx}$, respectively:

$$u_n^{k+1} = u_n^k - cu_x\Delta t + c^2u_{xx}\Delta t^2/2$$

Discretizing $u_x$ and $u_{xx}$ with central differences results in the Lax–Wendroff scheme:

$$u_n^{k+1} = u_n^k - \tfrac{\sigma}{2}\left(u_{n+1}^k - u_{n-1}^k\right) + \tfrac{\sigma^2}{2}\left(u_{n+1}^k - 2u_n^k + u_{n-1}^k\right), \tag{2.61}$$

where $\sigma = c\Delta t/\Delta x$ is known as the *Courant–Friedrichs–Lewy* (*CFL*) or *Courant number* [43] (for an English translation see [44]). If we introduce the Taylor expansions (2.49) into (2.61), we obtain

$$u_t + cu_x = \cancel{\left(-u_{tt} + c^2u_{xx}\right)}\Delta t/2 - u_{ttt}\Delta t^2/3! - cu_{xxx}\Delta x^2/3! + H.O.T.$$

The first term on the right is zero and the equation reduces to

$$u_t + cu_x = -\frac{c\Delta x^2}{3!}(1-\sigma^2)u_{xxx} + H.O.T. \tag{2.62}$$

This shows that the scheme is consistent and second-order accurate. Note that we solve (2.60) only in the limit $\Delta x \rightarrow 0$. We call (2.62) the *modified equation* which, in this case, is a KdV-like equation, (2.2).

The scheme proposed by MacCormack in 1969 [134], and widely used thereafter, performs the integration over a time interval, $\Delta t$, in two successive steps: a *predictor*, in which a preliminary evaluation of the value of $\tilde{u}$ at level $k+1$ is obtained, via the first-order accurate scheme

$$\frac{\tilde{u}_n - u_n^k}{\Delta t} + c\frac{u_n^k - u_{n-1}^k}{\Delta x} = 0,$$

which uses a backward space derivative, and a *corrector*, in which a value of $u_n^{k+\frac{1}{2}}$ of $u$, averaged at time $t + \Delta t/2$ from $u_n^k$ and $\tilde{u}_n$, is updated over a half-interval using forward derivatives of $\tilde{u}$ in the manner of the following first-order accurate scheme:

$$\frac{u_n^{k+1} - u_n^{k+\frac{1}{2}}}{\Delta t/2} + c\frac{\tilde{u}_{n+1} - \tilde{u}_n}{\Delta x} = 0$$

with

$$u_n^{k+\frac{1}{2}} = \tfrac{1}{2}\left(u_n^k + \tilde{u}_n\right).$$

If we combine the *predictor* and *corrector* steps into one step, we find that

$$u_n^{k+1} = u_n^k - \tfrac{\sigma}{2}\left(u_{n+1}^k - u_{n-1}^k\right) + \tfrac{\sigma^2}{2}\left(u_{n+1}^k - 2u_n^k + u_{n-1}^k\right),$$

and that this is identical to the Lax–Wendroff scheme. The amplification factor for these two schemes is

$$\Omega = 1 - i\sigma \sin(\theta) - 2\sigma^2 \sin^2(\theta/2),$$

where $\theta = \alpha \Delta x$. In the complex plane $\Omega$ traces out an ellipse centered on the real axis at $1 - \sigma^2$ and having a semi-axis length of $\sigma^2$ along the abscissa and semi-axis length of $\sigma$ along the ordinate. The ellipse traced by $\Omega$ lies inside the unit circle if $\sigma \leq 1$. Hence, if the *CFL* number is less than or equal to 1 the scheme is stable. The requirement that the *CFL* number be less than 1 means that the characteristic, $dx/dt = c$, drawn back from $[x_n, t + \Delta t]$ lies within the footprint of the scheme, $[x_{n-1}, x_{n+1}]$, at time $t$. Hence, the domain of dependence of the finite-difference scheme should include the domain of dependence of the partial differential equation. The *phase error* defined as $\varepsilon_\theta = \omega/\alpha\, c\Delta t$ is given by

$$\varepsilon_\theta = \frac{\tan^{-1}\left[\sigma \sin(\phi)/\left(1 - 2\sigma^2 \sin^2(\theta/2)\right)\right]}{\sigma\theta}.$$

Plots of the amplification matrix, dissipation error, and phase error are shown in Figures 2.17, 2.18, and 2.19, respectively.

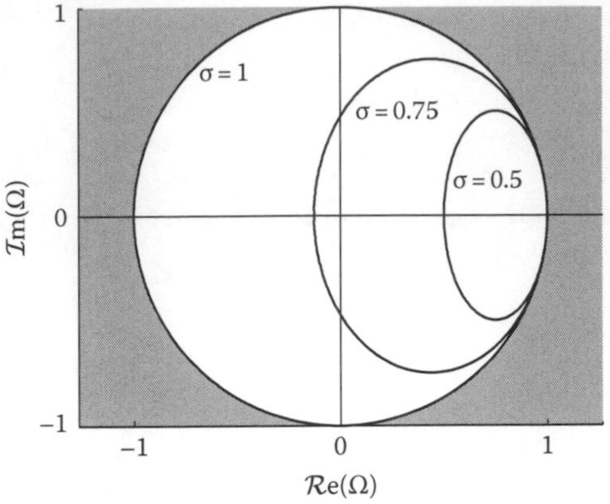


**FIGURE 2.17:** The amplification matrix for the Lax–Wendroff and MacCormack schemes. The shaded area is unstable.

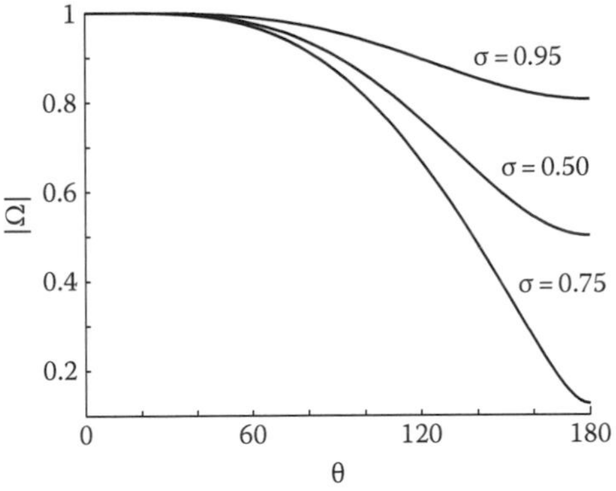


**FIGURE 2.18:** Dissipation error as a function of the phase angle in degrees.

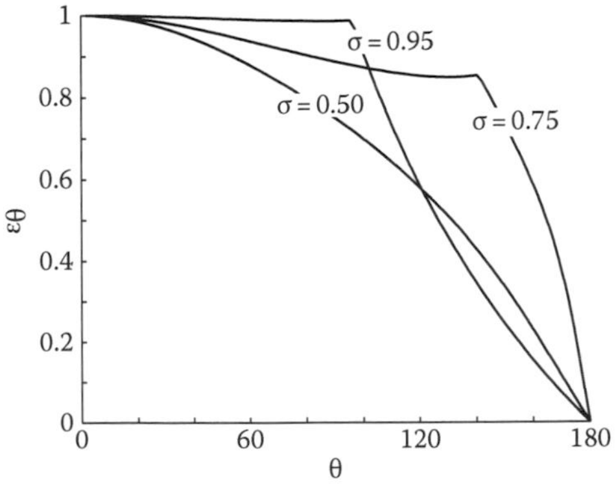


**FIGURE 2.19:** Phase error as a function of the phase angle in degrees.

### 2.3.8 Generalized Stability Analysis

The stability analysis of Section 2.3.4 lends itself to two-level schemes. However, many schemes involve more than two time levels. To study these schemes, we need a more general stability analysis. Consider the leap-frog scheme defined by

$$\frac{u_n^{k+1} - u_n^{k-1}}{2\Delta t} + c\frac{u_{n+1}^k - u_{n-1}^k}{2\Delta x} = 0.$$

Applying the analysis of Section 2.3.4 yields the equation

$$w^{k+1}(\alpha) + 2i\sigma \sin(\alpha\Delta x) w^k(\alpha) - w^{k-1}(\alpha) = 0, \tag{2.63}$$

which contains values of $w$ on three levels. The difficulty is overcome by developing a method to treat schemes of the form

$$w_n^{k+1} + Bw_n^k + Cw_n^{k-1} = 0,$$

including schemes where $w$ is a vector and the number of space variables is greater than 1. Let $H^k = w^{k-1}$, so that $H^{k+1} = w^k$. Therefore (2.63) can be written in the form

$$\begin{aligned} w^{k+1}(\alpha) &= -2i\sigma \sin(\alpha\Delta x) w^k(\alpha) + H^k(\alpha), \\ H^{k+1}(\alpha) &= w^k(\alpha), \end{aligned}$$

or, in the vector form

$$\Phi^{k+1}(\alpha) = \mathbf{g}\Phi^k(\alpha), \tag{2.64}$$

where

$$\Phi = \begin{vmatrix} w \\ H \end{vmatrix}, \quad \mathbf{g} = \begin{vmatrix} -B & -C \end{vmatrix}, \tag{2.65}$$

and $B = 2i\sigma \sin(\alpha\Delta x)$ and $C = -1$. The analysis can be reduced to determining the eigenvalues of $\mathbf{g}$, if the interest is mainly in stability (as opposed to phase errors). Indeed, we can write

$$\mathbf{g} = \mathbf{M}^{-1}\mathbf{D}\mathbf{M},$$

where

**D** is a diagonal matrix
**M** is a matrix which satisfies the equation
$\mathbf{Mg} = \mathbf{DM}$.

The latter is the same as (2.12) with all the left eigenvectors, $\vec{\mu}$ arranged in matrix form, and the scalars, $\lambda$ arranged on the diagonal of $\mathbf{D}$. Therefore, the nonzero elements of $\mathbf{D}$ are the eigenvalues of $\mathbf{g}$. Since the determinant of a product of matrices is the product of their determinants, and since $\det(\mathbf{M}^{-1}) = 1/\det \mathbf{M}$, we conclude that

$$\det \mathbf{g} = \det \mathbf{D},$$

that is, $\det \mathbf{g}$ is the product of all the eigenvalues of $\mathbf{g}$. For stability, the modulus of $\det \mathbf{g}$ must be less than 1; therefore, all eigenvalues of $\mathbf{g}$ must be within the unit circle. For the leap-frog scheme, the $\mathbf{g}$ is defined by (2.65) and the eigenvalues of $\mathbf{g}$ are the solutions of the equation

$$\Omega^2 + 2i\sigma \sin(\alpha\Delta x)\Omega - 1 = 0, \tag{2.66}$$

with $\Omega = X + iY$ and $\beta = \sigma \sin(\alpha\Delta x)$, (2.66) splits into the equations $Y = -\beta$ and $X^2 + \beta^2 - 1 = 0$, that is,

$$\begin{aligned} X^2 + Y^2 &= 1, \\ Y &= -\sigma \sin(\alpha\Delta x), \end{aligned} \tag{2.67}$$

and

$$\Omega(\alpha) = -i\sigma \sin(\alpha\Delta x) \pm \sqrt{1 - \sigma^2 \sin^2(\alpha\Delta x)}.$$

The fact that $\Omega(\alpha)$ has two roots means that the scheme has a spurious or parasitic solution. This is a typical problem associated with multiple step schemes. The positive root is the desired root. The negative root corresponds to a mode that oscillates in time and travels in the wrong direction, i.e., if $c > 0$, the parasitic wave travels to the left. Care must be taken in a numerical calculation not to excite this mode. Focusing our attention on (2.67), we find again that $\sigma \leq 1$ must be satisfied if the second of (2.67) has to be met. In this case, however, all eigenvalues lie on the unit circle, so that no damping occurs. In this regards, the leap-frog scheme seems optimal. The phase error is given by

$$\varepsilon_\theta = \pm \frac{\tan^{-1}\left[\sigma \sin(\theta)/\sqrt{1 - \sigma^2 \sin^2(\theta)}\right]}{\sigma\theta} = \pm \frac{\sin^{-1}(\sigma \sin(\theta))}{\sigma\theta}.$$

### 2.3.9 Interpolation

The Lax–Wendroff and MacCormack schemes belong to a class of schemes that treat the space and time differencing simultaneously. The reader should also have noticed that the only element of *physics* that comes into play with these schemes is in the stability requirement that the domain of dependence of the partial differential equation be satisfied, i.e., $\sigma \leq 1$. A different approach for developing a numerical scheme is to treat the variable $u$ as a time-continuous, space-discrete function,

$$u_n(t) \simeq U(x_n, t),$$

allowing us to write a system of ordinary differential equations of the form

$$\frac{\mathrm{d}u_n}{\mathrm{d}t} = \mathcal{L}_n\left[-U\frac{\partial U}{\partial x}\right], \tag{2.68}$$

where $\mathcal{L}_n$ is some discrete approximation of the flux. The system of equations represented by (2.68) is called a *semi-discretization* (or method of lines [73]) of the partial differential equation. In the next few sections, we will investigate how to discretize the flux while bringing some elements of the relevant physics into the approximation. In Section 2.3.14, the time integration is treated with a Runge–Kutta method.

The discretization of the flux difference can be viewed as a two-stage process. In the first stage, the discrete cell values of the conserved state variables are interpolated to the cell boundaries. This is called the *projection* stage. The space order of accuracy of the scheme is determined by the accuracy of the interpolation in the projection stage. For a one-dimensional problem, at the cell boundary the projected values from the left and the right create a jump in the state variable. This jump is resolved in the second stage, called the *evolution* stage, by finding a unique, physically meaningful, set of fluxes at the cell boundary. The evolution stage consists of solving a one-dimensional Riemann problem. This is in a sense a subgrid model of the physics representing the equation, or equations, we are solving. The Riemann problem can be solved exactly, as in Godunov's scheme, or approximately by one of the many "approximate Riemann solvers" available, see [220].

For the projection stage we consider the κ-*interpolation* proposed by van Leer [232], see also [236]. We begin by expressing the distribution of the state variable within a cell centered at $x_n$ with boundaries at $(x_{n-\frac{1}{2}}, x_{n+\frac{1}{2}})$ in terms of Legendre polynomials

$$u(x) = U_n + (x - x_n)\frac{\partial u}{\partial x} + \frac{3\kappa}{2}\left((x - x_n)^2 - \frac{\Delta x^2}{12}\right)\frac{\partial^2 u}{\partial x^2} + O(\Delta x^3), \qquad x_{n-\frac{1}{2}} < x < x_{n+\frac{1}{2}}, \tag{2.69}$$

where $U_n$ is the volume average of $u$ over the $n$th cell:

$$U_n = \frac{1}{\Delta x_n}\int_{x_n-\frac{\Delta x}{2}}^{x_n+\frac{\Delta x}{2}} u\mathrm{d}x. \tag{2.70}$$

If we expand in a Taylor series about $u_n$, taking $x_n$ as the origin and denoting derivatives with respect to $x$ by primes, we find

$$U_n = \frac{1}{\Delta x_n} \int_{\frac{-\Delta x}{2}}^{\frac{\Delta x}{2}} \left( u_n + u_n' x + u_n'' \frac{x^2}{2!} + u_n''' \frac{x^3}{3!} + \ldots \right) \mathrm{d}x,$$
$$= u_n + \frac{1}{12} u_n'' \frac{\Delta x^2}{2!} + O(\Delta x^4). \tag{2.71}$$

This expression relates the cell averaged value, $U_n$, to the point value, $u_n$. More generally, we have

$$U_{n+m} = \frac{1}{\Delta x_{n+m}} \int_{\frac{\Delta x}{2}+(m-1)\Delta x}^{\frac{\Delta x}{2}+m\Delta x} \left( u_n + u_n' x + u_n'' \frac{x^2}{2!} + u_n''' \frac{x^3}{3!} + \ldots \right) \mathrm{d}x,$$
$$= u_n + m u_n' \Delta x + \left( m^2 + \frac{1}{12} \right) u_n'' \frac{\Delta x^2}{2!}$$
$$+ \left( m^3 + \frac{m}{4} \right) u_n''' \frac{\Delta x^3}{3!} + O(\Delta x^4). \tag{2.72}$$

It follows that

$$\begin{aligned} U_{n+1} &= u_n + u_n' \Delta x + \left( 1 + \frac{1}{12} \right) u_n'' \frac{\Delta x^2}{2} + O(\Delta x^3), \\ U_{n-1} &= u_n - u_n' \Delta x + \left( 1 + \frac{1}{12} \right) u_n'' \frac{\Delta x^2}{2} + O(\Delta x^3), \end{aligned} \tag{2.73}$$

and therefore

$$\begin{aligned} U_{n+1} - U_{n-1} &= 2 u_n' \Delta x + O(\Delta x^3), \\ U_{n+1} + U_{n-1} - 2U_n &= u_n'' \Delta x^2 + O(\Delta x^4). \end{aligned} \tag{2.74}$$

Within the order of accuracy of the interpolation defined by Equation 2.69, its derivatives can be approximated by centered differences of $U$ using values from nearby cells:

$$\begin{aligned} \frac{\partial U}{\partial x} &\approx \frac{U_{n+1} - U_{n-1}}{2\Delta x}, \\ \frac{\partial^2 U}{\partial x^2} &\approx \frac{U_{n+1} - 2U_n + U_{n-1}}{\Delta x^2}. \end{aligned} \tag{2.75}$$

The parameter $\kappa$ controls the order of the interpolation as indicated in Table 2.1.

**TABLE 2.1:** Options for κ-interpolation parameter.

| κ | Comments |
|---|---|
| −1 | One-sided, second-order upwind scheme |
| 0 | Upwind, second-order, Fromm's QUICK scheme |
| 1/3 | Upwind biased, third-order scheme |
| 1/2 | Quadratic upstream interpolation scheme |
| 1 | Central difference, second-order scheme |

To evaluate the projections on the left side of the cell boundary at $x_{n+\frac{1}{2}}$, let $x = x_{n+\frac{1}{2}}$:

$$U^L_{n+\frac{1}{2}} = U_n + \tfrac{1}{4}\{(1-\kappa)(U_n - U_{n-1}) + (1+\kappa)(U_{n+1} - U_n)\}. \tag{2.76}$$

Similarly, for the right side of this cell boundary we let again $x = x_{n+\frac{1}{2}}$ but with the interpolation shifted to $n+1$:

$$U^R_{n+\frac{1}{2}} = U_{n+1} - \tfrac{1}{4}\{(1+\kappa)(U_{n+1} - U_n) + (1-\kappa)(U_{n+2} - U_{n+1})\}. \tag{2.77}$$

A five-point stencil is needed to evaluate the projections at the boundaries of a cell. This presents some problems near the boundaries of the computational domain. If $n$ corresponds to a cell centered at a shock wave that is treated as a boundary of the flow, a *"ghost point"* at $n+1$ will be needed in order to evaluate the projections for the computation of the $n-1$ cell.

### 2.3.10 Total Variation Diminishing Property

In 1970, Glimm and Lax [84] showed that the weak solution of a scalar one-dimensional conservation law is *monotonicity preserving*: (1) no new local extrema in $x$ may be created; and (2) the value of the local minimum is nondecreasing, while the value of a local maximum is nonincreasing. It follows from this property that the total variation of $u(x,t)$ at some time $t$ defined by

$$TV(u(t)) = \int_{-\infty}^{\infty} \left|\frac{\mathrm{d}u}{\mathrm{d}x}\right| \mathrm{d}x \tag{2.78}$$

is not increasing in time, i.e.,

$$TV(u(t_2)) \leq TV(u(t_1)), \quad \text{for all } t_2 \geq t_1. \tag{2.79}$$

This can be shown by tracing back from $t_2$ to $t_1$ the solution along characteristics. At shock waves, extrema are consumed, hence the total variation might decrease.

Between successive extrema you get $u_{\max} - u_{\min}$ contributions to the integral in (2.78). Over all extrema you get

$$TV(u(t)) = 2\sum_{-\infty}^{\infty} u_{\max} - 2\sum_{-\infty}^{\infty} u_{\min}. \tag{2.80}$$

It follows that an initially monotone profile remains monotone, hence we say that the conservation law is *monotonicity preserving*.

This property appears as a fundamental property of the governing equation, and it seems reasonable to search for numerical schemes satisfying this property. However, Godunov [86] had shown in 1959 that *if an advection scheme preserves the monotonicity of the solution it cannot be better than first-order accurate* (Godunov's theorem). The possibility of high-resolution monotonicity preserving schemes was not completely shut, since in his analysis Godunov had assumed that the numerical scheme was based on a linear discretization. It was thus possible to look for high-resolution schemes that preserve monotonicity using a nonlinear discretization. Research in this direction was pursued in the early 1970s by J. P. Boris with the development of the flux-corrected transport method [20] and by B. van Leer who developed a nonoscillatory version of the Lax–Wendroff scheme [230]. Harten [95] in 1983, motivated by Glimm and Lax's paper, initiated a search for a class of explicit second-order schemes that had the property of being *total variation diminishing* (TVD). During the 1980s, an extraordinary amount of work was focused on high-resolution schemes, much of it conducted at ICASE,* the epicenter of numerical analysis during this period. This body of work is documented in the compendium [104].

The discrete analog of Equation 2.78 is

$$TV(u_n) = \sum_{n=-\infty}^{\infty} |u_{n+1} - u_n|. \tag{2.81}$$

Osher and Chakravarthy [159] showed that a semi-discretization of (2.4) is TVD if it can be written in the form

$$\frac{\mathrm{d}u_n}{\mathrm{d}t} = c^{+}_{n+\frac{1}{2}}(u_{n+1} - u_n) - c^{-}_{n-\frac{1}{2}}(u_n - u_{n-1}), \tag{2.82}$$

where the $c$'s are nonnegative. The proof by Osher and Chakravarthy is rather involved, and we will not discuss it here. Some details of a simpler proof by Jameson and Lax [106] are worth considering, since it deals with a point of interest to this book. The Jameson–Lax proof depends on a theorem that says that for the semi-discretization

$$\frac{\mathrm{d}u}{\mathrm{d}t} = \sum_{j} c_j u_{i-j}, \tag{2.83}$$

* The Institute for Computer Applications in Science and Engineering, NASA Langley Research Center, 1972–2002.

where the coefficients $c_j$ may depend on $i$ and $t$ either directly or through a dependence on $u$, the function $|u(t)|$ in the $L_1$ norm is a nonincreasing function of $t$ if and only if for all $h$ and $t$

$$c_0(h) + \sum_{j\neq 0} |c_j(h+j)| \leq 0. \tag{2.84}$$

Thus, what they want to show is that going from $t_1$ to $t_2$, where $t_2 > t_1$, $\mathrm{d}|u(t)|_1/\mathrm{d}t \leq 0$. To prove this they write

$$|u(t)|_1 = \sum_i s_i(u_i(t))u_i(t), \tag{2.85}$$

where $s_i$ is the sign of $u_i(t)$. They then argue that since $s_i$ is piecewise constant, it follows that

$$\frac{\mathrm{d}}{\mathrm{d}t}|u(t)|_1 = \sum_i s_i(t)\frac{\mathrm{d}u_i}{\mathrm{d}t}. \tag{2.86}$$

Introducing (2.83) into (2.86)

$$\frac{\mathrm{d}}{\mathrm{d}t}|u(t)|_1 = \sum_i s_i \sum_j c_j(i)u_{i-j} = \sum_h w_h u_h$$

where, after letting $h = i - j$,

$$w_h = \sum_j c_j(h+j)s_{h+j}. \tag{2.87}$$

Assume $u_h \neq 0$, then multiply (2.87) by $s_h$ and, since $c_j(h+j)s_h s_{h+j} \leq \left|c_j(h+j)\right|$, using assumption (2.84), it follows that

$$s_h w_h = c_0(h) + \sum_{j\neq 0} c_j(h+j)s_h s_{h+j} \leq 0.$$

Since by definition $s_h$ and $u_h$ have the same sign, we have for all $h$

$$u_h w_h \leq 0;$$

since this relation holds also for $u_h = 0$, it follows that

$$\frac{\mathrm{d}}{\mathrm{d}t}|u(t)|_1 \leq 0,$$

which is the desired proof. The problem, of course, is that the differentiation of (2.85) is not correct. The sign function is defined by

$$s_i(u_i(t)) = 2H(u_i(t)) - 1,$$

where $H$ is a Heaviside function and (2.86) should read

$$\frac{\mathrm{d}}{\mathrm{d}t}|u(t)|_1 = \sum_i \left[ u_i(t)\frac{\mathrm{d}s_i}{\mathrm{d}t} + s_i(t)\frac{\mathrm{d}u_i}{\mathrm{d}t} \right]. \tag{2.88}$$

In what follows, we do not evaluate $\mathrm{d}u_i/\mathrm{d}t$, since the intention in the proof is to replace this term with (2.83). To proceed with the differentiation, we have to examine four possible cases. These are shown in Figure 2.20. The first case is trivial, $u$ is a continuous function that does not go through zero, and hence $s$ is constant and $\mathrm{d}s_i/\mathrm{d}t = 0$. For case 2, $u$ is continuous but goes through zero at $t_1$, therefore we have

$$u_i(t)\frac{\mathrm{d}s_i}{\mathrm{d}t} = 2\delta(u_i)u_i\frac{\mathrm{d}u_i}{\mathrm{d}t}.$$

However, since $u_i(t_1) = 0$, we have $u_i(t)\mathrm{d}s_i/\mathrm{d}t = 0$. For case 3, $u$ is discontinuous at $t_1$, but does not change sign, therefore, $\mathrm{d}s_i/\mathrm{d}t = 0$. Finally, for case 4, $u$ is discontinuous and changes sign at $t_1$:

$$\begin{aligned} u_i(t)\frac{\mathrm{d}s_i}{\mathrm{d}t} &= 2\delta(u_i)(u_1 + [u_2 - u_1]H(t - t_1))\frac{\mathrm{d}u_i}{\mathrm{d}t} \\ &= (u_1 + u_2)\delta(t_1)\frac{\mathrm{d}u_i}{\mathrm{d}t}. \end{aligned} \tag{2.89}$$

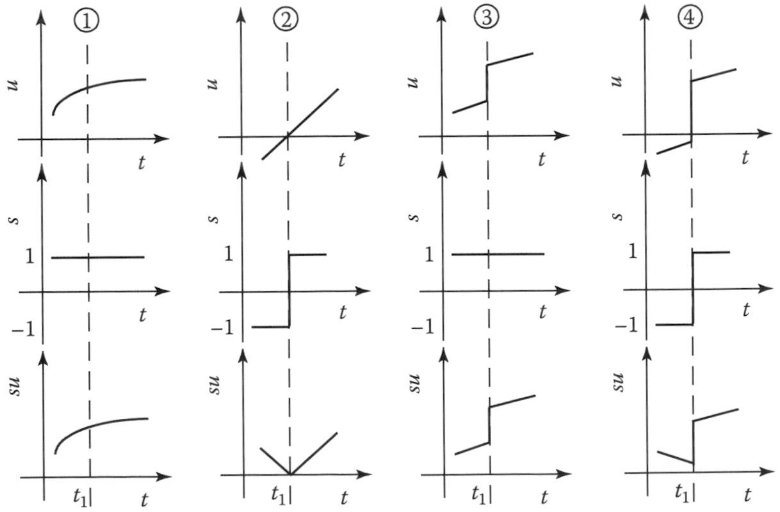


**FIGURE 2.20:** Four cases illustrating the behavior of $u$ needed to be considered in the proof by Jameson and Lax.

We note that in (2.89) we have replaced $\delta(u_i)$ with $\delta(t_1)$, since $u_i$ changes sign at $t_1$, and we have made the substitution $\delta H \approx \frac{1}{2}\delta$, further more $u_1 = u(t_1^-)$ and $u_2 = u(t_1^+)$. For case 4, the proof breaks down. Unlike cases 1 through 3, case 4 involves the multiplication of two generalized functions. As we have already discussed, this is not a trivial operation. Indeed, if we were to evaluate $du_i/dt$, we will end up with objects like $\delta^2$ in (2.89).

In the same paper for the more general case (ignoring the problem with case 4 just discussed), Jameson and Lax have proven that explicit and semi-discrete schemes are TVD for scalar conservation laws if and only if the coefficients of the differences $u_{n-m} - u_{n-m-1}$ have the same sign as $u$ for $m \geq 0$ (points on the upwind side); and the opposite sign for $m < 0$ (points on the downwind side). To be more specific, consider an explicit scheme for $u_n^{k+1}$, where $k$ is the time index:

$$u_n^{k+1} = u_n^k + \sum_{m=-M}^{M-1} c_m(n)\Delta_{n-m-\frac{1}{2}}(u^k), \tag{2.90}$$

where

$$\Delta_{n+\frac{1}{2}}(u^k) = u_{n+1}^k - u_n^k, \tag{2.91}$$

then the scheme is TVD if and only if,

$$\begin{aligned} &c_{-1}(n-1) \geq c_{-2}(n-2) \ldots \geq c_{-M}(n-M) \geq 0, \\ &-c_0(n) \geq -c_1(n+1) \geq \ldots \geq -c_{M-1}(n+M-1) \geq 0. \end{aligned} \tag{2.92}$$

This means that not only must the coefficients have the proper sign but their influence must diminish as they get further away from $n$. In addition, the following condition must be satisfied,

$$-c_0(n) + c_{-1}(n-1) \leq 1. \tag{2.93}$$

Since we have written this for the full discretization (space and time) the coefficients $c_0$ and $c_1$ contain the Courant number. Hence, (2.93) imposes an additional constrain on the Courant number for the TVD property to hold.

In general, no second-order (or higher) scheme is TVD if the scheme is linear (i.e., the $c$'s are not functions of the $u_n$'s). Therefore, any linear TVD scheme is at most first-order. High-order TVD schemes degenerate to first-order accuracy at local extrema. The constant damping of local extrema makes the global error of high-order TVD schemes $O(h)$ in the $L_\infty$ norm and $O(h^{3/2})$ in the $L_2$ norm [96]. Furthermore, Goodman and LeVeque [89] have shown that every (even nonlinear schemes!) TVD scheme for scalar conservation laws in two space dimensions is at most first-order accurate.

The TVD property is too strong. It needs to be relaxed. We do this by looking for a scheme whose coefficients have the proper sign and are uniformly bounded (i.e., there exists an $\varepsilon$, $\varepsilon > 0$, such that for $\forall$ $c$'s $|c| \leq \varepsilon$). And here is where limiters come in.

### 2.3.11 Limiter

For a second-order or higher scheme, let us write the interpolation function for values at cell faces located at $x_n \pm \Delta x/2$ as

$$\begin{aligned} U^L_{n+\frac{1}{2}} &= U_n + \frac{1}{2}\psi(\rho_n)(U_n - U_{n-1}), \\ U^R_{n-\frac{1}{2}} &= U_n - \frac{1}{2}\psi\left(\frac{1}{\rho_n}\right)(U_{n+1} - U_n), \end{aligned} \tag{2.94}$$

where

$$\rho_n = (U_{n+1} - U_n)/(U_n - U_{n-1}). \tag{2.95}$$

Here $\psi$ is called the limiter and $\rho$ is the ratio of consecutive gradients. If we write the $\kappa$-interpolation in this form we get,

$$\begin{aligned} \psi^L_\kappa &= \frac{1-\kappa}{2} + \frac{1+\kappa}{2}\rho_n, \\ \psi^R_\kappa &= \frac{1-\kappa}{2} + \frac{1+\kappa}{2}\frac{1}{\rho_n}. \end{aligned} \tag{2.96}$$

If we impose the TVD constraints of Equations 2.92 and 2.93 and require the $c$'s to be uniformly bounded, then it can be shown [216] that for a second-order scheme to be TVD it must lie in the shadowed region of Figure 2.21. Furthermore, all second-order accurate interpolations must pass through the point (1,1) in the ($\psi$, $\rho$) plane. This follows from the fact that $\rho = 1$ corresponds to $u_{xx} = 0$ and $\psi = 1$ corresponds to the proper interpolation for a linear variation. For the $\kappa$-interpolation with

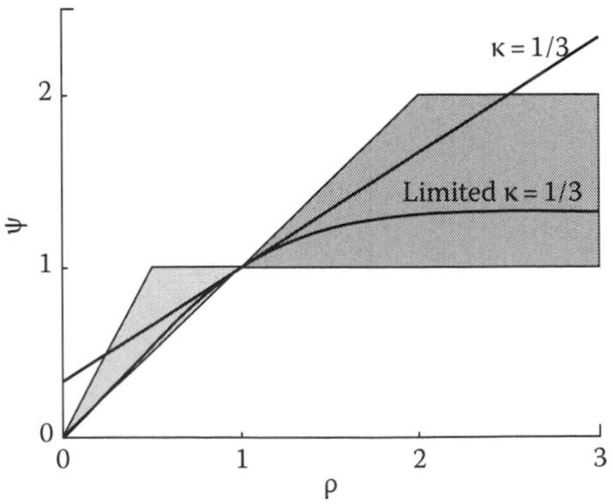


**FIGURE 2.21:** Second-order TVD region.

$\kappa = 1/3$, the line $\psi_{\kappa=\frac{1}{3}}$ passes through (1,1) as expected, but extends outside the shadowed (non-TVD) region.

This can be fixed by the introduction of a gradient limiter. Of the many limiters that have been proposed [208,216,220,236], we will consider only the van Albada limiter [227].

The van Albada limiter replaces the definitions of $\psi_\kappa^L$ and $\psi_\kappa^R$ with,

$$\begin{aligned} \psi_\kappa^L &= s(1 - \kappa s + (1 + \kappa s)\rho_n)/4, \\ \psi_\kappa^R &= s(1 - \kappa s + (1 + \kappa s)/\rho_n)/4, \end{aligned} \tag{2.97}$$

where

$$s = \frac{2\rho_n}{\rho_n^2 + 1}. \tag{2.98}$$

As shown in Figure 2.21 for $\kappa = 1/3$, the van Albada limiter restricts the $\kappa$-interpolation to the TVD region.

### 2.3.12 Riemann's Problem

The Riemann problem is to find the flow that evolves from an initial state where the flow conditions to the left and right of $x = 0$ are in constant but different states. Riemann studied this problem in 1860 [176] using what we know today as Riemann variables and found solutions to the Euler equations for several examples, but did not give a general solution. Since for the problem just formulated there is no reference length, we look for a similarity solution in the variable $z = x/t$. Introducing this variable in (2.4), we get

$$(u - z)u_z = 0.$$

Hence two solutions are possible,

$$u_z = 0, \rightarrow u(z) = \text{constant},$$

and

$$u(z) = z.$$

For the initial data

$$u = \begin{cases} u_l & x < 0, \\ u_r & x > 0, \end{cases}$$

the solution is one of the following:

1. $u(z) = \text{constant}$.

2. For $u_l \leq u_r$, the similarity solution corresponding to $u = z$ is a rarefaction wave and is given by

$$u = \begin{cases} u_l & x/t \leq u_l, \\ x/t & u_l \leq x/t \leq u_r, \\ u_r & x/t \geq u_r. \end{cases}$$

3. For $u_l \geq u_r$, the other similarity solution corresponds to a shock wave with $w = \langle u \rangle$ and satisfies the entropy condition $u_l \geq w \geq u_r$, see Equation 2.32. The solution is given by

$$u = \begin{cases} u_l & x/t < w, \\ u_r & x/t > w. \end{cases}$$

We are now ready to solve the Riemann problem. There are only two significant solutions, namely 2 and 3. Let us consider 3 first, the case of a shock, see Figure 2.11a:

Case (1): $u_l > u_r > 0,\ u = u_l,\quad f = \frac{1}{2}u_l^2$.

Case (2): $u_l > 0,\ u_r < 0$,

(a) $w = \frac{1}{2}(u_l + u_r) < 0,\ u = u_r,\quad f = \frac{1}{2}u_r^2$,

(b) $w = \frac{1}{2}(u_l + u_r) > 0,\ u = u_l,\quad f = \frac{1}{2}u_l^2$.

Solution 2 consists of a *fan* with constant state $u(x,t) = u_l$ to the left of the characteristic defined by $x = u_l t$ and constant state $u(x,t) = u_r$ to the right of the characteristic defined by $x = u_r t$. Within the fan, $u_l \leq x/t \leq u_r$, the solution is given by $u(x/t) = x/t$.

Case (3): $u_l \leq 0 \leq u_r,\ u = 0, f = 0$.

There are other combinations, but they can all be summarized as follows:

$$f(u_l, u_r) = \begin{cases} \frac{1}{2}u_l^2 & \begin{cases} u_l, u_r > 0, \text{ or} \\ u_l > w \geq 0 \geq u_r \end{cases} \\ \frac{1}{2}u_r^2 & \begin{cases} u_l, u_r < 0, \text{ or} \\ u_l \geq 0 > w > u_r \end{cases} \\ 0 & \quad u_l \leq 0 \leq u_r \end{cases} . \qquad (2.99)$$

### 2.3.13 Approximate Riemann Solvers

If we resolve the jump in $u$ at the cell face using the exact solution to the Riemann problem given in the previous section, we will be using what is known as the Godunov scheme [86]. The difference between the exact Riemann solution and Godunov's is due to the cell averaging used in Godunov's scheme. However, it is possible to use an approximation to the exact Riemann solution without significant deterioration of the results. The arguments in favor of an approximate Riemann solver are (1) the exact Riemann solver provides more information than is needed to resolve the jump (particularly for the Euler equations); (2) the extrapolation of $u$ to the cell faces is not exact, hence an exact solution based on inexact initial conditions is not necessary; (3) the exact Riemann solver requires too much effort (for the Euler equations); (4) there is no exact Riemann solver for the multidimensional problem; and (5) averaging of $u$ over the cell introduces errors. A number of approximate Riemann solvers have been introduced and we will consider two: Roe's [181] and Engquist–Osher's [66].

Roe approximates the solution to the Riemann problem by locally linearizing the flux Jacobian matrix, i.e., he solves the equation

$$u_t + \hat{\mathbf{A}} u_x = 0,$$

where $\hat{\mathbf{A}}$ is an approximation to $\mathbf{A} = \mathrm{d}f/\mathrm{d}u$. Roe imposes the following conditions on $\hat{\mathbf{A}}$:

1. $\hat{\mathbf{A}}(u_\mathrm{l}, u_\mathrm{r})(u_\mathrm{r} - u_\mathrm{l}) = f(u_\mathrm{r}) - f(u_\mathrm{l})$
2. $\hat{\mathbf{A}}$ is diagonalizable with real eigenvalues
3. $\hat{\mathbf{A}}(u_\mathrm{l}, u_\mathrm{r}) \rightarrow f_u(u)$ as $(u_\mathrm{l}, u_\mathrm{r}) \rightarrow u$

Condition 1 captures isolated discontinuities exactly (in multidimensional flows it only works for discontinuities that align with cell faces). Condition 2 is to ensure that the approximate problem remains hyperbolic. Condition 3 is the consistency condition.

Consider the hyperbolic linear system represented by

$$u_t + \hat{\mathbf{A}} u_x = 0,$$

with initial conditions,

$$u(x, 0) = \begin{cases} u_l, x < 0, \\ u_r, x > 0, \end{cases}$$

where

$\hat{\mathbf{A}}$ is some constant matrix
$u$ is an $m$ component vector

We want to represent the jump in $u$ at a cell face as a linear combination of the $m$ real distinct right eigenvectors, $v_m$, of $\hat{\mathbf{A}}$:

$$u_\mathrm{r} - u_\mathrm{l} = \sum_i^m \alpha_i v_i.$$

It follows that

$$f_r - f_l = \hat{\mathbf{A}} \sum_i^m \alpha_i \nu_i,$$

but since we are assuming that $\hat{\mathbf{A}}$ is diagonalizable with real eigenvalues, we can write, see Section 2.1.5, $\hat{\mathbf{A}} = \vec{\nu}\Lambda\vec{\nu}^{-1}$ and $\hat{\mathbf{A}}\vec{\nu} = \vec{\nu}\Lambda$ (or expanded $\hat{\mathbf{A}}\nu_i = \lambda_i \nu_i$), therefore,

$$f_r - f_l = \sum_i^m \alpha_i \lambda_i \nu_i. \tag{2.100}$$

Each term in the sum of Equation 2.100 represents one wave in the Riemann solution. For example, for a system of three equations, $m = 3$, we have three waves, with speeds: $\lambda_1$, $\lambda_2$, and $\lambda_3$. $\alpha_i$ is the amplitude of the $i$th wave and $\nu_i$ represents how the effect of the $i$th wave is distributed. Across a discontinuity with speed $\lambda_i$, we have $[\![u]\!] = \alpha_i \nu_i$, $[\![f]\!] = \alpha_i \lambda_i \nu_i$, and therefore $[\![f]\!] = \lambda_i [\![u]\!]$.

On $x/t = 0$,

$$f = f_l + \sum^{(-)} \alpha_i \lambda_i \nu_i \quad \text{for } \forall i \text{ such that } \lambda_i < x/t.$$

Similarly,

$$f = f_r - \sum^{(+)} \alpha_i \lambda_i \nu_i \quad \text{for } \forall i \text{ such that } \lambda_i > x/t.$$

If we add these two results and average,

$$f = \tfrac{1}{2}(f_l + f_r) - \tfrac{1}{2}\sum \alpha_i |\lambda_i| \nu_i. \tag{2.101}$$

For Burgers' equation (2.4), the linearized Jacobian matrix is

$$\hat{\mathbf{A}}(u_l, u_r) = \tfrac{1}{2}(u_l + u_r) = w,$$

which is the exact Rankine–Hugoniot shock speed for this equation. The jump $[\![u]\!]$ propagates along a characteristic with speed $w$. The two possible solutions are

$$u = \begin{cases} u_l & x/t < w, \\ u_r & x/t > w, \end{cases}$$

$$f(u_l, u_r) = \begin{cases} \frac{1}{2}u_l^2 & w \geq 0, \\ \frac{1}{2}u_r^2 & w < 0, \end{cases}$$

or in compact form, as (2.101),

$$f(u_l, u_r) = \tfrac{1}{2}\left(\tfrac{1}{2}u_l^2 + \tfrac{1}{2}u_r^2\right) - \tfrac{1}{2}\left|\tfrac{1}{2}(u_l + u_r)\right|(u_r - u_l).$$

The Roe scheme for Burgers' equation is equivalent to the exact Riemann solver except that it allows for expansion shocks. This can be fixed in a number of ways. Roe's solution is to let $f(u_l, u_r) = \frac{1}{2}u_l u_r$ when $u_l \leq 0 \leq u_r$ [231].

In the usual Riemann problem, we start with a jump in $u$ at $x = 0$, $t = 0$ and look for the solution at a later time. Engquist–Osher instead ask: what was the solution before $t = 0$ that gave rise to the jump in $u$? There is no unique answer to this question. Thus, Engquist–Osher take advantage of the ambiguity and look for a solution that involves only *isentropic* expansions and compressions. The resulting solution to the Riemann problem can be expressed in closed form (this is quite an accomplishment for Euler's equations).

Any scalar function $f(u)$ can be represented as the sum of a nonincreasing function $f^-(u)$ and a nondecreasing function $f^+(u)$, such that $f(u) = f^-(u) + f^+(u)$. The Jacobian matrix $\mathbf{A}$ can also be decomposed into an $\mathbf{A}^-$ and an $\mathbf{A}^+$, such that $\mathbf{A} = \mathbf{A}^- + \mathbf{A}^+$, as follows. Define two eigenvalue matrices

$$\Lambda^{\pm} = \begin{vmatrix} \lambda_1^{\pm} & & \\ & \lambda_2^{\pm} & \\ & & \ddots \end{vmatrix},$$

where $\lambda_i^{\pm} = \frac{1}{2}[\lambda_i \pm |\lambda_i|]$. Then $\Lambda = \Lambda^+ + \Lambda^-$ and $|\Lambda| = \Lambda^+ - \Lambda^-$. Now, since $\mathbf{A} = \vec{v}\Lambda\vec{v}^{-1}$, we have that $\mathbf{A}^+ = \vec{v}\Lambda^+\vec{v}^{-1}$, $\mathbf{A}^- = \vec{v}\Lambda^-\vec{v}^{-1}$, $\mathbf{A} = \mathbf{A}^+ + \mathbf{A}^-$, and $|\mathbf{A}| = \mathbf{A}^+ - \mathbf{A}^-$. With $\mathbf{A}^-$ and $\mathbf{A}^+$, Engquist–Osher define the flux as

$$f(u_l, u_r) = f(u_l) + \int_{u_l}^{u_r} \mathbf{A}^- du, \tag{2.102}$$

which can also be written as

$$f(u_l, u_r) = f(u_r) - \int_{u_l}^{u_r} \mathbf{A}^+ du. \tag{2.103}$$

The paths of integration are taken parallel to the right eigenvectors of $\mathbf{A}$. If we sum and average (2.102) and (2.103), we get

$$f(u_l, u_r) = \tfrac{1}{2}(f(u_l) + f(u_r)) - \tfrac{1}{2}\int_{u_l}^{u_r} |\mathbf{A}| du.$$

For Burgers' equation, $\mathbf{A} = u$, therefore let

$$\mathbf{A}^+ = \max(u, 0) \text{ in the interval } [u_l, u_r], \tag{2.104}$$

$$\mathbf{A}^- = \min(u, 0) \text{ in the interval } [u_l, u_r], \tag{2.105}$$

and $|\mathbf{A}| = \mathbf{A}^+ - \mathbf{A}^-$. Therefore, the Engquist–Osher flux is, in compact form,

$$f(u_l, u_r) = \tfrac{1}{2}\left(\tfrac{1}{2}u_l^2 + \tfrac{1}{2}u_r^2\right) - \tfrac{1}{2}\int_{u_l}^{u_r} |\mathbf{A}|\mathrm{d}u.$$

Recall the exact solution, (2.99):

$$f(u_l, u_r) = \begin{cases} \tfrac{1}{2}u_l^2 & u_l, u_r > 0, \\ \tfrac{1}{2}u_r^2 & u_l, u_r < 0, \end{cases}$$
$$f(u_l, u_r) = 0 \; u_l \leq 0 \leq u_r.$$

For these two cases, the Engquist–Osher scheme has the same solution. For the remaining case,

$$f(u_l, u_r) = \begin{cases} \tfrac{1}{2}u_l^2 & u_l > w \geq 0 \geq u_r \\ \tfrac{1}{2}u_r^2 & u_l \geq 0 > w > u_r, \end{cases}$$

in the domain $u_l \geq x/t \geq u_r$, $u$ folds over and consists of three branches, as shown in Figure 2.22. For the exact Riemann solution these folded branches are replaced by a compression shock, but for Engquist–Osher we get,

$$\begin{aligned} f(u_l, u_r) &= \tfrac{1}{2}u_l^2 - \tfrac{1}{2}(x/t)^2 + \tfrac{1}{2}u_r^2 \\ &= \tfrac{1}{2}\left(u_l^2 + u_r^2\right) \text{ at } x/t = 0. \end{aligned}$$

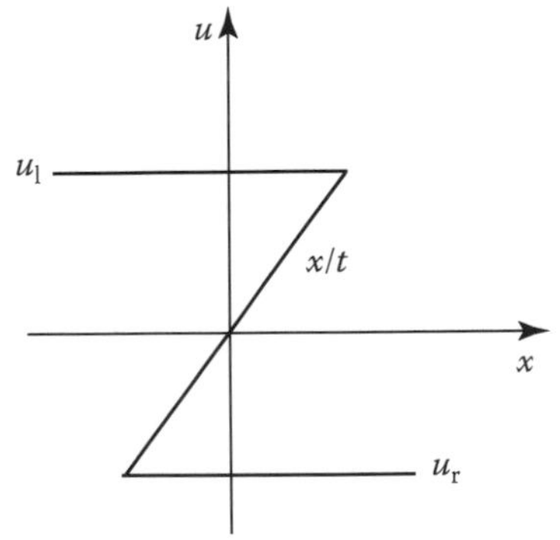


**FIGURE 2.22:** Approximate Riemann solution of Engquist–Osher.

This is the difference between Engquist–Osher and the exact Riemann solution.

### 2.3.14 Time Integration: Runge–Kutta Methods

Let

$$\dot{y} = f(t, y), \quad y(t_0) = y_0.$$

If we integrate from $t_0$ to $t_1 = t_0 + \Delta t$,

$$y(t_1) = y_0 + \int_{t_0}^{t_1} f(t, y(t))\mathrm{d}t.$$

If we now approximate the integral by the trapezoidal rule, we find

$$y(t_1) = y_0 + (f(t_0, y_0) + f(t_1, y_1))\Delta t/2.$$

This was the starting point for Carl David Tolmé Runge [185], who approximated the term $f(t_1, y_1)$ with $f(t_1, y_0 + \Delta t f(t_0 + y_0))$ taking the following steps:

$$\begin{aligned} k_1 &= f(t_0, y_0), \\ k_2 &= f(t_0 + \Delta t, y_0 + \Delta t k_1), \\ y_1 &= y_0 + (k_1 + k_2)\Delta t/2. \end{aligned}$$

Figure 2.23 illustrates how the method works. The dashed line represents the exact curve $y(t)$. At $t_0$, we know $y_0$ and the slope $k_1$. The slope $k_1$ is used to find an approximate $y_1$, $\tilde{y}_1$, which is used to predict a new slope $k_2$. The final value of $y_1$ is $y_0$ plus an average of the slope times $\Delta t$. The symmetrization or averaging cancels out

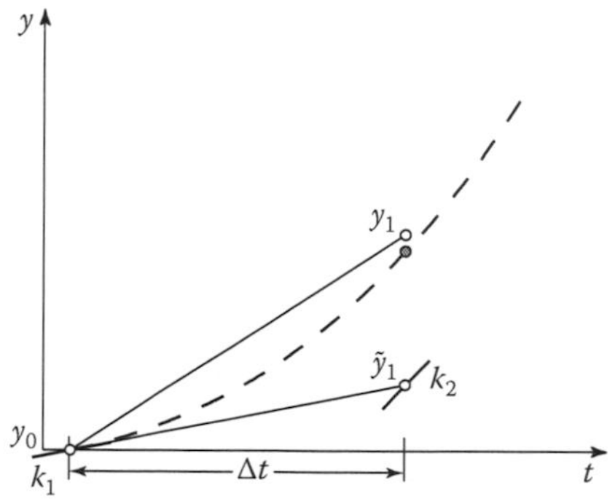


**FIGURE 2.23:** Geometrical illustration of Runge–Kutta method. Exact solution represented by dashed line.

the first-order error, making the method second-order accurate. This line of thinking led to the idea of evaluating $f(t, y)$ to first order with various slopes $k_i$, each evaluation having different coefficients of higher order terms. The trick then is to find the right combination to eliminate the error terms and achieve higher order.

Referring to (2.68), a $K$-stage Runge–Kutta method is given by

$$\begin{aligned} U^{(0)} &= U^{(k)}, \\ U^{(1)} &= U^{(0)} + \alpha_1 \mathcal{L}(U^{(1)})\Delta t, \\ U^{(2)} &= U^{(0)} + \alpha_2 \mathcal{L}(U^{(2)})\Delta t, \\ &\vdots \\ U^{(K-1)} &= U^{(0)} + \alpha_{K-1} \mathcal{L}(U^{(K-2)})\Delta t, \\ U^{(K)} &= U^{(0)} + \Delta t \sum_{k=1}^{K} \beta_k \mathcal{L}(U^{(k)}), \end{aligned} \tag{2.106}$$

where for consistency, it is required that

$$\sum_{k=1}^{K} \beta_k = 1.$$

$\mathcal{L}$ represents the space semi-discretization operator which is repeatedly evaluated using the values of $U$ obtained in the previous stage. At each stage, the solution is advanced by the fractional step $\alpha_i \Delta t$. On the last stage, all previous space semi-discretizations are added up to get the final value. This standard Runge–Kutta *last step* has been found to require too much memory, and has led to the design of *low-memory* schemes [228] of the following type:

$$\begin{aligned} U^{(0)} &= U^{(k)}, \\ U^{(1)} &= U^{(0)} + \alpha_1 \mathcal{L}(U^{(1)})\Delta t, \\ U^{(2)} &= U^{(0)} + \alpha_2 \mathcal{L}(U^{(2)})\Delta t, \\ &\vdots \\ U^{(K)} &= U^{(0)} + \Delta t \mathcal{L}(U^{(K-1)}). \end{aligned}$$

The $\alpha$ coefficients for a low-memory third-order Runge–Kutta scheme are $\alpha_1 = \frac{1}{3}$, $\alpha_2 = \frac{1}{2}$; and for a fourth-order accuracy they are $\alpha_1 = \frac{1}{4}$, $\alpha_2 = \frac{1}{3}$, and $\alpha_3 = \frac{1}{2}$.

For a low-memory third-order Runge–Kutta scheme, we have

$$\begin{aligned} U^{(1)} &= \left(1 + \tfrac{1}{3}\Delta t \mathcal{L}\right) U^{(k)}, \\ U^{(2)} &= \left(1 + \tfrac{1}{2}\Delta t \mathcal{L} + \tfrac{1}{6}(\Delta t \mathcal{L})^2\right) U^{(k)}, \\ U^{(3)} &= \left(1 + \Delta t \mathcal{L} + \tfrac{1}{2}(\Delta t \mathcal{L})^2 + \tfrac{1}{6}(\Delta t \mathcal{L})^3\right) U^{(k)}, \end{aligned}$$

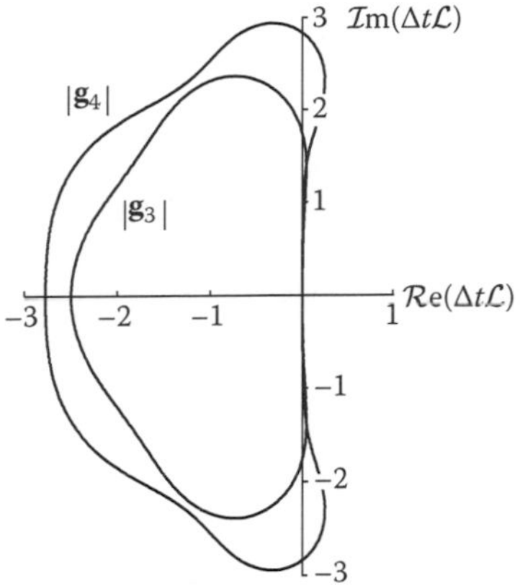


**FIGURE 2.24:** Stability boundaries for third- and fourth-order low-memory Runge–Kutta schemes.

and, therefore, referring to (2.64), the amplification matrix, $\mathbf{g}$, is given by

$$\mathbf{g}_3 = \left(1 + \Delta t\mathcal{L} + \tfrac{1}{2!}(\Delta t\mathcal{L})^2 + \tfrac{1}{3!}(\Delta t\mathcal{L})^3\right).$$

Similarly, for a fourth-order accurate scheme, we get

$$\mathbf{g}_4 = \left(1 + \Delta t\mathcal{L} + \tfrac{1}{2!}(\Delta t\mathcal{L})^2 + \tfrac{1}{3!}(\Delta t\mathcal{L})^3 + \tfrac{1}{4!}(\Delta t\mathcal{L})^4\right).$$

For stability, we must require that $x = 0$. The stability boundaries for these two schemes are shown in Figure 2.24. For central difference semi-discretization, the intercept of the locus of points of $|\mathbf{g}_K| = 1$ with the imaginary axis bounds the allowable *CFL* number; for the third-order and the fourth-order schemes this corresponds to $\sigma \leq 1.8$ and $\sigma \leq 2\sqrt{2}$, respectively. If time accuracy is not a concern (for example, if only the steady state is of interest), schemes can be designed to maximize the stability limit. Van der Houwen [228] has shown that the maximum stability interval along the imaginary axis for an odd $K$-stage scheme is $K - 1$, and Sonneveld and van Leer [207] have obtained the coefficients for the even schemes. For example, for $K = 4$ they find $\alpha_1 = \frac{1}{3}$, $\alpha_2 = \frac{4}{15}$, and $\alpha_3 = \frac{5}{9}$ with a stability region reaching about 3 along the imaginary axis.

### 2.3.15 Verification and Validation

Verification and Validation (V&V) are conceptually very simple ideas which today suffer from overexposure, see [179] and references therein. Overexposure breeds confusion. We will strive not to add to the confusion.

To understand a physical event and hence to be able to predict behavior man *measures and models*. Protagoras had it wrong. Man is not the measure, but the measurer of all things. Modeling is a process of abstraction. Both measurements and modeling involve errors. This is not to say that man cannot grasp reality. On the contrary, man can grasp reality and also know to within what precision. The Navier–Stokes equations are mathematical models of certain physical processes. The Euler equations are mathematical models of similar processes in the limit of vanishing viscosity. With farther abstraction, we say that the potential equation is a mathematical model of a similar process in the absence of viscosity, entropy, and vorticity. The inviscid Burgers' equation is such a highly abstracted model that it has no direct bearing on a physical process. To be more precise, by a mathematical model we mean the full mathematical apparatus that describes an event, which might or might not be physical. Thus, typically in a mathematical model we would have a set of differential equations, some constitutive relations, and boundary and initial conditions as part of our model. To express the fact that there are many mathematical models for the same physical event, we say that the models have a range of fidelity. When we test a *model* of an airplane in a wind tunnel, we are using a *physical model* to simulate a physical process. Like our mathematical models, physical models have a range of fidelity. All models, both mathematical and physical, have modeling errors. Depending on the process being modeled and the model being used, the modeling errors might be insignificant, or not. The modeling error of the Euler equations when used to predict the pressure over a slender wing might be insignificant at low angles of attack, but not at high angles of attack. The modeling error created by the proximity of wind tunnel walls might be insignificant at low speeds and unacceptable at transonic speeds. All measurements have errors and, like modeling errors, their magnitude could be significant or not.

Part of the process of finding a solution to a mathematical model is *verifying* that the solution satisfies the governing equations and the boundary and initial conditions. For example, the motion of a steady, two-dimensional, irrotational, isentropic flow with perturbation velocities that are small compared to the free stream velocity is modeled by the following equation:

$$\beta^2 \frac{\partial^2 \varphi}{\partial x^2} + \frac{\partial^2 \varphi}{\partial y^2} = 0, \tag{2.107}$$

where

$\varphi$ is the velocity potential

$\beta = \sqrt{1 - M_\infty^2}$

If we look for a solution to this equation for flow over a wavy wall defined by

$$y_{\text{wall}} = h \cos(2\pi x), \quad h \ll 1, \tag{2.108}$$

we might consider the following expression for the velocity potential

$$\varphi = \frac{V_\infty}{\beta} h \sin(2\pi x) e^{-2\pi\beta y}. \tag{2.109}$$

To verify that (2.109) is a solution of (2.107) for flow over a wall defined by (2.108) we have to show that (2.109) satisfies (2.107) and the tangency condition at the wall. We do this by inserting (2.109) into the governing equation and the equation for the tangency condition and showing that the equations are satisfied. If we had solved the same problem using a finite-difference scheme, the verification process would be quite different. First, with a finite-difference scheme we do not actually solve the partial differential equation (2.107). What we solve is a discrete model of (2.107), that approximates the *modified equation*, see Section 2.3.7. Therefore, the first step is to verify that the finite-difference solution satisfies the discrete model. This is done by showing that for each grid the iterative (or temporal) process converges and that as the grid is refined the discrete solution is convergent in some norm with some order of accuracy. If we are successful in this, we have verified the discrete model. In general, to verify that the discrete solution satisfies the partial differential equation, we appeal to Lax's equivalence theorem: *a stable, consistent, finite-difference solution converges as the grid size goes to zero to the solution of the differential equation*, see Section 2.3.6. However, there is a caveat; the Lax equivalence theorem is valid only for *linear* difference schemes. Therefore, it is possible to find discrete solutions, using a nonlinear finite-difference scheme, that are not solutions of the differential equation. The carbuncle solution for a supersonic blunt body flow is an example of this anomaly [38]. For problems for which we know the exact solution of the differential equation, an example is Ringleb Flow [177], we verify the validity of the discrete solution by showing that in some norm the error between the exact and discrete solutions vanishes as the grid size goes to zero. We will use this approach throughout this chapter. The reader should note that in the verification process we have outlined, we have discussed the verification of solutions, not codes. It would be reasonable to expect that in some neighborhood of a verified solution, where the solution depends continuously on the initial and boundary data, other solutions from the same code would also satisfy the differential equation, however this is just a reasonable expectation that can be easily violated by either poor or clever programming. The reader should also note that we have not discussed how the solutions relate to the physical process. This is the job of the *validation* process, which we discuss next.

Suppose that we have *exact* data describing a physical event. *Validation* then consists of comparing the verified numerical solution to the physical data with the purpose of establishing how the *differential* equations we have solved describe the physical event, and to what accuracy. Thus, if we have a verified finite-difference solution of the Euler equations for flow over a slender wing at some small angle of attack and we compare the pressure distribution over the wing obtained by finite-difference to the physical pressure distribution and find that the error is within acceptable bounds, we might conclude that we have validated the solution. However, it will be unreasonable to try to validate the velocity profile from the Euler solution with the velocity profile within the boundary layer of the experiment. Of course, we never have exact data from a physical experiment. Therefore, part of the validation process is to establish error bounds for the experiment and then to show that the numerical solution is within those error bounds.

### 2.3.16 Application of Background Scheme

The results of a shock capturing simulation using the background scheme previously described (Roe, R-K) are shown in Figures 2.25 and 2.26. The van Albada limiter was used to reduce Gibbs' oscillations [108] at the shock. The MATLAB code (SHOCKCAP01.M) is available in the companion CD. The calculation used 400 mesh points, with the left boundary set at $-1.5$ and the right boundary set at 2.8, corresponding to a mesh interval of 0.0107. The Courant number was set equal to 1. Figure 2.25a compares the exact $u$-profile to the $u$-profile predicted by the simulation using the limiter. Figure 2.25b shows the Gibbs' oscillations that develop if the limiter is not used.

The computed and exact shock paths are compared in Figure 2.26a. In the simulation the shock location was defined as the midway point in the interval corresponding to max $|\Delta u|$. Various other ways to define the shock locations were tried, but none gave consistently better results. The log of the error defined by

$$\varepsilon = \frac{1}{N}\sum_{1}^{N} |u_n - u_{\text{exact}}|, \tag{2.110}$$

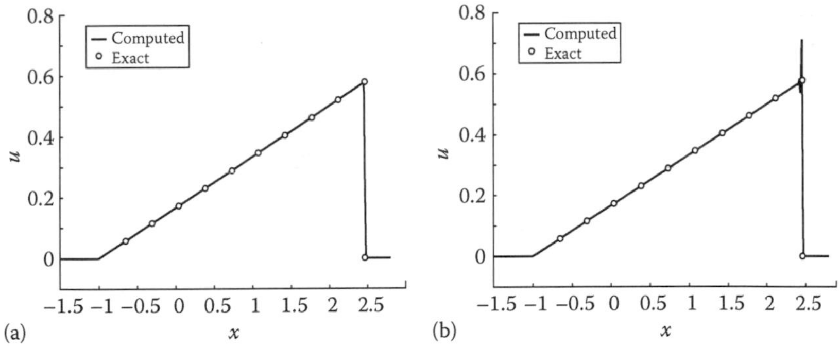


**FIGURE 2.25:** Computed and exact $u$-profile at $t = 5$ for one-saw-tooth problem, 400 mesh points, with (a) limiter on, and (b) limiter off.

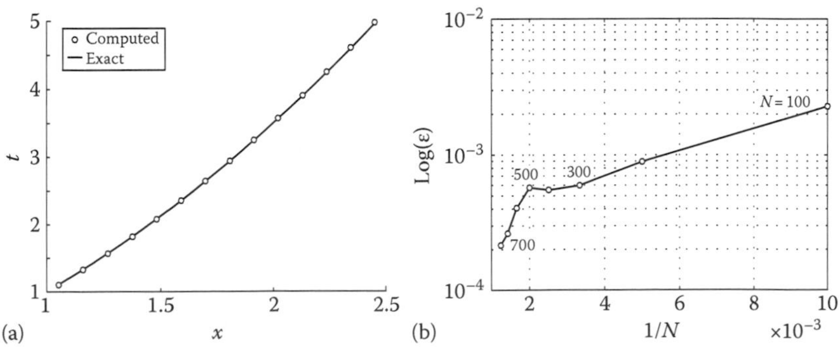


**FIGURE 2.26:** (a) Computed and exact shock wave path of *one-saw-tooth* problem using 400 mesh points; (b) convergence of log(ε) as a function of $1/N$.

where $N$ is the number of mesh points, is shown as a function of $1/N$ in Figure 2.26b. $N$ ranges from 100 to 800 mesh points. Overall, the results look impressive. However, the "one-saw-tooth" function consists of four straight line segments which could be described with five points!

## 2.4 Mappings, Conservation Form, and Transformation Matrices

It will be advantageous to retain the conservation form of the equation after performing a coordinate mapping. To show how this can be accomplished, consider the following mapping from physical coordinates $(t, x)$ to computational coordinates $(\zeta,\tau)$:

$$\begin{aligned} \tau &= t, & t &= \tau, \\ \zeta &= \zeta(x,t), & x &= x(\zeta,\tau). \end{aligned} \tag{2.111}$$

The transformation from one set of coordinates to another is achieved by

$$\begin{bmatrix} \dfrac{\partial}{\partial \tau} \\ \dfrac{\partial}{\partial \zeta} \end{bmatrix} = \begin{bmatrix} 1 & x_\tau \\ 0 & x_\zeta \end{bmatrix} \begin{bmatrix} \dfrac{\partial}{\partial t} \\ \dfrac{\partial}{\partial x} \end{bmatrix}. \tag{2.112}$$

The determinant of this transformation is $D = x_\zeta$ and the Jacobian is $\boldsymbol{J} = D^{-1} = \zeta_x$. The inverse transformation is given by

$$\begin{bmatrix} \dfrac{\partial}{\partial t} \\ \dfrac{\partial}{\partial x} \end{bmatrix} = \begin{bmatrix} 1 & \zeta_t \\ 0 & \zeta_x \end{bmatrix} \begin{bmatrix} \dfrac{\partial}{\partial \tau} \\ \dfrac{\partial}{\partial \zeta} \end{bmatrix}. \tag{2.113}$$

Now consider the conservation law

$$u_t + f_x = 0. \tag{2.114}$$

If we write this equation in terms of computational coordinates, we get

$$u_\tau + \zeta_t u_\zeta + \zeta_x f_\zeta = 0, \tag{2.115}$$

which we can write as

$$(u/J)_\tau - u(1/J)_\tau + (\zeta_t u/J)_\zeta - u(\zeta_t/J)_\zeta + f_\zeta = 0. \tag{2.116}$$

Since

$$\begin{aligned}(1/J)_\tau &= x_{\zeta\tau},\\ (\zeta_t/J)_\zeta &= -x_{\tau\zeta},\end{aligned} \tag{2.117}$$

we can write the conservation law in terms of computational coordinates as

$$U_\tau + F_\zeta = 0, \tag{2.118}$$

where

$$\begin{aligned}U &= u/J\\ F &= (\zeta_t u + \zeta_x f)/J\end{aligned} \tag{2.119}$$

Let $\vec{u}$ be a vector of conservative dependent variables of a conservative system

$$\vec{u}_t + \vec{f}_x = 0.$$

For a system of three equations we have $\vec{u} = [u_1, u_2, u_3]^{\mathrm{T}}$ and $\vec{f} = [f_1, f_2, f_3]^{\mathrm{T}}$. Let $\vec{v}$ be another vector of dependent variables for the same system of equations, for the three equation example we have $\vec{v} = [v_1, v_2, v_3]^{\mathrm{T}}$. The two sets of variables are related by the Jacobian matrix

$$J = \frac{\partial \vec{u}}{\partial \vec{v}}. \tag{2.120}$$

For the three equation example, $J = [\partial\vec{u}/\partial v_1 \; \partial\vec{u}/\partial v_2 \; \partial\vec{u}/\partial v_3]$. The relation between the conservative and nonconservative Jacobian matrices can be expressed through a similarity transformation with the matrix $J$. For the conservative system, we have

$$\vec{u}_t + \mathbf{A}_u \vec{u}_x = 0, \tag{2.121}$$

where $\mathbf{A}_u = \partial\vec{f}/\partial\vec{u}$. From (2.121)

$$\frac{\partial\vec{u}}{\partial\vec{v}}\frac{\partial\vec{v}}{\partial t} + \mathbf{A}_u \frac{\partial\vec{u}}{\partial\vec{v}}\frac{\partial\vec{v}}{\partial x} = 0,$$

which, using (2.120), we write as

$$J\vec{v}_t + \mathbf{A}_u J v_x = 0. \tag{2.122}$$

Now multiply (2.122) by the inverse transformation $J^{-1} = \partial\vec{v}/\partial\vec{u}$ to obtain

$$\begin{aligned} J^{-1}J\vec{v}_t + J^{-1}A_u J\vec{v}_x &= 0, \\ \vec{v}_t + A_v\vec{v}_x &= 0, \end{aligned}$$

that is, $\mathbf{A}_v = J^{-1}\mathbf{A}_u J$. Similarly, it is easy to show that $\mathbf{A}_u = J\mathbf{A}_v J^{-1}$. Note that if $\vec{v}$ are characteristic variables, then $\Lambda = J^{-1}A_u J$ and $J$ is the right eigenvector matrix.

---

## 2.5 Boundary Shock-Fitting

We introduce boundary shock-fitting by considering it in the context of a problem we can solve analytically. It will be the foundation for our numerical work.

Since for the *one-saw-tooth* problem the flow to the right of the shock is at rest, we only need to compute the flow to the left of the shock. By making the shock a boundary, we can reduce the computational region to $-1 \le x \le x_s(t)$. We begin by introducing a new set of coordinates

$$\begin{aligned} \tau &= t, \\ \zeta &= \frac{x - x_l}{x_s(t) - x_l}, \end{aligned} \tag{2.123}$$

where $x_l$, a constant, is the location of the left boundary. It follows that

$$\begin{aligned} \frac{\partial}{\partial t} &= \frac{\partial}{\partial \tau} + \zeta_t \frac{\partial}{\partial \zeta}, \\ \frac{\partial}{\partial x} &= \zeta_x \frac{\partial}{\partial \zeta}, \\ \zeta_t &= -\zeta\zeta_x w, \quad \zeta_x = 1/(x_s - x_l). \end{aligned} \tag{2.124}$$

The equation for the characteristic reaching the shock on the left side is given by Equation 2.7, which written in terms of our new coordinates is

$$u_\tau + (u - w)\zeta_x u_\zeta = 0. \tag{2.125}$$

We emphasize that this equation is valid along the shock by labeling $u$ as $u_l$:

$$u_{l\tau} + (u_l - w)\zeta_x u_{\zeta l} = 0. \tag{2.126}$$

With the shock jump condition, we can replace $u_l$ with $2w$, thus

$$w_\tau + \tfrac{1}{2} w\zeta_x u_{\zeta l} = 0. \tag{2.127}$$

This is the equation that governs the shock motion. The equation controls the motion of the shock through the term $u_{\zeta 1}$ which represents the conditions in the neighborhood of the shock. If we were solving this problem numerically, the term $u_{\zeta 1}$ would be approximated by a one-sided derivative and $w_\tau$ would be advanced in time by a time integration scheme, for example a Runge–Kutta scheme. We proceed to show that we can recover the exact shock path. We carry out the integration analytically:

$$\begin{aligned} \zeta_x u_{\zeta 1} &= \frac{1}{1+\tau}, \\ \int_{\frac{1}{2}}^{w} \frac{dw}{w} &= -\tfrac{1}{2} \int_{1}^{\tau} \frac{d\tau}{1+\tau}, \\ w &= \tfrac{1}{2}\left(\tfrac{1}{2}(1+\tau)\right)^{-1/2.} \end{aligned} \tag{2.128}$$

Now integrating the last of (2.128), we find

$$x_x(\tau) = \sqrt{2(1+\tau)} - 1. \tag{2.129}$$

As a side note we need to point out that for this simple problem, the compatibility equation and the governing equation are the same. This, of course, will not be true in general.

Here is the snippet of the MATLAB code (SHOCKFIT01.M) that performs the shock path integration. The code is available in the companion CD.

```
uz=(u(NP)-u(N))/dz;
R=-.5*w*zx*uz;
xs=xso+a(k)*w*dt;
w=wo+a(k)*R*dt;
usL=2.*w;
zx=1./(xs-xl);
```

NP is the mesh point index corresponding to the left side of the shock. The coefficients for the three-stage Runge–Kutta scheme are stored in a(k). z stands for $\zeta$.

Note that the order is important. The shock location, xs, is integrated before the shock speed, w, is updated. The velocity on the left side of the shock, usL, is not found from the integration of the conservation law, as is the case at the other mesh points, but follows from the jump condition.

Figure 2.27a shows the results of a calculation with the shock fitted as a boundary. The problem is the same as the one computed in Section 2.3.16, but this time it is computed with only 10 mesh points, resulting in a mesh interval of 0.10. With an equivalent mesh interval, the shock capturing scheme of Section 2.3.16 gives the results shown in Figure 2.27b. Here we used 40 mesh points, for a mesh interval of 0.107. Figure 2.28 shows the convergence history of the shock-fitting scheme using

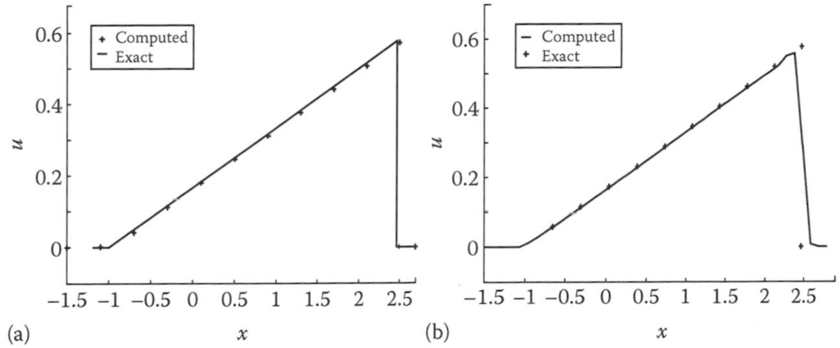


**FIGURE 2.27:** (a) Shock-fitted with 10 mesh points, $\Delta\zeta = 0.10$; (b) Shock captured with 40 mesh points, $\Delta x = 0.107$.

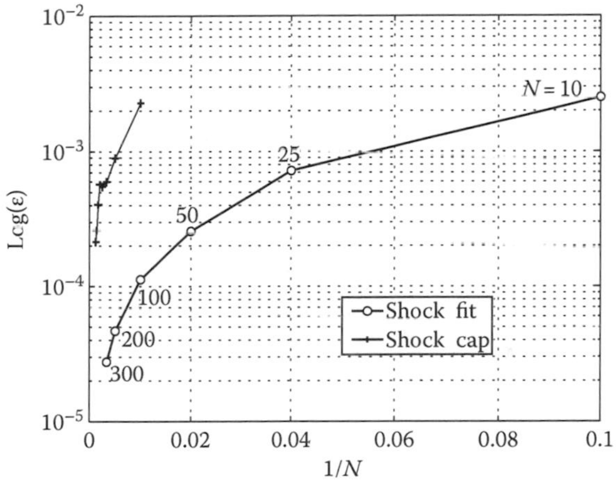


**FIGURE 2.28:** Convergence of the log(ε) as a function of $1/N$. The shock captured results of Figure 2.27(b) are included for reference.

the error defined by (2.110) and with the number of mesh points, $N$, ranging from 10 through 300. For comparison, we have included in the figure the convergence results for the shock capturing scheme, previously shown in Figure 2.26b.

Table 2.2 lists the error in the shock location at $t = 5$ for the shock-fitting and shock capturing schemes. This error is defined by

$$\varepsilon_s = 1 - x_{sc}/x_{s,\text{exact}}, \tag{2.130}$$

where

$x_{sc}$ is the computed shock location

$x_{s,\text{exact}}$ is the exact shock location

**TABLE 2.2:** Error in shock location at $t = 5$ as a function of the number of mesh intervals $N$, with shock-fitting, ε(sf) and with shock capturing, ε(sc).

| $N$ | ε(sf) | ε(sc) |
|---|---|---|
| 10 | −7.7444e–006 | 0.29995 |
| 25 | 3.1955e–005 | 0.038189 |
| 50 | 1.1033e–005 | 0.006159 |
| 100 | 3.2219e–006 | 0.012013 |
| 200 | 8.6915e–007 | 0.007650 |

Keep in mind that this is not a one-to-one comparison, since the physical mesh interval is different for each scheme. For the shock capturing scheme the computational interval is fixed at $[-1.5, 2.8]$, while for the shock-fitting scheme it is $[-1.5, x_s(t)]$, and therefore it changes with the shock motion.

## 2.6 Gaussian Pulse Problem

A disadvantage, as a test for numerical schemes, of the previous Cauchy problem is that the initial data is only piecewise continuous. We can remedy this by considering instead the following initial conditions:

$$u_0(x) = e^{-x^2}, \quad -\infty < x < \infty. \tag{2.131}$$

The initial data is now infinitely smooth (i.e., all its derivatives are smooth). However, as we will see, this problem is, in some ways, harder; both numerically and analytically.

The solution to this problem is given by

$$u(x,t) = e^{-(x-ut)^2}. \tag{2.132}$$

The time evolution of the solution is shown on the right panel of Figure 2.29. We see that the "compression" side of the wave steepens and eventually topples over. To determine where this first happens, we look for the place where $u_x \to \infty$. If we take $\partial/\partial x$ of Equation 2.132, we find that

$$u_x = -2gu/(1 - 2tug), \tag{2.133}$$

where $g(u) = \sqrt{-\ln(u)}$. The solution $u(x,t)$, as a function of $x$, has vertical tangents where the denominator $(1–2tug)$ vanishes. We can draw the curve satisfying this

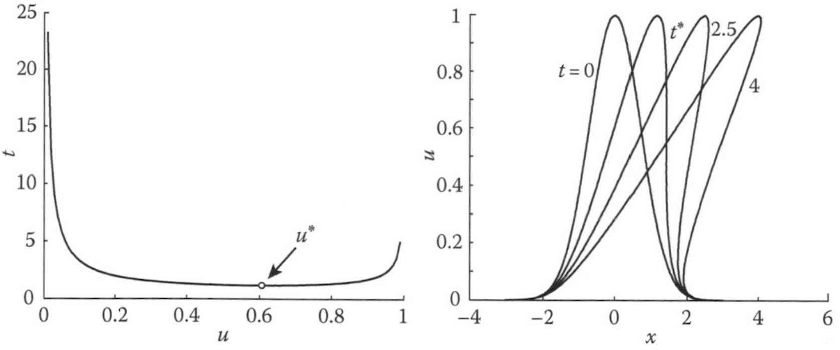


**FIGURE 2.29:** Points where (1–2$tug$) vanishes as a function of ($u$, $t$), shown on the left panel; triple-valued behavior of $u$, shown on the right panel.

condition by considering $t$ as a function of $u$, $t = 1/(2ug)$. This curve is shown on the left panel of Figure 2.29. The curve has a minimum at $t^* = \sqrt{e/2}$ corresponding to $u^* = 1/\sqrt{e}$ and $x^* = \sqrt{2}$. For all values of $t > t^*$, $u$ as a function of $x$ is somewhere triple-valued, as shown in the right panel of Figure 2.29, and has no "physical" meaning. A shock wave has to be inserted in the region where $u$ folds over. The characteristic line where the shock forms is given by

$$\frac{\mathrm{d}x}{\mathrm{d}t} = u^*. \tag{2.134}$$

If we trace this characteristic back to the initial data, we find its foot at $x_0 = 1/\sqrt{2}$. This corresponds to a minimum for $u_x(x, 0)$. That is, the shock forms on the characteristic originating from the point where the initial $u$-profile has the most negative slope. Figure 2.30 shows how the singularity in $u_x$ develops in time; the left panel is a carpet plot of $-u_x$ and the right panel shows contours of $u_x$. Note how $\max |u_x|$ propagates along the characteristic curve (2.134), shown as the solid line.

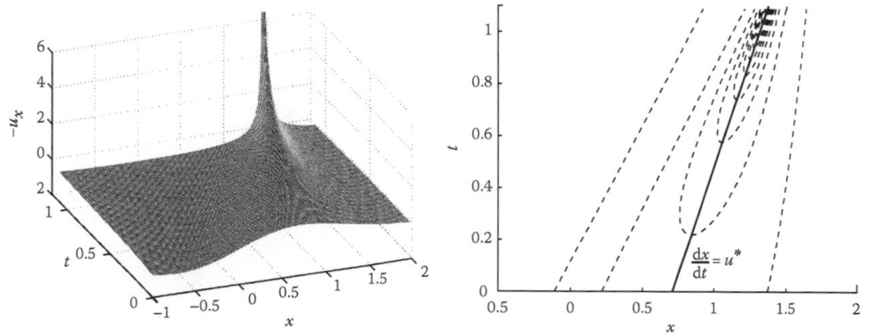


**FIGURE 2.30:** Carpet plot of $-u_x(x, t)$, shown on the left panel; contours of $u_x$, shown on the right panel.

### 2.6.1 Shock Formation and Branch-Points

The inviscid Burgers' equation reproduces the observations made by Gabriel Stokes in 1848, and is discussed in Section 1.2. The pilling up of compression waves brought about by the nonlinearity of the flux term results first in the development of a vertical tangent in the $u$-profile, followed by the subsequent formation of a shock wave. There are however other mathematical subtleties taking place, sort of *in the background,* that are worth exploring. These will provide a better understanding of the shock formation process and make us aware of some additional numerical difficulties. Let us begin by considering the following two functions:

$$u_1 = e^{-x^2},$$
$$u_2 = (1 + x^2)^{-1}.$$

We recognize the first as the initial conditions to the Gaussian pulse problem, Equation 2.131. As already stated, $u_1$ is an infinitely smooth function of $x$. The second function is also a very well-behaved function of $x$ and superficially similar to the first, as shown in Figure 2.31.

If we Taylor* expand these functions, we find

$$u_1(x) = 1 - x^2 + \frac{x^4}{2} - \frac{x^6}{6} + \frac{x^8}{24} - \ldots = 1 + \sum_{i=1}^{\infty} (-1)^i \frac{x^{2i}}{(2i-2)!},$$

$$u_2(x) = 1 - x^2 + x^4 - x^6 + x^8 - \ldots = 1 + \sum_{i=1}^{\infty} (-1)^i x^{2i}.$$

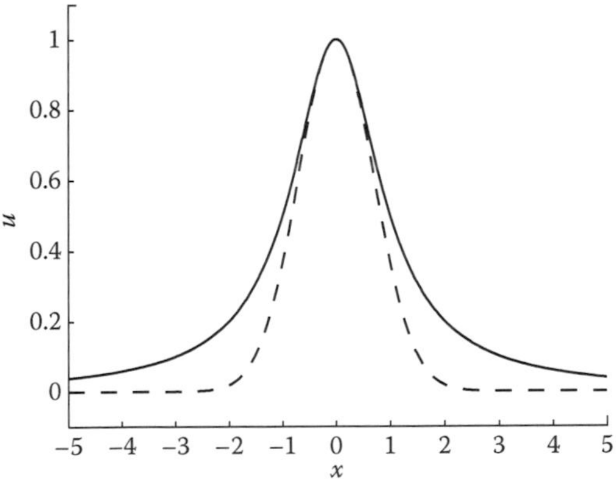


**FIGURE 2.31:** Plot of functions $u_1$ (dashed line) and $u_2$ (solid line).

* The MATLAB function *taylortool* is a GUI for the study of Taylor expansions. Use it to evaluate these Taylor expansions about $x = 2$ with 24 terms.

Using now the ratio test, we find for $u_1$,

$$\lim_{i\to\infty} \frac{(2i-2)!x^{2(i+1)}}{(2i)!x^{2i}} = \lim_{i\to\infty} \frac{x^2}{2i(2i-1)} = 0,$$

and for $u_2$,

$$\lim_{i\to\infty} \frac{x^{2(i+1)}}{x^{2i}} = x^2 \to \quad \begin{cases} |x| < 1 \text{ converges,} \\ |x| \geq 1 \text{ diverges.} \end{cases}$$

Why is this? Both functions are well behaved on the real axis; why should one converge while the other diverges if $|x| \geq 1$? To understand what is going on, let $x \to z = x + iy$, so that $u_1$ and $u_2$ are now functions of the complex variable $z$. An investigation of $u_1$ reveals that it is well behaved over the entire complex plane, however, for $u_2$ we find simple pole singularities at $z = \pm i$ as shown in Figure 2.32. The behavior of the series expansion of $u_2$ is now clear, since for any power series about some point $z_o$ there is a circle in the complex plane, centered at $z_o$, called the *circle of convergence*, with the property that if the complex number $z$ lies inside the circle the series converges for that value of $z$, while if it lies outside the circle, it diverges for that value of $z$. The radius of this circle is the distance in the complex plane from $z_o$ to the nearest singularity [171]. The function $u_2(z)$ is simple enough that we can identify the singularities by just inspecting the function. For more complex functions, we can find the singularities by looking for the points where $du/dz \to \infty$ or alternatively where $dz/du = 0$. For $u_2$ we find

$$\frac{dz}{du} = -\frac{(1+z^2)^2}{2z} = 0,$$

hence, the function is singular at $z = \pm i$ and at $z = \pm\infty$.

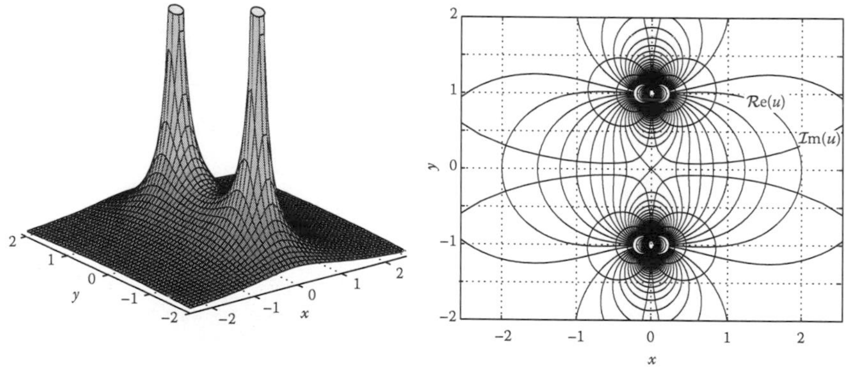


**FIGURE 2.32:** Left pane shows the surface $|u_2|$, while the right pane shows lines $\mathcal{R}e(u_2) =$ constant and $\mathcal{I}m(u_2) =$ constant. A geometrical property of an analytic function $u$ is that the lines $\mathcal{R}e(u) =$ constant and $\mathcal{I}m(u) =$ constant form an orthogonal net, i.e., the lines intercept at right angles.

Let us now look at the function defined by Equation 2.132. First let us extend the function to the complex plane by letting $x \to z = x + yi$. Next, we look for the points where $dz/du = 0$, treating $t$ as a fixed parameter, we find

$$\begin{aligned} \frac{dz}{du} &= t - \frac{1}{2u(z-ut)} = 0, \\ u &= \frac{z \pm \sqrt{z^2-2}}{2t}. \end{aligned} \tag{2.135}$$

Now substituting the last of Equation 2.135 into Equation 2.132, we find, by Newton iteration, the singular points of Equation 2.132 as the roots of the following equation:

$$F(z) = z \pm \sqrt{z^2-2} - 2te^{-\left(z-\left(z\pm\sqrt{z^2-2}\right)/2\right)^2} = 0.$$

The roots lying on the upper half of the complex plane correspond to the negative branch of $\sqrt{z^2-1}$, and are shown in Figure 2.33. Denote these roots as $z_b(t)$. Note that the curve intercepts the real axis at $t = \sqrt{e/2}$, the time at which the shock wave is formed. The function defined by Equation 2.132 when extended to the complex plane is a multivalued function of $z$. In order to treat a multivalued function as a single-valued function, we have to break the function into branches that are analytic.* To accomplish this we introduce a curve, called a *branch-cut*, in the complex plane to separate the branches of a multivalued function into singled-valued analytic branches. Any point of a multivalued function $u(z)$ that cannot be an interior point of the region of definition of a single-valued branch of $u$ is a singular point called a *branch-point*. The singular points depicted in Figure 2.33 are the branch-points of Equation 2.132 at different times. The characteristic feature of a branch-point is that

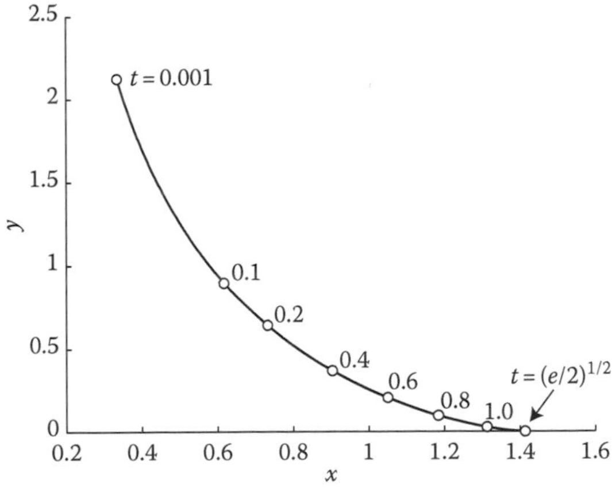


**FIGURE 2.33:** Locus of branch-points $z_b$ on the upper half of the complex plane.

* A function is analytic in a region $\mathcal{R}$ of the complex plane if it has a finite derivative at each point of $\mathcal{R}$ and if it is single-valued in $\mathcal{R}$.

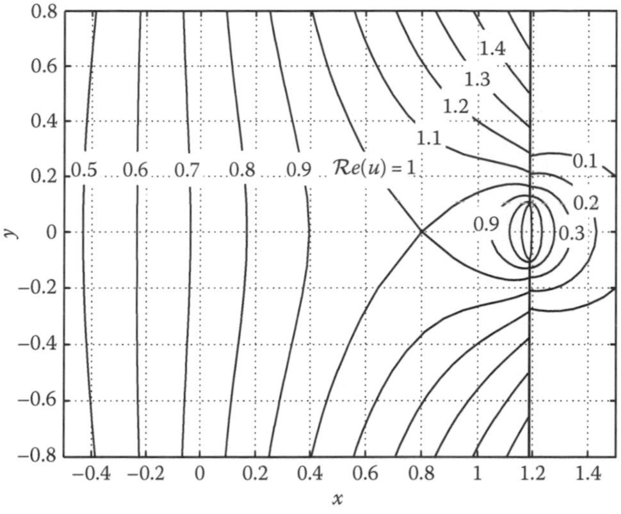


**FIGURE 2.34:** Lines of $\mathcal{R}e(u) = $ constant on the complex plane for the branch of Equation 2.132 of interest at $t = 0.8$.

if we take a closed circuit about such a point, the value of the function at the end of the circuit is different from its initial value. Figure 2.34 shows the single-valued branch of Equation 2.132 at $t = 0.8$. The lines $\mathcal{R}e(u) = $ constant are shown. On the abscissa, the values of $\mathcal{R}e(u)$ coincide with the exact solution (2.132) at $t = 0.8$. The two heavy lines, symmetric about the abscissa, are the branch-cuts, which extend from the branch-points, $z_b(0.8) = 1.185 \pm 0.095i$ to $y = \pm\infty$. The velocity of the branch-point $z_b$ is given by*

$$\dot{z}_b(t) = u(z_b, t).$$

Figure 2.35 shows the two velocity components of $\dot{z}_b$.

One of the major shortcomings of the analysis of consistency, as outlined in Section 2.3.2, is that it relies on Taylor expansions. An evaluation of the order of accuracy of a given scheme makes sense only if the Taylor series converges monotonically, so that each additional term contributes to the sum by less than the preceding one. This is not necessarily the case in most problems of practical interest, when singularities such as shocks are about to appear within a flow which is continuous and differentiable, as in the present example. Say that we want to find $u(x_0 + \Delta x)$ by expanding $u$ about $x_0 = 1.185$ in a Taylor series with $\Delta x = 0.1$ at $t = 0.8$. Let $u^{(i)}$ be the $i$th derivative of $u$, where $u$ is given by (2.132). Let $S_i$ be the partial sums of the Taylor expansion, i.e., $S_i = u_0 + \sum_{k=1}^{i} u^{(k)} \Delta x^k / k!$ and let $\varepsilon_i$ be the error, $u_{\text{exact}}(x_0 + \Delta x) - S_i$. The results of this exercise are shown on Table 2.3. The radius of convergence from $x_0 = 1.185$ at $t = 0.8$ is 0.095; with $\Delta x = 0.1$ the series does not converge, and increasing the order of accuracy makes the result worse.

* Let $x \to z$ in Equation 2.132, then take $d/dt$.

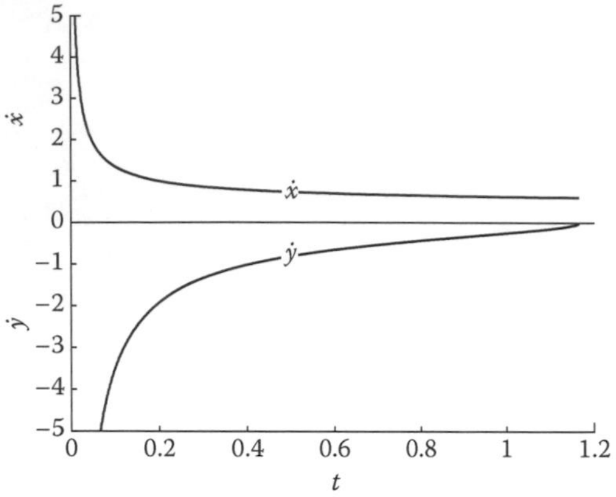


**FIGURE 2.35:** The two velocity components of $\dot{z}_b$.

**TABLE 2.3:** Results of Taylor expansion of $u$ about $x_0$, with $\Delta x = 0.1$.

| $i$ | 1 | 2 | 3 | 4 |
|---|---|---|---|---|
| $u^{(i)}$ | 2.724 | 2.615 | $3.563 \times 10^2$ | $1.157 \times 10^5$ |
| $S_i$ | 0.354 | 0.341 | 0.400 | −0.082 |
| $\varepsilon_i$ | 0.031 | 0.044 | −0.015 | 0.467 |

### 2.6.2 Gaussian Pulse Shock Path

For this problem, it is difficult to find the shock path by integrating the shock speed relation; however it is fairly easy to find the shock path using the area conservation rule. The solution is in two parts depending on whether $t > x_s$ or not. For $t > x_s$, the area conservation rule results in the following three simultaneous equations:

$$\begin{aligned} x_s &= u_l t - g(u_l), \\ x_s &= u_r t + g(u_r), \\ x_s &= t\langle u\rangle + \frac{\sqrt{\pi}\langle \mathrm{erf}(g)\rangle - 2\langle ug\rangle}{[\![u]\!]}. \end{aligned} \tag{2.136}$$

In the range $t^* \leq t \leq x_s$ the shock path is defined by

$$\begin{aligned} x_s &= u_l t + g(u_l), \\ x_s &= u_r t + g(u_r), \\ x_s &= t\langle u\rangle + \frac{[\![ug]\!] - \frac{1}{2}\sqrt{\pi}[\![\mathrm{erf}(g)]\!]}{[\![u]\!]}. \end{aligned} \tag{2.137}$$

The condition $t = x_s$ occurs when $u_l = 1$, corresponding approximately to $t = 1.7421$.

Equations 2.136 and 2.137 can be simplified to

$$\begin{aligned} t &= 2\langle g\rangle/[\![u]\!], \\ 2\langle g\rangle\langle u\rangle &= \sqrt{\pi}\langle \mathrm{erf}(g)\rangle \end{aligned} \tag{2.138}$$

and

$$\begin{aligned} t &= -[\![g]\!]/[\![u]\!], \\ 2[\![g]\!]\langle u\rangle &= \sqrt{\pi}[\![\mathrm{erf}(g)]\!], \end{aligned} \tag{2.139}$$

respectively. These constitute a set of two simultaneous equations for $u_\mathrm{l}$ and $u_\mathrm{r}$ at a given time $t$. A MATLAB program to solve these equations is available on the companion CD (`EXACTGAUSS.M`). In the neighborhood of $t = t^*$ the first equation of (2.139) is difficult to solve because both the numerator and denominator vanish. To solve this problem, we Taylor expand the numerator about $u = u^*$,

$$\begin{aligned} [\![g]\!] &\approx g_u(u^*)[\![\Delta u]\!] + \frac{g_{uuu}(u^*)}{3!}[\![\Delta u^3]\!] \\ &+ \frac{g_{uuuu}(u^*)}{4!}[\![\Delta u^4]\!] + O\left([\![\Delta u^5]\!]\right), \\ \Delta u_\mathrm{l} &= u_\mathrm{l} - u^*, \quad \Delta u_\mathrm{r} = u_\mathrm{r} - u^*, \end{aligned} \tag{2.140}$$

thus,

$$\begin{aligned} t \approx\, & t^* + e\sqrt{2e}\left(\Delta u_\mathrm{r}^2 + \Delta u_\mathrm{l}^2 + \Delta u_\mathrm{r}\Delta u_\mathrm{l}\right)/3! \\ & - e^2\sqrt{2}(\Delta u_\mathrm{r}^2 + \Delta u_\mathrm{l}^2)(\Delta u_\mathrm{r} + \Delta u_\mathrm{l})/4! + O(\Delta u^4) \end{aligned} \tag{2.141}$$

replaces the first equation in (2.139) near $t = t^*$.

Finally, for large $t$ we find

$$\begin{aligned} u_\mathrm{r} &\approx 0, \\ u_\mathrm{l} &\approx \sqrt{2\sqrt{\pi}/t}, \\ w &\approx \tfrac{1}{2}\sqrt{2\sqrt{\pi}/t}, \\ x_\mathrm{s} &\approx \sqrt{2\sqrt{\pi}}\sqrt{t}. \end{aligned} \tag{2.142}$$

The exact shock path and characteristic lines for the initial conditions given by (2.131) are shown in Figure 2.36a. Figure 2.36b shows a carpet plot of $u$ for the same conditions. Values along the shock path are tabulated in Table 2.4.

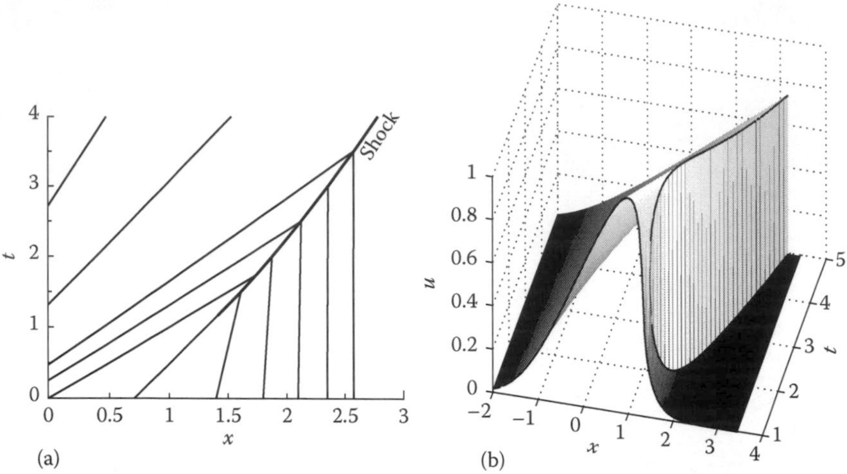


**FIGURE 2.36:** Shock wave path and characteristics for initial conditions given by Equation 2.131.

**TABLE 2.4:** Exact shock wave parameters for initial conditions corresponding to Equation 2.131.

| $t$ | $x_s$ | $u_l$ | $u_r$ |
|---|---|---|---|
| $\sqrt{e/2}$ | $\sqrt{2}$ | $1/\sqrt{e}$ | $1/\sqrt{e}$ |
| 1.5000 | 1.6091 | 0.9814 | 0.1430 |
| 1.7421 | 1.7421 | 1.0000 | 0.0742 |
| 2.0000 | 1.8776 | 0.9897 | 0.0393 |
| 2.5000 | 2.1256 | 0.9452 | 0.0124 |
| 3.0000 | 2.3576 | 0.8962 | 0.0041 |
| 3.5000 | 2.5765 | 0.8509 | 0.0013 |
| 4.0000 | 2.7843 | 0.8106 | 4.3396e–004 |
| 5.0000 | 3.1722 | 0.7434 | 1.3758e–004 |

## 2.7 Boundary Shock-Fitting Revisited

In the last problem the flow field is changing on both sides of the shock wave, therefore it is worthwhile to reconsider how to fit the shock. We start from Equation 2.126. Using the definition of the shock speed, Equation 2.27, we rewrite Equation 2.126 as follows:

$$2w_\tau - u_{r\tau} + (u_l - w)\zeta_{xl} u_{\zeta l} = 0. \tag{2.143}$$

Carrying out this operation, Equation 2.143 now reads

$$w_\tau + \tfrac{1}{2}(u_\mathrm{r} - w)\zeta_{x\mathrm{r}} u_{\zeta r} + \tfrac{1}{2}(u_\mathrm{l} - w)\zeta_{x\mathrm{l}} u_{\zeta \mathrm{l}} = 0. \quad (2.144)$$

This equation describes the motion of the shock using information about the nature of the flow on both sides of the shock. The implementation, using two regions, is illustrated in the following code snippet from the two-region shock-fitting code (SHOCKFIT04.M) available in the companion CD (compare it to the single region code of Section 2.5).

```
 1. uzL=(usL-u(1,N(1)))/dz(1);
 2. RuL=-(usL-w)*zx(1)*uzL;
 3. uzR=(u(2,3)-usR)/dz(2);
 4. RuR=-(usR-w)*zx(2)*uzR;
 5. R=.5*(RuL+RuR);
 6. xs=xso+a(k)*w*dt;
 7. w=wo+a(k)*R*dt;
 8. if INIT==1 || INIT==2
 9.   usL=usLo+a(k)*RuL*dt;
10.   usR=2.*w-usL;
11. elseif INIT==4 || INIT==3
12.   usR=usRo+a(k)*RuR*dt;
13.   usL=2.*w-usR;
14. end
15. zx(1)=1./(xs-xl);
16. zx(2)=1./(xr-xs);
```

We now compute both a $u_{\zeta \mathrm{l}}$ (line 1) and a $u_{\zeta \mathrm{r}}$ (line 3). Note that N, dz, etc., are now functions of the region counter. Thus uzL ($= u_{\zeta \mathrm{l}}$) is computed with information from region 1 and uzR ($= u_{\zeta \mathrm{r}}$) with information from region 2. The shock position is updated on line 6 and the shock speed is updated on the next line. The index INIT, lines 8 and 9, determines which problem is being computed (i.e., either a Gaussian pulse running to the left or right, or the saw-tooth wave running to the left or right), and in which order to apply the Rankine–Hugoniot jump. This minor detail makes little difference for Burgers' equation. The Jacobians for regions 1 and 2 are updated on lines 15 and 16.

We begin the numerical study of this problem with the shock capturing code based on the Runge–Kutta Roe-solver SHOCKCAP01.M. We compute first the Cauchy problem (2.131) up to a time $t = 1$; just prior to the formation of the shock. The computational domain is $[-3, 4]$, we use a $\kappa$-interpolation parameter of 1/3, a Courant number of 1, no limiter is applied, and we use 800 mesh points. The $u$-profile and the local error at $t = 1$ are plotted in Figure 2.37a. For plotting convenience, the error is scaled as follows:

$$\bar{\varepsilon} = \frac{u_n - u_\mathrm{exact}}{2 \max |u_n - u_\mathrm{exact}|}.$$

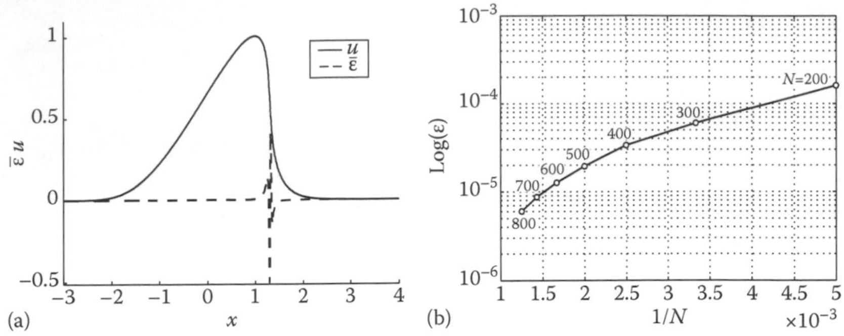


**FIGURE 2.37:** Computed solution at $t = 1$ for "Gaussian" initial conditions, using a shock capturing scheme, panel (a); the convergence of the code for this problem is shown on panel (b).

As expected, and as indicated by the figure, the error is large where the gradients are large. The convergence behavior of the code, with meshes ranging from 200 to 800 points, is shown in Figure 2.37b. Again, this is for a computation that stops at $t = 1$.

Now we repeat the calculation, but go to a time $t = 5$, use the van Albada limiter, and 600 mesh points. The computed $u$-profile is shown on the left panel of Figure 2.38 and the shock path is shown on the right panel.

Finally, we perform a grid convergence study using grids ranging from 200 to 900 mesh points and with the limiter both on and off. The results are shown in Figure 2.39. At the higher grid densities the convergence is very poor for both limiter on and limiter off.

The poor convergence behavior of the Runge–Kutta Roe-solver suggests trying the much simpler nonconservation MacCormack scheme as the background scheme with the shock-fitting technique. This option, along with the Runge–Kutta Roe-solver, is

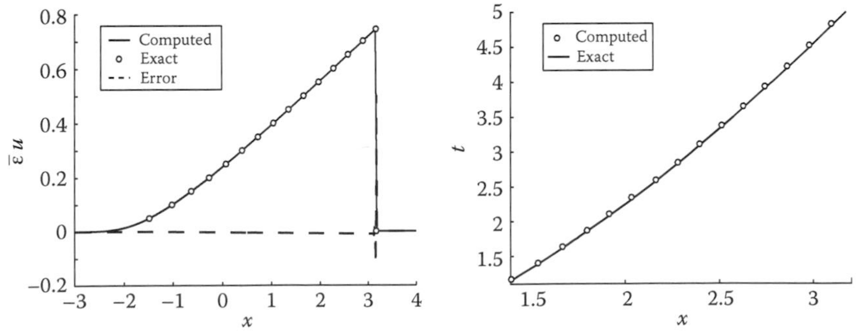


**FIGURE 2.38:** Computed $u$-profile with shock capturing code, $t = 5$, with 600 mesh, is shown on left panel; the computed shock path is shown on the right panel.

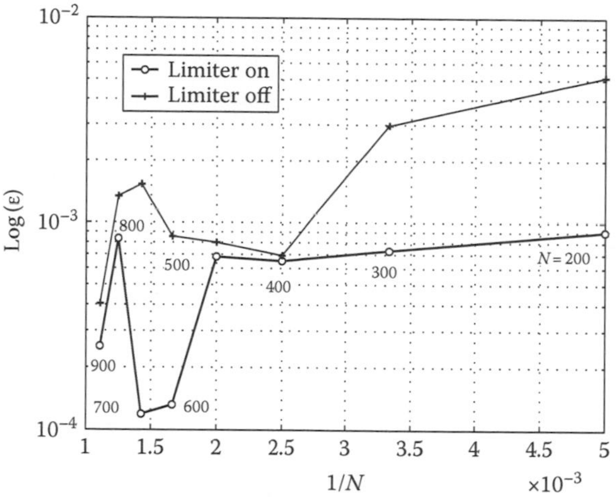


**FIGURE 2.39:** Grid convergence study for shock capturing scheme, Gaussian initial conditions and $t = 5$.

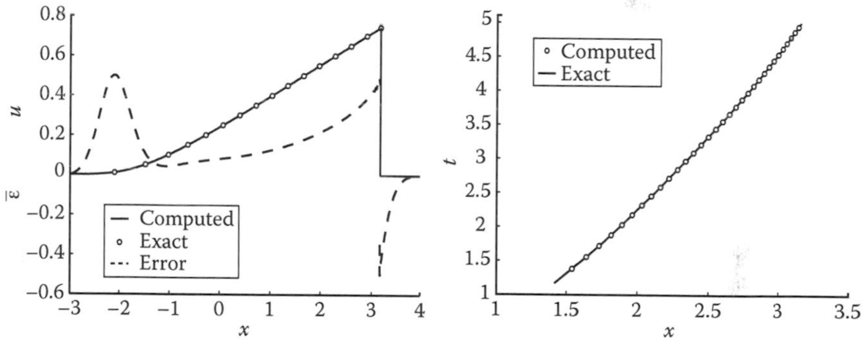


**FIGURE 2.40:** Computed and exact $u$-profile with shock-fitting code using two regions, nonconservative MacCormack scheme, with 300 mesh points in each region, shown on the right panel; computed and exact shock paths, shown on the left panel.

available in the code SHOCKFIT04.M, available in the companion CD. The results for this problem using 300 mesh points on each region are shown in Figure 2.40. Results from a convergence study for both the nonconservative MacCormack scheme and the Runge–Kutta Roe-solver, both using shock-fitting, are shown in Figure 2.41. As shown in the figure, with the shock fitted, both schemes show good convergence characteristics; the MacCormack scheme giving slightly better results. The computed and exact $u_l$, $u_r$, $u_{\varsigma l}$, and $u_{\varsigma r}$ are shown in Figure 2.42. The computation was made with the nonconservative MacCormack scheme and 400 mesh points.

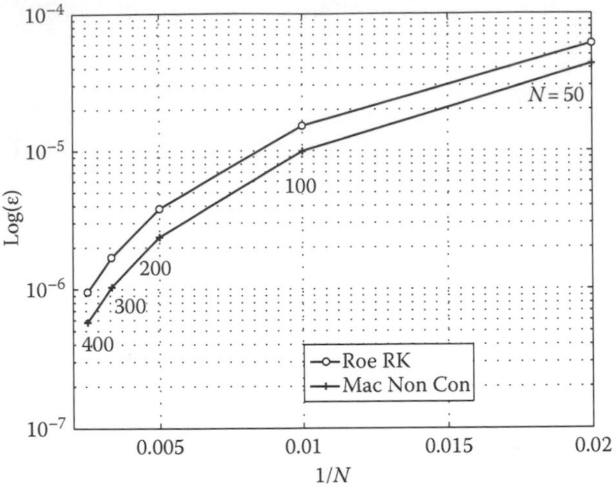


**FIGURE 2.41:** Convergence study using Roe–Runge–Kutta conservative and MacCormack nonconservative schemes, both schemes use shock-fitting.

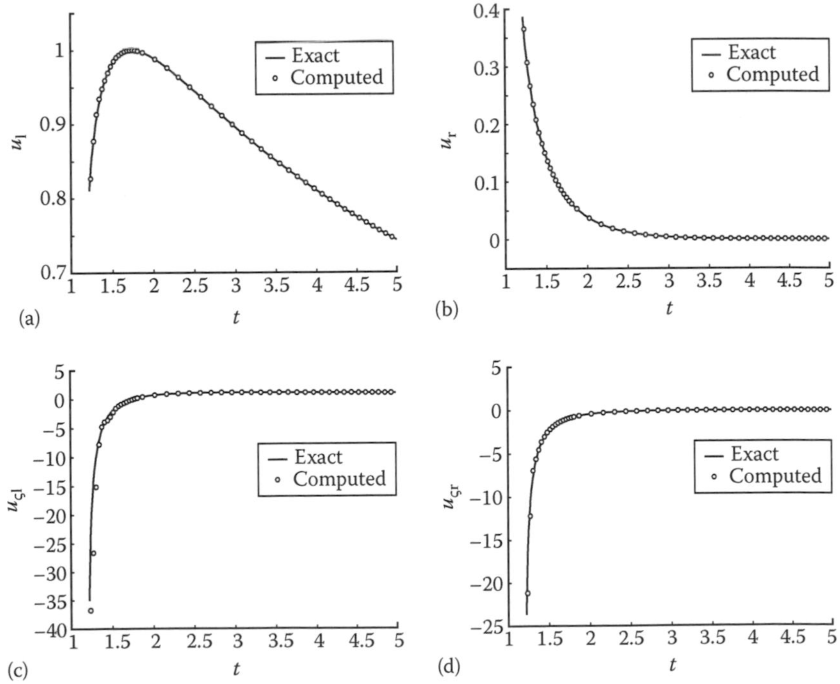


**FIGURE 2.42:** Comparison of computed and exact for $u_l$, $u_r$, $u_{\varsigma l}$, and $u_{\varsigma r}$ with MacCormack nonconservative code, with 400 mesh points.

## 2.8 Floating Shock-Fitting

The high-water mark for the calculation of shock waves as boundaries of the flow was set by Marconi and Salas in the early 1970s with the publication of [136,137]. Those two references reported on a method for solving high-speed flow over aircraft- and spacecraft-like configurations. Figure 2.43 illustrates the level of complexity that was achieved. The configuration represents a high-speed research aircraft that was of interest to NASA at that time. The flight conditions correspond to $M_\infty = 6$, $\gamma = 1.2$, and $\alpha = 0°$. In the simulation, the aircraft bow shock, canopy shock, wing shock, and vertical tail shock are treated as boundaries. The solution of the Euler equation required approximately 1 h of CPU time on an IBM 370/168 computer using a cross-sectional grid consisting of 25 by 30 mesh points.

This computer was among the fastest computers available in the early 1970, running at about 3.5 MIPS (million instructions per second). It became evident with this work that the technique of creating shock-bounded regions was a limiting factor for multidimensional complex flows. This led Moretti to develop a technique where "the shocks float among mesh points" [147] which he labeled *floating shock-fitting*.

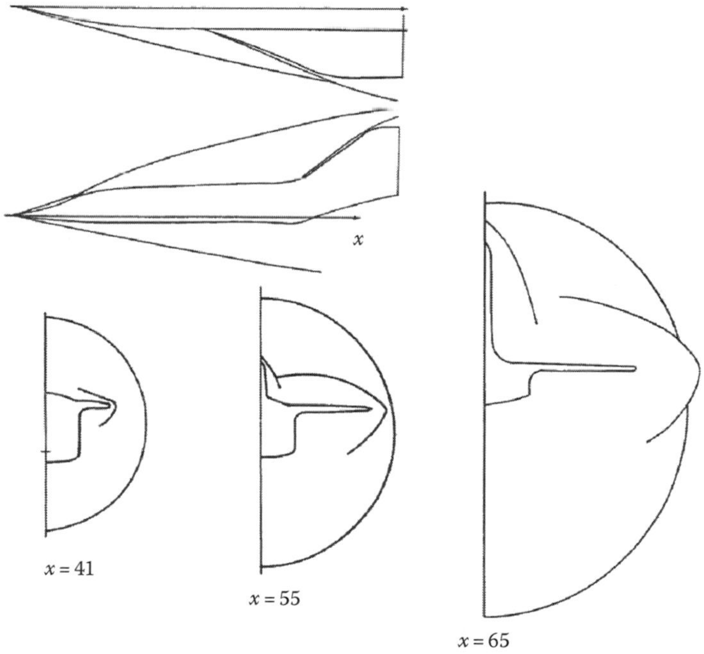


**FIGURE 2.43:** Shock pattern calculated for a high-speed research aircraft. Free stream conditions corresponding to $M_\infty = 6$, $\gamma = 1.2$, and $\alpha = 0°$. (Adapted from Marconi, F. et al., Development of a computer code for calculating the steady super/hypersonic inviscid flow around real configurations, Vol. I, Computational technique, NASA CR-2675, 1976.)

For one-dimensional problems, floating shock-fitting has the advantage over shocks-as-boundaries in that it does not require adding and subtracting mesh points as regions enlarge or contract. This is not just the cost in additional logic required with shocks-as-boundaries, but also the elimination of interpolation errors associated with the addition and subtraction of mesh points. The disadvantage of floating shock-fitting, however, is the need for additional difference expressions for mesh points in the vicinity of shocks (both in space and time) and the possibility of inconsistencies between the truncation errors from these expressions and those of the *background scheme*. In what follows, we present the details of floating shock-fitting, as described in [188], with the nonconservative MacCormack scheme as the *background scheme*. We chose this scheme because of its simplicity and, as we have seen, better overall results. The extension to the conservative version of the MacCormack scheme is trivial. The extension to the Roe scheme with κ-interpolation requires some additional logic because of the larger stencil used in this scheme.

With floating shock-fitting, we do not have to introduce a new set of computational coordinates for the computation of field points, although we can just as easily introduce floating shock-fitting within a shock-as-a-boundary scenario. Thus for field points, we consider computational coordinates $(x,t)$ which coincide with the physical $(x,t)$ coordinates. The general problem is illustrated in Figure 2.44. We have a shock wave, identified by the index $k$, between mesh points $[n, n+1]$ at time $t$ and the shock remains within this cell at time $t+\Delta t$. The computation of mesh points $n-1$ and $n+2$ is not affected by the presence of the shock and proceeds without any changes. However, the computation of mesh points $n$ and $n+1$ has to be modified. Let us say that we write the MacCormack scheme such that in the predictor stage $u_x$ is approximated by the forward difference,

$$u_x = (u_{n+1} - u_n)/\Delta x.$$

In floating shock-fitting this is replaced with,

$$u_x = \big((\varepsilon - 2)u_{n-1} + u_1^k + (1-\varepsilon)u_n\big)/2\Delta x, \qquad (2.145)$$

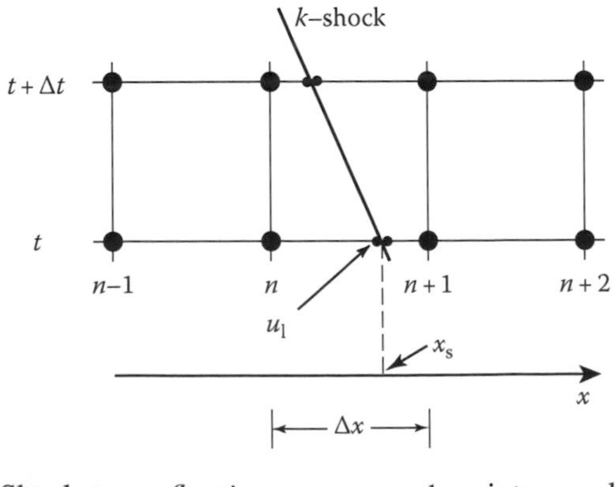


**FIGURE 2.44:** Shock wave *floating* among mesh points $n$ and $n+1$.

where

$$\varepsilon = \left(x_s^k - x_n\right)/\Delta x, \tag{2.146}$$

$u_l^k$ is the value of $u$ on the left side of the $k$th shock wave and $x_s^k$ is the abscissa of the $k$th shock wave at time $t$. The truncation error for (2.145) is

$$(\varepsilon + 2)(\varepsilon - 1)u_{xx}\Delta x/4.$$

If there is a shock wave, identified by index $k-1$, in cell $[n-1, n]$, then (2.145) is replaced with

$$u_x = \left(u_l^k - u_r^{k-1}\right)/\left(x_s^k - x_s^{k-1}\right). \tag{2.147}$$

The equivalent expressions are used for the calculation of mesh point $n+1$ in the corrector stage. Before we discuss the calculation of the shock itself, let us see how this is implemented in an actual code.* First, at any time $t$, all shock waves are ordered from left to right and each is associated with an integer $k = 1, 2, \ldots, k_{\max}$. Then, we introduce at each mesh point $n$ and index $I(n, j)$ defined as follows, see Figure 2.45:

$$\begin{aligned} I(n,1) &= I(n,2) = 0 \text{ if no shocks in cell } [n, n+1], \\ I(n,1) &= \min(k) \text{ of any shock in cell } [n, n+1], \\ I(n,2) &= \max(k) \text{ of any shock in cell } [n, n+1]. \end{aligned}$$

If, as in Figure 2.44, there is only one shock wave in cell $[n, n+1]$, then $I(n,1) = I(n,2) = k$. If shocks $k, k+1$, and $k+2$ reside within cell $[n, n+1]$, then $I(n,1) = k$ and $I(n,2) = k+2$. A snippet of MATLAB for the

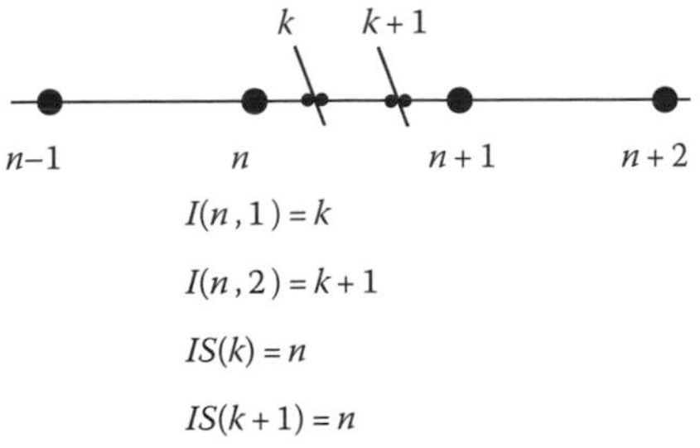


**FIGURE 2.45:** Indexes used to track shock waves.

* The MATLAB implementation of the floating shock-fitting method is file SHOCKFIT07.M in the CD.

implementation of this procedure with the MacCormack scheme (see function `mac_non_conservative7`) is listed below:

```
 1. for j=1:2
 2.   fac=j/2;
 3.   np=2-j;
 4.   nm=1-j;
 5.   for n=2:N
 6.     sigma=dts(n)/dx;
 7.     ux=u(n+np)-u(n+nm);
 8.     if j== 1 && I(n,1) ~=0
 9.       if I(n-1,1)== 0
10.         k=I(n,1);
11.         eps1= (xs(k)-x(n))/dx;
12.         ux= ((eps1-2)*u(n-1)+usL(k)+(1-eps1)*u(n))/2;
13.     else
14.       k=I(n,1);
15.       ux= (usL(k)-usR(k-1))*dx/(xs(k)-xs(k-1));
16.     end
17.   elseif j== 2 & I(n-1,1) ~=0
18.     if I(n,1)== 0
19.       k=I(n-1,2);
20.       eps1= (x(n)-xs(k))/dx;
21.       ux=-((eps1-2)*u(n+1)+usR(k)+(1-eps1)*u(n))/2;
22.     else
23.       k=I(n,1);
24.       ux= (usL(k)-usR(k-1))*dx/(xs(k)-xs(k-1));
25.     end
26.   end
27.   ut=-u(n)*ux;
28.   if j== 1
29.     U(n)=u(n)+sigma*ut;
30.   else
31.     U(n)=.5*(Uo(n)+u(n)+sigma*ut);
32.   end
33.   end
34. .
35. . (shock point computation)
36. .
37. end
```

Line 1 is the beginning of the predictor–corrector loop. At line 6 we compute the ratio of $\Delta t/\Delta x$ which is used later in the loop. The reason for storing $\Delta t$ (i.e., `dts(n)`) in an array will be explained later. At line 7, $u_x$ is computed for each mesh point regardless of the possible presence of a shock. Line 8 is true if we are computing the predictor stage and there are one or more shock waves within cell

$[n, n+1]$, in that case we continue with line 9, otherwise we jump to line 17. At line 9, we check for the presence of shocks within cell $[n-1, n]$. If there are no shocks in this cell, we evaluate $u_x$ using (2.145) in line 12, otherwise in line 15 we use (2.147). The $k$th index needed in these expressions is extracted from the $I(n, j)$ index in lines 10 and 14. Lines 17 through 26 perform the equivalent operations for the corrector stage. At line 27, $u_t$ is computed. In line 29, the predicted value of $u$ is computed. Similarly, in line 31 the corrected value of $u$ is computed. The predictor–corrector loop ends at line 37.

At the shock it is necessary to introduce a coordinate transformation to ensure that the calculation is done along the shock surface. Thus, at the $k$th shock wave introduce the transformation

$$\tau = t,$$
$$\varsigma^k = x - x_s^k(t).$$

Then, for example, the governing equation on the left side of the shock is

$$u_{\tau l} + (u_l - w^k)u_{\varsigma l} = 0.$$

Since with this transformation $\partial/\partial x = \partial/\partial\varsigma$, in what follows we use the notation $u_x$. To solve the equations at the shock, we need to evaluate $u_x$ on the left and right sides of the shock. The expressions we use are

$$u_{xl} = \left(2\left(u_l^k - u_n\right)/(1+\varepsilon) + 0.5(u_{n-2} - u_n)\right)/\Delta x, \quad \text{if no shock } \in [n-2, n], \tag{2.148}$$

$$u_{xl} = \left((2\varepsilon - 1)u_{n-1}/(1+\varepsilon) + 3u_l^k/(1+\varepsilon) - 2u_n\right)/\Delta x, \quad \text{if shock } \in [n-2, n-1], \tag{2.149}$$

$$u_{xl} = \left(u_l^k - u_r^{k-1}\right)/\Delta x_s, \quad \text{if shock } \in [n-1, n] \text{ and } \Delta x_s \geq \Delta x, \tag{2.150}$$

$$u_{xl} = 0 \begin{cases} \text{if } k\text{-shock not first } \in [n, n+1], \\ \text{or if } \Delta x_s < \Delta x, \end{cases} \tag{2.151}$$

where

$$\Delta x_s = x_s^k - x_s^{k-1}.$$

For Equations 2.148 and 2.149 $\varepsilon$ is defined by Equation 2.146. The truncation error for (2.148) is $O(\Delta x^2)$ and for (2.149) it is $(\varepsilon - 1)u_{xx}\Delta x/2$. Equivalent expressions are used for the evaluation of $u_{xr}$. In order to implement these equations, we introduce an index $IS(k)$ defined by $IS(k) = n$, where $n$ is the mesh point to the left of the $k$th shock, see Figure 2.45. The snippet of code corresponding to the evaluation of $u_x$ at the shock is shown below. Line 1 is the beginning of the predictor–corrector loop. In line 5, we enter the shock computation if $k_{max} > 0$. In line 6, we start a loop over all the shocks. Line 7 defines the index for the mesh point to the left of the $k$th shock. Line 8 checks if the $k$th shock is the first shock within cell $[n, n+1]$. If this is true, we proceed with the evaluation of Equations 2.148 through 2.150, as required. If the statement in line 8 is false, then we evaluate (2.151) at line 27. The rest of the shock computation follows below line 30.

```
 1. for j=1:2
 2. .
 3. . (regular mesh point computation)
 4. .
 5.   if kmax>0
 6.     for k=1:kmax
 7.       n=IS(k);
 8.       if I(n,1)==k
 9.         if I(n-1,1)==0 && I(n-2,1)==0
10.           epsL=(xs(k)-x(n))/dx;
11.           c1=2/(1+epsL);
12.           uxL=(c1*(usL(k)-u(n-1))+.5*(u(n-2)-u(n)))/dx;
13.       elseif I(n-1,1)==0 && I(n-2,1) ~=0
14.           epsL=(xs(k)-x(n))/dx;
15.           c1=(2*epsL-1)/(1+epsL);
16.           c2=3/(1+epsL);
17.           uxL=(c1*u(n-1)+c2*usL(k)-2*u(n))/dx;
18.       elseif I(n-1,1) ~=0
19.         epsL=xs(k)-xs(k-1);
20.           if epsL>=dx
21.             uxL=(usL(k)-usR(k-1))/epsL;
22.           else
23.             uxL=0;
24.           end
25.         end
26.       else
27.           uxL=0;
28.       end
29. .
30. . (evaluation of uxR)
31. . (continue with shock computation)
32. end
```

When a shock wave crosses a mesh line, the computation of the mesh point crossed by the shock is modified as follows. Consider Figure 2.46, the $k$th shock is shown crossing the mesh line $n$ between $[t, t+\Delta t]$. The value of $u_n(t)$ cannot be used to evaluate $u_n(t+\Delta t)$. Instead, we must use the value of $u$ at point $b$ which is interpolated from $u$ at $c$ and $d$ along the shock. Similarly, the $u_x$ derivative is

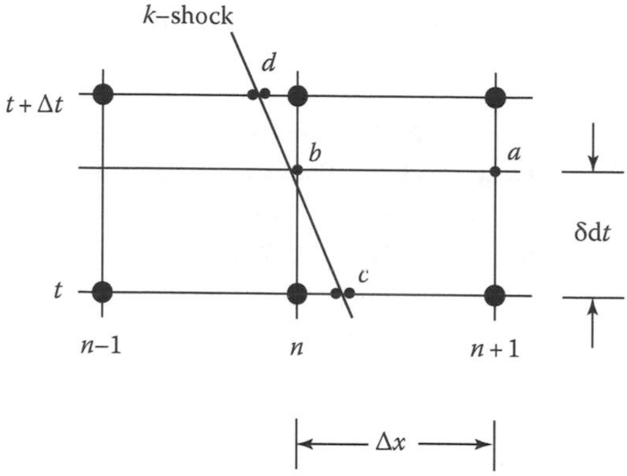


**FIGURE 2.46:** Details of shock crossing mesh line.

approximated from values of $u$ at $a$ and $b$. Let us review the snippet of code where this is implemented. This snippet occurs following the predictor stage of the MacCormack scheme within function mesh_xing.

```
for k=1:kmax
  if ISO(k) < IS(k)
.
. (computation for shock crossing mesh line to the right)
.
  elseif ISO (k) > IS(k)
    n=IS(k) +1;
    if I(n,1) == 0
      kk=I(n-1,2);
      delta= (x(n)-xso(kk))/(xs(kk)-xso(kk));
      ua=Uo(n+1) +delta*(U(n+1)-Uo(n+1));
      ub=usRo(kk) +delta*(usR(s)-usRo(kk));
      U(n) =ub-ub*(ua-ub)*(1-delta)*dt/dx;
      Uo(n) =ub;
      dts(n) = (1-delta)*dt
    else
      delta= (x(n)-xso(k))/(xs(k)-xso(k));
      U(n) =usL(k+1)*(x(n)-xs(k))/(xs(k+1)-xs(k)) ...
      + usR(k)*(x(n)-xs(k+1))/(xs(k)-xs(k+1));
      Uo(n) =usRo(k) +delta*(usR(k)-usRo(k));
      dts(n) = (1-delta)*dt;
    end
...
  end
...
end
```

Line 1 is the beginning of a loop that will sweep through all the shocks. Recall that `IS(k)` defines the mesh point $n$ to the left of the $k$th shock. Here `IS(k)` has the value of $n-1$ at time $t+\Delta t$ and `ISO(k)` has the value of $n$ at $t$. Line 2 checks to see if the $k$th shock has crossed the mesh line to its right, if this is not the case, we check at line 6 for a crossing to the left. If this is the case, see Figure 2.46, we define $n$ at line 7. At line 8, we check that there are no shocks in the interval $[n, n+1]$. At line 9, we define $k$. If multiple shocks have crossed the mesh line $n$ during this time interval, $k$ could be different from the value defined at line 1. To ensure that our computation is based on the last shock in the interval $[n-1, n]$, we use the index `I(n-1,2)` to define $k$.* Line 10 evaluates the fraction of $\Delta t$ corresponding to point b of Figure 2.46. Lines 11 and 12 find $u_a$ and $u_b$ by interpolation. Line 13 predicts $u$ and line 14 stores the old value $u$. The value of $\Delta t$ needed to do the correction step of the MacCormack scheme is saved at line 15. This ends the computation. However, if there is a shock in the interval $[n, n+1]$, line 8 would have sent us to line 16. Here, lines 17 through 21 do the interpolations using the values of the $k+1$ shock instead of the values on line $n+1$.

The first test of the code is the one-saw-tooth problem of Section 2.2. An advantage of the floating shock-fitting code is that we can, just as easily as fitting a shock wave, fit the gradient discontinuity that occurs at $x=-1$. The gradient discontinuity is treated as an infinitesimally weak shock by initializing the jump to zero. We have done that in a series of runs to evaluate the convergence properties of the code. As in Section 2.3.16, the boundaries of the computational region are set at $[-1.5, 2.8]$. The results are shown in Figure 2.47 where the log of the error in $u$

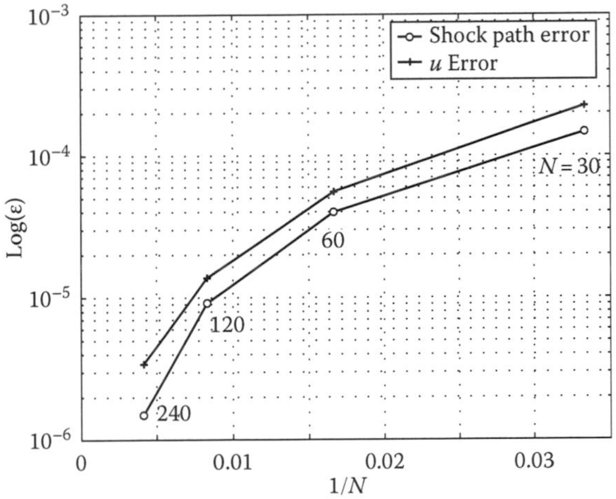


**FIGURE 2.47:** Convergence history for the one-saw-tooth problem calculated by floating shock-fitting.

* The MATLAB language allows for the counter variable to be redefined within a "for" loop without affecting the counter value, however you get a warning. We use a different counter, $kk$, to avoid confusion.

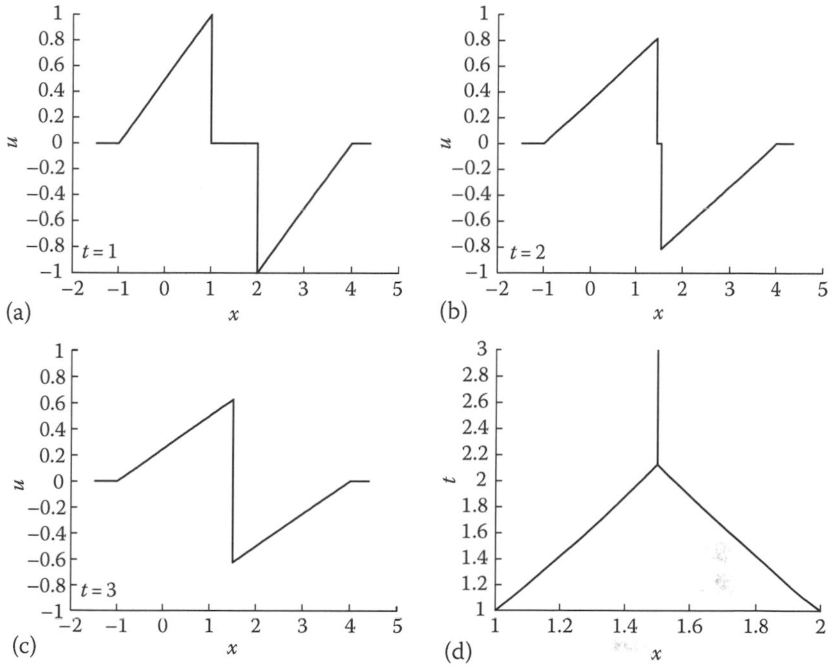


**FIGURE 2.48:** (a) Initial $u$-profile, $t = 1$; (b) computed $u$-profile at $t = 2$; (c) computed $u$-profile at $t = 3$; and (d) computed shock wave paths.

defined by (2.110), along with the log of the error in shock location defined by 2.130 are plotted for $N = [30,60,120,240]$ mesh points. This plot should be compared to the one in Figure 2.26b and the one in Figure 2.28. As is evident from the figure, the results are significantly better now that we have also fitted the gradient singularity.

Next, we consider a double-saw-tooth problem with the initial profile at $t = 1$ shown in the first plot of Figure 2.48a. The first saw-tooth generates a right-running shock; the second, inverted, saw-tooth generates a shock of equal strength running to the left. The computational interval is from $x_l = -1.5$ to $x_r = 4.5$. A total of 60 mesh points are used for the calculation. In addition to fitting the two shock waves, the gradient discontinuities at $x = -1$ and $x = 4$ are also fitted. Figure 2.48b and c shows the computed $u$-profiles at $t = 2$ and at $t = 3$. The computed shock paths are shown in Figure 2.48d. As expected from the symmetry of the problem, a stationary shock emerges at $x = 1.5$ from the interaction of the left- and right-running shocks.

As a final example, we have some fun solving a Cauchy problem with the following *staircase* initial conditions:

$$u_0(x) = u_l - [\![u]\!] \sum_{1}^{k\max} H(x - x_k),$$

$$x_k = x_{k-1} + (k + 1)/10, \quad k = 2, 3, \ldots, k_{\max},$$

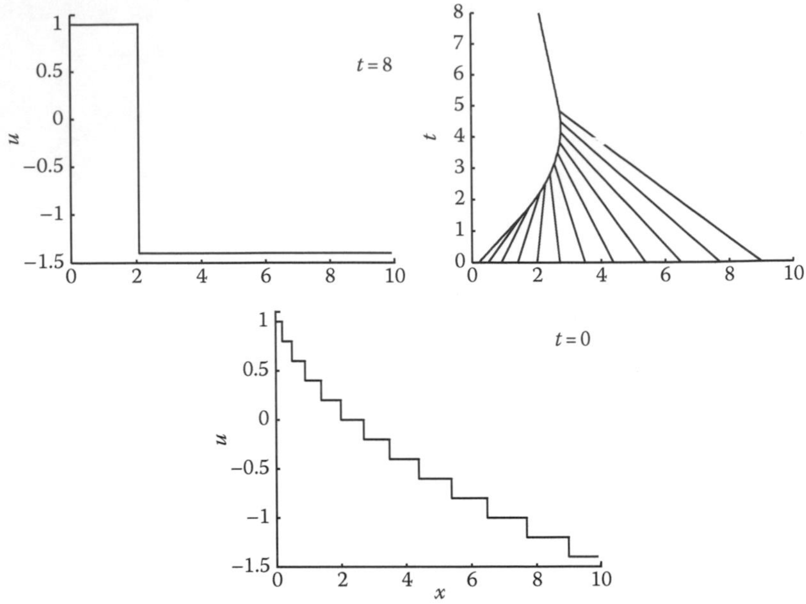


**FIGURE 2.49:** Staircase problem; lower panel shows initial $u$-profile at $t = 0$; right upper panel shows computed shock wave paths; and left upper panel shows $u$-profile at $t = 8$.

with $u_l = 1$, $[\![u]\!] = -0.2$, $x_1 = 0.2$, $k_{max} = 12$. The computational region is the $x$ interval [0, 10] and the time interval [0, 8]. For this problem we use a total of 90 mesh points, giving an average of seven mesh points per shock wave. The initial profile is shown on the lower panel of Figure 2.49; the upper panel shows the computed $u$-profile at $t = 8$, which agrees with the exact solution. The computed shock wave paths for all 12 shocks are shown on the middle panel. Essentially exact solutions are obtained for these patterns with zero gradients between shocks.

## 2.9 Detection of Shock Formation

The detection of shock formation requires that we predict the location in space and time where a shock forms. To avoid introducing wiggles, the shock location should be found within one mesh interval in space and a few time steps before its actual occurrence. The early detection of a weak shock is inconsequential compared to the problems created by a shock detected too late. We can take advantage of two events that signal the formation of a shock: (1) characteristics intercept; and (2) the gradient $u_x \to -\infty$. Of course, the second event is a direct consequence of the first.

In what follows, we will outline two shock detection techniques each based on one of these two events.

For Burgers' equation, shocks form only in regions where $u_x$ is negative, thus we begin by restricting the detection of shock formation to these regions. The idea for using $u_x \to -\infty$ for shock detection was first applied by Moretti in [146]. It begins by tracking the location of the maximum $|u_x|$ at each time step. However, a search for $|u_x|$ becoming infinite is impractical. Instead, let us consider $x$ as a function of $u$ and look for

$$x_u = 0. \tag{2.152}$$

Assume that the maximum $|u_x|$ occurs in the interval $[n,\ n+1]$ as shown in Figure 2.50. In order to capture the typical behavior of $u(x)$, we approximate $x(u)$ by a third-order polynomial $x(u) = au^3 + bu^2 + cu + d$, and look for the point $(x^*, t^*)$ when the condition (2.152)

$$x_u = 3au^2 + 2bu + c = 0 \tag{2.153}$$

is satisfied by one real root. This takes place when

$$b^2 - 3ac = 0. \tag{2.154}$$

Thus, at each time step (2.154) is evaluated until we detect a change in sign. When this happens, we introduce a shock wave into the computation. The MATLAB program named `DETECTSHOCK.M` that comes in the accompanying CD, shows how this detection method works on a test function, Equation 2.132 that behaves like an incipient shock wave. Figure 2.51 shows how (2.154) changes sign at the point where the shock forms. This simulation detects the shock formation at $t = 1.1547$, while the exact time is given by $t^* = \sqrt{e/2} = 1.1658$. The ragged behavior of the line is a result of max $|u_x|$ jumping discretely from one mesh point to another. At the point where (2.154) changes sign, Equation 2.153 is satisfied by a single real value of $u$ given by $u = -b/3a$. The $x$ location where the shock forms follows by using this

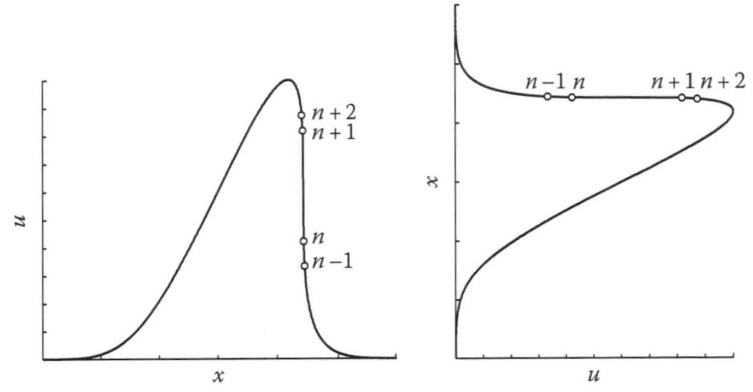


**FIGURE 2.50:** Behavior of $u(x)$ and $x(u)$ in the neighborhood of an incipient shock wave.

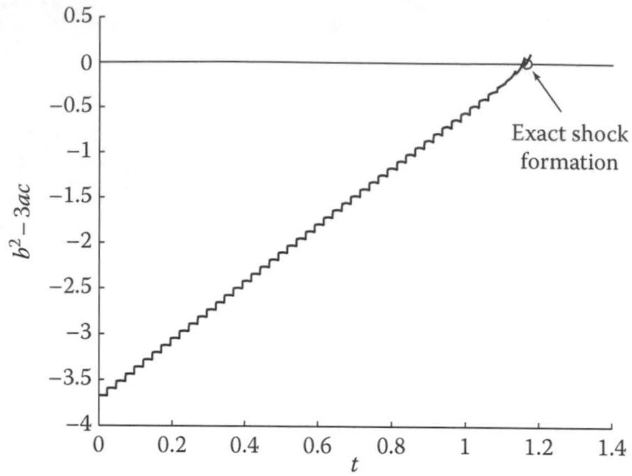


**FIGURE 2.51:** Prediction of shock formation by Equation 2.154.

value of $u$ in the polynomial fit. The computed and exact values are compared in Table 2.5. In a real computation, it would be advisable to insert the shock before the point at which (2.154) changes sign, say when $b^2 - 3ac > -0.5$. This shock detection technique is implemented in the code discussed in Section 2.10.

The other approach used to detect the formation of a shock wave consists of checking for the interception of characteristics. This technique was introduced in [188]. This approach uses a smaller stencil (2 mesh points, instead of 4) and requires fewer operations. Consider the nondimensional quantity defined by

$$\Gamma = -\frac{x_n - x_{n-1}}{(u_n - u_{n-1})\Delta t}. \tag{2.155}$$

This quantity is the number of time steps that must be taken before the characteristics emanating from mesh points $n$ and $n-1$ intercept. In regions of the flow where $u_x$ is negative, we look for the mesh interval $[n, n-1]$ where $\Gamma$ is a minimum, and when this minimum reaches a threshold value a shock wave is injected into the flow field. The studies of several problems indicate a threshold between 4 and 5 time steps. In the calculations that follow, we use $\Gamma \leq 4$.

Figure 2.52 shows the behavior of $\Gamma$ for the Gaussian problem studied in Section 2.6. As the $u_x$-profile becomes more and more steep, $\Gamma$ decreases reaching

**TABLE 2.5:** Comparison of computed and exact shock formation time, shock location, and $u$ at the shock.

| | Computed | Exact |
|---|---|---|
| $t^*$ | 1.1547 | 1.1658 |
| $x^*$ | 1.4226 | 1.4142 |
| $u^*$ | 0.5861 | 0.6065 |

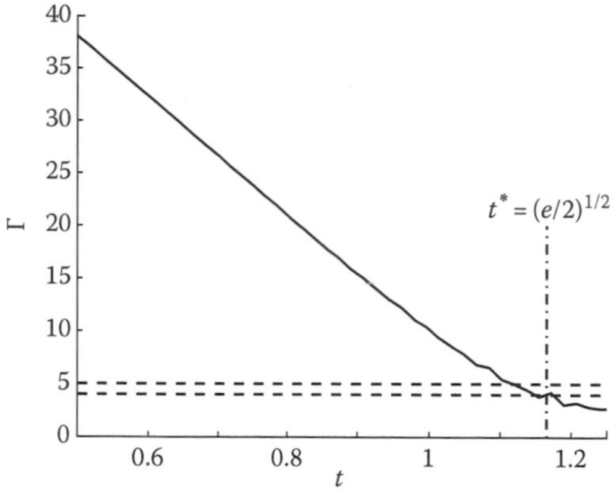


**FIGURE 2.52:** Computed behavior of Γ for the formation of a shock wave from a Gaussian distribution.

a value of 4 shortly before the theoretical value for shock formation. At this point a shock is inserted into the flow field. The point of insertion is

$$x^* = .5(x_n + x_{n-1}).$$

Figure 2.53 shows the computed shock path for the Gaussian initial conditions as discussed in Section 2.6. The calculation was carried out with 200 mesh points. As shown in the figure, the shock wave is detected just ahead of the point predicted by theory. The computed shock path nicely tracks the theoretical path.

We consider next a more difficult problem defined by the initial conditions:

$$u(x) = 1, \quad x < 0,$$
$$u(x) = \cos(\pi x), \quad 0 \le x \le 1,$$
$$u(x) = -1, \quad x > 1.$$

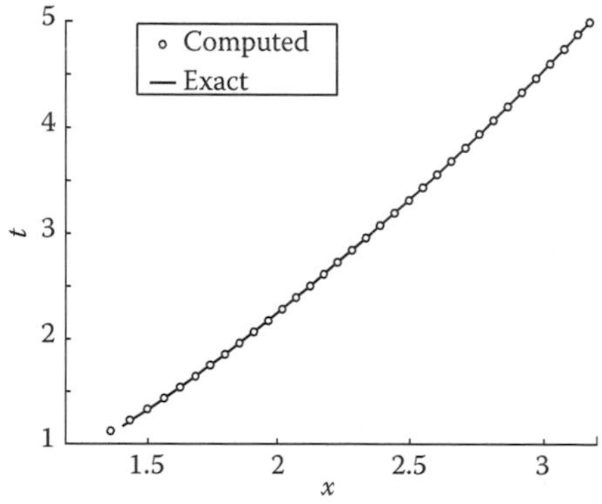


**FIGURE 2.53:** Computed shock path for a shock detected with $\Gamma = 4$.

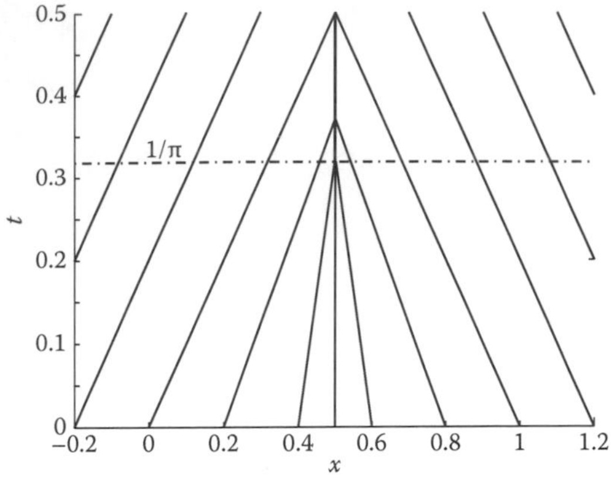


**FIGURE 2.54:** Characteristics coalescing to form a shock at $t = 1/\pi$.

This initial profile is more challenging because it is steeper and it is symmetric. Symmetrical problems tend to magnify computational problems. The exact solution for this problem is given by $u(x, t) = \cos(\pi(x - ut))$. From the exact solution we find the time for the formation of the shock wave to be $t^* = 1/\pi \approx 0.3183$. The exact characteristic pattern is shown in Figure 2.54. The resulting shock wave is stationary at $x = 0.5$. The initial $u$-profile and the computed $u$-profile at $t = 0.5$, using 400 mesh points in the $x$ interval $[-0.5, 1.5]$, are shown in Figure 2.55.

A study of the shock detection process is shown in Figure 2.56. Note that the region shown in Figure 2.56 is a blowup of the region around the shock in Figure 2.54. With a 100 mesh points, left panel of Figure 2.56, two symmetric weak shocks are detected at the same time step. A few steps later, a third shock is detected, upsetting the symmetric pattern. The appearance of the third shock is a result of lack of resolution. The first two shocks come together at approximately $1/\pi$. After the third shock intercepts with the shock created by the previous interaction, the

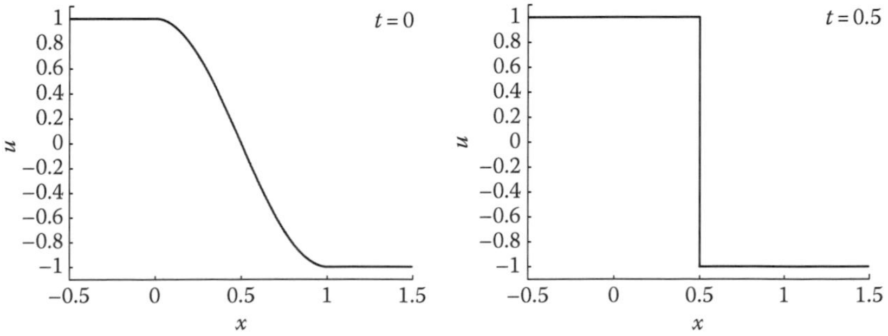


**FIGURE 2.55:** Initial profile, $u(x) = \cos(\pi x)$ and computed profile at $t = 0.5$, using 400 mesh points.

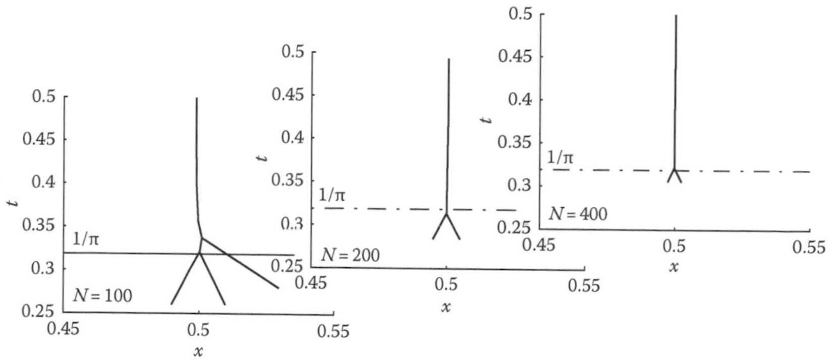


**FIGURE 2.56:** Computed shock paths with 100, 200, and 400 mesh points for cosine profile.

remaining shock settles in its stationary position at $x = 0.5$. The results are greatly improved with 200 mesh points, middle panel. Here two weak symmetric shocks are detected slightly ahead of the exact time for shock formation. The two shocks coalesce at approximately $1/\pi$ forming a stationary shock at $x = 0.5$. With 400 mesh points, right panel, we see the same behavior as with 200 mesh points, but the pattern is tighter around $[0.5, 1/\pi]$.

The numerical implementation of the detection of shock formation using the interception of characteristics can be found in the MATLAB code `SHOCKFIT07.M` in functions *detect_shock* and *inject_shock*.

## 2.10 Application of Colombeau's Generalized Functions to a Nonconservative System of Equations

The generalized function theory of Colombeau allows us to find the Rankine–Hugoniot jumps of *strictly nonconservative systems*. By *strictly nonconservative* we mean a system of equations that cannot be expressed in conservative form without transforming it through multiplication by a generalized function. For example,

$$u_t + uu_x = 0$$

is not a strictly nonconservative equation, since it can be written in conservation form without multiplying by a generalized function. On the other hand, the entropy equation

$$S_t + uS_x = 0$$

is strictly nonconservative. It can be written in conservative form by multiplying by $\rho$ and adding the conservation of mass

$$\rho_t + (u\rho)_x = 0$$

to the system, see Section 3.7.5 for more details. However, in doing so, we change the jump conditions of the original equation. As an example of a strictly nonconservative system, we will consider in this section the system of equations discussed by Oberguggenberger in [157], and Cauret, Colombeau, and Le Roux in [29], namely:

$$\begin{aligned} u_t + uu_x &= \sigma_x, \\ \sigma_t + u\sigma_x &= u_x. \end{aligned} \tag{2.156}$$

In [29] the claim is made that (2.156) follows from the more general one-dimensional elasto-plastic equations if the density is assumed constant. This is not correct, since constant density implies $u_x = 0$. The system can be written in conservation form with the transformation $q = \sigma + u^2/2$:

$$\begin{aligned} u_t + (u^2 - q)_x &= 0, \\ q_t + (u^3/3 - u)_x &= 0. \end{aligned}$$

However, this transformation requires multiplication by distribution functions, and the resulting equations do not have the same Rankine–Hugoniot jumps as those of (2.156). As written, it is difficult to see the *range of influence* and *domain of dependence* associated with (2.156). Thus unlike the work of [29,157], we begin by transforming (2.156) into its characteristic form, see Section 2.1.4. We want to write (2.156) such that the directional derivative of some combination of $u$ and $\sigma$, let us call it $f$, with respect to $t$ along a line with slope $\lambda$ is given by $f_t + \lambda f_x$. To this end, we multiply the first equation of (2.156) by some factor $\mu_1$, and the second by some other factor $\mu_2$ and add:

$$\mu_1 \left[ u_t + \frac{\mu_1 u - \mu_2}{\mu_1} u_x \right] + \mu_2 \left[ \sigma_t + \frac{\mu_2 u - \mu_1}{\mu_2} \sigma_x \right] = 0. \tag{2.157}$$

The coefficients of $u_x$ and $\sigma_x$ take the desired value $\lambda$, if the following two equations are satisfied:

$$\begin{aligned} (\lambda - u)\mu_1 + \mu_2 &= 0, \\ \mu_1 + (\lambda - u)\mu_2 &= 0. \end{aligned} \tag{2.158}$$

The system (2.158) has a nontrivial solution if the determinant of the coefficients vanishes,

$$\begin{vmatrix} \lambda - u & 1 \\ 1 & \lambda - u \end{vmatrix} = 0, \tag{2.159}$$

that is, if

$$\lambda = u \pm 1.$$

The factors $\mu_1$ and $\mu_2$, the left eigenvectors, are proportional to the complementary minors of any row of the determinant (2.159). Thus, we may take $\mu_1 = \pm 1$ and $\mu_2 = 1$. The directions

$$\frac{dx}{dt} = \lambda^{\pm} = u \pm 1$$

are called characteristic slopes, and a line tangent to the characteristic slope at all its points is called a characteristic. The two equations that we can now obtain from (2.157) are the compatibility equations

$$\begin{aligned} u' + \sigma' &= 0, \text{ on } \lambda^-, \\ u' - \sigma' &= 0, \text{ on } \lambda^+. \end{aligned} \tag{2.160}$$

It should be clear that $(\ )'$ means differentiation with respect to $t$ along a line $dx/dt = \lambda^{\pm}$, i.e., $(\ )' = \partial(\ )/\partial t + \lambda^{\pm}\partial(\ )/\partial x$. Let us introduce the notation

$$\begin{aligned} r &= u + \sigma, \\ s &= u - \sigma. \end{aligned}$$

With this notation, (2.160) reduces to

$$\begin{aligned} r' &= 0, \text{ on } \lambda^-, \\ s' &= 0, \text{ on } \lambda^+. \end{aligned} \tag{2.161}$$

The system of equations (2.161) is similar to the system studied by Riemann in 1860 [176], and the variables $s$, and $r$ are called Riemann invariants, because they are constant along the characteristics $\lambda^+$ and $\lambda^-$, respectively.

To summarize, let us write the results in matrix form. Let $\vec{q} = [u, \sigma]^T$. The system (2.156) is given by $\vec{q}_t + \mathbf{A}\vec{q}_x = 0$, where the matrix $\mathbf{A}$ is

$$\mathbf{A} = \begin{bmatrix} u & -1 \\ -1 & u \end{bmatrix}.$$

The compatibility equations are $\vec{p}_t + \Lambda\vec{p}_x = 0$, where $\vec{p} = [r, s]^T$ and

$$\Lambda = \begin{bmatrix} u - 1 & 0 \\ 0 & u + 1 \end{bmatrix}.$$

The right eigenvectors matrix $\boldsymbol{\nu}$ and its inverse are

$$\boldsymbol{\nu} = \begin{bmatrix} \frac{1}{2} & \frac{1}{2} \\ \frac{1}{2} & -\frac{1}{2} \end{bmatrix}, \ \boldsymbol{\nu}^{-1} = \begin{bmatrix} 1 & 1 \\ 1 & -1 \end{bmatrix}.$$

The characteristic variables are related to the original variables by the transformations $\vec{p} = \boldsymbol{\nu}^{-1}\vec{q}$ and $\vec{q} = \boldsymbol{\nu}\vec{p}$, see Section 2.4. The left eigenvectors matrix $\boldsymbol{\mu}$ and its inverse are

$$\boldsymbol{\mu} = \begin{bmatrix} 1 & 1 \\ -1 & 1 \end{bmatrix}, \ \boldsymbol{\mu}^{-1} = \begin{bmatrix} \frac{1}{2} & -\frac{1}{2} \\ \frac{1}{2} & \frac{1}{2} \end{bmatrix}.$$

We are now ready to study the Rankine–Hugoniot jumps of (2.161). We look for solutions of the form

$$\begin{aligned} r(x,t) &= r_l + (r_r - r_l)H(\xi), \\ s(x,t) &= s_l + (s_r - s_l)H(\xi), \end{aligned} \tag{2.162}$$

where

$r_l$, $r_r$, $s_l$, and $s_r$ are constants
$H$ is the classical Heaviside function
$\xi = x - wt$
$w$ is the shock wave speed

In writing (2.162), we have made the tacit assumption that the jumps of the functions $r$ and $s$ can be represented by the same Heaviside function. Differentiating (2.162) we obtain

$$\begin{aligned} r_t &= -w(r_r - r_l)\delta(\xi), \\ r_x &= (r_r - r_l)\delta(\xi), \end{aligned}$$

and similar expressions for $s_t$ and $s_x$. Introducing these expressions into (2.161), we obtain

$$\begin{aligned} -w\delta + \left[\frac{r_l + s_l}{2} - 1\right]\delta + \frac{1}{2}(r_r - r_l + s_r - s_l)H\delta = 0, \\ -w\delta + \left[\frac{r_l + s_l}{2} + 1\right]\delta + \frac{1}{2}(r_r - r_l + s_r - s_l)H\delta = 0. \end{aligned} \tag{2.163}$$

The troublesome term $H\delta$ is evaluated as $H\delta \approx \frac{1}{2}\delta$, see Section 2.1.8, and (2.163) yields

$$\begin{aligned} w &= \frac{1}{4}(r_l + s_l + r_r + s_r) - 1, \quad \text{for an } r\text{-shock}, \\ w &= \frac{1}{4}(r_l + s_l + r_r + s_r) + 1, \quad \text{for an } s\text{-shock}. \end{aligned} \tag{2.164}$$

An $r$-shock is a shock where the Riemann invariant $r$ jumps, and similarly an $s$-shock is one where the Riemann invariant $s$ jumps. Furthermore, along an $r$-shock the Riemann invariant $s$ is $C^0$ continuous and similarly on an $s$-shock the Riemann invariant $r$ is $C^0$ continuous. This trivially follows, since $r(x,t) = r_l + (r_r - r_l)H(\xi)$, $s(x,t) = \text{constant}$ is a solution of (2.161). We can also show that the slopes of $r$ and $s$ are continuous across a shock of the opposite family. To show this, consider the trial solution

$$\begin{aligned} s &= \frac{s_l + s_r}{2} + s_{xr}\xi H(\xi) - s_{xl}\xi H(-\xi), \\ r &= r_l + (r_r - r_l)H(\xi), \end{aligned} \tag{2.165}$$

where

$s_l = s_r$

$\xi = x - wt$

$s_{xr}$ and $s_{xl}$ are the slopes of the Riemann invariant $s$ to the right and left of the shock, respectively

We would like to show that (2.165) satisfy the second of (2.161) only if $s_{xr} = s_{xl}$. Taking derivatives of the first of (2.165) and inserting in the second of, (2.161), we find

$$\begin{aligned} &-ws_{xr}H(\xi) - \cancel{ws_{xr}\xi\delta(\xi)} + ws_{xl}H(-\xi) - \cancel{ws_{xl}\xi\delta(-\xi)} \\ &+ \left[\frac{r_l + (r_r - r_l)H(\xi) + \dfrac{s_l + s_r}{2} + \cancel{s_{xr}\xi H(\xi)} - \cancel{s_{xl}\xi H(-\xi)}}{2} + 1\right] \\ &\times (s_{xr}H(\xi) + \cancel{s_{xr}\xi\delta(\xi)} - s_{xl}H(-\xi) + \cancel{s_{xl}\xi\delta(-\xi)}) \stackrel{?}{=} 0. \end{aligned}$$

The terms $\xi H$ and $\xi\delta \to 0$ as $\xi \to 0$, while $H \to \frac{1}{2}$ as $\xi \to 0$.* Hence,

$$-\frac{ws_{xr}}{2} + \frac{ws_{xl}}{2} + \left[\frac{r_r + r_l + s_r + s_l}{4} + 1\right]\left(\frac{s_{xr}}{2} - \frac{s_{xl}}{2}\right) \stackrel{?}{=} 0.$$

The term inside the square brackets equals $w + 2$, and, therefore, the expression is satisfied only if $s_{xr} = s_{xl}$.

To apply Colombeau's theory to (2.161) we repeat the previous analysis in terms of generalized Heaviside functions. A generalized Heaviside function is an element of $\mathcal{G}$ if it is bounded, i.e., its representatives $y(\varphi_\varepsilon)$ are bounded as $\varepsilon \to 0$ for any $\varphi$, and if any of its representatives $y(\varphi_\varepsilon)$ converges pointwise almost everywhere to the

---

* $H(-\xi) = 1 - H(\xi)$ and $\delta(-\xi) = \delta(\xi)$.

classical Heaviside function $H$ as $\varepsilon \to 0$. Therefore, we look for solutions $(R, S) \in \mathcal{G}$ of the form

$$R(x,t) = R_\mathrm{l} + (R_\mathrm{r} - R_\mathrm{l})Y(\xi),$$
$$S(x,t) = S_\mathrm{l} + (S_\mathrm{r} - S_\mathrm{l})Z(\xi),$$

to the system of equations,

$$R_t + \Lambda^- R_x \approx 0,$$
$$S_t + \Lambda^+ S_x \approx 0,$$
$$\Lambda^\pm = \tfrac{1}{2}(R+S) \pm 1,$$

where $Y$ and $Z$ are generalized Heaviside functions. From this point on, the analysis follows as the previous one. However, in order to complete the analysis a term $YZ'$ must be evaluated by assuming that $Y = Z$. Once this is done, we find the same Rankine–Hugoniot jumps we found before.

We can show that the Rankine–Hugoniot jumps, (2.164), are consistent with those obtained by the standard analysis, see Section 2.1.10, by considering the following degenerate case of (2.161). Let $u = \sigma$, then from (2.161) we find

$$r_t + (r/2 - 1)r_x = 0,$$
$$s = 0,$$

and, therefore,

$$w = \frac{[\![\frac{1}{4}r^2 - r]\!]}{[\![r]\!]} = \tfrac{1}{4}(r_\mathrm{r} + r_\mathrm{l}) - 1$$

in agreement with (2.164). A similar result is obtained from the $s$ Riemann variable equation if we let $u = -\sigma$. Note that by assuming that $u = \pm\sigma$, we are saying that the jumps of $r$ and $s$ are defined by the same Heaviside function. Therefore, $H\delta \approx \frac{1}{2}\delta$ is consistent with the results of the standard analysis. To have the $r$ and $s$ jumps defined by different Heaviside functions, we would need to supplement the governing equations, (2.156), with additional information from experimental observations or a more complete theory. Mathematical reasoning alone cannot solve this problem.

For this system, shock waves are valid, see Section 2.1.11, if they satisfy the condition

$$\begin{aligned} &\lambda_\mathrm{l}^+ \geq w \geq \lambda_\mathrm{r}^+, \quad \text{for an } s\text{-shock, and} \\ &\lambda_\mathrm{l}^- \geq w \geq \lambda_\mathrm{r}^-, \quad \text{for an } r\text{-shock.} \end{aligned} \tag{2.166}$$

Both result in the requirement that $u_\mathrm{l} \geq u_\mathrm{r}$.

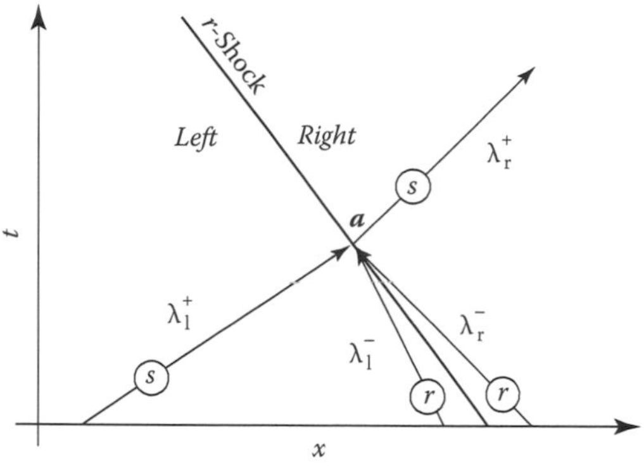


**FIGURE 2.57:** Domain of dependence of a point $a$ on an $r$-shock.

In order to compute the paths of the shock waves associated with this system of equations, we have to consider the *domain of dependence* of a shock point. For example, point $\boldsymbol{a}$ on the $r$-shock of Figure 2.57 has two characteristics bringing information from the left side of the shock. Along the $\lambda_l^+$ characteristic the Riemann invariant $s$ is constant, while along the $\lambda_l^-$ characteristic the Riemann invariant $r$ is constant. Knowing $r$ and $s$ on the left side of the shock provides all the information we need on the left side. The shock has no influence on the state of the left side. For this reason, we refer to the left side as the *supersonic* side of the shock. On the right side of the shock, we only have one characteristic reaching point $\boldsymbol{a}$. Along this characteristic, $\lambda_r^-$, $r$ is constant. Thus, we know $r$ on the right side of the shock at point $\boldsymbol{a}$. To find $s$, we use the Rankine–Hugoniot jump condition, which across an $r$-shock is given by $s_r = s_l$. Since the right side of point $\boldsymbol{a}$ influences the flow to the right of the shock, along the $\lambda_r^+$ characteristic, we refer to the right side as the *subsonic* side. For an $s$-shock, we have a similar situation, except that for an $s$-shock the left side is the subsonic side and the right side is the supersonic side. These notions are incorporated into the calculation of the shock wave speed as follows. The shock speed is defined by the second expression in (2.164). We differentiate this expression with respect to a time coordinate $\tau$ that follows the shock to obtain

$$w_\tau = \tfrac{1}{4}(2s_{\tau l} + r_{\tau l} + r_{\tau r}). \tag{2.167}$$

The derivatives $r_{\tau l}$, $s_{\tau l}$, and $r_{\tau r}$ follow from (2.161). Note that $s_{\tau l}$ is multiplied by two, reflecting the fact that $s_r = s_l$. Integrating $w_\tau$, we update the shock speed.

We are now ready to investigate the numerical solution of (2.161). Let us consider the initial value problem corresponding to the *one-saw-tooth* condition:

$$\begin{aligned} r(x,0) &= 2(1+x)[H(x+1) - H(x)] \\ &\quad + 2(1-x)[H(x) - H(x-1)], \\ s(x,0) &= s_0. \end{aligned} \tag{2.168}$$

The exact solution to this problem is given by

$$r_1(x,t) = 2\left(1+\frac{x-s_0\,t/2}{1+t}\right)[H(x+1+(1-s_0/2)t) - H(x-s_0\,t/2)]+$$
$$+2\left(1-\frac{x-s_0\,t/2}{1-t}\right)[H(x-s_0\,t/2) - H(x-1+(1-s_0/2)t)],$$
$$s_1(x,t) = s_0,$$

if $t \le \hat{t}$, where $\hat{t}$ is the time at which the shock forms. An examination of the above solution reveals that the shock wave, an $r$-shock, forms at $\hat{t} = 1$ and $x_s = s_0/2$. This is the point at which the gradient discontinuity originating at $x = 0$, $t = 0$ meets the gradient discontinuity originating at $x = 1$, $t = 0$. For $t \ge \hat{t}$, the exact solution is given by

$$r_2(x,t) = 2\left(1+\frac{x-s_0\,t/2}{1+t}\right)[H(x+1+(1-s_0/2)t) - H(x_s(t))],$$
$$s_2(x,t) = s_0$$

The shock path, $x_s(t)$, is obtained from

$$w = \tfrac{1}{4}(r_l + 2s_0) - 1.$$

From the exact solution, we have $r_l = 2(1 + (x - s_0\,t/2)/(1+t))$. Therefore, the ordinary differential equation defining the shock path is

$$\dot{x}_s - x_s/2(1+t) = (s_0/2 - 1)/2 + s_0/4(1+t).$$

The solution of this ordinary differential equation is easily obtained by standard methods [21] or with MATLAB*:

$$x_s(t) = \sqrt{2(1+t)} + (s_0/2 - 1)t - 1.$$

The numerical solution is obtained with the code `SHOCKFIT12.M` available on the accompanying CD. This code is a modification of `SHOCKFIT07.M`. It solves system (2.161) for both $r$- and $s$-shocks. For this problem, the three gradient discontinuities in the initial profile, (2.168), are treated as infinitesimally weak $r$-shocks. The numerical results at $t = 3$ are compared to the exact solution in Figure 2.58.

---

* With MATLAB:

```
syms x t s0
x=dsolve('Dx-x/(1+t)/2=(s0/2-1)/2+s0/(1+t)/4', 'x(1)=s0/2')
```

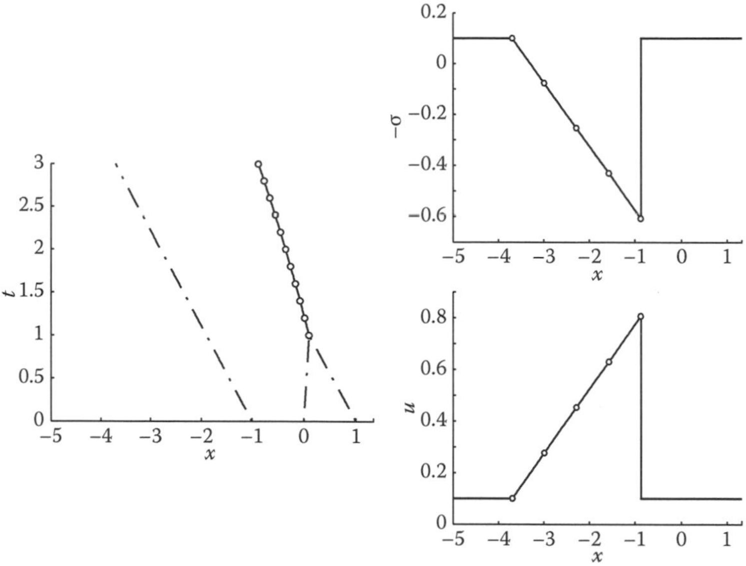


**FIGURE 2.58:** The left panel shows the computed paths of the three gradient discontinuities, dashed lines, and the computed path of the $r$-shock, solid line. The exact shock path is indicated by the circles. The right panel shows the computed, solid line, $u$ and $-\sigma$ profiles at $t = 3$, the exact values are shown as circles.

Similar results are obtained for the initial value problem defined by

$$\begin{aligned} s(x,0) &= 2(1+x)[H(x+1) - H(x)] \\ &\quad + 2(1-x)[H(x) - H(x-1)], \\ r(x,0) &= r_0, \end{aligned}$$

corresponding to a one-saw-tooth problem for the $s$ Riemann variable. In this case, the shock forms at $t = 1$, $x = 2 + r_0/2$, and the shock path is defined by

$$x_s(t) = (r_0/2 + 1)(3 + 2t) - \sqrt{(1+t)/2}(3 + 2r_0).$$

In this code, shock detection is implemented by finding when $x_r$ or $x_s \to 0$, as discussed in Section 2.9. A set of initial conditions corresponding to a Gaussian pulse is available in the code to test this method.

The MATLAB code `SHOCKFIT13.M` solves a similar system $\vec{q}_t + A\vec{q}_x = 0$, but with

$$\mathbf{A} = \begin{bmatrix} u & 1 \\ 1 & u \end{bmatrix}.$$

### 2.10.1 The Mask of Batman

The new element in this code is treating the interaction of $r$- and $s$-shocks. This interaction is illustrated in the physical plane and the Hugoniot plane in Figure 2.59. The implementation of these conditions is done by the function *shock_crossing*. Here is a snippet of the relevant lines:

```
 1. km=kcross-1;
 2. .
 3. .
 4. .
 5. if shock_type(kcross) == shock_type(km)
 6. .
 7. . (usual interaction of shocks of the same family)
 8. .
 9. else
10.    if shock_type(km) == 'r'
11.      shock_type(km) = 's';
12.      shock_type(km+1) = 'r';
13.      rR(km) = rL(km);
14.      sR(km) = sR(km+1);
15.      sL(km+1) = sR(km+1);
16.      rL(km+1) = rL(km);
17.      w(km) = (rL(km) + sL(km) + rR(km) + sR(km))/4 + 1;
18.      w(km+1) = (rL(km+1)+ sL(km+1)+ . . .
19.      rR(km+1)+ sR(km+1))/4-1;
20.    else
```

The variable `kcross` is the index of the incoming right shock and `km` is the index of the incoming left shock. The variable `shock_type(k)` defines the $k$th shock type, either r or s. If the shocks are of the same type, the interaction is handled by the omitted lines following line 5. If not, we go to line 9 which is followed by a test to

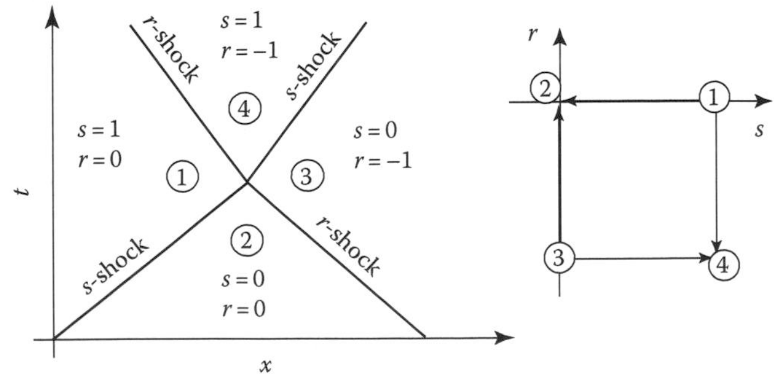


**FIGURE 2.59:** Figure depicts a typical interaction of an $s$- and $r$-shock in the $(x, t)$-plane and in the Hugoniot plane.

see if the left shock is of type $r$. If this is the case, the types are switched, i.e., the outgoing left shock is defined type $s$ and the outgoing right shock is defined type r. The variables `rR, rL, sR, sL` are the values of $r$ and $s$ at the left and right sides of the shock. These are updated in lines 13 through 16 according to the Hugoniot conditions illustrated in Figure 2.59. The new shock speeds are computed in lines 17 through 19. The lines following line 20 do the crossing if the incoming left shock is of type $s$.

To test this capability, consider the following problem:

$$s(x,0) = x[H(x) - H(x-1)],$$
$$r(x,0) = (x-3)[H(x-2) - H(x-3)].$$

These initial conditions correspond to an $s$-shock at $x = 1$ followed by an $s$-expansion in the interval $x \subset [0,1]$ and an $r$-shock at $x = 2$ followed by an $r$-expansion in the interval $x \subset [2,3]$. In addition to these two shocks, there is an $s$-gradient discontinuity ($s$-g.d.) at $x = 0$ and an $r$-gradient discontinuity ($r$-g.d.) at $x = 3$. The computed shock wave paths are shown in Figure 2.60. The shock interaction takes place at $t = 0.4$ and $x = 1.5$. The gradient discontinuities interact with the shocks at $t = 0.95$. At every step, a check is made to make sure that the shocks (and gradient discontinuities) satisfy the condition (2.166). After the gradient discontinuities cross the shock waves, this condition is not met by the gradient discontinuities and they are removed from the computation. The details of the computation are shown in Figure 2.61 where computed $\sigma$- and $u$-profiles are shown. The panels labeled $t = 0.4$ take place immediately after the shock cross. What appears as a delta function at the apex of the triangular $\sigma$-profile is a typical shock crossing profile. The delta-like behavior occurs because just after crossing, the region between the two shocks is infinitesimally thin. We would have seen a similar behavior, but with the spike pointing down, if we had plotted the profile just before the shock interaction. As the shocks move away, the delta-like pulse opens up into a rectangular-like profile ($t = 0.8$). After the shocks interact with the gradient discontinuities, they start moving into

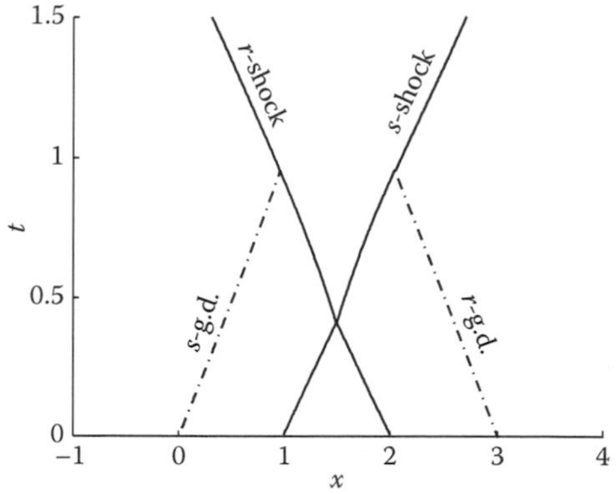


**FIGURE 2.60:** Computed shock wave and gradient discontinuity paths.

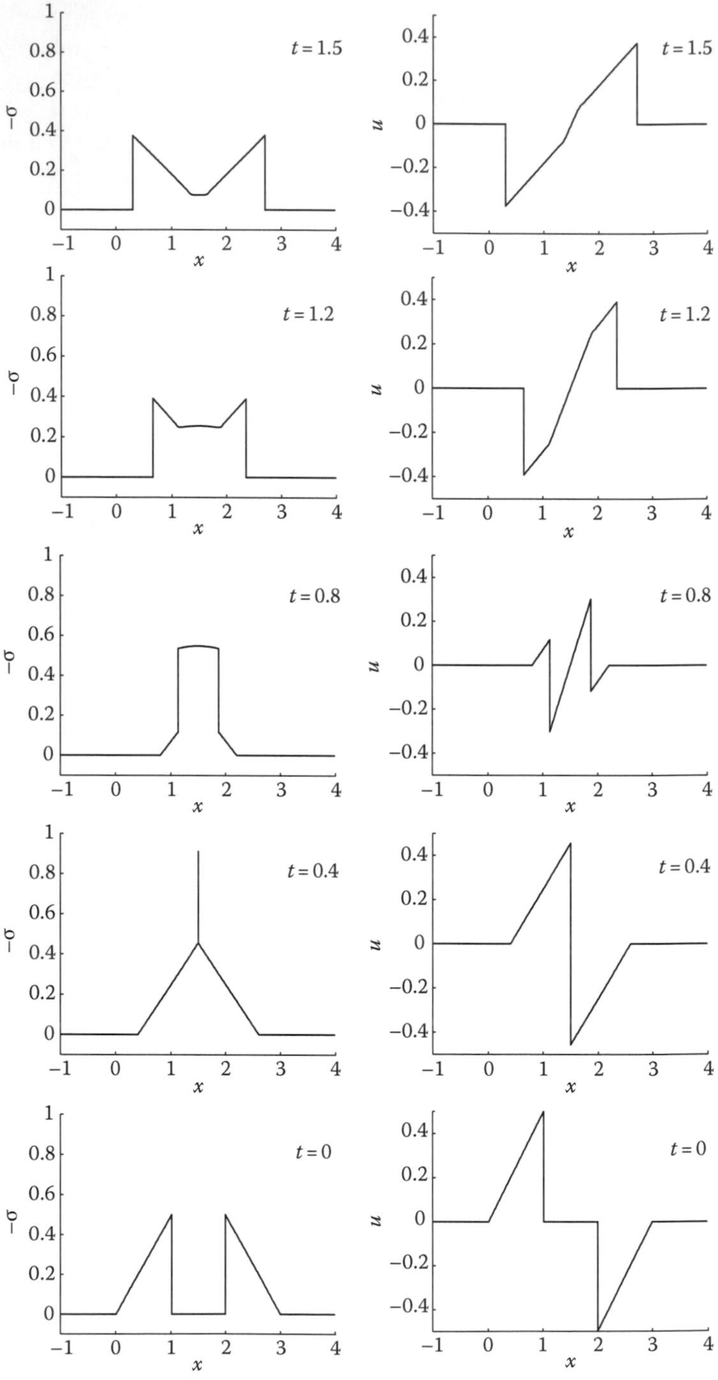


**FIGURE 2.61:** Computed σ- and $u$-profiles, note that $-\sigma$ is plotted.

regions of quiescent flow ($r = 0, s = 0$). This leads to the *Mask of Batman* $\sigma$-profile shown at $t = 1.2$. For large times, the expansions that are following the shocks eventually consume the shocks and the whole field becomes quiescent.

---

## Problems

**2.1** Let

$$H(x) = \lim_{\varepsilon \to 0} \frac{1}{1 + e^{-x/\varepsilon}},$$

and

$$\text{sgn}(x) = 2H(x) - 1.$$

(a) Plot $\sin|\pi x/2|$, $H(\sin(\pi x/2))$, $H(\sin|\pi x/2|)$ between $-4 \le x \le 4$. Plot $\{H(x) - H(-x)\}\cos(x)$ between $-\pi/2 \le x \le \pi/2$.
(b) Find the first and second generalized derivatives of $|x|$.
(c) Find the first and second generalized derivatives of $\sin|x|$.
(d) Find the first generalized derivative of $\text{sgn}(f(x))f(x)$, where $f(x) = \{H(x) - H(-x)\}\cos(x)$.

**2.2** For the Gaussian pulse problem defined by (2.131) show that the shock path is given by (2.137).

**2.3** For the following initial value problem

$$\frac{\partial u}{\partial t} + \frac{\partial}{\partial x}\left(\tfrac{1}{2}u^2\right) = 0, \qquad -\infty \le x \le \infty, \quad t > 0,$$
$$u(x,0) = \text{sech}(x),$$

show that the shock wave forms at $t = 2$. Find the location of the branch-point at $t = 1.9$.

**2.4** For the following initial value problem

$$\frac{\partial u}{\partial t} + \frac{\partial}{\partial x}\left(\tfrac{1}{2}u^2\right) = 0, \qquad -\infty \le x \le \infty, \quad t > 0,$$

$$u(x,0) = \begin{cases} 0 & x \le -1, \\ 1 + x & -1 < x \le 0, \\ 1 - x^2 & 0 < x < 1, \\ 0 & x \ge 1, \end{cases}$$

find the expression for the shock wave path.

**2.5** Study the problem

$$\frac{\partial u}{\partial t} + \frac{\partial}{\partial x}\left(\tfrac{1}{2}u^2\right) = u(1-u), \quad 0 \le x \le 1, \quad t > 0,$$
$$u(x,0) = g(x),$$

with characteristic boundary conditions:

$$\begin{aligned} u(0,t) &= \begin{cases} 0 & \text{if } u(\varepsilon,t) > 0, \\ \text{unspecified} & \text{if } u(\varepsilon,t) \le 0, \quad \varepsilon = 0+, \end{cases} \\ u(1,t) &= \begin{cases} 0 & \text{if } u(\delta,t) < 0, \\ \text{unspecified} & \text{if } u(\delta,t) \le 0, \quad \delta = 1-\varepsilon. \end{cases} \end{aligned} \tag{2.169}$$

Discuss the stability of a steady-state shock. Find the exact solution for

$$g(x) = b(1 - e^{-x}), \; b > 0.$$

Hint: see [197].

**2.6** Find the steady-state solution for

$$\frac{\partial u}{\partial t} + \frac{\partial}{\partial x}\left(\tfrac{1}{2}u^2\right) = \sin x \cos x, \quad 0 \le x \le \pi, \quad t > 0,$$
$$u(x,0) = g(x), \quad g(0) = g(\pi) = 0,$$

with boundary conditions (1.169) and $g(x) = \beta \sin x$, $0 > \beta > 1$.
Hint: see [197].

**2.7** Find the characteristic equations and jump conditions for the system

$$\begin{aligned} \rho_t + u\rho_x + \rho u_x &= 0, \\ u_t + uu_x - \sigma_x/\rho &= 0, \\ \sigma_t + u\sigma_x - \rho u_x/4 &= 0. \end{aligned}$$

# Chapter 3

## Fundamental Concepts and Equations

> Quelques sublimes que soient les recherches sur les fluides, dont nous sommes redevables à Mrs. Bernoullis, Clairaut, & d'Alembert, elles découlent si naturellement de mes deux formules générals: qu'on ne scauroit assés admirer cet accord de leurs profondes méditations avec la simplicité des principles, d'où j'ai tiré mes deux équations, & auxquels j'ai été conduit immédiatement par les premiers axioms de la Méchanique.*
>
> Leonhard Euler [68, p. 316]

The impetus for the development of the equations of fluid mechanics originated with the prize competition of 1750 of the Berlin Academy (Académie Royale des Sciences et Belles-Lettres de Prusse) for a theory explaining the resistance of fluids. As it turns out, the Academy was not happy with the quality of the submitted manuscripts.† The Academy decided that no contestant had earned the prize, and urged the contestants to compare their results with experiments. Sometime after this, on August 31, 1752, Euler read to the Academy his memoir on *De motu fluidorum in genere*.‡ This memoir is Euler's first attempt to develop a theory of fluid motion. It is of great historical value because in it Euler derived the incompressible equation of continuity, see Figure 3.1. A few years later Euler derived the inviscid compressible equations (mass and momentum conservation, not energy) from Newton's laws of mechanics. The original work was presented to the Berlin Academy on September 4, 1755. The work was published in 1757 as three separate, but consecutive, articles in the journal of the Academy. The equations appeared on the second article under the title *General Principles Concerning the Motion of Fluids* [69]. These were among the first nonlinear field equations ever formulated. For a brief history of fluid mechanics covering the period from 1687 to 1845, see [222]; for an in-depth account, see [52] and [223].

* However sublime the researches on fluids that we owe to Messrs Bernoullis, Clairaut, and d'Alembert may be, they follow so naturally from my two general formulas that one could not cease to admire this agreement of their profound thoughts with the simplicity of the principles from which I have drawn my two equations and to which I have been immediately driven by the first axioms of Mechanics.

† The fact that the leading entry was written by d'Alembert and that Euler was the director of the Mathematics department of the Academy surely played into the decision.

‡ In Truesdell's introduction to Euler's *Opera Omnia* Series II, Volume 12, the claim is made that the presentation date was established by C.G.J. Jacobi. The lecture was later published as a three-part treatise in 1761 [70].

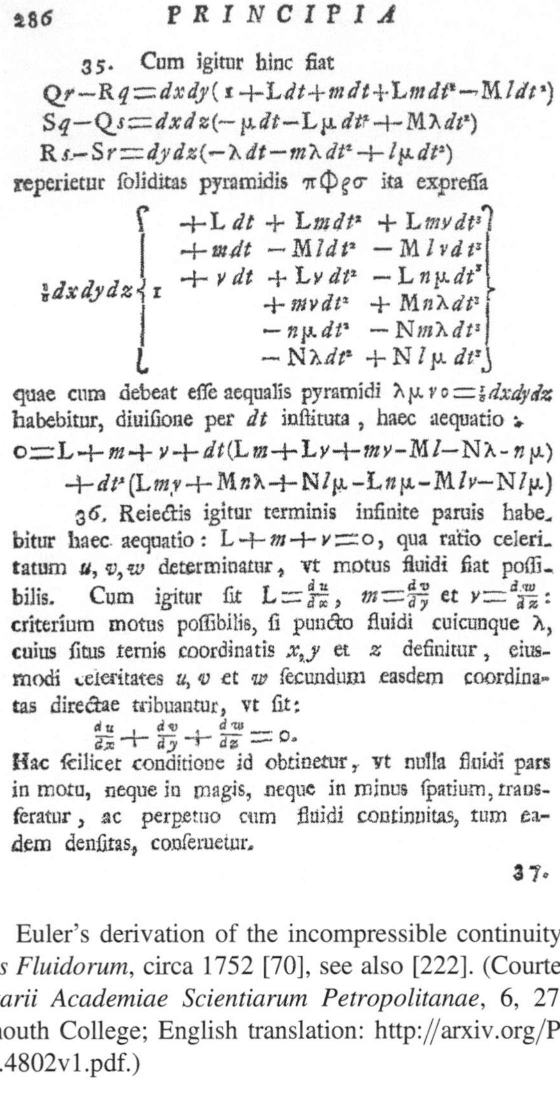

286 PRINCIPIA

35. Cum igitur hinc fiat

$$Qr - Rq = dxdy(1 + Ldt + mdt + Lmdt^2 - Mldt^2)$$
$$Sq - Qs = dxdz(-\mu dt - L\mu dt^2 + M\lambda dt^2)$$
$$Rs - Sr = dydz(-\lambda dt - m\lambda dt^2 + l\mu dt^2)$$

reperietur soliditas pyramidis $\pi\Phi\varrho\sigma$ ita expressa

$$\tfrac{1}{6}dxdydz\left\{1 \begin{array}{lll} +L\,dt & +Lmdt^2 & +Lm\nu dt^3 \\ +mdt & -Mldt^2 & -Ml\nu dt^3 \\ +\nu\,dt & +L\nu dt^2 & -Ln\mu dt^3 \\ & +m\nu dt^2 & +Mn\lambda dt^3 \\ & -n\mu dt^2 & -Nm\lambda dt^3 \\ & -N\lambda dt^2 & +Nl\mu dt^3 \end{array}\right\}$$

quae cum debeat esse aequalis pyramidi $\lambda\mu\nu o = \tfrac{1}{6}dxdydz$ habebitur, diuisione per $dt$ instituta, haec aequatio:

$$o = L + m + \nu + dt(Lm + L\nu + m\nu - Ml - N\lambda - n\mu)$$
$$+ dt^2(Lm\nu + Mn\lambda + Nl\mu - Ln\mu - Ml\nu - Nl\mu)$$

36. Reiectis igitur terminis infinite paruis habebitur haec aequatio: $L + m + \nu = o$, qua ratio celeritatum $u, v, w$ determinatur, vt motus fluidi fiat possibilis. Cum igitur sit $L = \frac{du}{dx}$, $m = \frac{dv}{dy}$ et $\nu = \frac{dw}{dz}$: criterium motus possibilis, si puncto fluidi cuicunque $\lambda$, cuius situs ternis coordinatis $x, y$ et $z$ definitur, eiusmodi celeritates $u, v$ et $w$ secundum easdem coordinatas directae tribuantur, vt sit:

$$\frac{du}{dx} + \frac{dv}{dy} + \frac{dw}{dz} = o.$$

Hac scilicet conditione id obtinetur, vt nulla fluidi pars in motu, neque in magis, neque in minus spatium, transferatur, ac perpetuo cum fluidi continuitas, tum eadem densitas, conseruetur.

37.

**FIGURE 3.1:** Euler's derivation of the incompressible continuity equation from *Principia Motus Fluidorum*, circa 1752 [70], see also [222]. (Courtesy of Euler, L., *Novi Commentarii Academiae Scientiarum Petropolitanae*, 6, 271, 1761, Euler Archive, Dartmouth College; English translation: http://arxiv.org/PS_cache/arxiv/pdf/0804/0804.4802v1.pdf.)

---

## 3.1 Physical Problem

We assume that the reader has a basic knowledge of fluid dynamics. Therefore, we confine my remarks to a recapitulation of fundamental equations and definitions of symbols used in the rest of the book. We do not attempt to justify such definitions or assumptions. We ask the reader to find justifications in books that deal with the theoretical aspects of fluid dynamics. In addition, we do not attempt to start from the equations of motion in general form, since our interest in this chapter and Chapter 4 is limited to one-dimensional, mostly inviscid flows.

### 3.1.1 Thermodynamics of the Medium

To formulate a model of the gas, let us begin by postulating three assumptions.

(a) *The medium is a continuum*, in the sense that physical quantities associated with matter contained in a small volume are regarded as being spread uniformly over that volume, see Section 2.1.1 and [13]. The assumption is considered acceptable for all gases, if not extremely rarefied; but is commonplace even in astrophysics, as a first approximation to stellar environments. Consequently, at each point in space occupied by the gas certain quantities can be defined which describe the thermodynamical state of the gas:

$p$, the pressure

$\rho$, the density

$\Theta$, the absolute temperature

$s$, the specific entropy (entropy per unit mass)

$e$, the specific internal energy, and

$h$, the specific enthalpy, defined by

$$h = e + p/\rho. \tag{3.1}$$

It is known from thermodynamics that only two of the above variables are independent. In addition to (3.1), thus, we should be able to find three more relations among them. The first is the definition of internal energy as a quantity which increases with heat addition and with work done on the medium by the pressure:

$$\mathrm{d}e = \Theta \mathrm{d}s + (p/\rho^2)\mathrm{d}\rho. \tag{3.2}$$

This formula holds for all gases. The remaining two relations, instead, take different forms according to the assumed nature of the gas. In the present context, they will be defined by the following assumptions.

(b) *The gas is ideal*, that is, a medium which satisfies the *equation of state*

$$p = \mathcal{R}\rho\Theta, \tag{3.3}$$

where $\mathcal{R}$ is a constant (the universal gas constant divided by the molecular weight of the gas). A brilliant inductive result of primitive experiments (by Boyle, Gay-Lussac, and Charles), (3.3) is justified by the kinetic theory of gases [32, p. 37], except in ranges of low temperature and high pressure, which lie outside the scope of this book. An important corollary follows from (3.3): *The internal energy is a function of the temperature alone* [41, p. 8]. Obviously the same can be said of the enthalpy, because of (3.1).

(c) *The gas is polytropic*. This more restrictive assumption consists of choosing the relationship between $e$ and $\Theta$ in the very simple form

$$e = c_v \Theta, \tag{3.4}$$

where $c_v$ is a constant, the specific heat at constant volume. Today, we can accept (3.4) as a first approximation only if the temperature excursion is relatively small and no chemical effects take place. We accept it mostly because it allows any thermodynamical variable to be expressed explicitly as a function of any two others. Indeed, note first that from (3.1), (3.3), and (3.4),

$$h = c_p \Theta, \tag{3.5}$$

where $c_p$, the specific heat at constant pressure, is given by

$$c_p = c_v + \mathcal{R}. \tag{3.6}$$

We will also write (3.6) in the form

$$\mathcal{R} = c_v(\gamma - 1), \tag{3.7}$$

where

$$\gamma = c_p / c_v. \tag{3.8}$$

For a monatomic gas $\gamma = \frac{5}{3}$, and for a diatomic gas it is $\frac{7}{5}$. Air is a mixture of 1:100 parts monatomic to diatomic, therefore, $\gamma = \frac{7}{5}$ is a good approximation of air. If (3.4) is used in (3.2), the latter can be integrated to give a closed-form relationship between $s$, $\Theta$, and $\rho$:

$$(s - s_1)/c_v = \ln(\Theta/\Theta_1) - (\gamma - 1)\ln(\rho/\rho_1), \tag{3.9}$$

where $s_1$, $\Theta_1$, and $\rho_1$ are suitable integration constants.

More complicated thermodynamical models are available to solve problems of current interest in ionized gases, plasma physics, and combustion.

### 3.1.2 Mechanical Assumptions

Assumption (a) (the medium is a continuum) allows us to define a vector **q**, the velocity of the gas at any given point. Two more assumptions are introduced now to specify the nature of the forces acting on the gas.

(d) *There are no body forces*. In particular, gravity effects and electromagnetic effects are negligible.

(e) *The gas is inviscid*. That is, the stress tensor reduces to its diagonal elements, all equal to the pressure (the medium exerts no tangential stress). The absence of tangential stresses implies the absence of internal friction. An argument based on the kinetic theory of gases [88, p. 10] shows that, if a gas is assumed as inviscid, each volume element is also *adiabatic*, that is, no heat can be exchanged between volume elements or transmitted through walls. The assumption is amply justified in a large number of practical applications.

### 3.1.3 Conservation Laws

Whatever the model of the gas, conservation of mass, momentum, and energy is postulated. The three conservation laws provide two scalar equations (for the conservation of mass and energy) and one vector equation (for the conservation of momentum). Hence, the number of equations available equals the number of unknowns (the velocity vector and two thermodynamical variables). The conservation laws can be expressed by considering a finite volume, bounded by material points, and following such points in their motion. The enclosed volume changes in shape, but the total mass contained in the volume does not change (unless the volume contains sources or sinks). This is the law of mass conservation which (in more general form) could be considered as expressing the balance between variations of mass and the effects of sources and sinks. The momentum conservation law states that the change in time of the total momentum of the matter contained in the volume equals the vectorial integral, in space and time, of the external forces applied to the volume. Thus, it expresses the balance between variations of momentum and force actions. The energy conservation law states that the change in time of the total (internal plus kinetic) energy of the matter contained in the volume equals the integrated flux of energy (by heat exchange or mechanical work) through the volume boundary (balance of energy and energy exchanges). Consistently with the definition of $e$ as the specific energy, we define a quantity $E$ as the specific total energy,

$$E = e + q^2/2, \tag{3.10}$$

and similarly, we define a quantity $H$ as the specific total enthalpy,

$$H = h + q^2/2. \tag{3.11}$$

### 3.1.4 Boundary Conditions

The physical picture of a flow is completed by a description of the boundaries and the nature of the flow at infinity. The physical boundaries can be

(a) *Rigid walls* (which in turn can be either fixed or moving according to a given law), or *flexible walls*

(b) *Free boundaries*, that is, surfaces which separate the gas from another fluid, with which the gas does not mix

In the case of walls, their geometry (either invariable or varying in time) is assumed known. We assume that the wall is impermeable by definition. This means that every material element in contact with the wall has the same velocity component normal to the wall as the wall itself. In particular, if the wall geometry does not change in time, the normal velocity component of the gas vanishes at the wall. Such a *kinematical boundary condition* is true for all gases considered above. Since we assume that the gas is inviscid, the condition is also sufficient to determine the flow pattern, assuming that the nature of the gas at infinity is known. The geometry of a free surface cannot be the datum of a problem. It results, instead, as part of the solution. A free surface is a boundary typical of inviscid flows, since mixing would occur between two adjacent fluids otherwise. Two conditions must be satisfied at each and every point of a free surface:

(1) *The surface is a locus of particle paths*. This is the same condition satisfied at a wall (the material elements can slide along the free surface but cannot penetrate through it). This condition, however, is not sufficient to determine the flow field and the surface geometry. Therefore, in addition,

(2) *The pressure must be the same on both sides of the free surface* or, if surface tension develops, the pressure jump across the surface must balance such tension.

### 3.1.5 Discontinuities

The assumption that the medium is a continuum does not imply that the physical parameters are continuous and differentiable functions of space and time everywhere. Quite to the contrary, we must expect sudden jumps to occur (either in the parameters themselves or in their derivatives) across certain surfaces. Three different cases may occur in an inviscid, compressible flow:

(a) Surfaces across which all physical parameters are continuous, but some of their derivatives are not; such surfaces will be called *gradient discontinuities*.

(b) Surfaces across which the tangential velocity component jumps (and so will all thermodynamical parameters, except pressure); we will say in this case that the surface is a *contact discontinuity*.

(c) Surfaces across which the normal velocity component jumps (and so will all thermodynamical parameters); we will say in this case that the surface is a *shock wave*.

---

## 3.2 Mathematical Formulation

### 3.2.1 Nondimensionalization

In this section, we stipulate a convenient way of writing the equations in a nondimensional form. Let us choose a pressure, $p_1$, as a reference pressure and a density, $\rho_1$, as a reference density (in a way conveniently defined for each problem). Consistently, we define a reference temperature, $\Theta_1$, according to (3.3):

$$\Theta_1 = p_1/(\rho_1 \mathcal{R}). \tag{3.12}$$

All pressures, densities, and temperatures will be expressed as multiples of their reference values. It is also convenient to introduce a symbol for the logarithm of pressure (after scaling by $p_1$):

$$P = \ln p. \tag{3.13}$$

Therefore, (3.3) is replaced by

$$p = \rho\Theta, \tag{3.14}$$

or

$$\Theta = \exp\,(P - \ln \rho). \tag{3.15}$$

The reference energy, in mechanical units, is chosen as

$$e_1 = p_1/\rho_1 = \mathcal{R}\Theta_1. \tag{3.16}$$

Consequently, (3.4) and (3.5) take the forms

$$e = \frac{1}{\gamma - 1}\Theta = \frac{1}{\gamma - 1}\frac{p}{\rho}, \qquad h = \frac{\gamma}{\gamma - 1}\Theta = \frac{\gamma}{\gamma - 1}\frac{p}{\rho}, \tag{3.17}$$

respectively. Finally, if in (3.9) $S$ (a nondimensional quantity) is written in lieu of $(s - s_1)/c_v$, (3.9) becomes

$$S = \ln \Theta - (\gamma - 1) \ln \rho. \tag{3.18}$$

Other useful equivalent relations are

$$S = P - \gamma \ln \rho \tag{3.19}$$

and

$$S = \gamma \ln \Theta - (\gamma - 1)P. \tag{3.20}$$

Since energy, in mechanical units, is the square of a velocity, a convenient choice for a reference velocity is

$$u_1 = \sqrt{e_1}. \tag{3.21}$$

Next, let us choose a reference length, $x_1$. A reference time follows:

$$t_1 = x_1/u_1. \tag{3.22}$$

### 3.2.2 Conservation Laws in Integral Form

Mass, momentum, and total energy of the matter contained in a volume $\tau$ are expressible as volume integrals, by writing

$$F = \int_\tau f \mathrm{d}\tau, \tag{3.23}$$

where $F$ is
- the mass if $f = \rho$
- the momentum if $f = \rho\mathbf{q}$
- the total energy if $f = \rho E$

Since the integrand is a continuous function for any volume $\tau$ or, at most, it undergoes finite jumps, $F$ is a continuous function.

Let $\tau$ move with the matter, and let $\Delta F$ be the total change in $F$ during a finite interval of time, $\Delta t$. The conservation laws expressed in Section 3.1.3 imposed three relations between $\Delta F$ and certain events occurring across the surface $\sigma$ bounding $\tau$. Specifically, we can write

$$\Delta F = \Delta \int_\tau f \mathrm{d}\tau = \int_t^{t+\Delta t} \mathrm{d}t \int_\sigma \psi \mathrm{d}\sigma, \tag{3.24}$$

where
- $\psi = 0$ in the mass conservation equation
- $\psi = p\boldsymbol{n}$ in the momentum equation (where $\boldsymbol{n}$ is the unit vector normal to $\mathrm{d}\sigma$, pointing inwards)
- $\psi = p\mathbf{q} \cdot \boldsymbol{n}$ in the energy equation

So long as $\psi$ is bounded, $\int \psi \mathrm{d}\sigma$ is a continuous function of space and time. Consequently, (3.23) may be replaced by

$$\frac{D}{Dt} \int_\tau f \mathrm{d}\tau = \int_\sigma \psi \mathrm{d}\sigma, \tag{3.25}$$

where $D/Dt$ means, as usual, that the derivative is taken following the volume elements in their motion (*material*, or *Lagrangian* derivative, see Section 2.1.2). The mathematical formulation (3.23) or (3.24) of the conservation laws is indeed the natural expression of the physical facts stated in Section 3.1.3, but it is not necessarily the best representation for all practical purposes.

### 3.2.3 Conservation Laws in Differential Form

Wherever the flow parameters are continuous and differentiable, the conservation laws are more conveniently expressed in differential form. To eliminate the integrals in (3.24), two steps have to be taken: (1) The surface integral on the right-hand side of (3.24) must be transformed into a volume integral, and (2) the differential and integral operators on the left-hand side must be interchanged. For both steps we must make use of the integral calculus formula

$$\int_{\sigma} f n_i \mathrm{d}\sigma = -\int_{\tau} \partial f/\partial x_i \mathrm{d}\tau, \tag{3.26}$$

where

$n_i$ $(i=1,2,3)$ are the Cartesian components of the unit vector $\boldsymbol{n}$
$x_i$ are Cartesian coordinates

The importance of (3.26) stems from its physical meaning: the global effect of certain events within a volume (right-hand side) must have an influence on the global effect of related events on the surrounding surface (left-hand side). Of course, for (3.26) to be true, $f$ must be continuously differentiable in $\tau$.

A number of vector equations can be obtained from (3.26) through simple manipulations. For example,

$$\int_{\sigma} f\boldsymbol{n} \mathrm{d}\sigma = -\int_{\tau} \nabla f \mathrm{d}\tau, \tag{3.27}$$

and

$$\int_{\sigma} fg \cdot \boldsymbol{n} \mathrm{d}\sigma = -\int_{\tau} \nabla \cdot (fg) \mathrm{d}\tau. \tag{3.28}$$

For the first step mentioned above, we use (3.27) or (3.28) to write

$$\int_{\sigma} \psi \mathrm{d}\sigma = \int_{\tau} \chi \mathrm{d}\tau, \tag{3.29}$$

where

$\chi = 0$ for the mass conservation equation
$\chi = -\nabla p$ for the momentum equation
$\chi = -\nabla p\mathbf{q}$ for the energy equation

The second step can be taken by observing that the material rate of change of $F$ (in a moving volume $\tau$) results from the local rate of change of $F$ in a fixed volume $\tau^*$ (which coincides with $\tau$ initially),

$$\frac{\partial F}{\partial t} = \frac{\partial}{\partial t} \int_{\tau^*} f \mathrm{d}\tau^* = \int_{\tau^*} \frac{\partial f}{\partial t} \mathrm{d}\tau^*, \tag{3.30}$$

plus the balance between the inflow and outflow of $f$ across the surface $\sigma^*$ surrounding $\tau^*$, with the elements on $\sigma^*$ moving with velocity $\mathbf{q}$,

$$-\int_{\sigma^*} f\mathbf{q}\cdot\boldsymbol{n}\,\mathrm{d}\sigma^*, \tag{3.31}$$

or, by applying (3.27) or (3.28),

$$\int_{\tau^*} \nabla(f\mathbf{q})\mathrm{d}\tau^*. \tag{3.32}$$

Note that, for the momentum equation, $f$ is a vector and the integrand in the last expression becomes $\nabla\cdot(\rho\mathbf{q}\mathbf{q})$ which can be given meaning, consistent with the usual notations of vector calculus, if the latter is considered as a subset of general tensor calculus. In this context, it will suffice to say that, in Cartesian coordinates, the expression is a vector, defined by

$$\nabla\cdot(\rho\mathbf{q}\mathbf{q}) = \mathbf{q}\nabla\cdot(\rho\mathbf{q}) + \rho\mathbf{q}\nabla\cdot\mathbf{q},$$

whose component along the $x_i$-axis ($i=1,2,3$) is $\nabla\cdot(\rho u_i\mathbf{q})$ where $u_i$ is the component of $\mathbf{q}$ along the same axis. In conclusion, the material rate of change of $F$ can be expressed in the form

$$\frac{DF}{Dt} = \int_{\tau^*}\left[\frac{\partial f}{\partial t} + \nabla\cdot(f\mathbf{q})\right]\mathrm{d}\tau^*. \tag{3.33}$$

On the other hand, the integral on the right-hand side of (3.29) is evaluated at a given instant of time; its domain of integration, thus, can be denoted by $\tau$ or $\tau^*$, indifferently. Consequently, (3.29) can be replaced by

$$\int_{\tau}\left[\frac{\partial f}{\partial t} + \nabla\cdot(f\mathbf{q}) - \chi\right]\mathrm{d}\tau = 0. \tag{3.34}$$

This equation, however, must be satisfied regardless of the choice of $\tau$; therefore, the integrand must vanish identically:

$$\frac{\partial f}{\partial t} + \nabla\cdot(f\mathbf{q}) = \chi. \tag{3.35}$$

In this way, a differential form of the conservation equations has been obtained. As said earlier, its validity is restricted to those regions where $f$ is continuous and differentiable. If $f$ undergoes a sudden jump somewhere, the derivatives in the integrand of (3.34) can still be accepted in a generalized way as representing δ-functions whose integrals may be, and are in fact, well-behaved functions. However, all meaning is lost if the integral signs are removed, as in (3.35). It is clear that, in the presence of discontinuities, the equations in the differential form (3.35) must be supplemented by other equations, expressing the same physical conservation laws in a vanishingly thin layer draped about the discontinuity surface. A solution of (3.35), having discontinuities across which the conservation laws in integral form are satisfied, is called a *weak solution* of (3.35).

## 3.3 Explicit Form of the Equations of Motion

Starting from the basic formula (3.35), we now proceed to write the partial differential equations that express the three conservation laws separately, and we introduce alternative forms that are commonly used as fundamental sets in numerical fluid dynamics. Let us restate that, because of their origin, all equations in this section are valid only wherever the flow parameters are differentiable.

### 3.3.1 Divergence Form

With $f$ and $\chi$ defined in the preceding sections, the following three equations are obtained from (3.35):

$$\frac{\partial \rho}{\partial t} + \nabla \cdot (\rho \mathbf{q}) = 0, \tag{3.36}$$

$$\frac{\partial \rho \mathbf{q}}{\partial t} + \nabla \cdot (\rho \mathbf{q}\mathbf{q} + p\mathbf{I}) = 0, \tag{3.37}$$

$$\frac{\partial \rho E}{\partial t} + \nabla \cdot ((p + \rho E)\mathbf{q}) = 0, \tag{3.38}$$

where $\mathbf{I}$ is a $3 \times 3$ unit matrix. The set above is known as the divergence (or conservation) form of the equations of motion (the mass conservation equation, the momentum conservation equation, and the energy conservation equation, in that order). The mass conservation equation is also commonly called the "continuity equation," a name that stuck, as it happens, without any apparent reason.* There is reason, instead for saying that the equations above are written in divergence form. The divergence of a vector, in Cartesian coordinates, is the sum of three terms, each being a first partial derivative of a quantity, not multiplied by any coefficient. It is evident that the same can be said of all the terms appearing in (3.36) through (3.38)

* The term is probably due to Euler. In his derivation of the equation, see Figure 3.1, last sentence, he writes: "This condition expresses that through the motion no part of the fluid is carried into a greater or lesser space, but perpetually *the continuity of the fluid* as well as the identical density is conserved."

when explicitly written in Cartesian coordinates. For unsteady flows, the evolution of the flow in time will be found by integrating the equations with respect to time. The unknowns of the problem are thus:

$$\boldsymbol{U} = [\rho,\ \rho\mathbf{q},\ \rho E]^{\mathrm{T}}. \tag{3.39}$$

We note that (3.38) may be simplified using (3.11):

$$\frac{\partial \rho E}{\partial t} + \nabla \cdot (\rho H \mathbf{q}) = 0. \tag{3.40}$$

If we define the flux vector $\mathcal{F}$ to be

$$\mathcal{F} = \begin{bmatrix} \rho\mathbf{q} \\ \rho\mathbf{q}\mathbf{q} + p\mathbf{I} \\ (p + \rho E)\mathbf{q} \end{bmatrix},$$

then the system can be expressed in compact form as

$$\frac{\partial \boldsymbol{U}}{\partial t} + \nabla \cdot \mathcal{F}(\boldsymbol{U}) = 0, \tag{3.41}$$

since $p = (\gamma - 1)(\rho E - (\rho\mathbf{q})^2/2)/\rho$ and $\mathbf{q} = \rho\mathbf{q}/\rho$.

Also in compact form, we write the quasilinear form of (3.41) as

$$\frac{\partial \boldsymbol{U}}{\partial t} + \mathbf{A}(\boldsymbol{U}) \cdot \nabla \boldsymbol{U} = 0,$$

where $\mathbf{A}(\boldsymbol{U}) = \partial\mathcal{F}/\partial\boldsymbol{U}$. If the pressure can be expressed as some function of the internal energy times the density, $p = \rho f(e)$, as is the case with our model, then the components of the flux vector $\mathcal{F}$ are homogeneous functions of degree one with respect to the unknown vector $\boldsymbol{U}$ [211]. This means that for any scalar $\alpha$, $\mathcal{F}(\alpha\boldsymbol{U}) = \alpha\mathcal{F}(\boldsymbol{U})$. By applying Euler's homogeneity relation [42, p. 11] [72], we obtain the remarkable relation

$$\mathcal{F} = \mathbf{A}\boldsymbol{U}.$$

### 3.3.2 Euler's Form

Let us now make use of the well-known relation between material and local derivatives:

$$\frac{Df}{Dt} = \frac{\partial f}{\partial t} + \mathbf{q} \cdot \nabla f. \tag{3.42}$$

Note that the formula holds if $f$ is a vector, provided that the Cartesian components of the vector $\mathbf{q} \cdot \nabla f$ are defined as

$$(\mathbf{q} \cdot \nabla f)_i = \sum_j u_j \frac{\partial f_i}{\partial x_j}.$$

By using (3.42), (3.36) becomes

$$\frac{D\rho}{Dt} + \rho \nabla \cdot \mathbf{q} = 0. \tag{3.43}$$

If (3.43) is multiplied by $\mathbf{q}$ and then subtracted from (3.37), (3.42) can be applied again (this time, for a vector) and the following new form of the momentum equation is obtained:

$$\rho \frac{D\mathbf{q}}{Dt} + \nabla p = 0. \tag{3.44}$$

Similarly, if (3.43) is multiplied by $E$ and subtracted from (3.38), the energy equation takes the form

$$\rho \frac{DE}{Dt} + \nabla \cdot (p\mathbf{q}) = 0. \tag{3.45}$$

This, however, is not the form we will commonly use. To obtain the form we prefer, first subtract from it (3.44), dot multiply the result by $\mathbf{q}$; then replace $e$ with $p/(\gamma - 1)\rho$ and simplify the resulting $D\rho/Dt$ with (3.43). The new form of the energy equation so obtained is

$$\frac{Dp}{Dt} + \gamma p \nabla \cdot \mathbf{q} = 0. \tag{3.46}$$

We repeat here the three equations of motion as obtained above; in the second, though, the term $D\mathbf{q}/Dt$ has been replaced by an equivalent expression which can be expanded more easily in any system of coordinates:

$$\begin{aligned} &\frac{D\rho}{Dt} + \rho \nabla \cdot \mathbf{q} = 0, \\ &\frac{\partial \mathbf{q}}{\partial t} - \mathbf{q} \times (\nabla \times \mathbf{q}) + \nabla \frac{q^2}{2} + \frac{1}{\rho} \nabla p = 0, \\ &\frac{Dp}{Dt} + \gamma p \nabla \cdot \mathbf{q} = 0. \end{aligned} \tag{3.47}$$

We denote this set as Euler's equations of motion, only because they are referred to under that name in textbooks. For unsteady flows, the unknowns of the problem are $\rho$, $\mathbf{q}$, and $p$. No other parameters need to be evaluated for the numerical integration of the equations. In this sense, (3.47) is simpler than (3.41). Most of the terms, however, contain variable coefficients.

### 3.3.3 Modified Euler's Form: Conservation of Entropy

If the first equation of (3.47) is multiplied by $\gamma/\rho$ and the last equation is divided by $p$, and the two equations so obtained are subtracted from each other, it follows that

$$\frac{DP}{Dt} - \gamma\frac{D\ln\rho}{Dt} = 0 \tag{3.48}$$

or, recalling Equation 3.19,

$$\frac{DS}{Dt} = 0. \tag{3.49}$$

This result, remarkable in its simplicity, can be expressed in words as a new conservation law: *The entropy of a material element is conserved in an inviscid flow*. It is obvious that this law must be respected in a numerical computation, even if it is not used explicitly. It must be noted, however, that (3.48) is not exempt of the limitations mentioned at the opening of this section. The new conservation law and the related conservation of entropy law are valid so long as the flow variables, $P$, $\rho$, and $S$ are differentiable. The conservation laws of the preceding section are unrestricted because they have been obtained in integral form. The law of conservation of entropy is limited to differentiable flows because it has been obtained from the other conservation equations only after such equations had been reduced to a differential form.

In the *absence of discontinuities*, (3.49) can replace the energy equation and the third of (3.47), divided by the pressure, can be used instead of the continuity equation, to obtain the following set:

$$\begin{aligned}
&\frac{DP}{Dt} + \gamma\nabla\cdot\mathbf{q} = 0,\\
&\frac{\partial\mathbf{q}}{\partial t} - \mathbf{q}\times(\nabla\times\mathbf{q}) + \nabla\frac{q^2}{2} + \Theta\nabla P = 0,\\
&\frac{DS}{Dt} = 0.
\end{aligned} \tag{3.50}$$

The unknowns of the problem are $P$, $\mathbf{q}$, and $S$. The equations are nonlinear and one of the coefficients, $\Theta$, has to be evaluated from $P$ and $S$ using (3.20). The interesting features of these equations are

1. The conservation of energy law is explicitly replaced by the conservation of entropy law.

2. The entropy is completely defined by the third equation and is not differentiated in any of the other equations; it only affects the coefficient $\Theta$.

### 3.3.4 Crocco's Equation

The vorticity vector is defined by

$$\boldsymbol{\omega} = \nabla\times\mathbf{q}. \tag{3.51}$$

With (3.51), we can write the momentum equation in (3.50) in the form

$$\frac{\partial \mathbf{q}}{\partial t} - \mathbf{q} \times \boldsymbol{\omega} + \nabla \frac{q^2}{2} + \Theta \nabla P = 0. \tag{3.52}$$

If we let $\boldsymbol{a}$ be the acceleration vector, then

$$\boldsymbol{a} = \frac{\partial \mathbf{q}}{\partial t} - \mathbf{q} \times \boldsymbol{\omega} + \nabla \frac{q^2}{2}. \tag{3.53}$$

In an inviscid flow it is equal to $-\Theta \nabla P$. The term $-\mathbf{q} \times \boldsymbol{\omega}$ is the centripetal acceleration. If we take the curl of the acceleration,* we find

$$\nabla \times \boldsymbol{a} = \frac{\partial \boldsymbol{\omega}}{\partial t} - \boldsymbol{\omega} \cdot (\nabla \mathbf{q}) + \boldsymbol{\omega}(\nabla \cdot \mathbf{q}).$$

Now if we take the curl of $-\Theta \nabla P$,[†]

$$\nabla \times \boldsymbol{a} = -\nabla \Theta \times \nabla P.$$

Combining the last two expressions, we obtain the equation governing the distribution of vorticity:

$$\frac{\partial \boldsymbol{\omega}}{\partial t} + \boldsymbol{\omega}(\nabla \cdot \mathbf{q}) - \boldsymbol{\omega}(\nabla \mathbf{q}) + \nabla \Theta \times \nabla P = 0. \tag{3.54}$$

Now combining (3.2) with (3.1), nondimensionalizing the resulting expression, and taking the gradient, we arrive at the following equation:

$$\frac{\Theta}{\gamma - 1} \nabla S = \nabla h - \Theta \nabla P. \tag{3.55}$$

If we now use (3.51), (3.55), and (3.11), we can write the momentum equation in (3.50) in the form

$$\frac{\partial \mathbf{q}}{\partial t} - \mathbf{q} \times \boldsymbol{\omega} = \frac{\Theta}{\gamma - 1} \nabla S - \nabla H. \tag{3.56}$$

Equation 3.56 was derived by Crocco in 1937 [48] and is known as Crocco's equation. For steady flows, it establishes a relation between vorticity and entropy, since from (3.38) $\nabla H = 0$.

* To obtain this result we used the following two identities: $\nabla \times \nabla \phi = 0$, $\nabla \times (\boldsymbol{\alpha} \times \boldsymbol{\beta}) = (\boldsymbol{\beta} \nabla)\boldsymbol{\alpha} - (\boldsymbol{\alpha} \nabla)\boldsymbol{\beta} - \boldsymbol{\beta}(\nabla \boldsymbol{\alpha}) + \boldsymbol{\alpha}(\nabla \boldsymbol{\beta})$, where $\phi$ is a scalar, and $\boldsymbol{\alpha}$ and $\boldsymbol{\beta}$ are vectors.

† Here we used the identities: $\nabla \times \nabla \phi = 0$, $\nabla \times (\phi \boldsymbol{\alpha}) = \nabla \phi \times \boldsymbol{\alpha} + \phi(\nabla \times \boldsymbol{\alpha})$, where $\phi$ is a scalar and $\boldsymbol{\alpha}$ is a vector.

## 3.4 Orthogonal Curvilinear Coordinates

In previous sections we have expressed the governing equations in vectorial form. From the vectorial form, we will need to write the equations in some orthogonal coordinate system specifically suited to the problem under investigation. The general properties of orthogonal curvilinear coordinates are as follows.

Let $\varsigma$, $\eta$, $\xi$ be a set of orthogonal curvilinear coordinates and let $\vec{i}_\varsigma$, $\vec{i}_\eta$, $\vec{i}_\xi$ be the unit vectors corresponding to each coordinate. Formally, we express the relationship between the curvilinear coordinates and the Cartesian coordinates by

$$x = x(\varsigma, \eta, \xi),$$
$$y = y(\varsigma, \eta, \xi),$$
$$z = z(\varsigma, \eta, \xi).$$

At points where the Jacobian matrix $\partial(x, y, z)/\partial(\varsigma, \eta, \xi)$ does not vanish, we can write the inverse transformation

$$\varsigma = \varsigma(x, y, z),$$
$$\eta = \eta(x, y, z),$$
$$\xi = \xi(x, y, z).$$

In Cartesian coordinates the differentials d$x$, d$y$, and d$z$ corresponds to elemental distances along each of the Cartesian coordinates. Let $\mathrm{d}l_\varsigma$, $\mathrm{d}l_\eta$, and $\mathrm{d}l_\xi$ be elements of distance measured along each of the orthogonal curvilinear coordinates and define the metric coefficients $h_\varsigma = |\mathrm{d}\varsigma/\mathrm{d}l_\varsigma|$, $h_\eta = |\mathrm{d}\eta/\mathrm{d}l_\eta|$, and $h_\xi = |\mathrm{d}\xi/\mathrm{d}l_\xi|$. If $\boldsymbol{r}(\varsigma,\eta,\xi)$ is the position vector, then $\vec{i}_k = \partial \boldsymbol{r}/\partial l_k$, where $k = \varsigma,\eta,\xi$. Therefore,

$$\vec{i}_\varsigma = \frac{\partial \boldsymbol{r}}{\partial l_\varsigma} = \frac{\partial \boldsymbol{r}}{\partial \varsigma}\frac{\partial \varsigma}{\partial l_\varsigma} = \frac{\partial \boldsymbol{r}}{\partial \varsigma} h_\varsigma,$$

and similarly

$$\vec{i}_\eta = \frac{\partial \boldsymbol{r}}{\partial \eta} h_\eta, \quad \vec{i}_\xi = \frac{\partial \boldsymbol{r}}{\partial \xi} h_\xi.$$

An infinitesimal displacement of the position vector is given by

$$\begin{aligned} \mathrm{d}\boldsymbol{r} &= \frac{\partial \boldsymbol{r}}{\partial \varsigma}\mathrm{d}\varsigma + \frac{\partial \boldsymbol{r}}{\partial \eta}\mathrm{d}\eta + \frac{\partial \boldsymbol{r}}{\partial \xi}\mathrm{d}\xi, \\ &= \vec{i}_\varsigma \frac{\mathrm{d}\varsigma}{h_\varsigma} + \vec{i}_\eta \frac{\mathrm{d}\eta}{h_\eta} + \vec{i}_\xi \frac{\mathrm{d}\xi}{h_\xi}. \end{aligned}$$

An element of distance between two points in Cartesian coordinates is given by $dl^2 = dx^2 + dy^2 + dz^2$. In orthogonal curvilinear coordinates it is given by

$$\begin{aligned} dl^2 &= |d\boldsymbol{r}|^2 = \frac{d\varsigma^2}{h_\varsigma^2} + \frac{d\eta^2}{h_\eta^2} + \frac{d\xi^2}{h_\xi^2} + 2\vec{i}_\varsigma \cdot \vec{i}_\eta \frac{d\varsigma d\eta}{h_\varsigma h_\eta} + 2\vec{i}_\xi \cdot \vec{i}_\eta \frac{d\xi d\eta}{h_\xi h_\eta} + 2\vec{i}_\xi \cdot \vec{i}_\varsigma \frac{d\xi d\varsigma}{h_\xi h_\varsigma} \\ &= \frac{d\varsigma^2}{h_\varsigma^2} + \frac{d\eta^2}{h_\eta^2} + \frac{d\xi^2}{h_\xi^2}. \end{aligned}$$

The operator $\nabla\varphi$ is defined by the relation $d\varphi = d\boldsymbol{r} \cdot \nabla\varphi$. From this it follows that, if $\varphi$ is a scalar, the gradient operator is defined by

$$\nabla\varphi = \vec{i}_\varsigma h_\varsigma \frac{\partial\varphi}{\partial\varsigma} + \vec{i}_\eta h_\eta \frac{\partial\varphi}{\partial\eta} + \vec{i}_\xi h_\xi \frac{\partial\varphi}{\partial\xi}.$$

If we take the gradient of $\varsigma$,

$$\nabla\varsigma = h_\varsigma \vec{i}_\varsigma, \quad \text{therefore, } h_\varsigma = |\nabla\varsigma|,$$

similarly, $h_\eta = |\nabla\eta|$, and $h_\xi = |\nabla\xi|$. If we write the gradient operator in Cartesian coordinates, then

$$h_\varsigma^2 = \left(\frac{\partial\varsigma}{\partial x}\right)^2 + \left(\frac{\partial\varsigma}{\partial y}\right)^2 + \left(\frac{\partial\varsigma}{\partial z}\right)^2,$$

With similar expressions for $h_\eta$, and $h_\xi$.

If $\boldsymbol{v}$ is a vector defined by

$$\boldsymbol{v} = v_\varsigma \vec{i}_\varsigma + v_\eta \vec{i}_\eta + v_\xi \vec{i}_\xi,$$

then the divergence of $\boldsymbol{v}$ is

$$\nabla \cdot \boldsymbol{v} = h_\varsigma h_\eta h_\xi \left[\frac{\partial}{\partial\varsigma}\left(\frac{v_\varsigma}{h_\eta h_\xi}\right) + \frac{\partial}{\partial\eta}\left(\frac{v_\eta}{h_\varsigma h_\xi}\right) + \frac{\partial}{\partial\xi}\left(\frac{v_\xi}{h_\eta h_\varsigma}\right)\right],$$

and the curl of $\boldsymbol{v}$ is

$$\nabla \times \boldsymbol{v} = h_\varsigma h_\eta h_\xi \begin{vmatrix} \dfrac{\vec{i}_\varsigma}{h_\varsigma} & \dfrac{\vec{i}_\eta}{h_\eta} & \dfrac{\vec{i}_\xi}{h_\xi} \\ \dfrac{\partial}{\partial\varsigma} & \dfrac{\partial}{\partial\eta} & \dfrac{\partial}{\partial\xi} \\ \dfrac{v_\varsigma}{h_\varsigma} & \dfrac{v_\eta}{h_\eta} & \dfrac{v_\xi}{h_\xi} \end{vmatrix}.$$

The Jacobian matrix relating the Cartesian coordinates to the orthogonal curvilinear coordinates is given by

$$\frac{\partial(x,y,z)}{\partial(\varsigma,\eta,\xi)} = \frac{1}{\partial(\varsigma,\eta,\xi)/\partial(x,y,z)} = \frac{1}{h_\varsigma h_\eta h_\xi}.$$

---

## 3.5 Differential Geometry of Singular Surfaces

In this section, we will explore the properties associated with *gradient discontinuities*. We will refer to any surface across which all flow parameters are continuous, but some of their space derivatives are discontinuous as a *singular surface*. We study these surfaces using geometrical and physical concepts. For a different treatment see [44, pp. 75–78].

Let $\Sigma$ be a singular surface and let $A$ be one of its points. The location of $A$ is defined by the position vector $\boldsymbol{x} = \alpha\boldsymbol{e}_x + \beta\boldsymbol{e}_y + \delta\boldsymbol{e}_z$, where the $\boldsymbol{e}_i$'s are the unit vectors along the respective coordinate axes. The surface $\Sigma$ is defined in space and time by

$$\mathbb{F}(\boldsymbol{x},t) = 0. \tag{3.57}$$

$\Sigma$ is a geometrical locus. When we describe the motion of a point $A$ on $\Sigma$ we do not describe the motion of a fluid particle (material element) which happens to be at $A$ at time $t$, instead we describe the motion of a geometrical entity. Displacements of $A$ along $\Sigma$ are of no interest, since they do not describe changes in the shape of $\Sigma$. The motion of $\Sigma$, thus, is defined solely by its speed $w$ at each of its points. At any point $A$, the direction of $w$ is along the normal $\boldsymbol{n}$ at $A$, see Figure 3.2. From (3.57), $\boldsymbol{n}$ is defined by

$$\boldsymbol{n} = \frac{\nabla\mathbb{F}}{|\nabla\mathbb{F}|}. \tag{3.58}$$

Let $[\![g]\!]$ denote the jump of any scalar function $g$ across $\Sigma$. By our definition of $\Sigma$, if $g$ is a physical variable then $[\![g]\!] = 0$, but $[\![\nabla g]\!]$ may be different from zero. We would like to find expressions for $[\![Dg/Dt]\!]$, $[\![\nabla g]\!]$, as well as, $[\![\nabla\cdot\boldsymbol{g}]\!]$ for a vector $\boldsymbol{g}$. In what follows, we make use of the following relations:

$$\text{if } g = a + b \text{ then } [\![g]\!] = [\![a]\!] + [\![b]\!],$$
$$\text{if } g = a\cdot b \text{ then } [\![g]\!] = [\![a]\!]\langle b\rangle + \langle a\rangle[\![b]\!],$$

where $\langle a\rangle$ and $\langle b\rangle$ are the averages of $a$ and $b$ between the two sides of $\Sigma$.

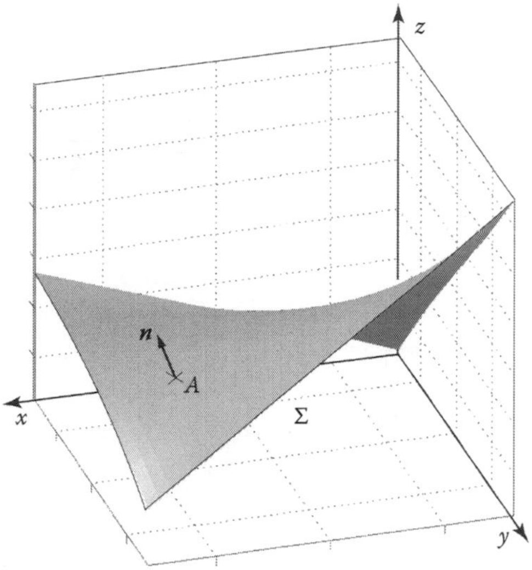


**FIGURE 3.2:** Singular surface orientation.

At a point $A$, moving with velocity, $w\boldsymbol{n}$, $\mathbb{F}$ remains zero in time. We describe this by the total derivative

$$\frac{D\mathbb{F}}{Dt} = \frac{\partial \mathbb{F}}{\partial t} + w\boldsymbol{n} \cdot \nabla \mathbb{F} = 0,$$

or with the help of (3.58) by

$$\frac{\partial \mathbb{F}}{\partial t} = -w|\nabla \mathbb{F}|. \tag{3.59}$$

Now let $d\vec{x}$ be an infinitesimal displacement along the surface $\Sigma$. $d\vec{x}$ is a geometric displacement uniquely defined at each point of $\Sigma$; a quantity, thus, that does not jump across $\Sigma$. The change in $g$ following the motion of $A$ along the surface is

$$dg = \frac{\partial g}{\partial t} dt + \nabla g \cdot d\vec{x}. \tag{3.60}$$

It is important to observe that since $g$ does not jump across $\Sigma$, $dg$ is the same for any two points infinitely close to each other on either side of $\Sigma$, in other words $[\![dg]\!] = 0$, and from (3.60) we have

$$\begin{aligned} \left[\!\!\left[\frac{\partial g}{\partial t} dt + \nabla g \cdot d\vec{x}\right]\!\!\right] &= 0, \\ \left[\!\!\left[\frac{\partial g}{\partial t}\right]\!\!\right] dt + [\![\nabla g]\!] \cdot d\vec{x} &= 0. \end{aligned} \tag{3.61}$$

For the function $\mathbb{F}$, since it always vanishes on $\Sigma$, we can write the same expression, without brackets:

$$\frac{\partial \mathbb{F}}{\partial t} \mathrm{d}t + \nabla \mathbb{F} \cdot \mathrm{d}\vec{x} = 0. \tag{3.62}$$

Comparing (3.61) and (3.62), and since d$t$ and d$\vec{x}$ are independent, we can write

$$\left[\!\left[ \frac{\partial g}{\partial t} \right]\!\right] = \bar{\mu} \frac{\partial \mathbb{F}}{\partial t}, \quad [\![ \nabla g ]\!] = \bar{\mu} \nabla \mathbb{F}, \tag{3.63}$$

where $\bar{\mu}$ is the same in both. If we let $\bar{\mu} = \mu / |\nabla \mathbb{F}|$ and use (3.59), the first of (3.63) becomes

$$\left[\!\left[ \frac{\partial g}{\partial t} \right]\!\right] = -\mu w, \tag{3.64}$$

while the second is

$$[\![ \nabla g ]\!] = \mu \boldsymbol{n}. \tag{3.65}$$

We now have the ingredients to evaluate the jump on $Dg/Dt$. The material derivative of $g$ is given by

$$\frac{Dg}{Dt} = \frac{\partial g}{\partial t} + \mathbf{q} \cdot \nabla g,$$

where $\mathbf{q}$ is the particle velocity, which by the definition of $\Sigma$ is continuous across $\Sigma$. Then,

$$\left[\!\left[ \frac{Dg}{Dt} \right]\!\right] = \left[\!\left[ \frac{\partial g}{\partial t} \right]\!\right] + \mathbf{q} \cdot [\![ \nabla g ]\!],$$

and with (3.64) and (3.65) we obtain

$$\left[\!\left[ \frac{Dg}{Dt} \right]\!\right] = \mu (\mathbf{q} \cdot \boldsymbol{n} - w). \tag{3.66}$$

The extension to a vector is a straightforward process. Let $\boldsymbol{g} = g_x \boldsymbol{e}_x + g_y \boldsymbol{e}_y + g_z \boldsymbol{e}_z$ and $\boldsymbol{\mu} = \mu_x \boldsymbol{e}_x + \mu_y \boldsymbol{e}_y + \mu_z \boldsymbol{e}_z$, where the subscripts denote components in the $(x, y, z)$ directions, respectively. Since

$$\nabla \boldsymbol{g} = \nabla g_x \boldsymbol{e}_x + \nabla g_y \boldsymbol{e}_y + \nabla g_z \boldsymbol{e}_z,$$

applying (3.65) and (3.66) to each component we obtain

$$\left[\!\left[\frac{Dg_i}{Dt}\right]\!\right] = \mu_i(\mathbf{q}\cdot\boldsymbol{n} - w),$$
$$[\![\nabla g_i]\!] = \mu_i \cdot \boldsymbol{n}, \tag{3.67}$$

for each component $i = 1, 2, 3$. Therefore

$$\left[\!\left[\frac{Dg}{Dt}\right]\!\right] = \boldsymbol{\mu}(\mathbf{q}\cdot\boldsymbol{n} - w),$$
$$[\![\nabla \mathbf{g}]\!] = \boldsymbol{\mu}\cdot\boldsymbol{n}. \tag{3.68}$$

Now let $\Gamma$, $\boldsymbol{\Omega}$, and $\Pi$ be the values of $\mu$ associated with $P$ (recall $P = \ln p$), $\mathbf{q}$, and $S$, respectively. We write the governing equations,

$$\frac{DP}{Dt} + \gamma\nabla\cdot\mathbf{q} = 0,$$
$$\frac{D\mathbf{q}}{Dt} + \Theta\nabla P = 0,$$
$$\frac{DS}{Dt} = 0,$$

on both sides of $\Sigma$ at some point $A$ and take their difference across $\Sigma$. Then applying (3.64), (3.65), and (3.67) we get

$$\begin{aligned}
(\mathbf{q}\cdot\boldsymbol{n} - w)\Gamma + \quad & \gamma\boldsymbol{n}\cdot\boldsymbol{\Omega} & & = 0,\\
\Theta\boldsymbol{n}\Gamma + \quad & (\mathbf{q}\cdot\boldsymbol{n} - w)\boldsymbol{\Omega} & & = 0,\\
& & (\mathbf{q}\cdot\boldsymbol{n} - w)\boldsymbol{\Pi} & = 0.
\end{aligned}$$

A nontrivial solution to this homogeneous system follows if the determinant is zero,

$$(\mathbf{q}\cdot\boldsymbol{n} - w)((\mathbf{q}\cdot\boldsymbol{n} - w)^2 - \gamma\Theta) = 0.$$

This requires that

1. $\Gamma$ and $\Omega$ are both different from zero and $\Pi = 0$, or,
2. $\Pi \neq 0$, but $\Gamma = \boldsymbol{\Omega} = 0$.

For case (1), nontrivial solutions are obtained if

$$(\mathbf{q}\cdot\boldsymbol{n} - w)^2 = \gamma\Theta. \tag{3.69}$$

For case (2), nontrivial solutions are obtained if

$$(\mathbf{q}\cdot\boldsymbol{n} - w) = 0. \tag{3.70}$$

### 3.5.1 Speed of Sound and Propagation of Small Disturbances

In a gas at rest $\mathbf{q} = 0$. If a small perturbation, expressed by a discontinuity in $\nabla P$, but not in $\nabla S$, is initially found on a surface $\Sigma$, such surface will subsequently move with a speed (magnitude) defined by (3.69), that is,

$$a = \sqrt{\gamma\Theta}. \tag{3.71}$$

This is what is commonly called the *speed of sound* (because sound itself means very small pressure perturbations). Naturally, the concept of "speed of sound" is not limited to the case of a gas at rest. In general, in a moving gas, small perturbations expressed by discontinuities in the space derivatives of $P$ and $\mathbf{q}$, but not $S$, will be carried by surfaces moving with speed equal to

$$w = \mathbf{q} \cdot \boldsymbol{n} \pm a. \tag{3.72}$$

We can also say that such small perturbations *propagate* along the surfaces defined by (3.72). A disturbance expressed by a discontinuity in $\nabla S$, but not in $\nabla P$ or $\nabla \cdot \mathbf{q}$, is instead carried on a surface defined by (3.70), which is nothing else but the locus of material elements moving with a velocity $\mathbf{q}$. At any point where the flow speed is smaller than the speed of sound, the flow is called *subsonic*; at any point where the flow speed is greater than the speed of sound, the flow is called *supersonic*. If a flow is subsonic in certain regions and supersonic in others, it is said to be transonic. The dimensionless ratio that compares the speed of the flow to the speed of sound is called the *Mach number*,

$$M = |\mathbf{q}|/a; \tag{3.73}$$

it provides a measure of the compressibility of the flow. In an adiabatic flow, an increase in $|\mathbf{q}|$ always results in an increase in $M$.

---

## 3.6 Finite Discontinuities

We now turn our attention to the other two types of discontinuities mentioned above, viz. contact discontinuities and shock waves. In both cases, some of the physical parameters jump across the surface. As before, we refer to the locus of points representing the surface discontinuity as $\Sigma$, defined by (3.57). The unit normal is still represented by (3.58) and the motion of $\Sigma$ is still characterized by the speed, $w$, directed along $\boldsymbol{n}$.

### 3.6.1 Standard Analysis

The standard analysis of a finite discontinuity was presented in Section 2.1.10. After writing the conservation laws in divergence form, (3.36) through (3.38), the jump conditions follow directly from (2.26). They are

$$-w[\![\rho]\!] + [\![\rho\mathbf{q}\cdot\boldsymbol{n}]\!] = 0, \tag{3.74}$$

$$-[\![\rho\mathbf{q}]\!]w + [\![\rho\mathbf{q}\mathbf{q}\cdot\boldsymbol{n} + p\boldsymbol{n}]\!] = 0, \tag{3.75}$$

$$-[\![\rho E]\!]w + [\![(p+\rho E)\mathbf{q}\cdot\boldsymbol{n}]\!] = 0. \tag{3.76}$$

### 3.6.2 Nonstandard Analysis

In smooth regions of the flow, the differential form of the Euler equations representing the conservation of mass, momentum, and energy may be written in nonconservation form:

$$\begin{aligned}\frac{Dv}{Dt} - v\nabla\cdot\mathbf{q} &= 0,\\ \frac{D\mathbf{q}}{Dt} + v\nabla p &= 0,\\ \frac{Dp}{Dt} + \gamma p\nabla\cdot\mathbf{q} &= 0,\end{aligned} \tag{3.77}$$

where $v$ is the specific volume, $v = 1/\rho$. (We will consider the mass conservation equation in terms of $\rho$ after we complete this analysis). However in the presence of discontinuities, current mathematical orthodoxy rejects (3.77) as a starting point for obtaining the jump conditions across discontinuities. The argument is that while the integral across a discontinuity of a term like $\nabla\ \mathbf{q}/v$ can be defined, this is not true if the same term is written in the form $\mathbf{q}\cdot\nabla v - v\nabla\cdot\mathbf{q}$. In what follows, we show that this is not the case. We begin by looking for solutions of the form (at this point, we strongly recommend a review of Sections 2.1.7 through 2.1.8)

$$\begin{aligned}v &= v_1 + [\![v]\!]H(\mathbb{F}),\\ \mathbf{q} &= \mathbf{q}_1 + [\![\mathbf{q}]\!]K(\mathbb{F}),\\ p &= p_1 + [\![p]\!]L(\mathbb{F}),\end{aligned} \tag{3.78}$$

where $H$, $K$, and $L$ are generalized Heaviside functions (note that $\mathbb{F}$ is zero on $\Sigma$). The macroscopic properties of $H$, $K$, and $L$ are the same, but we do not assume that they have the same microscopic behavior. For now we only require that

$$H, K, L = \begin{cases} 0 & \text{for } \mathbb{F} \to -, \\ 1 & \text{for } \mathbb{F} \to +, \end{cases}$$

and $[\![g]\!] = g_r - g_l$. The functions $H$, $K$, and $L$ provide a description of the flow in the immediate vicinity of $\Sigma$ with constant end conditions $g_l$ at $\mathbb{F} \to -$ and $g_r$ at $\mathbb{F} \to +$, where $g$ stands for $v$, $\mathbf{q}$, and $p$.

If we let $T$ represent the appropriate Heaviside function, then in the vicinity of $\Sigma$ we can write

$$\frac{\partial g}{\partial t} = [\![g]\!]\frac{\partial T}{\partial t}, \ \nabla g = [\![g]\!]\nabla T. \tag{3.79}$$

Obviously,

$$\nabla T = \frac{\mathrm{d}T}{\mathrm{d}\mathbb{F}}\nabla\mathbb{F},$$

and with (3.59)

$$\frac{\partial T}{\partial t} = -\frac{\mathrm{d}T}{\mathrm{d}\mathbb{F}}w|\nabla\mathbb{F}|.$$

Therefore, we can write that

$$\frac{\partial g}{\partial t} = -[\![g]\!]w|\nabla\mathbb{F}|\frac{\mathrm{d}T}{\mathrm{d}\mathbb{F}}, \quad \nabla g = [\![g]\!]\boldsymbol{n}|\nabla\mathbb{F}|\frac{\mathrm{d}T}{\mathrm{d}\mathbb{F}}.$$

Furthermore, proceeding as in Section 3.5, we can also find an expression for $\nabla\cdot\boldsymbol{g}$, where $\boldsymbol{g}$ is a vector:

$$\nabla\cdot\boldsymbol{g} = [\![\boldsymbol{g}\cdot\boldsymbol{n}]\!]|\nabla\mathbb{F}|\frac{\mathrm{d}T}{\mathrm{d}\mathbb{F}}.$$

Let us apply these results to the first of (3.77), in the immediate vicinity of $\Sigma$:

$$-w[\![v]\!]H' + \mathbf{q}\cdot\boldsymbol{n}[\![v]\!]H' - v[\![\mathbf{q}\cdot\boldsymbol{n}]\!]K' \approx 0, \tag{3.80}$$

where the primes denote differentiation with respect to the argument. We can rewrite (3.80) as

$$\frac{H'}{(v_l + [\![v]\!]H)} \approx \frac{K'}{(a+K)[\![v]\!]}, \tag{3.81}$$

where

$$a = (\mathbf{q}_l\cdot\boldsymbol{n} - w)/[\![\mathbf{q}\cdot\boldsymbol{n}]\!]. \tag{3.82}$$

Integrating (3.81) we obtain

$$H \approx \frac{-v_l}{[\![v]\!]} + b(a+K),$$

where $b$ is a constant of integration, which we can evaluate by letting $\mathbb{F} \to -$, where both $H$ and $K$ vanish, hence $b = v_1/(a[\![v]\!])$. Therefore,

$$H \approx \frac{v_1}{[\![v]\!]}\frac{K}{a}. \tag{3.83}$$

But since for $\mathbb{F} \to +$ both $H$ and $K \to 1$, it follows that

$$\frac{v_1}{a[\![v]\!]} = 1. \tag{3.84}$$

From this last relation we find (in terms of $\rho$) that

$$-[\![\rho]\!]w + [\![\rho\mathbf{q} \cdot \boldsymbol{n}]\!] = 0, \tag{3.85}$$

which is the Rankine–Hugoniot jump for the mass conservation equation. From (3.83) and (3.84) it follows that $H \approx K$.

Now consider the second equation of (3.77). Proceeding as before, we find

$$-w[\![\mathbf{q}]\!]K' + \mathbf{q}[\![\mathbf{q} \cdot \boldsymbol{n}]\!]K' + v[\![p\boldsymbol{n}]\!]L' \approx 0. \tag{3.86}$$

With (3.84), (3.82), (3.78), and since, as we have shown, $H \approx K$, we can write the above equation as

$$L' + \frac{[\![\mathbf{q}]\!][\![\mathbf{q} \cdot \boldsymbol{n}]\!]}{[\![v]\!][\![p\boldsymbol{n}]\!]}K' \approx 0,$$

which upon integration gives

$$L + \frac{[\![\mathbf{q}]\!][\![\mathbf{q} \cdot \boldsymbol{n}]\!]}{[\![v]\!][\![p\boldsymbol{n}]\!]}K \approx d, \tag{3.87}$$

where $d$ is a constant of integration. By taking $\mathbb{F} \to -$, we find $d = 0$, and by taking $\mathbb{F} \to +$, we find

$$\frac{[\![\mathbf{q}]\!][\![\mathbf{q} \cdot \boldsymbol{n}]\!]}{[\![v]\!][\![p\boldsymbol{n}]\!]} = -1. \tag{3.88}$$

The jump condition represented by (3.88) is related to *Prandtl's relation* [34, p. 135]. For example, with the aid of (3.85), for a one-dimensional flow, it reduces to

$$(u_r - w)(u_l - w) = \frac{p_r - p_l}{\rho_r - \rho_l}.$$

With (3.87) and (3.88), we can assert that $K \approx L$. Thus, all three generalized Heaviside functions are equal (in a weak sense). Now we can eliminate the δ-functions from (3.86) and write

$$-w\rho[\![\mathbf{q}]\!] + \rho\mathbf{q}[\![\mathbf{q}\cdot\boldsymbol{n}]\!] + [\![p\boldsymbol{n}]\!] = 0. \tag{3.89}$$

(Note that we have replaced the association sign with an equal sign.) Let us multiply (3.85) by $\mathbf{q}$ and add the resulting equation to (3.89):

$$-w(\rho[\![\mathbf{q}]\!] + \mathbf{q}[\![\rho]\!]) + \rho\mathbf{q}\cdot\boldsymbol{n}[\![\mathbf{q}]\!] + \mathbf{q}[\![\rho\mathbf{q}\cdot\boldsymbol{n}]\!] + [\![p\boldsymbol{n}]\!] = 0. \tag{3.90}$$

Now evaluate (3.90) on both sides of the surface $\Sigma$ and add the resulting two equations:

$$-w(\langle\rho\rangle[\![\mathbf{q}]\!] + \langle\mathbf{q}\rangle[\![\rho]\!]) + \langle\rho\mathbf{q}\cdot\boldsymbol{n}\rangle[\![\mathbf{q}]\!] + \langle\mathbf{q}\rangle[\![\rho\mathbf{q}\cdot\boldsymbol{n}]\!] + [\![p\boldsymbol{n}]\!] = 0.$$

This last expression can be written in the form

$$-[\![\rho\mathbf{q}]\!]w + [\![\rho\mathbf{q}\mathbf{q}\cdot\boldsymbol{n}]\!] + [\![p\boldsymbol{n}]\!] = 0. \tag{3.91}$$

Consider the last of (3.77). As with the previous equations, but since we have established that $K \approx L$, it follows that

$$([\![p]\!](\mathbf{q}_l\cdot\boldsymbol{n} - w) + \gamma p_l[\![\mathbf{q}\cdot\boldsymbol{n}]\!])K' + [\![p]\!][\![\mathbf{q}\cdot\boldsymbol{n}]\!](\gamma+1)KK' \approx 0.$$

Integrating from $\mathbb{F}-$ to $\mathbb{F}+$,

$$([\![p]\!](\mathbf{q}_l\cdot\boldsymbol{n} - w) + \gamma p_l[\![\mathbf{q}\cdot\boldsymbol{n}]\!])\int_0^1 \mathrm{d}K + (\gamma+1)[\![p]\!][\![\mathbf{q}\cdot\boldsymbol{n}]\!]\int_0^1 K\mathrm{d}K \approx 0,$$

we find the third Rankine–Hugoniot jump,

$$\frac{\mathbf{q}_l\cdot\boldsymbol{n} - w}{\mathbf{q}_r\cdot\boldsymbol{n} - w} = \frac{(\gamma+1)p_r + (\gamma-1)p_l}{(\gamma-1)p_r + (\gamma+1)p_l}, \tag{3.92}$$

which may also be expressed as

$$[\![(\gamma-1)\langle p\rangle(\mathbf{q}\cdot\boldsymbol{n} - w)]\!] + [\![p(\mathbf{q}\cdot\boldsymbol{n} - w)]\!] = 0. \tag{3.93}$$

In the previous analysis, we found the jumps for the conservation laws using the specific volume as a variable instead of the density. The reason for doing the analysis this way is that the Heaviside function for the density variable is not equal (in a weak sense) to the Heaviside functions representing the volume, velocity, and pressure. This can be shown by considering the behavior of the density and the volume near a discontinuity. By definition, we have $\rho v = 1$. Therefore, near a discontinuity

$$[\rho_1 + [\![\rho]\!] I_\varepsilon(\mathbb{F})]\,[\nu_1 + [\![\nu]\!] H_\varepsilon(\mathbb{F})] \approx 1, \tag{3.94}$$

where
$H_\varepsilon = e^{\mathbb{F}/\varepsilon}/(1 + e^{\mathbb{F}/\varepsilon})$ is a classical Heaviside function
$\varepsilon$ is a small parameter (see Section 2.1.8)

From (3.94) we find $I_\varepsilon \approx -\nu_r H_\varepsilon/(\nu_l + [\![\nu]\!] H_\varepsilon)$, which can be written as

$$I_\varepsilon \approx \frac{-H_\varepsilon}{1 - 2[\![\nu]\!]/\{([\![\nu]\!] + 2\langle\nu\rangle)(1 + e^{\mathbb{F}/\varepsilon})\}}. \tag{3.95}$$

The profiles of $H_\varepsilon$ and $-I_\varepsilon$ are shown in Figure 3.3. The two functions have the same macrostructure (away from $\mathbb{F} = 0$), but different microstructure. For example, $H_\varepsilon(0) \to \frac{1}{2}$, while $-I_\varepsilon(0) \to \frac{1}{2}\left(1 + \frac{1}{2}[\![\nu]\!]/\langle\nu\rangle\right)$. Thus, in the Mach number range $[1, \infty]$, $-I_\varepsilon(0)$ is in the range $\left[\frac{1}{2}, (\gamma - 1)/2\gamma\right]$. We want to show that $I \approx -H$, i.e., that $\int_{-\infty}^{\infty} (I - (-H))\phi(\mathbb{F})\mathrm{d}\mathbb{F} \approx 0$. Formally, we look for

$$\lim_{\varepsilon \to 0} \int_{-\infty}^{\infty} (I_\varepsilon(\varepsilon, \mathbb{F}) - (-H_\varepsilon(\varepsilon, \mathbb{F})))\phi(\mathbb{F})\mathrm{d}\mathbb{F},$$

with $I_\varepsilon$, $\mathrm{H}_\varepsilon \in \mathcal{G}(\Omega)$ and $\phi \in \mathcal{D}(\Omega)$. Since

$$\begin{aligned}
\lim_{\varepsilon \to 0} \int_{-\infty}^{\infty} (I_\varepsilon(\varepsilon, \mathbb{F}) - (-H_\varepsilon(\varepsilon, \mathbb{F})))\phi(\mathbb{F})\mathrm{d}\mathbb{F} &= \int_{-\infty}^{\infty} \lim_{\varepsilon \to 0} (I_\varepsilon(\varepsilon, \mathbb{F}) - (-H_\varepsilon(\varepsilon, \mathbb{F})))\phi(\mathbb{F})\mathrm{d}\mathbb{F} \\
&= \int_{-\infty}^{\infty} (-H + H)\phi(\mathbb{F})\mathrm{d}\mathbb{F} \\
&\approx 0,
\end{aligned}$$

we conclude that $I \approx -H$.

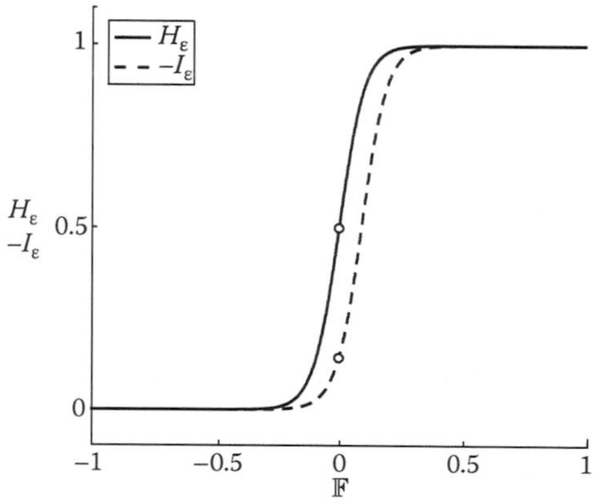


**FIGURE 3.3:** Illustration of the behavior of $H_\varepsilon$ and $-I_\varepsilon$; the case shown corresponds to $M_\infty \to \infty$ and $\varepsilon = 0.1$. The circles indicate $H_\varepsilon(0)$ and $-I_\varepsilon(0)$.

With this as background, we are now ready to study the jump of the mass conservation law

$$\frac{D\rho}{Dt} + \rho\nabla \cdot \mathbf{q} = 0. \tag{3.96}$$

As before, we look for a solution in the form $\rho = \rho_1 + [\![\rho]\!] I(\mathbb{F})$, where $I$ is a generalized Heaviside function. Proceeding as before, we find,

$$-w[\![\rho]\!]I' + \mathbf{q} \cdot \boldsymbol{n}[\![\rho]\!]I' + \rho[\![\mathbf{q} \cdot \boldsymbol{n}]\!]K' \approx 0, \tag{3.97}$$

which can be integrated to obtain

$$(\rho_1 + [\![\rho]\!]I)(a + K) \approx e. \tag{3.98}$$

The constant of integration, $e$, is found by letting $\mathbb{F} \to -$: $e = a\rho_1$. Now using (3.82) and (3.84), we find

$$I \approx -\frac{a+1}{a+K}K. \tag{3.99}$$

An analysis of (3.99), following the same steps used to show that $-I \approx H$, reveal that $-I \approx K$. Therefore, we can rewrite (3.97) as

$$(\mathbf{q}_1 \cdot \boldsymbol{n} - w + [\![\mathbf{q} \cdot \boldsymbol{n}]\!]K)[\![\rho]\!](-K') + (\rho_1 + [\![\rho]\!]I)[\![\mathbf{q} \cdot \boldsymbol{n}]\!](-I') \approx 0.$$

Now integrating from $\mathbb{F}-$ to $\mathbb{F}+$, we find, after some simplification,

$$-[\![\rho]\!]w + [\![\rho\mathbf{q} \cdot \boldsymbol{n}]\!] = 0.$$

To summarize, we have found the following jump conditions:

$$\begin{aligned} [\![(\mathbf{q} \cdot \boldsymbol{n} - w)\rho]\!] &= 0, \\ [\![(\mathbf{q} \cdot \boldsymbol{n} - w)\rho\mathbf{q}]\!] + [\![p]\!]\boldsymbol{n} &= 0, \\ [\![((\gamma - 1)\langle p\rangle + p)(\mathbf{q} \cdot \boldsymbol{n} - w)]\!] &= 0. \end{aligned} \tag{3.100}$$

The first two jumps of (3.100) are the same as those we found from the divergence form of the equations of motion, (3.74) and (3.75). The last jump, although different from 3.76, is an equally valid jump.

### 3.6.3 Contact Discontinuities

An obvious solution to (3.100) is obtained by assuming that $\mathbf{q} \cdot \boldsymbol{n}$ does not jump: $[\![\mathbf{q} \cdot \boldsymbol{n}]\!] = 0$. Therefore, $\mathbf{q} \cdot \boldsymbol{n}$ has a single value at every point on the surface. If the surface moves with a velocity $w = \mathbf{q} \cdot \boldsymbol{n}$ and $[\![p]\!] = 0$, the system (3.100) is identically satisfied, regardless of any jump in ρ or **q**. Surfaces which move with a velocity $w$, and such that the pressure is the same on both sides, are called *contact surfaces*, or *contact discontinuities*. If we let $S = S_1 + [\![S]\!]T$, where $T(\mathbb{F})$ has the following properties:

$$T \to 0, \quad \text{if } \mathbb{F} \to \mathbb{F}-,$$
$$T \to 1, \quad \text{if } \mathbb{F} \to \mathbb{F}+,$$

it is easy to find, from (3.48) and (3.79), that

$$\{[\![S]\!]T'\}(\mathbf{q} \cdot \boldsymbol{n} - w) = 0. \tag{3.101}$$

(We will discuss in more detail in Section 3.7.5 the nature of the function $T$). Equation 3.101 implies that $[\![S]\!]$ is not necessarily equal to zero across a contact surface. In fact, $S$ is a function of $p$ and ρ, and $p$ does not jump across Σ but ρ generally does. More generally, all thermodynamical parameters may jump across a contact surface, except the pressure. In particular, from (3.71) and (3.20) the important relation follows:

$$2\gamma[\![\ln a]\!] = [\![S]\!]. \tag{3.102}$$

### 3.6.4 Shock Waves

Surfaces across which the pressure jumps are called shock waves. Equations 3.100 are still valid, but they do not define the shock velocity, $w$; the latter, indeed, is a consequence of the environment around the shock, as we will see later. Let us write **q** as the sum of two vectors, one in the direction of the normal to the shock surface and the other perpendicular to the former,

$$\mathbf{q} = \tilde{u}\boldsymbol{n} + \tilde{\boldsymbol{v}}, \tag{3.103}$$

and let a normal component of velocity relative to the shock be defined by

$$\tilde{u}_{\text{rel}} = \tilde{u} - w. \tag{3.104}$$

From the first of (3.100) we obtain

$$[\![\rho\tilde{u}_{\text{rel}}]\!] = 0. \tag{3.105}$$

From the second of (3.100) we obtain two scalar equations:

$$[\![\rho\tilde{u}\tilde{u}_{\text{rel}} + p]\!] = 0,$$

which can be also written as

$$[\![\rho\tilde{u}_{\mathrm{rel}}^2 + p]\!] = 0, \tag{3.106}$$

if (3.105) is taken into account; and

$$[\![\tilde{v}]\!] = 0, \tag{3.107}$$

again, having used (3.105). From the third of (3.100) we obtain, taking (3.1) and (3.106) into account,

$$[\![\rho\left(h + \tilde{u}^2/2\right)\tilde{u}_{\mathrm{rel}} + \rho\tilde{v}\tilde{u}_{\mathrm{rel}}/2 - \rho\tilde{u}_{\mathrm{rel}}^2 w]\!] = 0.$$

The common factor, $\rho\tilde{u}_{\mathrm{rel}}$, can be dropped because of (3.105); the same can be said for the term containing $\tilde{v}^2$. Using (3.104) the remaining terms can be rearranged to yield

$$[\![h + \tilde{u}_{\mathrm{rel}}^2/2]\!] = 0. \tag{3.108}$$

Equations 3.105 through 3.108 are called *Rankine–Hugoniot conditions*. Three important remarks should be made at this point: (1) The velocity component tangent to the shock wave is the same on both sides. (2) All jumps are governed by the normal velocity component, relative to the shock. Therefore, the nature of the shock wave at any of its points can be investigated through a *one-dimensional analysis* (transverse velocities do not appear in the equations). (3) The Rankine–Hugoniot conditions are sufficient to determine all values on one side of the shock, if the values on the other side are known; but they give no information on $w$, which is hidden within $\tilde{u}_{\mathrm{rel}}$. Many other equations can be obtained from the Rankine–Hugoniot conditions by simple algebraic manipulations; some of them are important for practical applications, others provide further insight into the physics of a shock wave. Starting from (3.105) to (3.107), using $h = \gamma p/(\gamma-1)\rho$ and eliminating all $\tilde{u}_{\mathrm{rel}}$, we obtain

$$p_r/p_l = \frac{(\gamma+1)(\rho_r/\rho_l) - (\gamma-1)}{(\gamma+1) - (\gamma-1)(\rho_r/\rho_l)}. \tag{3.109}$$

This relation can also be obtained from (3.92). If the entropy were the same on both sides of the shock, a simple application of (3.19) shows that $p_r/p_l$ and $\rho_r/\rho_l$ should be related by the equation

$$p_r/p_l = (\rho_r/\rho_l)^{\gamma}, \tag{3.110}$$

which is obviously different from (3.109). It should be noted that at $\rho_r/\rho_l = 1$, the two curves (3.109) and (3.110), have a contact of the second order; this is easy to verify by Taylor expanding about $\rho_r/\rho_l = 1$. Therefore, weak shocks are practically isentropic. The following formulas are of practical interest, particularly for numerical applications:

$$\frac{\Theta_r}{\Theta_l} = \frac{a_r^2}{a_l^2} = \frac{p_r/p_l}{\rho_r/\rho_l}, \tag{3.111}$$

$$[\![S]\!] = \ln p_r/p_l - \gamma \ln \rho_r/\rho_l, \tag{3.112}$$

$$[\![p]\!]/[\![\rho]\!] = \gamma\langle p\rangle/\langle\rho\rangle, \tag{3.113}$$

and Prandtl's relation,

$$\tilde{u}_{r,\text{rel}}\tilde{u}_{l,\text{rel}} = a_*^2 - \frac{\gamma - 1}{\gamma + 1}\tilde{v}^2, \tag{3.114}$$

where $a_*$ is the speed of sound at sonic conditions. In terms of the low pressure side *shock Mach number* defined by

$$M_l = \frac{\tilde{u}_l - w}{a_l}, \tag{3.115}$$

we have

$$p_r/p_l = \frac{2\gamma M_l^2 - (\gamma - 1)}{\gamma + 1}, \tag{3.116}$$

$$\rho_r/\rho_l = \frac{(\gamma + 1)M_l^2}{(\gamma - 1)M_l^2 + 2}, \tag{3.117}$$

$$M_r^2 = \frac{(\gamma - 1)M_l^2 + 2}{2\gamma M_l^2 - (\gamma - 1)}, \tag{3.118}$$

### 3.6.4.1 Vorticity Jump across a Shock Wave

The vorticity jump across a shock wave can be derived purely from kinematical considerations, i.e., the result is independent of the equation of state and thermodynamics of the flow. The derivation for the general case can be found in [98]. Here we confine ourselves to the results for steady and unsteady flow with uniform (irrotational) flow in front of the shock. The vorticity vector is defined by (3.51). Let us decompose the vorticity into a component directed along the normal to the shock and a component on the tangent plane, thus, $\boldsymbol{\omega} = \boldsymbol{n}\omega_n + \boldsymbol{\omega}_t$. If the flow is steady and uniform in front of the shock, the jump in vorticity is given by

$$\begin{aligned} [\![\omega_n]\!] &= 0, \\ [\![\boldsymbol{\omega}_t]\!] &= -\frac{(1-\varepsilon)^2}{\varepsilon}\boldsymbol{n} \times \tilde{\boldsymbol{v}}\mathrm{K}, \end{aligned} \tag{3.119}$$

where $\varepsilon = \rho_r/\rho_l$ and K is the two-dimensional curvature tensor defined by $\mathrm{K} = -\nabla\boldsymbol{n}$. The curvature is defined positive if the surface is concave on the side from which $\boldsymbol{n}$ is

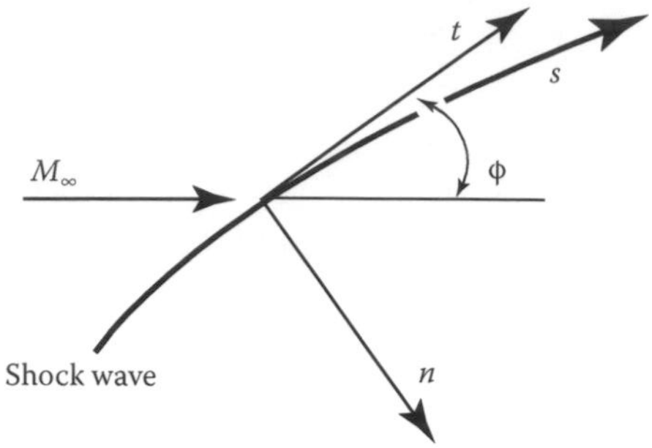


**FIGURE 3.4:** Shock wave geometry for 2D and axisymmetric flows.

directed. For plane or axisymmetric flow, the second relation in (3.119) provides the following relation for the vorticity at the shock:

$$\boldsymbol{\omega} = -\frac{(1-\varepsilon)^2}{\varepsilon}\kappa\tilde{v}\boldsymbol{k}, \tag{3.120}$$

where, with reference to Figure 3.4, $\boldsymbol{t}$ is the unit vector tangent to the shock; $\boldsymbol{k}$ is the unit vector pointing perpendicularly to the plane, $\boldsymbol{k} = \boldsymbol{n} \times \boldsymbol{t}$; $\tilde{v}$ is the magnitude of the velocity component tangent to the shock, $\tilde{v} = \sqrt{\gamma} M_\infty \cos(\phi)$; and $\kappa$ is the shock curvature, $\kappa = -\mathrm{d}\phi/\mathrm{d}s$, where $s$ is the arc length along the shock.

If the shock wave is moving with a velocity $w\boldsymbol{n}$ and the flow in front of the shock is uniform, the jump in vorticity is given by

$$[\![\omega_n]\!] = 0,$$

$$[\![\boldsymbol{\omega}_t]\!] = -\frac{(1-\varepsilon)^2}{\varepsilon}\boldsymbol{n} \times (\tilde{v}\mathrm{K} + \nabla w).$$

## 3.7 Shock Wave Structure

### 3.7.1 Exact Solution to the Shock Layer

The Navier–Stokes equations can be derived from Boltzmann's equation as a first-order approximation in terms of the mean free path. Although, the Navier–Stokes equations agree well with experimental profiles for shock layers, particularly for weak shocks, attempts were made in the early 1950s to derive a more accurate set of equations by retaining higher order terms. The results obtained for shock layer profiles from these higher order equations did not show better agreement with experiments than those obtained with the Navier–Stokes equation. It was not until the work of Bird [17], in the early 1970s, with direct Monte Carlo simulation of

Boltzmann's equation, that it became clear that improving the Navier–Stokes equation by retaining higher order terms would require so many terms that it makes this approach impracticable [91].

In this and the next section, we want to show that the inviscid Rankine–Hugoniot jumps are outer limits of the Navier–Stokes equations. The steady state conservation of mass, momentum, and energy for a one-dimensional viscous, heat-conducting fluid written in nondimensional form are [143]

$$\begin{aligned} \frac{\mathrm{d}(\rho u)}{\mathrm{d}x} &= 0, \\ \frac{\mathrm{d}\left(\rho u^2 + p - \frac{\frac{4}{3}\sqrt{\gamma}M_\infty}{R_e}\frac{\mathrm{d}u}{\mathrm{d}x}\right)}{\mathrm{d}x} &= 0, \\ \frac{\mathrm{d}\left(u\,\left(p + \rho\left(e + \frac{1}{2}u^2\right)\right) - \frac{\frac{4}{3}\sqrt{\gamma}M_\infty u}{R_e}\frac{\mathrm{d}u}{\mathrm{d}x} - \frac{\gamma\sqrt{\gamma}M_\infty}{(\gamma-1)R_eP_r}\frac{\mathrm{d}\Theta}{\mathrm{d}x}\right)}{\mathrm{d}x} &= 0. \end{aligned} \tag{3.121}$$

Here we have assumed that the coefficients of viscosity, $\mu$, and heat conduction, $\kappa$, are constant. This is an over simplification, done to facilitate a closed-form solution. Both coefficients are functions of temperature [124]. A consequence of this simplification is that the predicted shock wave thickness is smaller than if we account for the variation of the coefficients with temperature. At $M_\infty = 2$ the difference is about 30%, but it gets larger at higher Mach numbers [82]. The nondimensionalization is done using reference values taken in the free stream as $x \to -\infty$, where the Mach number, $M_\infty$, is assumed greater than one. The dimensionless parameters $R_e$ and $P_r$ are the Reynolds number and Prandtl number, respectively. They are defined by

$$R_e = \frac{\rho_\infty u_\infty \lambda}{\mu}, \quad P_r = \frac{\mu c_p}{\kappa},$$

where the reference length, $\lambda$, is the average distance between molecular collisions, i.e., the mean free path.* The Reynolds number measures the relative magnitude of the nonviscous and viscous effects. Osborne Reynolds first introduced it in 1883 while studying turbulence [175]. The Prandtl number measures the relative magnitude of the viscous and heat-conducting effects.

Integrating the first of (3.121) yields

$$\rho u = \rho_\infty u_\infty = \sqrt{\gamma}M_\infty. \tag{3.122}$$

Using (3.122) to replace $\rho$ in the equation of state of an ideal gas, (3.3), we get

$$p = \frac{\sqrt{\gamma}M_\infty \Theta}{u}. \tag{3.123}$$

* At sea level, the mean free path in air is of the order of $6 \times 10^{-8}$ m.

Now we eliminate $e$ from the third of (3.121) using the polytropic gas relation, (3.4), and using (3.122) and (3.123), and integrating once,

$$\frac{u^2}{2}+\frac{\gamma}{\gamma-1}\Theta-\frac{4}{3}\frac{u}{R_e}\frac{\mathrm{d}u}{\mathrm{d}x}-\frac{\gamma}{\gamma-1}\frac{1}{R_eP_r}\frac{\mathrm{d}\Theta}{\mathrm{d}x}=c_1, \tag{3.124}$$

where $c_1$ is a constant of integration. For Prandtl number of 3/4,* we can rewrite (3.124) in the form

$$\frac{u^2}{2}+\frac{\gamma}{\gamma-1}\Theta-\frac{4}{3}\frac{1}{R_e}\frac{\mathrm{d}}{\mathrm{d}x}\left(\frac{u^2}{2}+\frac{\gamma}{\gamma-1}\Theta\right)=c_1. \tag{3.125}$$

(3.125) can be integrated to yield

$$\frac{u^2}{2}+\frac{\gamma}{\gamma-1}\Theta=c_1+c_2\exp\left(\tfrac{3}{4}R_e\textstyle\int\mathrm{d}x\right),$$

where $c_2$ is a constant of integration. We look for a solution with uniform flow as $x\to\infty$ and, hence, we require that $c_2=0$. If we evaluate $c_1$ at $x\to-\infty$, we obtain the well-known exact integral of the one-dimensional energy equation

$$\frac{u^2}{2}+\frac{\gamma}{\gamma-1}\Theta=\frac{\gamma}{2}M_\infty^2+\frac{\gamma}{\gamma-1}. \tag{3.126}$$

Now using (3.122) in the second of (3.121) and integrating once, we get

$$\sqrt{\gamma}M_\infty u+p-\frac{4}{3}\frac{\sqrt{\gamma}M_\infty}{R_e}\frac{\mathrm{d}u}{\mathrm{d}x}=c_3. \tag{3.127}$$

Again we evaluate $c_3$ at $x\to-\infty$, remembering that $\mathrm{d}u/\mathrm{d}x\to 0$ as $x\to-\infty$, to obtain $c_3=1+\gamma M_\infty^2$. Using (3.123) and (3.126) in (3.127) yields

$$\frac{4}{3}\frac{\nu}{R_e}\frac{\mathrm{d}\nu}{\mathrm{d}x}-\frac{\gamma+1}{2\gamma}\left[\nu^2-\frac{2}{\gamma+1}\left(\gamma+\frac{1}{M_\infty^2}\right)\nu+\nu_r\right]=0, \tag{3.128}$$

where $\nu=u/\sqrt{\gamma}M_\infty$ and

$$\nu_r=\frac{\gamma-1}{\gamma+1}+\frac{2}{\gamma+1}\frac{1}{M_\infty^2},\quad M_\infty\geq 1,\ \frac{\gamma-1}{\gamma+1}\leq\nu_r\leq 1. \tag{3.129}$$

We recognize $\nu_r$ as the inverse of the Rankine–Hugoniot density jump, (3.117). The term in square brackets in (3.128) can be factored to obtain

* The Prandtl number is related to $\gamma$ by *Eucken's relation*: $P_r=4\gamma/(9\gamma-5)$ [233]. For air, we get $P_r=0.737$, which is close to the $\frac{3}{4}$ value we are using.

$$\frac{4}{3}\frac{\nu}{R_e}\frac{d\nu}{dx} - \frac{\gamma+1}{2\gamma}[(\nu-1)(\nu-\nu_r)] = 0. \tag{3.130}$$

Integrating (3.130) results in

$$\frac{1-\nu}{(\nu-\nu_r)^{\nu_r}} = c_4 \exp\left(\frac{3(\gamma+1)}{8\gamma}R_e(1-\nu_r)x\right).$$

The constant of integration, $c_4$, is evaluated by defining the origin, $x=0$, to be the point where $\nu$ has an inflection point, $d^2\nu/dx^2=0$. Carrying out this operation we find that it occurs at $\nu = \sqrt{\nu_r}$*, and that

$$\frac{1-\nu}{(\nu-\nu_r)^{\nu_r}} = \frac{1-\sqrt{\nu_r}}{\left(\sqrt{\nu_r}-\nu_r\right)^{\nu_r}} \exp\left(\frac{3(\gamma+1)}{8\gamma}R_e(1-\nu_r)x\right), \quad -\infty \le x \le \infty. \tag{3.131}$$

Equation 3.131 may also be written as

$$x = \frac{8\gamma}{3R_e(\gamma+1)(1-\nu_r)} \ln\left[\left(\frac{\sqrt{\nu_r}-\nu_r}{\nu-\nu_r}\right)^{\nu_r}\frac{1-\nu}{1-\sqrt{\nu_r}}\right], \quad \nu_r \le \nu \le 1. \tag{3.132}$$

### 3.7.2 Description of the Shock Layer

Von Kármán [234] derived the following relation by means of the kinetic theory of gases. It relates the Reynolds number, based on a reference length $L$, to the Mach number through the Knudsen number,

$$K_n = \sqrt{\frac{\pi\gamma}{2}}\frac{M_\infty}{R_{eL}}, \tag{3.133}$$

where the Knudsen number is the ratio of the mean free path to the reference length $L$. Since our reference length is the mean free path $K_n=1$, we have

$$R_e = \sqrt{\frac{\pi\gamma}{2}}M_\infty.$$

Using this expression in (3.132) we get

$$x = \frac{\frac{8}{3}\sqrt{\frac{2\gamma}{\pi}}}{M_\infty(\gamma+1)(1-\nu_r)} \ln\left[\left(\frac{\sqrt{\nu_r}-\nu_r}{\nu-\nu_r}\right)^{\nu_r}\frac{1-\nu}{1-\sqrt{\nu_r}}\right], \quad \nu_r \le \nu \le 1. \tag{3.134}$$

* At the origin, $\nu = \sqrt{\nu_r}$, represents the geometric mean of the end values $[1, \nu_r]$.

This last expression gives the dependence of $\nu$ on the free stream Mach number. We can describe the flow through the shock layer as a function of the free stream Mach number with (3.129), (3.134), and the following relations:

$$
\begin{aligned}
u &= \sqrt{\gamma} M_\infty \nu, \\
\rho &= 1/\nu, \\
\Theta &= 1 + \frac{\gamma - 1}{2} M_\infty^2 (1 - \nu^2), \\
p &= \frac{1}{\nu}\left(1 + \frac{\gamma - 1}{2} M_\infty^2 (1 - \nu^2)\right), \\
S &= \ln p - \gamma \ln \rho.
\end{aligned}
$$

The distribution of these variables across the viscous shock layer is shown in Figure 3.5.

### 3.7.3 Shock Wave Thickness

The shock wave thickness, $\Delta$, in multiples of the mean free path, can be defined in a number of ways. Here we consider the definition proposed by Prandtl [170]:

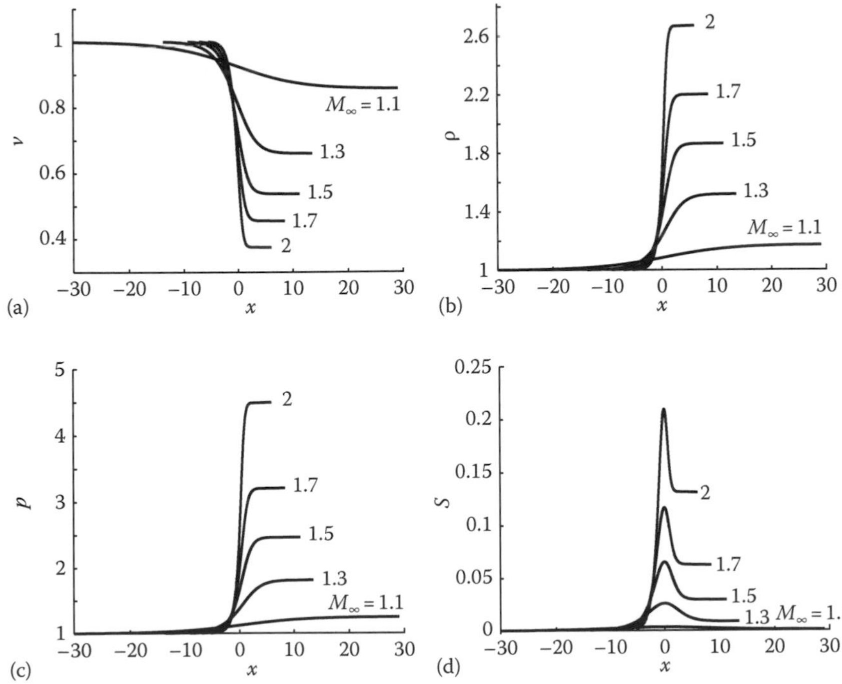


**FIGURE 3.5:** Distribution of (a) specific volume, (b) density, (c) pressure, and (d) entropy across a viscous shock layer.

$$\Delta = \frac{-(1-\nu_r)}{\left.\frac{d\nu}{dx}\right|_0}.$$

With this definition, the shock thickness is the interval in the $(\nu, x)$-plane between the interception of a straight line drawn from the origin with slope $d\nu/dx$, evaluated at $x=0$, and the lines $\nu=1$ and $\nu=\nu_r$. From (3.130), since $\nu(0)=\sqrt{\nu_r}$, we find

$$\left.\frac{d\nu}{dx}\right|_0 = \frac{-3}{8}\frac{\gamma+1}{\gamma}R_e\left(\sqrt{\nu_r}-1\right)^2,$$

and, therefore,

$$\Delta = \frac{8}{3}\frac{\gamma}{\gamma+1}\frac{1}{R_e}\frac{1+\sqrt{\nu_r}}{1-\sqrt{\nu_r}}.$$

Using von Kármán's relation, (3.133), we can write $R_e = K_n R_{eL} = \sqrt{\pi\gamma/2}M_\infty$ to obtain

$$\Delta = \frac{8}{3}\frac{\gamma}{\gamma+1}\frac{1}{\sqrt{\pi\gamma/2}M_\infty}\frac{1+\sqrt{\nu_r}}{1-\sqrt{\nu_r}}, \tag{3.135}$$

where $\nu_r$ is given by (3.129). The variation of $\Delta$ with free stream Mach number is shown in Figure 2.1. The relation between shock wave thickness, Mach number, Reynolds number, and Knudsen number is shown in Figure 3.6. For $M_\infty=1$, the

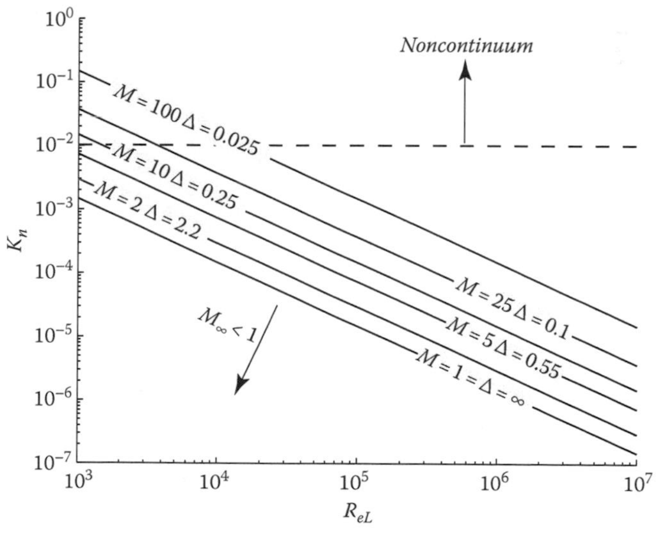


**FIGURE 3.6:** Relation between shock wave thickness, $\Delta$, and Mach number, Reynolds number, and Knudsen number.

shock wave thickness $\rightarrow\infty$. The continuum hypothesis starts to fail for $K_n > 0.01$ with the onset of slip flow at solid surfaces [198]. In addition, we have to consider that for $M_\infty > 4$ some of our gas model assumptions breakdown as nonequilibrium effects become important. Nevertheless, from Figure 3.6 we can estimate that to resolve a Navier–Stokes shock layer, with just 10 mesh points within the layer, at a free stream Mach number of two for flow over a wing with a 1 m chord would require approximately 500 million mesh points in the stream-wise direction.

### 3.7.4 Shock Relations in the Limit of Vanishing Viscosity

In this section, we consider the limit of the exact solution represented by (3.131) as $R_e \rightarrow \infty$. Then we look at the jump conditions across a shock wave starting from the Navier–Stokes equations and taking the limit $R_e \rightarrow \infty$. We expect that in the limit of vanishing viscosity we will recover the inviscid Rankine–Hugoniot jumps. We also want to investigate the jump across a shock associated with entropy and we would like to *derive* the so-called entropy condition.

Equation 3.131 can be written in the form*

$$\left(\frac{1-\nu}{1-\sqrt{\nu_r}}\right)^2\left(\frac{\sqrt{\nu_r}-\nu_r}{\nu-\nu_r}\right)^{2\nu_r}=\frac{\tanh(-3(\gamma+1)[\![\nu]\!]R_e x/8\gamma)+1}{\tanh(3(\gamma+1)[\![\nu]\!]R_e x/8\gamma)+1},$$

where $[\![\nu]\!]=\nu_r-1\leq 0$. In the limit $R_e\rightarrow\infty$, the above relation takes the form

$$\left(\frac{1-\nu_i}{1-\sqrt{\nu_r}}\right)^2 H(x)-\left(\frac{\nu_i-\nu_r}{\sqrt{\nu_r}-\nu_r}\right)^{2\nu_r} H(-x)=0,$$

where $\nu_i$ is the value that $\nu$ takes as $R_e\rightarrow\infty$, see (2.30). Examining the last expression for $x<0$ and $x>0$, we see that it is equivalent to

$$\nu_i = 1+[\![\nu]\!]H(x). \tag{3.136}$$

Equation 3.136 is the inviscid limit of (3.131).

The steady Navier–Stokes conservation laws, in nondimensional form, follow from Equation 3.121:

$$\begin{aligned}
[\![\rho u]\!]_{-\varepsilon}^{\varepsilon}&=0,\\
\left[\!\!\left[\rho u^2+p-\frac{\frac{4}{3}\sqrt{\gamma}M_\infty}{R_e}\frac{\mathrm{d}u}{\mathrm{d}x}\right]\!\!\right]_{-\varepsilon}^{\varepsilon}&=0,\\
\left[\!\!\left[u\left(p+\rho\left(e+\frac{1}{2}u^2\right)\right)-\frac{\frac{4}{3}\sqrt{\gamma}M_\infty u}{R_e}\frac{\mathrm{d}u}{\mathrm{d}x}-\frac{\gamma\sqrt{\gamma}M_\infty}{(\gamma-1)R_eP_r}\frac{\mathrm{d}\Theta}{\mathrm{d}x}\right]\!\!\right]_{-\varepsilon}^{\varepsilon}&=0.
\end{aligned}\tag{3.137}$$

* If $z=e^y$, then $\tanh(y)=(z^2-1)/(z^2+1)$, etc.

The three conservation laws are written assuming a steady shock wave transition about the point $x=0$. The analysis is also valid for a shock moving with a constant velocity, thus we could assume a reference frame moving with a constant velocity so that in this frame the shock is at rest at $x=0$. Equations 3.137 represent the integrals for the conservation of mass, momentum, and energy evaluated between $-\varepsilon$ and $\varepsilon$, such that $-\varepsilon<0<\varepsilon$. To these equations we add the entropy equation

$$S_x = \frac{\frac{4}{3}(\gamma-1)}{R_e}\frac{u_x^2}{\Theta} + \frac{\gamma}{R_eP_r}\frac{\Theta_{xx}}{\Theta}. \tag{3.138}$$

Equation 3.138 follows from (3.2) and (3.121), see [128, pp. 261, 293]. The two terms on the right-hand side are sources of entropy. The first represents the entropy created by dissipation; the second represents the entropy created by heat transfer. Since

$$\Theta_{xx}/\Theta = (\Theta_x/\Theta)^2 + \mathrm{d}(\Theta_x/\Theta)/\mathrm{d}x,$$

we can also write (3.138) as

$$[\![S]\!]_{-\varepsilon}^{\varepsilon} = \int_{-\varepsilon}^{\varepsilon} \frac{\frac{4}{3}(\gamma-1)}{R_e}\frac{u_x^2}{\Theta}\mathrm{d}x + \int_{-\varepsilon}^{\varepsilon} \frac{\gamma}{R_eP_r}\frac{\Theta_x^2}{\Theta^2}\mathrm{d}x + \left[\!\!\left[\frac{\gamma}{R_eP_r}\frac{\Theta_x}{\Theta}\right]\!\!\right]_{-\varepsilon}^{\varepsilon}. \tag{3.139}$$

Recall that the exact solution we found for the shock profile is valid only for $P_r=3/4$, therefore we will consider the limit $R_e \to \infty$, with $P_r$ constant. At this point, the usual argument (see, for example, [41, pp. 136–137]) is to say that since the quantities in (3.137) are all bounded away from $x=0$, in the limit $R_e \to \infty$, we obtain

$$[\![\rho u]\!]_{-\varepsilon}^{\varepsilon} = 0,$$
$$[\![\rho u^2 + p]\!]_{-\varepsilon}^{\varepsilon} = 0,$$
$$\left[\!\!\left[u\left(p+\rho\left(e+\tfrac{1}{2}u^2\right)\right)\right]\!\!\right]_{-\varepsilon}^{\varepsilon} = 0.$$

where $\varepsilon$ can be arbitrarily small, but fixed while taking the limit $R_e \to \infty$. Now, the argument continues, we let $\varepsilon \to 0$ and we obtain the familiar Rankine–Hugoniot jumps. The argument sounds reasonable, but it is misleading! Consider the second equation of (3.137). The term $\mathrm{d}u/\mathrm{d}x \sim \mathrm{d}v/\mathrm{d}x$, and from (3.130) we know that $\mathrm{d}v/\mathrm{d}x \sim R_e$. Therefore, the term represented by $\mathrm{d}u/\mathrm{d}x\,/R_e$ does not vanish with $R_e \to \infty$. The same is true of the terms with $\mathrm{d}u/\mathrm{d}x$ and $\mathrm{d}\Theta/\mathrm{d}x$ in the third equation of (3.137) and the source terms in (3.138) and (3.139). *The Rankine–Hugoniot jump conditions do not depend on the Reynolds number!* What is important is to take $\varepsilon$ sufficiently large for the gradients $\mathrm{d}u/\mathrm{d}x$ and $\mathrm{d}\Theta/\mathrm{d}x$ to vanish.

Let us look more critically at the entropy equation (3.138). But first, let us label the source terms as follows

$$I_1 = \frac{4}{3}\frac{(\gamma-1)}{R_e}\frac{u_x^2}{\Theta},$$
$$I_2 = \frac{\gamma}{R_eP_r}\frac{\Theta_{xx}}{\Theta},$$
$$I_3 = \frac{\gamma}{R_eP_r}\frac{\Theta_x^2}{\Theta^2},$$
$$I_4 = \frac{\gamma}{R_eP_r}\frac{\mathrm{d}}{\mathrm{d}x}\left(\frac{\Theta_x}{\Theta}\right).$$

Let us integrate $S_x$ from $x \to \pm\infty$, but the integration is done in terms of $v$ rather than $x$ through a change of variables, using $\mathrm{d}x = \mathrm{d}v/v_x$, and replacing the limits with $v_- = 1$ and $v_+ = v_r$. Using the exact solution, (3.130), and letting $M_\infty = 2$ and $R_e = 1$, we obtain the result shown in Figure 3.7. We are showing the result of the integration as a function of $x$. As previously shown in Figure 3.5, the entropy rise across the shock layer is not monotonic. It reaches a peak value at $x = 0$ and then decreases. This is always the case as long as heat conduction is present (i.e., $P_r > 0$) [144]. The decrease in entropy is due to the behavior of $\Theta_{xx}$ as shown in Figure 3.8. The term $I_2$, which represents $\Theta_{xx}$, is positive ahead of $x = 0$, but becomes negative just before $x = 0$. This term represents the heat added to the fluid by conduction. Thus, when this term is negative, the fluid loses heat. When this loss exceeds the heat added by dissipation by $I_1$, the entropy decreases. The only effect of Reynolds number on this result is a squeezing of the profile around $x = 0$ as the Reynolds number increases. The end point values of $S$ at $x \to \pm\infty$ are independent of the Reynolds number, as is the peak value at $x = 0$. Aside from the fact that this is manifested in the exact

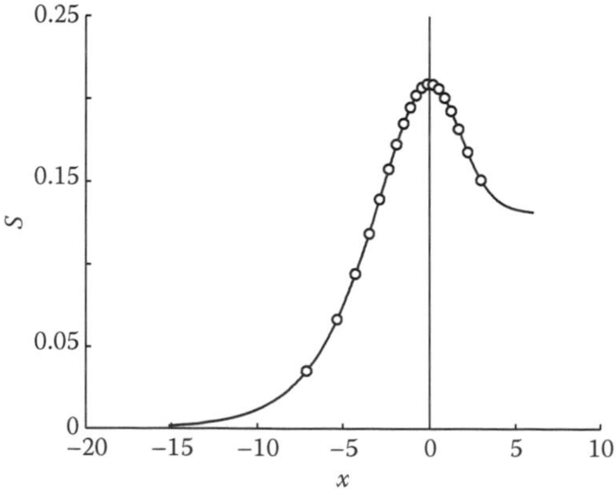


**FIGURE 3.7:** Evaluation of $S$ by integrating $S_x$ using the exact solution, solid line, and from $S = \ln p - \gamma \ln \rho$, circles.

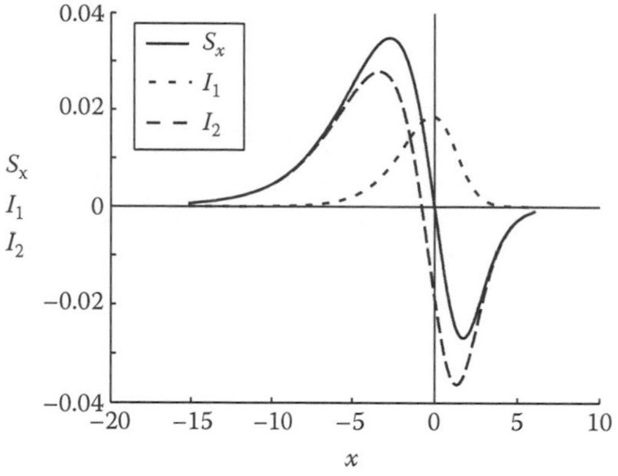


**FIGURE 3.8:** Evaluation of $S_x$ by integrating $I_1$ and $I_2$.

solution, it can also be deduced from the fact that *the entropy is a function of state*, and, therefore, its value depends only on the end state, not the path the fluid particle takes. This is the reason why *the drag generated by a shock wave can be evaluated by integrating the jump in entropy without reference to the Reynolds number of the flow* [161]. It is easy to show* that the peak entropy value is given by

$$S_0 = [\![S]\!] + \ln\left(\frac{\rho_r}{\langle\rho\rangle}\frac{\langle p\rangle}{p_r}\rho_r^{(\gamma-1)/2}\right), \tag{3.140}$$

where $S_0$ is the peak value, the subscript $r$ refers to values at $x \to \infty$, and the log term is the amount of entropy lost due to heat conduction. Equation 3.140 may also be written as

$$S_0 = \frac{1}{2}[\![S]\!] + \ln\frac{\sqrt{\rho_r}}{\langle\rho\rangle}\frac{\langle p\rangle}{\sqrt{p_r}}. \tag{3.141}$$

The log term now involves the ratios of geometric mean to arithmetic mean of the density and pressure values across the shock.

The integration of $S_x$ can also be done by adding the integrals corresponding to $I_1$, $I_3$, and $I_4$, as shown in Figure 3.9. This is the way usually presented in textbooks which aim to show that because $I_1$ and $I_3$ are squared terms and since $I_4$ vanishes at $x \to \pm\infty$, the entropy jump must be greater than zero. Now we can write $[\![S]\!] = I_1 + I_3$ and, therefore, the so-called entropy condition $[\![S]\!] \geq 0$ follows. The equal sign being true only for $M_\infty = 1$. This, however, can be misinterpreted to mean that the entropy rise is monotonic (the contribution of $I_4$ is hidden once we evaluate the integrals at $x \to \pm\infty$), which we know is not the case. Note, however, that both $I_1$ and $I_3$ vanish much faster than $I_4$, thus, it is important to take the limits at $x \to \pm\infty$ to obtain the proper jump. An important consequence of the entropy condition is that it rules out *expansion shocks* (i.e., shocks across which the pressure decreases) [243].

* First show that $S_x \sim (v-1)(v-v_r)(v^2-v_r)$, then evaluate $S_0$ using $v(0) = \sqrt{v_r}$.

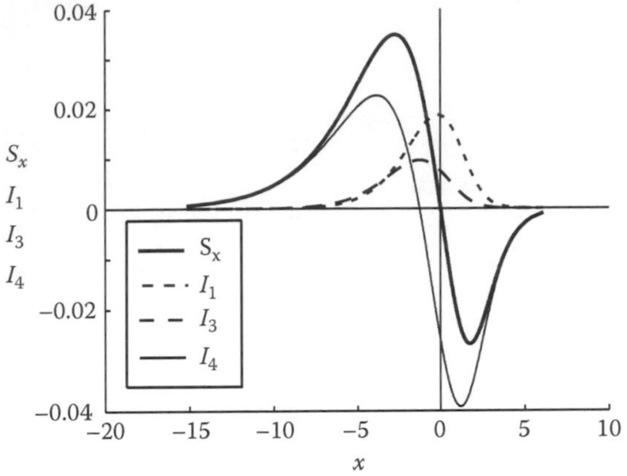


**FIGURE 3.9:** Evaluation of $S_x$ by integrating $I_1$, $I_3$, and $I_4$.

### 3.7.5 Entropy as a Generalized Function

We have shown in Section 3.7.4 that the entropy is not conserved across a shock wave and that its stationary and turning point values are independent of the Reynolds number. Here we want to show that a valid entropy jump can be obtained from the entropy equation if the entropy is represented by a proper generalized function and that the entropy shock-microstructure is build into the entropy–pressure–density relation.

The relation between entropy, pressure, and specific volume is given by $\mathrm{d}S=\mathrm{d}p/p+\gamma\mathrm{d}v/v$. Therefore, Equation 3.49 may be written as

$$\frac{DS}{Dt}=\frac{p_t}{p}+\mathbf{q}\frac{\nabla p}{p}+\gamma\left(\frac{v_t}{v}+\mathbf{q}\frac{\nabla v}{v}\right)=0. \tag{3.142}$$

Let

$$S=S_1+[\![S]\!]T(\mathbb{F}), \tag{3.143}$$

where, at this point, we only require $T$ to have the following behavior:

$$\begin{aligned} T&\to 0, \quad \text{if } \mathbb{F}\to\mathbb{F}-\,, \\ T&\to 1, \quad \text{if } \mathbb{F}\to\mathbb{F}+\,. \end{aligned}$$

Introducing (3.143) and (3.78) into (3.142) results in the following equation:

$$\{[\![S]\!]T'\}(\mathbf{q}\cdot\boldsymbol{n}-w)\approx\left\{\frac{[\![p]\!]L'}{(p_1+[\![p]\!]L)}+\frac{\gamma[\![v]\!]H'}{(v_1+[\![v]\!]H)}\right\}(\mathbf{q}\cdot\boldsymbol{n}-w)\approx 0. \tag{3.144}$$

If $\mathbf{q}\cdot\boldsymbol{n}=w$, the surface on which (3.144) is valid is a contact discontinuity and (3.144) is satisfied for any value of the expression inside the curly brackets, i.e., for any entropy jump. If $\mathbf{q}\cdot\boldsymbol{n} \neq w$, the surface is a shock. For this case, we eliminate the common term $\mathbf{q}\cdot\boldsymbol{n} - w$, make use of the fact that $L \approx H$, and integrate, to find

$$[\![S]\!]T + c_1 \approx \ln\left((p_l + [\![p]\!]H)(v_l + [\![v]\!]H)^{\gamma}\right), \tag{3.145}$$

where $c_1$ is a constant of integration. At $\mathbb{F}-$, $H = 0$, $T = 0$ and, therefore, $c_1 = s_1 = \ln\ p_1 v_1^{\gamma}$, thus

$$[\![S]\!]T \approx \ln\left(((p_1 + [\![p]\!]H)/p_1)((v_1 + [\![v]\!]H)/v_1)^{\gamma}\right). \tag{3.146}$$

Now evaluating (3.146) at $\mathbb{F}+$, $H = 1$, $T = 1$:

$$\begin{aligned} [\![S]\!] &= \ln\left((p_r/p_1)(v_r/v_1)^{\gamma}\right), \\ &= \ln\left(\left(\frac{p_1 + [\![p]\!]}{p_1}\right)\left(\frac{v_1 + [\![v]\!]}{v_1}\right)^{\gamma}\right). \end{aligned} \tag{3.147}$$

Equation 3.147 defines the entropy jump across a shock. From (3.146), we find the *definition of the function T*:

$$T \approx \ln\left(((p_1 + [\![p]\!]H)/p_1)((v_1 + [\![v]\!]H)/v_1)^{\gamma}\right)/[\![S]\!]. \tag{3.148}$$

Now we want to show that the microstructure of $T$ is equivalent to the structure we found for $S$ in the viscous shock layer. This was first shown in [196]. To simplify the notation, let $p_1 = v_1 = 1$. Evaluate $\mathrm{d}T/\mathrm{d}\mathbb{F}$:

$$\frac{\mathrm{d}T}{\mathrm{d}\mathbb{F}} \approx \frac{\mathrm{d}T}{\mathrm{d}H}H' = \frac{((\gamma+1)[\![p]\!]\,[\![v]\!]H + [\![p]\!] + \gamma[\![v]\!])}{[\![S]\!]pv}H'. \tag{3.149}$$

From (3.109) the jump in $p$ is given by

$$[\![p]\!] = -\frac{2\gamma[\![v]\!]}{(\gamma+1)[\![v]\!] + 2}. \tag{3.150}$$

Introducing (3.150) into (3.149) and simplifying, we get

$$\frac{\mathrm{d}T}{\mathrm{d}\mathbb{F}} \approx \frac{1}{[\![S]\!]pv}\frac{\gamma(\gamma+1)[\![v]\!]^2}{(\gamma+1)[\![v]\!]+2}(1-2H)H'.$$

Since $(1 - 2H)H' \approx 0$,* we have $\mathrm{d}T/\mathrm{d}\mathbb{F} \approx 0$; hence there is a turning point at the origin. Note that we know this is the case for the viscous shock layer (see Section 3.7.2) and that this behavior is independent of the Reynolds number. Here, we have again confirmed the Reynolds number independence by showing it as the inviscid limit of Heaviside functions (3.78). We can confirm that the turning point is a maximum by looking at the second derivative of $S$ or by looking at the behavior

* $2HH' \approx H'$, see Equation 2.20.

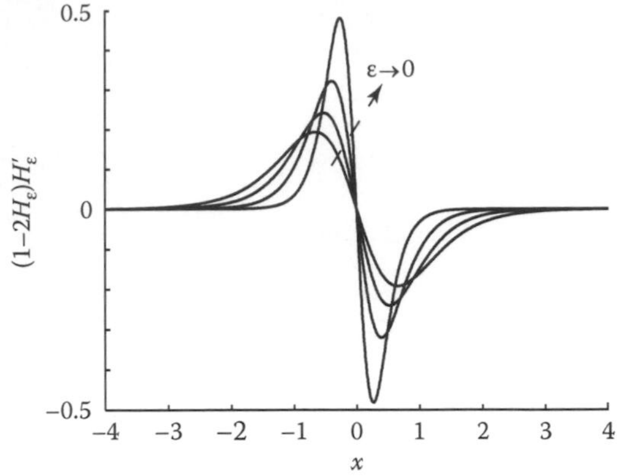


**FIGURE 3.10:** Behavior of $(1-2H_\varepsilon)H'_\varepsilon$ as $\varepsilon \to 0$. Compare this figure with Figure 3.8.

of $(1-2H_\varepsilon)H'_\varepsilon$ as $\varepsilon \to 0$ as shown in Figure 3.10 for $H_\varepsilon = e^{x/\varepsilon}/\, 1 + e^{x/\varepsilon}$. Note that $H'_\varepsilon = (1-H_\varepsilon)H_\varepsilon/\varepsilon$ and that $H''_\varepsilon = (1-2H_\varepsilon)H'_\varepsilon/\varepsilon$. Therefore, $T'$ has the characteristics of a first derivative of the delta function, and $T$ is a generalized function sharing characteristics of both Heaviside and δ-functions.

A word of caution is in order. Although the function $T$ has similar characteristics to the microstructure we see in the viscous shock layer, it is different from it in some important ways. For example, $S$ from (3.143) has a maximum at the origin just as in the viscous shock layer, but the maximum value is different from the actual maximum given by (3.140). For these values to be the same would require that the value of the specific volume at the origin be equal to $\sqrt{v_r}$ when $v$ is represented by a step function which, of course, is not the case.*

The MATLAB code `SHOCK_LAYER.M` reproduces some of the figures appearing in this chapter.

## Problems

**3.1** Write the isentropic system,

$$\frac{D\rho}{Dt} + \rho\nabla\cdot\mathbf{q} = 0,$$
$$\frac{D\mathbf{q}}{Dt} + \frac{1}{\rho}\nabla p = 0,$$
$$p = \rho^\gamma,$$

* Similarly, the pressure at the origin, when represented by a step function, should take a value equal to $(1+(\gamma-1)M_\infty^2(1-v_r)/2)/\sqrt{v_r}$.

in conservation form. Find the jump conditions and discuss how they relate to the jumps of (3.77). For example, if we define the shock strength by $[\![\rho]\!]$ at a given $M_\infty$, which system of equations has the stronger shock?

**3.2** If we impose constant total enthalpy, we obtain the following system:

$$\frac{D\rho}{Dt} + \rho\nabla\cdot\mathbf{q} = 0,$$
$$\frac{D\mathbf{q}}{Dt} + \frac{1}{\rho}\nabla p = 0,$$
$$p = \frac{\gamma-1}{\gamma}\rho\left(\frac{\gamma}{\gamma-1} + \gamma M_\infty^2 - \frac{q^2}{2}\right).$$

Find the jumps for this system and compare them with those of (3.77).

**3.3** With $v(x, R_e)$ defined by (3.131), study the microstructure of the Heaviside function:

$$H_\varepsilon(x) = \lim_{\varepsilon\to 0}\frac{v-1}{[\![v]\!]}, \quad \varepsilon = 1/R_e,$$

at $M_\infty = 2$. The pressure across the shock layer is given by

$$p = \frac{1}{v}\left(1 + \frac{\gamma-1}{2}M_\infty^2(1-v^2)\right).$$

Let

$$L_\varepsilon(x) = \lim_{\varepsilon\to 0}\frac{p-1}{[\![p]\!]}.$$

Compare the microstructures of $L_\varepsilon$ and $H_\varepsilon$.

# Chapter 4

## *Euler Equations: One-Dimensional Problems*

> What I wish to put to you is that a physical idea founded on mathematics is like a good spring steel, which though you deflect it almost indefinitely from its original position will spring back unerringly into its correct shape.
>
> Sir James Lighthill [127]

A thorough understanding of one-dimensional flows provides a solid foundation for more complex multidimensional problems. If the flow is also isentropic, the governing equations can be integrated and the flow patterns can be studied using *Riemann invariants*. In the first half of this chapter, we will focus on flows resulting from the motion of a piston in a long tube of constant cross section. In the second half, we will consider flows in ducts with small variations in cross section.

### 4.1 Piston-Driven Flows

In this section, we consider the motion of a gas confined to a long frictionless tube of constant cross section. The tube is laid out along the $x$-axis, as shown in Figure 4.1. The motion that takes place in this tube is a result of the initial and boundary conditions imposed. In general, we will assume that the gas is initially at rest and that the ensuing motion will be a result of the boundary conditions imposed at the ends of the tube. The governing equations describing the motion follow from (3.50). They are

$$\begin{aligned} P_t + uP_x + \gamma u_x &= 0, \\ u_t + uu_x + \Theta P_x &= 0, \\ S_t + uS_x &= 0. \end{aligned} \tag{4.1}$$

We note that the only *continuous* steady state admissible is that describing a uniform flow, i.e., a flow with constant velocity, pressure, temperature, and entropy. If we allow *discontinuous* solutions, then the steady form of the equations admit a steady

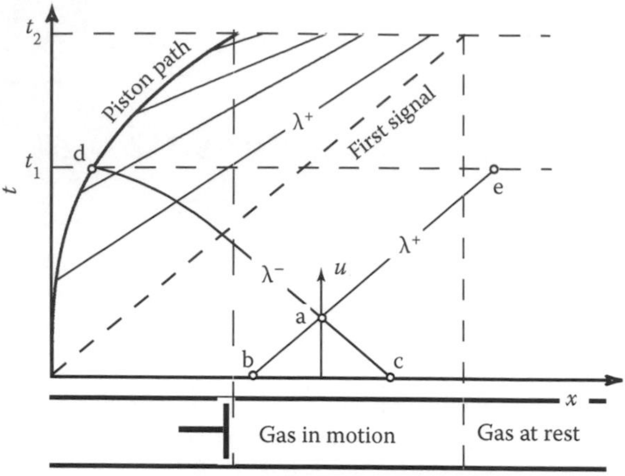


**FIGURE 4.1:** Illustration of the early time history of a piston-driven flow.

shock separating two uniform flows. This configuration is, however, unstable. The other admissible steady solution is that of a contact discontinuity separating two different gases at rest. Therefore, the repertoire of steady solutions is very limited for this type of flows and we must turn our attention to unsteady motions resulting from the actions of pistons at either end of the tube.

### 4.1.1 Characteristics

The physical features encapsulated in (4.1) are better understood if we rewrite (4.1) in characteristic form. As discussed in Section 2.1.5, the characteristics equations are obtained by linearly combining the first two equations of (4.1) such that the derivatives of $P$ and $u$ combine as directional derivatives in the same direction. To this end, we multiply the first of (4.1) by some factor $\mu_1$ and the second by some factor $\mu_2$ and add them, to find

$$\mu_1 \left[ P_t + \frac{\mu_1 u + \mu_2 \Theta}{\mu_1} P_x \right] + \mu_2 \left[ u_t + \frac{\mu_1 \gamma + \mu_2 u}{\mu_2} u_x \right] = 0. \tag{4.2}$$

In order to obtain the same directional derivative for $P$ and $u$, the coefficient of $P_x$ and $u_x$, which we will call $\lambda$, must satisfy the following conditions:

$$\begin{aligned} (\lambda - u)\mu_1 - \Theta\mu_2 &= 0, \\ -\gamma\mu_1 + (\lambda - u)\mu_2 &= 0. \end{aligned} \tag{4.3}$$

There exists a nontrivial solution to (4.3) only if the determinant of the coefficients vanishes,

$$\begin{vmatrix} \lambda - u & -\Theta \\ -\gamma & \lambda - u \end{vmatrix} = 0. \tag{4.4}$$

Equation 4.4 defines the characteristic slopes:

$$\lambda^{\pm} = \frac{dx}{dt} = u \pm a. \tag{4.5}$$

Since the values of $\mu_1$ and $\mu_2$ are proportional to the complementary minors of any row of the determinant of (4.4), it follows from (4.2) and (4.5) that the *compatibility equations* that replace the first two equations of (4.1) are

$$aP' \pm \gamma u' = 0 \text{ on } \lambda^{\pm}, \tag{4.6}$$

where $(\ )' = \partial(\ )/\partial t + \lambda^{\pm}\partial(\ )/\partial x$. Note that the third equation of (4.1) is already in characteristic form and may be written as

$$dS = 0, \quad \text{along the particle path } \frac{dx}{dt} = u. \tag{4.7}$$

Equations 4.6 and 4.7 only involve derivatives along the characteristic directions $\lambda^{\pm}$ and $u$ and, therefore, discontinuities in the derivatives of the flow variables normal to these characteristic directions may occur.

### 4.1.2 Riemann Invariants

If the entropy is zero, then $P$ can be expressed as a function of $a$, and the above equations, (4.6), can be integrated to obtain

$$\begin{aligned} \frac{2}{\gamma - 1}a + u &= R \text{ on } \lambda^{+}, \\ \frac{2}{\gamma - 1}a - u &= Q \text{ on } \lambda^{-}, \end{aligned} \tag{4.8}$$

where $R$ and $Q$, *the Riemann invariants*, are constants along the $\lambda^{+}$ and $\lambda^{-}$ characteristics, respectively.

If the entropy varies, we can write (4.6) in terms of the *Riemann variable*, $R$ and $Q$ (no longer invariants):

$$\begin{aligned} R_t + \lambda^{+}(R_x - aS_x/\gamma(\gamma - 1)) - aS_t/\gamma(\gamma - 1) &= 0, \\ Q_t + \lambda^{-}(Q_x - aS_x/\gamma(\gamma - 1)) - aS_t/\gamma(\gamma - 1) &= 0. \end{aligned} \tag{4.9}$$

Equations 4.9 will be the basis for one of the more advanced codes that we will develop later in this chapter.

### 4.1.3 Range of Influence and Domain of Dependence

A good understanding of the concepts of *range of influence* and *domain of dependence* is not only important to understand the results we obtain through a numerical simulation, but also to avoid programming behavior that is physically inconsistent with the governing equations. The latter is particularly important when imposing boundary conditions.

Let us consider the example illustrated in Figure 4.1. At $t=0$ a piston starts compressing a gas initially at rest. The first disturbance made by the piston travels along the characteristic $\lambda_0^+ = u_0 + a_0$, where the subscript denotes the values of the gas at rest. This disturbance corresponds to the dashed line originating at $t=0$, $x=0$ in the figure. Since $u_0=0$ and $a_0 = \sqrt{\gamma}$, $\lambda_0^+ = \sqrt{\gamma}$. From (4.8) it follows that $R$ is constant on this line. At any time $t$, to the right of this line the gas is at rest and to the left of this line the gas is in motion. Three characteristics pass through any point "a." The slopes of these characteristics are $\lambda^{\pm}$, and $u$. If we trace back the characteristics to the line $t=0$, we find that point "a" depends on the initial data given on the line segment $\overline{bc}$, and only on this data. This segment is the *domain of dependence* of point "a."

The CFL condition (see Section 2.3.7) is a direct consequence of the domain of dependence of the governing equations. If we want to evaluate point "a" in one time step, starting from $t=0$, the stencil we use to evaluate the $x$-derivatives must not be shorter than the segment $\overline{bc}$. If it is shorter, the scheme will not be stable. If we extend the characteristics passing through "a" forward in time, then any perturbation originating at "a" propagates along the characteristics and at time $t=t_1$ this perturbation would have influence the triangular region bounded by the segments $\overline{ad}$, $\overline{de}$, and $\overline{ea}$. This region is known as the *range of influence* of point "a." This limited dependence on data and limited range of influence is a characteristic feature of wave propagation phenomena. On one hand it simplifies the solution procedure by allowing the problem to be solved piecewise, on the other hand it allows for the solution to be nonanalytic.

The concepts of range of influence and domain of dependence apply to the propagation of small perturbations, not to shock waves. In the absence of shock waves, the concepts are reversible. For example, if we have data on line $\overline{de}$ at $t=t_1$ of Figure 4.1, we could integrate backward in time to find the state at point "a." That is, $\overline{de}$ is now the domain of dependence of point "a." If a shock wave is present, backward integration in time is no longer possible, i.e., the dependence is no longer reversible. Another way of thinking about this is that a state with a shock wave is not uniquely determined by data at a previous time.

### 4.1.4 Simple Waves

Let us return to the case where the entropy is constant throughout the flow. The governing equations are given by (4.8). In regions where the Riemann invariant $R$ is constant, then the $\lambda^-$-characteristics are straight lines. This follows since $Q$ is constant along the $\lambda^-$-characteristics, and, therefore, $a$ and $u$ are constant and $\lambda^-$ is constant. Conversely, if $Q$ is constant, the $\lambda^+$-characteristics are straight lines. Regions of the flow where either $R$ or $Q$ are constant are called *simple waves*.

Consider again the flow illustrated in Figure 4.1. The region bounded by the piston and the first signal is a simple wave region. Here $Q$ is a constant, since for any point in this region the $\lambda^-$-characteristics originate from some point along the $x$-axis at $t=0$. Therefore, in this region the $\lambda^+$-characteristics are straight lines. Note that $u$ and $a$ change along $\lambda^-$ but are constant along $\lambda^+$. This simple region corresponds to a compression. The slopes of the straight-line characteristics tend to coalesce. A simple wave can also represent an expansion, in which case the straight-line characteristics tend to spread out. In the example of Figure 4.1, the $\lambda^+$-characteristics would eventually intercept and a shock will be formed. Note that this is inevitable; no action by the piston can prevent the formation of the shock. Once a shock forms, the entropy is no longer zero and the concept of *simple waves* is no longer applicable. In regions of the flow that are not simple waves neither $R$ nor $Q$ are constant. In these regions perturbations are coupled, travel in both directions, and the solutions are strictly nonlinear.

### 4.1.5 Mathematical Background for Piston-Driven Flows

As previously discussed, the early phase of a piston-driven flow consists of a simple wave region that is amenable to analysis. Let $\tau$ be the value of $t$ at which the $\lambda^+$ characteristic passing through the point $(x, t)$ originates at the piston. Along such a line $u$, $a$, and $\lambda^+$ are constant. Hence integrating $\mathrm{d}x/\mathrm{d}t = u + a$ we find

$$x = b(\tau) + \left[\dot{b}(\tau) + a(\tau)\right](t - \tau), \tag{4.10}$$

where

$a(\tau)$ is the speed of sound on the piston at $t=\tau$
$\dot{b}$ is the speed of the piston

Since $Q$ has the same value inside the simple wave region and in the gas at rest, we find

$$Q = \frac{2\sqrt{\gamma}}{\gamma - 1}, \tag{4.11}$$

and therefore

$$a(\tau) = \sqrt{\gamma} + \frac{\gamma - 1}{2}\dot{b}(\tau). \tag{4.12}$$

Hence in the simple wave region we have along $\lambda^+$ that

$$x = b(\tau) + \left[\sqrt{\gamma} + \frac{\gamma + 1}{2}\dot{b}(\tau)\right](t - \tau). \tag{4.13}$$

For a given $b(t)$, the solution to the piston problem at some time $t_1$ is given by

$$u(x, t_1) = \dot{b}(\tau), \tag{4.14}$$

where $\tau$ follows from inverting (4.13) for $\tau = \tau(x, t_1)$. The derivatives $u_x$ and $u_t$ follow from $u_x = \ddot{b}(\tau)\frac{\partial\tau}{\partial x}$ and $u_t = \ddot{b}(\tau)\frac{\partial\tau}{\partial t}$. They are

$$
\begin{aligned}
u_x &= \frac{2\ddot{b}(\tau)}{(\gamma+1)\ddot{b}(\tau)(t-\tau) - 2\sqrt{\gamma} - (\gamma-1)\dot{b}(\tau)}, \\
u_t &= \frac{\ddot{b}(\tau)\left[2\sqrt{\gamma} + (\gamma+1)\dot{b}(\tau)\right]}{(\gamma+1)\ddot{b}(\tau)(t-\tau) - 2\sqrt{\gamma} - (\gamma-1)\dot{b}(\tau)}.
\end{aligned}
\tag{4.15}
$$

If the initial acceleration of the piston does not vanish, then as long as the denominator does not vanish the derivatives are continuous and differentiable, except along $x = \sqrt{\gamma}t$, i.e., the first signal.

Within the simple region, the slopes of the characteristics issuing from the piston steepen as time increases. As the characteristics coalesce, they define an envelope. Along this envelope, $\partial x/\partial\tau = 0$. From (4.13) we have, on the envelope,

$$
\frac{\gamma+1}{2}\ddot{b}(\tau)(t-\tau) - \sqrt{\gamma} - \frac{\gamma-1}{2}\dot{b}(\tau) = 0. \tag{4.16}
$$

Note that (4.16) is the same as the denominator in (4.15). The parametric equations for the envelope follow from (4.13) and (4.16):

$$
\begin{aligned}
x_e &= b(\tau) + \left[\sqrt{\gamma} + \frac{\gamma+1}{2}\dot{b}(\tau)\right]\frac{2\sqrt{\gamma} + (\gamma-1)\dot{b}(\tau)}{(\gamma+1)\ddot{b}(\tau)}, \\
t_e &= \tau + \frac{2\sqrt{\gamma} + (\gamma-1)\dot{b}(\tau)}{(\gamma+1)\ddot{b}(\tau)}.
\end{aligned}
\tag{4.17}
$$

If the initial acceleration of the piston is different from zero, the point at which the characteristics coalesce is given by

$$
\begin{aligned}
x_A &= \frac{2\gamma}{(\gamma+1)\ddot{b}(0)}, \\
t_A &= \frac{2\sqrt{\gamma}}{(\gamma+1)\ddot{b}(0)}.
\end{aligned}
\tag{4.18}
$$

This point is located along the first characteristic (first signal).

If the initial acceleration of the piston is zero, the coalescence occurs within the simple wave region. From the second of (4.17), we have $t$ on the envelope as a function of $\tau$. Since $\ddot{b}(0) = 0$, $t$ is infinite for $\tau = 0$. For all $\tau > 0$, $t$ is positive, since $\dot{b}(\tau) > 0$ and $\ddot{b}(\tau) > 0$ for an accelerating piston. From the second of (4.17) we see again that as $\tau \to \infty$, $t \to \infty$. It follows that $t(\tau)$ must have a minimum. Figure 4.2 shows this behavior for the case $b = \tau^3$. This minimum occurs at

$$
\left.\frac{dt}{d\tau}\right|_B = 2\gamma - \frac{2\sqrt{\gamma} + (\gamma-1)\dot{b}(\tau_B)}{\ddot{b}^2(\tau_B)}\dddot{b}(\tau_B) = 0. \tag{4.19}
$$

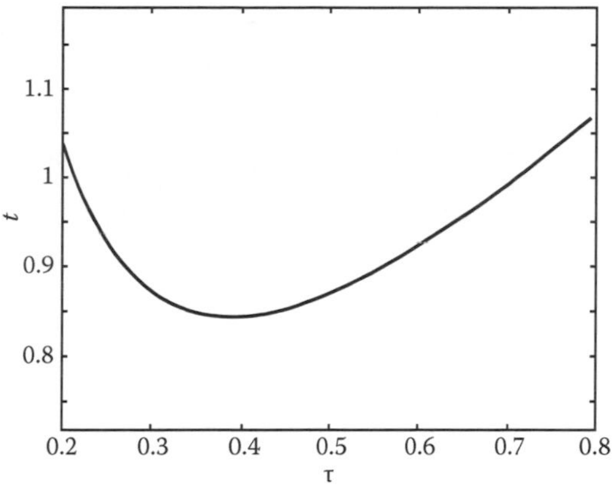


**FIGURE 4.2:** $t$ as a function of $\tau$ for $b=\tau^3$.

For the case defined by $b=\tau^3$, the minimum occurs at $\tau_B=0.3895$, corresponding to $t_B=0.8439$ and $x_B=0.8449$. Figure 4.3 shows the envelope for the case $b=\tau^3$.

Consider (4.13) again. For $b=\tau^3$ we have

$$x = \tau^3 + \left[\sqrt{\gamma} + \frac{\gamma+1}{2} 3\tau^2\right](t_1 - \tau),$$

which, using (4.14), may be written, after letting $t=t_1$, as

$$x - \sqrt{\gamma}t = \frac{\gamma+1}{2}u(x,t)t - \left[\sqrt{\gamma} + \frac{3\gamma+1}{6}u(x,t)\right]\sqrt{\frac{u(x,t)}{3}}.$$

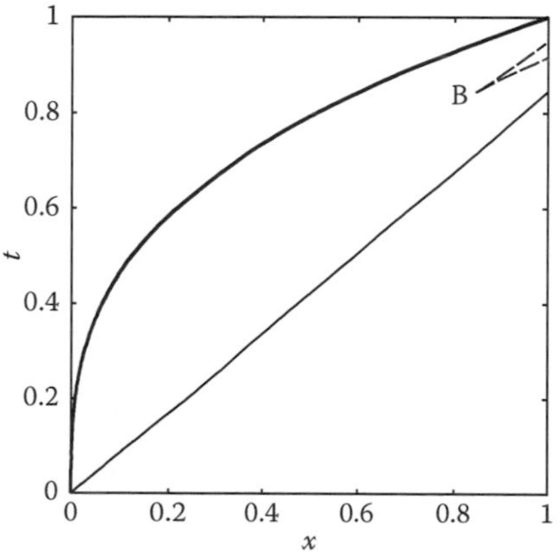


**FIGURE 4.3:** Envelope for $b=\tau^3$.

We are interested in the behavior of $u$ within the simple wave region. To this end, let $\xi = x - \sqrt{\gamma}t$; $\xi = 0$ is the locus of points along the first signal. Therefore

$$\xi = \frac{\gamma+1}{2}u(x,t)t - \left[\sqrt{\gamma} + \frac{3\gamma+1}{6}u(x,t)\right]\sqrt{\frac{u(x,t)}{3}}.$$

The region of interest at any time $t$ is then, $t(t^2 - \sqrt{\gamma}) < \xi < 0$. If we treat $\xi$ and $u$ as complex variables, we find that $u(\xi)$ has two branch-points in the $\xi$-plane. The branch-points (see Section 2.6.1 and [146]) are easily found by solving the equation,

$$\frac{d\xi}{du} = \frac{1}{6v}\left[3\frac{3\gamma+1}{2}v^2 - 3(\gamma+1)vt + \sqrt{\gamma}\right] = 0, \tag{4.20}$$

where $v = \sqrt{u/3}$. The locations of the branch-cuts in the complex plane at different times are shown in Figure 4.4. Note that as time increases the branch-cut moves toward the real axis, reaching it at $t = 0.8439$ when the shock forms. Figure 4.5 shows the branch of $u$ of interest at $t = 0.72$. There is a branch-point at "a" which forms a branch-cut with the branch-point at infinity, indicated by the heavy line. Lines of $\mathrm{Re}(u) = \text{constant}$ are plotted in the figure.

The proximity of the branch-cut to the real axis limits the convergence of any Taylor expansion of a real variable. Figure 4.6 shows the behavior of $u$ and its first five derivatives with respect to $\xi$ at $t = 0.77$. The behavior of $u$ does not show any obvious problem, but its derivatives clearly show the imminent analytic breakdown near $\xi = 0.2$. At $t = 0.77$, the branch-point is located at $\xi_0 = -0.1871 + 0.0210i$. If we expand $u$ in a Taylor series about $\xi = -0.1871$, the radius of convergence, $\mathfrak{R}$, is bounded by $\mathfrak{R} < 0.021$.

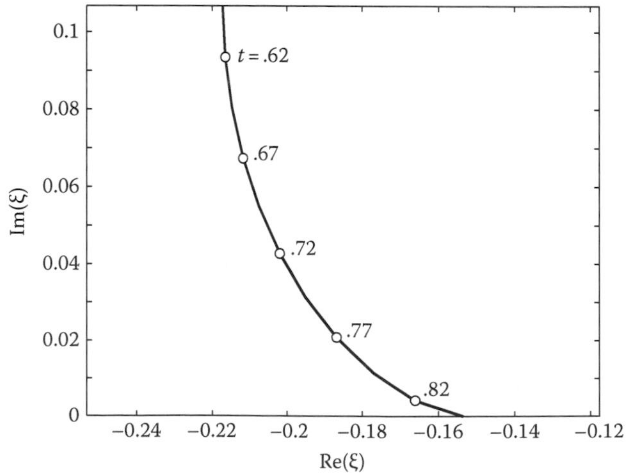


**FIGURE 4.4:** Branch-points in the complex plane at different times.

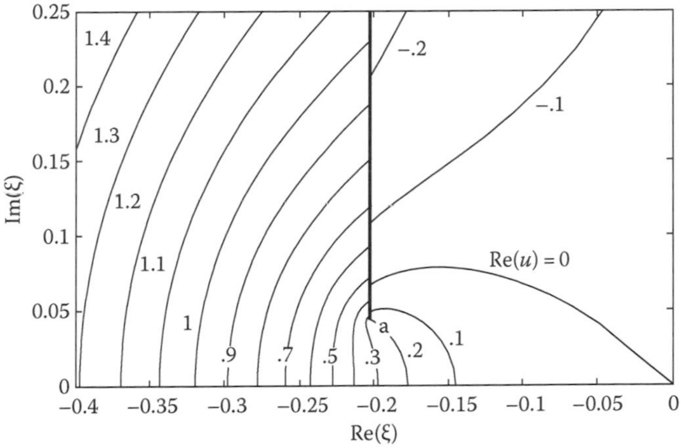


**FIGURE 4.5:** Branch of $u$ at $t = 0.72$.

The *remainder* of a Taylor series after $n$ terms is given by

$$R_n = u(\xi_0 + \Delta\xi) - \left[u(\xi_0) + \sum_{n=1}^{N} u^{(n)} \frac{\Delta\xi^n}{n!}\right].$$

In Figure 4.7 remainder of the Taylor series of $u$ is shown in $\Delta\xi$ interval of 0.01 and 0.03. The series does not converge with $\Delta\xi = 0.03$.

## 4.2 Numerical Analysis of a Simple Wave Region

We begin our numerical studies of the Euler equations with a very simple example: the study of a simple wave region. In Chapter 2, most of our numerical studies were based on the MacCormack scheme. Here we will focus on schemes that emphasize the wave propagation properties of the governing equations. The idea goes back to the seminal paper by Courant, Isaacson, and Rees [45] published in the early 1950s. As we shall see, several key properties of a numerical scheme (stability, domain of dependence, accuracy, simplicity) compete against each other and require some trade-offs and compromises.

Let $T$ be a time-like coordinate and $X$ be a space-like coordinate defined by

$$X = \frac{x - b(t)}{c(t) - b(t)}, \qquad T = t, \tag{4.21}$$

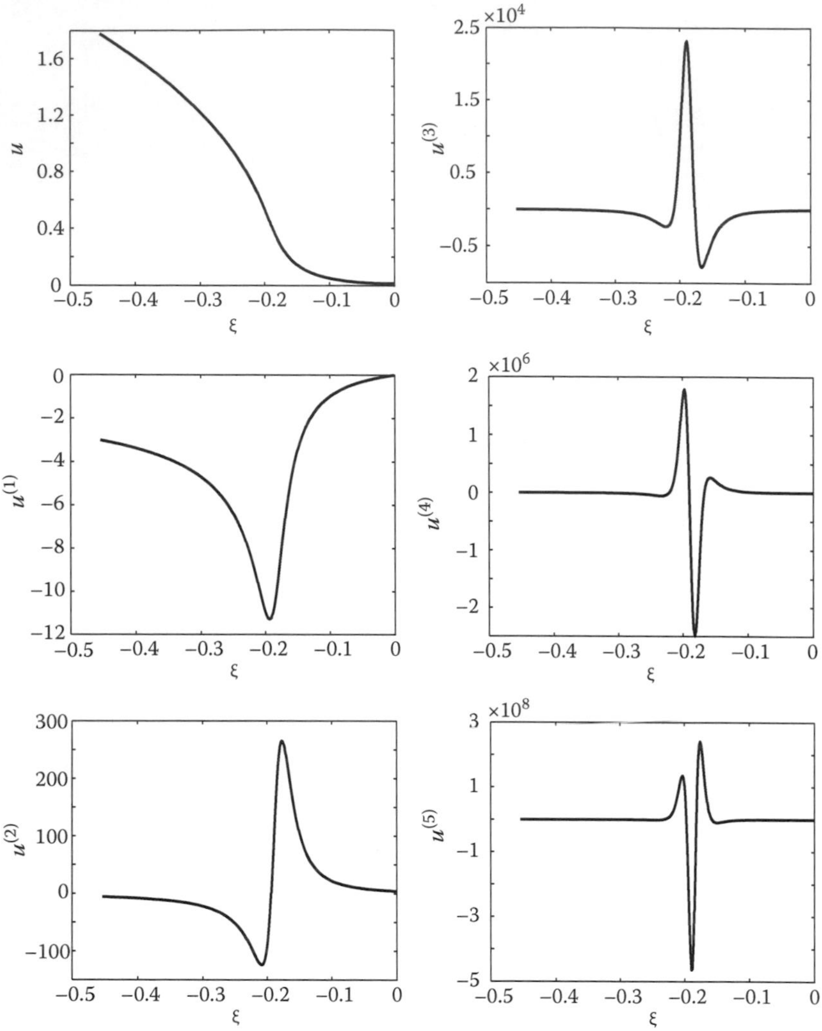


**FIGURE 4.6:** $u$ and its first five $x$-derivatives at $t=0.77$. $u$ shows a gentle compression, however its derivatives give an indication of the imminent analytical breakdown.

where $b(t)$ and $c(t)$ define the locus of a left and right boundary, respectively. These two boundaries could represent the paths of pistons or some other boundary condition. In the $X,T$ plane we will consider the following sets of governing equations representing an isentropic flow:

$$\begin{aligned} P_T + UP_X + \gamma X_x u_X &= 0, \\ u_T + Uu_X + \Theta X_x P_X &= 0, \end{aligned} \tag{4.22}$$

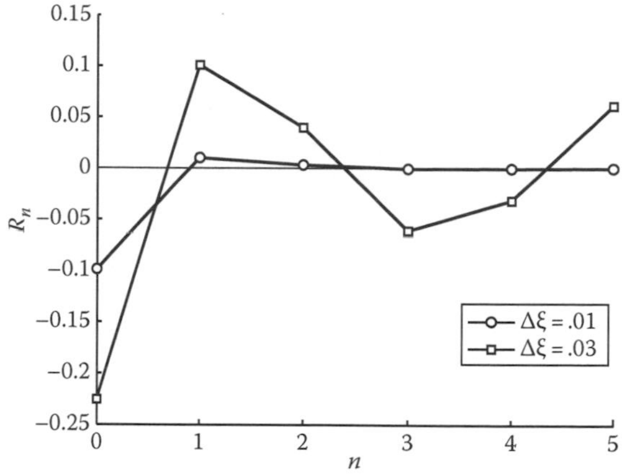


**FIGURE 4.7:** The remainder after $n$ terms.

$$\begin{aligned} R_T + \Lambda^+ R_X &= 0, \\ Q_T + \Lambda^- Q_X &= 0, \end{aligned} \tag{4.23}$$

$$\begin{aligned} a_T + (\gamma - 1)(\Lambda^+ R_X + \Lambda^- Q_X)/4 &= 0, \\ u_T + (\Lambda^+ R_X - \Lambda^- Q_X)/2 &= 0. \end{aligned} \tag{4.24}$$

In (4.22), $U$ is the contravariant velocity component defined by $U = X_t + uX_x$ and in (4.23) and (4.24), $\Lambda^\pm$ is the contravariant characteristic direction defined by $\Lambda^\pm = X_t + \lambda^\pm X_x$. The first set, (4.22), does not provide any obvious information on its domain of dependence or the direction of signal propagation. We will integrate this set using the MacCormack scheme introduced in Chapter 2. For the other two sets, we will use several variations of the λ-scheme.

The λ-scheme was developed in the late 1970s by Gino Moretti [150]. The motivation for developing the scheme was the strong belief that "a proper description of wave propagation relies on defining the domain of dependence of the wave and coding the discretized equations of motion accordingly" [155]. Several variations of the scheme have been proposed [30,79,155,245]. Here we will study three versions. First, let us consider the original version. We will apply it to (4.24). The application to (4.23) is trivial. The scheme is based on the predictor–corrector format of the MacCormack scheme. Let us represent discretely any function $f(X, T)$ by $f_n^k = f(n\Delta X, k\Delta T)$. Thus, in the predictor stage the $x$-derivatives are approximated by

$$\begin{aligned} R_X &= \left(2R_n^k - 3R_{n-1}^k + R_{n-2}^k\right)/\Delta X, \quad \Lambda^+ > 0, \\ R_X &= \left(R_{n+1}^k - R_n^k\right)/\Delta X, \quad \Lambda^+ < 0, \\ Q_X &= \left(2Q_n^k - 3Q_{n-1}^k + Q_{n-2}^k\right)/\Delta X, \quad \Lambda^- > 0, \\ Q_X &= \left(Q_{n+1}^k - Q_n^k\right)/\Delta X, \quad \Lambda^- < 0, \end{aligned}$$

according to the sign of $\Lambda^{\pm}$. The time derivatives are then given by

$$\begin{aligned} a_T &= -(\gamma - 1)(\Lambda^+ R_X + \Lambda^- Q_X)/4, \\ u_T &= -(\Lambda^+ R_X - \Lambda^- Q_X)/2, \end{aligned} \tag{4.25}$$

and predicted values are obtained from

$$\begin{aligned} \tilde{a}_n &= a_n^k + a_T \Delta T, \\ \tilde{u}_n &= u_n^k + u_T \Delta T. \end{aligned} \tag{4.26}$$

In the corrector step, the predicted values are used to approximate the $x$-derivatives as follows

$$\begin{aligned} \tilde{R}_X &= -\big(2\tilde{R}_n - 3\tilde{R}_{n+1} + \tilde{R}_{n+2}\big)/\Delta X, \quad \Lambda^+ < 0, \\ \tilde{R}_X &= \big(\tilde{R}_n - \tilde{R}_{n-1}\big)/\Delta X, \quad \Lambda^+ > 0, \\ \tilde{Q}_X &= -\big(2\tilde{Q}_n - 3\tilde{Q}_{n+1} + \tilde{Q}_{n+2}\big)/\Delta X, \quad \Lambda^- < 0, \\ \tilde{Q}_X &= \big(\tilde{Q}_n - \tilde{Q}_{n-1}\big)/\Delta X, \quad \Lambda^- > 0. \end{aligned}$$

New time derivatives are obtained from

$$\begin{aligned} \tilde{a}_T &= -(\gamma - 1)(\tilde{\Lambda}^+ \tilde{R}_X + \tilde{\Lambda}^- \tilde{Q}_X)/4, \\ \tilde{u}_T &= -(\tilde{\Lambda}^+ \tilde{R}_X - \tilde{\Lambda}^- \tilde{Q}_X)/2, \end{aligned}$$

and final values at $T = (k+1)\Delta T$ are given by

$$\begin{aligned} a_n^{k+1} &= \tfrac{1}{2}\big(a_n^k + \tilde{a}_n + \tilde{a}_T \Delta T\big), \\ u_n^{k+1} &= \tfrac{1}{2}\big(u_n^k + \tilde{u}_n + \tilde{u}_T \Delta T\big). \end{aligned}$$

A truncation error analysis of the $\lambda$-scheme, based on the simple equation $u_t + \lambda u_x = 0$, with $\lambda$ constant, shows that the leading error term is given by $\frac{1}{6} u_{ttt} \Delta t^2 + \frac{1}{3} \lambda u_{xxx} \Delta x^2$. In order to achieve second-order accuracy, the numerical stencil had to be enlarged; for MacCormack the stencil extends from $n-1$ to $n+1$, for the $\lambda$-scheme it extends from $n-2$ to $n+2$. This has two disadvantages: (1) the mesh point next to the boundary needs a special treatment, and (2) the numerical domain of dependence is broader than the physical domain of dependence. von Neumann's stability analysis shows that a Courant number less than $\frac{2}{3}$ is required for stability.

This is a somewhat more restrictive condition than for the MacCormack scheme and more restrictive than the physical range of influence of the scheme's footprint.

The first variation of the λ-scheme consists of a minor simplification of the logic. We will refer to this scheme as the λ-1 scheme. In this version, only two-point differences are made in the predictor stage:

$$\begin{aligned}
R_X &= \left(R_n^k - R_{n-1}^k\right)/\Delta X, \quad \Lambda^+ > 0,\\
R_X &= \left(R_{n+1}^k - R_n^k\right)/\Delta X, \quad \Lambda^+ < 0,\\
Q_X &= \left(Q_n^k - Q_{n-1}^k\right)/\Delta X, \quad \Lambda^- > 0,\\
Q_X &= \left(Q_{n+1}^k - Q_n^k\right)/\Delta X, \quad \Lambda^- < 0,
\end{aligned} \tag{4.27}$$

and only three-point differences are made in the corrector stage:

$$\begin{aligned}
\tilde{R}_X &= -\left(2\tilde{R}_n - 3\tilde{R}_{n+1} + \tilde{R}_{n+2}\right)/\Delta X, \quad \Lambda^+ < 0,\\
\tilde{R}_X &= \left(2\tilde{R}_n - 3\tilde{R}_{n-1} + \tilde{R}_{n-2}\right)/\Delta X, \quad \Lambda^+ > 0,\\
\tilde{Q}_X &= -\left(2\tilde{Q}_n - 3\tilde{Q}_{n+1} + \tilde{Q}_{n+2}\right)/\Delta X, \quad \Lambda^- < 0,\\
\tilde{Q}_X &= \left(2\tilde{Q}_n - 3\tilde{Q}_{n-1} + \tilde{Q}_{n-2}\right)/\Delta X, \quad \Lambda^- > 0.
\end{aligned} \tag{4.28}$$

The rest of the logic is the same as in the original form of the scheme. This variation has the same order of accuracy and stability constrain as the original scheme.

In 1983, Bruno Gabutti [79] developed a class of λ-like schemes that, by increasing the Courant number to three times that of the λ-scheme, removed the second disadvantage previously mentioned. The following variation of the λ-scheme is from this class of schemes. We will refer to this version as the λ-G scheme. The predictor stage consists of two parts. The first part uses (4.27), followed by (4.25) and (4.26), to obtain predicted values: $\tilde{a}_n$, $\tilde{u}_n$. The second part uses (4.28), but instead of using the predicted values, it evaluates (4.28) with values at time level $k$. Thus, a new time prediction is obtained:

$$\begin{aligned}
\widehat{a}_T &= -(\gamma - 1)(\Lambda^+ R_X + \Lambda^- Q_X)/4,\\
\widehat{u}_T &= -(\Lambda^+ R_X - \Lambda^- Q_X)/2.
\end{aligned}$$

The corrector stage uses (4.27) again, this time with the values predicted by (4.26) $(\tilde{a}_n, \tilde{u}_n)$ in the first part of the predictor, to obtain

$$\begin{aligned}
\tilde{a}_T &= -(\gamma - 1)\left(\tilde{\Lambda}^+ \tilde{R}_X + \tilde{\Lambda}^- \tilde{Q}_X\right)/4,\\
\tilde{u}_T &= -\left(\tilde{\Lambda}^+ \tilde{R}_X - \tilde{\Lambda}^- \tilde{Q}_X\right)/2.
\end{aligned}$$

The values at $T = (k+1)\Delta T$, follow from

$$a_n^{k+1} = a_n^k + \tfrac{1}{2}(\tilde{a}_T + \widehat{a}_T)\Delta T,$$
$$u_n^{k+1} = u_n^k + \tfrac{1}{2}(\tilde{u}_T + \widehat{u}_T)\Delta T.$$

The MATLAB code `PISTON.M`, available in the companion CD, implements these various schemes to solve the simple wave region produced by a piston moving according to the law $b(t) = t^3$. For these computations, the right boundary is taken along the first signal and the calculation is started at $t = 0.10$ with exact initial conditions. The results of various schemes approximating equation sets (4.22) through (4.24) are shown in Table 4.1. The table provides the $L_2$ and $L_\infty$ norms of the error at $t = 0.80$, first row, and at $t = 0.84$, second row, computed using 100 mesh points. The MacCormack scheme has lower values of the $L_2$ and $L_\infty$ norms at $t = 0.80$. All the λ-scheme variations shown have very similar behavior. As we get closer to the formation of the shock wave, we see better behavior with the λ-scheme variations than with the MacCormack scheme. This improved behavior persists after the discontinuity forms, as illustrated in Figure 4.8. Here we have extended the calculation to $t = 0.86$. Recall that the shock forms at $t = 0.8439$. The left panel shows the results obtained with the MacCormack scheme for equation set (4.22). The right panel shows the results obtained by the λ-scheme for equation set (4.23). Similar results are obtained with the other variants of the λ-scheme. The "exact" solution corresponds to the exact solution ignoring the entropy rise across the shock, which at $t = 0.86$ is small. The solution computed with the λ-scheme looks very good.

**TABLE 4.1:** $L_2$ and $L_\infty$ norms of the error for a simple wave region created by a piston moving according to the law. $b(t) = t^3$.

| Scheme | Equation Set | $L_2(u)$ | $L_\infty(u)$ |
|---|---|---|---|
| Mac | 4.22 | 0.002800 | 0.008994 |
| | | 0.009467 | 0.065503 |
| Mac | 4.23 | 0.002841 | 0.009301 |
| | | 0.008987 | 0.063117 |
| λ | 4.23 | 0.003581 | 0.017607 |
| | | 0.003268 | 0.027120 |
| λ-1 | 4.23 | 0.003617 | 0.017606 |
| | | 0.003289 | 0.026170 |
| λ-G | 4.24 | 0.003525 | 0.017003 |
| | | 0.003184 | 0.026491 |
| λ-G | 4.23 | 0.003525 | 0.017003 |
| | | 0.003184 | 0.026491 |

*Note:* The norms are evaluated at $t = 0.80$, first row, and at $t = 0.84$, second row.

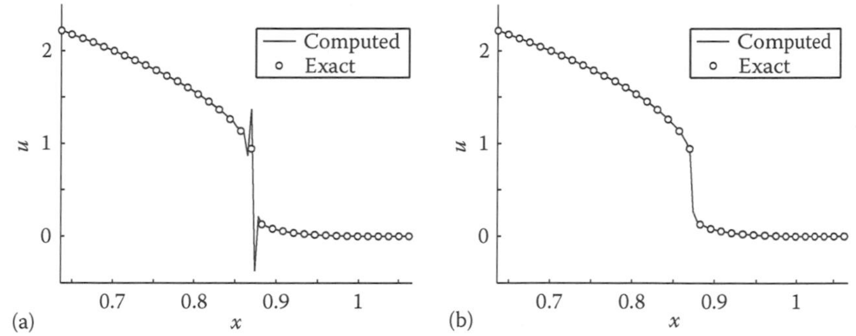


**FIGURE 4.8:** Compute $u$-profile at $t = 0.86$. Panel (a), MacCormack scheme; panel (b), λ-scheme.

However, there is a problem with it. The shock speed that it predicts is not correct and, consequently, the shock jumps are not correct either. To fix these problems we must either fit the shock or solve the equations in conservation form and introduce flux limiters and/or artificial dissipation to reduce the oscillations, and then deal with the ailments that come with these remedies. We will follow the former. The λ-scheme, or one of its variants, provides a clean (free of spurious oscillations) platform to do this work.

In the next section we will discuss the details of floating shock-fitting, for shocks treated as boundaries see [53], but before we dwell on that topic let us discuss why the λ-scheme produces wiggle-free shock profiles. For the case we are considering, a piston moving with the law $b(t) = t^3$, the Riemann variable $Q$ is constant throughout the computational region. Since we are modeling the flow as an isentropic flow, $Q$ remains constant even when a shock is present (as in panel (b) of Figure 4.8). If we look at the characteristic slopes in the computational plane, we find that to the left of the shock $\Lambda^+ > 0$ and $\Lambda^- < 0$, while to the right of the shock both $\Lambda^\pm < 0$. Therefore, the integration of the $\Lambda^+$-characteristic on the left side of the shock depends only on information to the left of the shock. The integration of the same characteristic to the right of the shock depends only on information to the right of the shock. Thus, the Riemann variable $R$ is evaluated without differentiating across the shock. A similar analysis for the $\Lambda^-$-characteristic shows that the integration of the mesh points to the left of the shock requires differentiating $Q$ across the shock. However, since $Q$ is constant throughout the region, no wiggles are created. Even when we take into account the entropy production at the shock, we find that when the Mach number relative to the shock is less than 4 the jump in $Q$ across the shock is so small that it can be differentiated without much harm to the calculation. This led Moretti [155] to suggest that the λ-scheme can be implemented by first doing an integration that ignores the presence of the shock wave. This first pass produces an accurate value of the jump in $R$ across the shock (since $R$ is computed without differentiating across the shock) between mesh points $n$, to the left of the shock, and $n+1$, to the right of the shock.

The jump in $R$ is used to obtain an accurate value of the relative shock Mach number, $M_{\text{rel}} = (u_{n+1} - w)/a_{n+1}$, which is related to the jump in $R$ through the relations

$$\Delta u = \frac{2\left(1 - M_{\text{rel}}^2\right)}{(\gamma + 1)M_{\text{rel}}} a_{n+1},$$

$$\Delta a = \left( \frac{2\sqrt{\left(\gamma M_{\text{rel}}^2 - (\gamma - 1)/2\right)\left(1 + (\gamma - 1)M_{\text{rel}}^2/2\right)}}{(\gamma + 1)M_{\text{rel}}} - 1 \right) a_{n+1},$$

where $a_{n+1}$ is the speed of sound at the mesh point immediately to the right of the shock, $\Delta u = u_n - u_{n+1}$ and $\Delta a = a_n - a_{n+1}$. With the shock Mach number known, and with accurate values computed to the right of the shock, the shock wave speed, $w$, can be computed. With the shock wave speed and the Rankine–Hugoniot jumps the values on the left side of the shock (that required differentiating $Q$ across the shock in the first pass) can be accurately updated. There are two aspects of this approach which we do not like: (1) the shock is resolved only to within one mesh point interval, and (2) it is bound to fail as the shock strength increases, since $Q$ and $S$ are differentiated across the shock.

## 4.3 Shock Wave Computation

When entropy is produced, equation set (4.23) must be replaced by

$$\begin{aligned} &R_T + \Lambda^+(R_X - aS_X/\gamma(\gamma - 1)) - aS_T/\gamma(\gamma - 1) = 0, \\ &Q_T + \Lambda^-(Q_X - aS_X/\gamma(\gamma - 1)) - aS_T/\gamma(\gamma - 1) = 0, \\ &S_T + US_X = 0. \end{aligned} \tag{4.29}$$

In what follows, we will use (4.29) and the $\lambda$-1 scheme for our numerical calculations. The reader is encouraged to try other variations of the $\lambda$-scheme. In the numerical implementation of (4.29), the derivatives inside the bracket multiplied by $\Lambda^+$ are discretized according to the sign of $\Lambda^+$, similarly for the derivatives associated with $\Lambda^-$. The time derivative of entropy that appears on the first, second, and third equations of (4.29) are evaluated from $S_T = -US_X$, where $S_X$ is discretized according to the sign of $U$ using the same rules as for $\Lambda^{\pm}$.

This set of equations, when integrated with one of the $\lambda$-scheme variants, provides a very stable and clean flow field for the detection of shock waves. In the code that we will develop* for the solution of the one-dimensional and quasi-one-dimensional Euler equations, we will use the shock detection method based on a

* The MATLAB code Q1D.M is available in the companion CD.

polynomial fit discussed in Section 2.9, for shock detection using characteristics see [58]. The polynomial fit is made for the variable $u=(R-Q)/2$ and is implemented in the function called `DETECT_SHOCK.M`. For the one-dimensional Euler equations, we have two families of shocks: shocks with the high pressure side on the left, and shocks with the high pressure side on the right. We will refer to these shocks as *l-shocks* and *r-shocks*, respectively. We will use a floating shock-fitting method for tracking and computing shock waves. The overall logic of the floating shock-fitting method is the same as described in Section 2.8. To distinguish between shock families, we introduce the string-variable `shock_type` defined as follows:

`shock_type(k)` = `'l'`, if the kth discontinuity is an *l-shock*,

`shock_type(k)` = `'r'`, if the kth discontinuity is an *r-shock*,

`shock_type(k)` = `'c'`, if the kth discontinuity is a contact surface.

Let us consider the computation of the shock wave created by a piston moving according to the law $b(t)=t^3$. After the shock is formed, the motion of the shock is governed by the signals reaching the shock along the $\lambda^+$ characteristic on the high pressure side of the shock, as illustrated in Figure 4.9. We want to find the flow variables at the shock (point $B$) at $t_0+\Delta t$. On the right side of the shock, the solution is known. It corresponds to the *gas at rest*. We can write three Rankine–Hugoniot conditions at $B$ to obtain the values (of say $P$, $u$, and $S$) on the high pressure side from the known values on the right. However, this introduces an additional unknown, the shock wave speed, $w$. To close the problem, we add the compatibility condition along $\lambda^+$, Equation 4.6, written in the form

$$\frac{a_B+a_A}{2}(P_B-P_A)+\gamma(u_B-u_A)=0,$$

where the subscripts refer to the values at points $A$ and $B$ as indicated in Figure 4.9. Note that the values at $A$ are interpolated from the values at mesh points $n$ and $n+1$

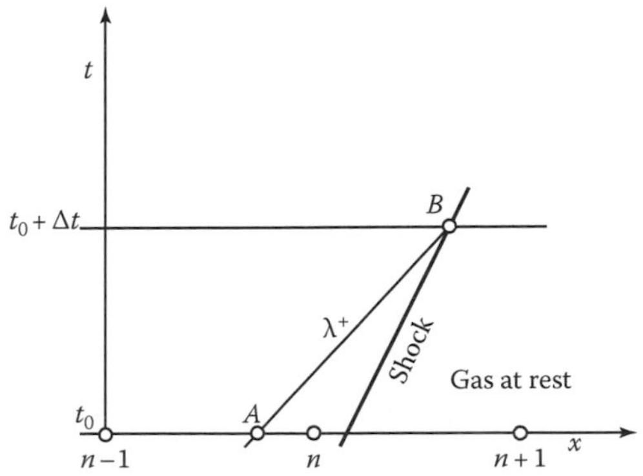


**FIGURE 4.9:** Details of shock computation.

from the known values at $t_0$. The location of point $A$ follows from the integration of the characteristic slope $\lambda^+$, namely

$$x_A = x_s + \left(w - \frac{u_A + u_B + a_A + a_B}{2}\right)\Delta t,$$

where $x_s$ is the known shock location at $t = t_0$. The set of equations we have described are solved by iteration. This approach has its origins in the method of characteristics, and it is rarely used today. It, however, clearly highlights the key points of shock-fitting:

1. Conditions on the low pressure side are known or can be computed with information available to the right of the shock.

2. Conditions on the high pressure side are related to the known conditions on the low pressure side through the Rankine–Hugoniot equations.

3. Perturbations reaching the shock from the high pressure side are embodied in the compatibility condition valid along the $\lambda^+$-characteristic. This equation and the three Rankine–Hugoniot conditions are sufficient to close the problem.

In 1970, Kentzer [113] reformulated this method in terms consistent with finite differences schemes. It is Kentzer's method that we will use for the computation of shocks. We begin by writing the following three Rankine–Hugoniot conditions:

$$\begin{aligned} p_2 &= p_1\left(\frac{2\gamma}{\gamma+1}M_{1\text{rel}}^2 - \frac{\gamma-1}{\gamma+1}\right), \\ \frac{u_2 - w}{u_1 - w} &= \frac{(\gamma-1)M_{1\text{rel}}^2 + 2}{(\gamma+1)M_{1\text{rel}}^2}, \\ M_{2\text{rel}}^2 &= \frac{(\gamma-1)M_{1\text{rel}}^2 + 2}{2\gamma M_{1\text{rel}}^2 - (\gamma-1)}, \end{aligned} \tag{4.30}$$

where
subscript 2 refers to the high pressure side
subscript 1 refers to the low pressure side

These conditions are valid *on the shock surface*. Therefore, in order to differentiate these conditions, we make a local coordinate transformation that makes the shock surface line up with one of our new coordinates,

$$\begin{aligned} \tilde{X} &= \frac{x - x_{s,k}(t)}{c(t) - b(t)}, \\ \tilde{T} &= t, \end{aligned}$$

where
$x_{s,k}(t)$ is the $x$-coordinate of the $k$th shock
$\tilde{X} = 0$ corresponds to the shock surface

Differentiating with respect to $\tilde{T}$ and, for convenience, dropping the tilde notation, we find

$$\begin{aligned}
P_{2T} &= P_{1T} + \frac{4\gamma}{\gamma+1}\frac{p_1}{p_2}M_{1\mathrm{rel}}M_{1\mathrm{rel}T}, \\
u_{2T} &= w_T + \frac{a_2 M_{2\mathrm{rel}}}{a_1 M_{1\mathrm{rel}}}(u_{1T} - w_T) - \frac{4a_1}{(\gamma+1)M_{1\mathrm{rel}}^2}M_{1\mathrm{rel}T}, \\
M_{i\mathrm{rel}T} &= \frac{u_{iT} - w_T}{a_i} - M_{i\mathrm{rel}}\frac{a_{iT}}{a_i}, \quad i = 1, 2, \\
a_{iT} &= \frac{\gamma - 1}{2\gamma}a_i\left(P_{iT} + \frac{S_{iT}}{\gamma - 1}\right), \quad i = 1, 2.
\end{aligned} \tag{4.31}$$

The compatibility condition along $\Lambda_2^+$ is

$$a_2\left(P_{2T} + \Lambda_2^+ P_{2X}\right) + \gamma\left(u_{2T} + \Lambda_2^+ u_{2X}\right) = 0. \tag{4.32}$$

Combining (4.31) and (4.32), then solving for $w_T$, we find

$$w_T = (B + C)/A,$$

where

$$\begin{aligned}
A &= \gamma\left(1 + \frac{4}{(\gamma+1)M_{1\mathrm{rel}}^2} - \frac{a_2}{a_1}\left(\frac{M_{2\mathrm{rel}}}{M_{1\mathrm{rel}}} + \frac{4}{\gamma+1}\frac{p_1}{p_2}M_{1\mathrm{rel}}\right)\right), \\
B &= -a_2\left(P_{1T} + \Lambda_2^+ P_{2X}\right) - \gamma\left(\frac{a_2 M_{2\mathrm{rel}}}{a_1 M_{1\mathrm{rel}}}u_{1T} + \Lambda_2^+ u_{2X}\right), \\
C &= -\frac{4\gamma}{\gamma+1}\left(\frac{p_1 a_2}{p_2 a_1}M_{1\mathrm{rel}} - \frac{1}{M_{1\mathrm{rel}}}\right)\left(u_{1T} - M_{1\mathrm{rel}}a_{1T}\right).
\end{aligned} \tag{4.33}$$

The second equation in (4.33) depends on the derivatives $P_{1T}$ and $u_{1T}$. If the low pressure side corresponds to a constant state, such as a gas at rest, these two derivatives would be zero; otherwise, these two derivatives are obtained by solving the governing equations, (4.22). The other two derivatives in the second of (4.33) are $P_{2X}$ and $u_{2X}$. They provide information about the local environment on the high pressure side of the shock. They are evaluated with one-sided derivatives with values on the left side of the shock. Similar comments apply to the third equation of (4.33). This approach is implemented in the function called `DISCONTINUITY.M` within the function `LAMBDA_Q1D.M` in the MATLAB code `Q1D.M` found in the companion CD.

Contact discontinuities are evaluated in a similar fashion. Along a contact we have

$$\begin{aligned}
p_\mathrm{l} &= p_\mathrm{r}, \\
u_\mathrm{l} &= u_\mathrm{r},
\end{aligned}$$

where the subscript l and r refer to conditions on the left and the right of the contact, respectively. In addition, since the contact is a particle path, the entropy on either side is constant:

$$S_l = \text{constant},$$
$$S_r = \text{constant}.$$

From these it follows, see (3.102), that

$$\frac{a_l}{a_r} = \exp\left(\frac{S_l - S_r}{2\gamma}\right)$$

is a constant along the contact. We have four conditions and six unknowns. To close the problem, we evaluate $R_l$ from the first of (4.29) and $Q_r$ from the second of (4.29). The remaining variables are given by

$$\begin{aligned} a_r &= \frac{(\gamma - 1)(R_l + Q_r)}{2(1 + a_l/a_r)}, \\ a_l &= a_r(a_l/a_r), \\ u_l &= u_r = (R_l - Q_r)/2 - (a_l - a_r)/(\gamma - 1), \\ P_l &= P_r = \frac{\gamma \ln(a_l a_r/\gamma) - (S_l + S_r)/2}{\gamma - 1}. \end{aligned}$$

This computation is also implemented in the function `DISCONTINUITY.M`.

### 4.3.1 Shock Interactions

The study of shock waves and shock wave interactions became very active during the Second World War. John von Neumann, Richard Courant, Kurt O. Friedrichs, and a few others were responsible for most of the theoretical work. In 1944, Courant and Friedrichs, under contract with the Office of Scientific Research and Development,* prepared a *manual* on the mathematical theory of nonlinear wave motion entitled *Supersonic Flow and Shock Waves* [40]. The manual was classified confidential and only 163 copies were made. Most copies went to various departments of the Navy; a few made their way to England; eight copies were given to NACA. Figure 44 appearing on page 116 of the manual is reproduced in Figure 4.10. In this particular chapter, Courant and Friedrichs are discussing the interaction of a shock with a contact surface. They correctly indicate that, assuming that $\gamma$ is the same for both gases, when a shock traveling on a gas of low density impinges on a gas of higher density the result of the interaction is that a shock is reflected back into the gas of low density and another shock is transmitted into the

* Many of the documents created under the OSRD can be found at www.archives.gov. In particular, reference [40] can be found at www.archives.gov/details/supersonicflowsh00cour.

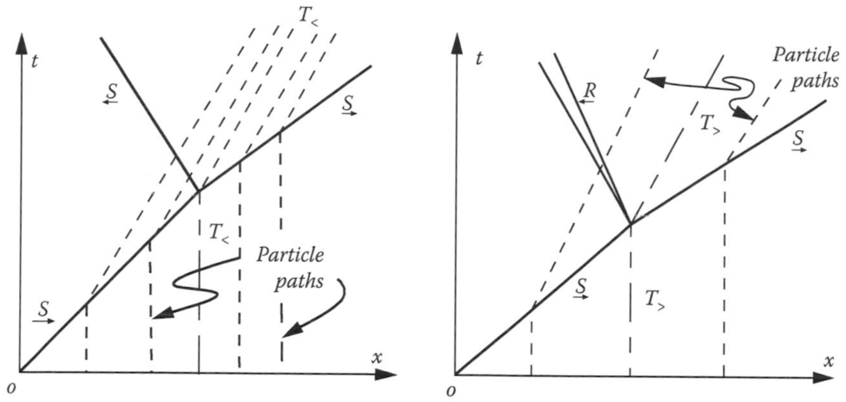


**FIGURE 4.10:** Figure 44, page 116, of *Supersonic Flow and Shock Waves*, 1944 version. (Adapted from Courant, R. and Friedrichs, K.O., Supersonic flow and shock waves, a manual on the mathematical theory of nonlinear wave motion, *App. Math. Panel, Nat. Def. Res. Comm.*, AMP Report 38.2R, 1944.)

gas of higher density. The *manual* was declassified after the war and was published in the open literature in 1948 [41]. It is now regarded as a classic treatise on applied mathematics. The topic of discussion in Figure 4.10 takes place on page 179 of the 1948 version. The text now reads: "...if a shock wave in one gas hits a second gas with a higher sound speed, at a discontinuity surface, a reflected and a transmitted shock wave result." This is not correct; since the pressures are equal across the contact, the speed of sound is inversely proportional to the square root of the density and, therefore, the speed of sound of the first gas should be greater than that of the second gas. The error found its way into Volume II of Shapiro's text book on compressible fluid flow [201, p. 1024], and has survived many reprints of both books. The more complex case, where the isentropic exponent of the two gases is different, was treated by Polachek and Seeger [169]. They show that a shock is reflected if

$$a_1/(\gamma_1((\gamma_1+1)+(\gamma_1-1)\xi))^{\frac{1}{2}} > a_2/(\gamma_2((\gamma_2+1)+(\gamma_2-1)\xi))^{\frac{1}{2}},$$

where, if the impinging shock is moving from left to right, subscript 1 refers to conditions for the gas immediately to the left of the contact, subscript 2 refers to conditions for the gas immediately to the right of the contact, and $\xi$ is the pressure ratio of the incoming shock. If this condition is not true, then an expansion is reflected.

Figure 4.11 illustrates the results of a head-on collision of two shocks of opposite families and the merging of two shocks of the same family. The interactions take place at $t = t^*$. Note that the results of these interactions are the same as those resulting from the interaction of a shock and a contact surface, as shown in Figure 4.10. Therefore, the solution of these problems is very similar. Let us consider the solution of the head-on collision of two shocks of opposite families, Figure 4.11a. The result of this interaction

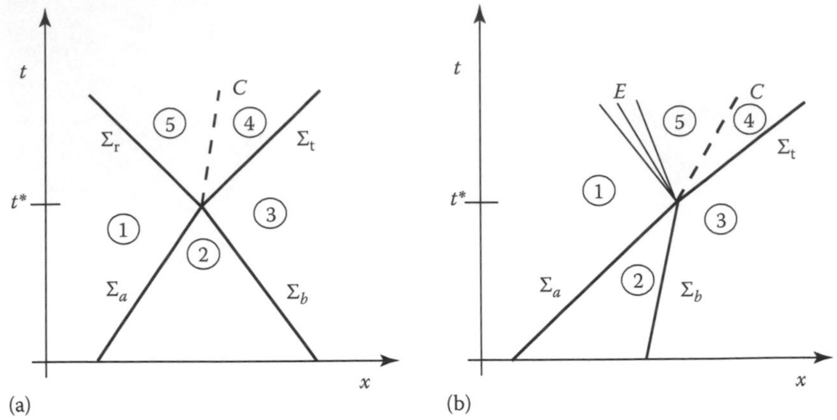


**FIGURE 4.11:** Illustration of (a), head-on collision of two shocks of opposite families; and (b), coalescence of two shocks of the same families.

is a transmitted shock, $\Sigma_t$, a contact surface, $C$, and a reflected shock, $\Sigma_r$. At the contact surface, we know that

$$\begin{aligned} p_4 &= p_5, \\ u_4 &= u_5. \end{aligned} \tag{4.34}$$

To solve the problem, we need to find the speeds of the reflected, $w_r$, and transmitted, $w_t$, shocks. The Mach number relative to shock $M_a$ is given by

$$M_{a,\,\mathrm{rel}} = \frac{u_2 - w_a}{a_2}.$$

A guess for $w_r$ is given by

$$w_r' = \left[1 - \frac{2(\gamma - 1)}{\gamma + 1}\left(1 - \frac{1}{M^2_{a,\,\mathrm{rel}}}\right)\right] w_b. \tag{4.35}$$

With the guessed value of $w_r$, we can find, using the Rankine–Hugoniot jumps, the values of $p_5$ and $u_5$; and with (4.34), we obtain $p_4$ and $u_4$. With $p_4$, $u_4$, and the Rankine–Hugoniot jumps, we can find the relative shock Mach number for shock $\Sigma_t$, $M_{t,\mathrm{rel}}$, and its speed, $w_t$:

$$\begin{aligned} M^2_{t,\,\mathrm{rel}} &= ((\gamma + 1)p_4/p_3 + (\gamma - 1))/2\gamma, \\ \rho_4/\rho_3 &= (\gamma + 1)M^2_{t,\,\mathrm{rel}}/((\gamma - 1)M^2_{t,\,\mathrm{rel}} + 2), \\ w_t &= (u_4\rho_4/\rho_3 - u_3)/(\rho_4/\rho_3 - 1). \end{aligned}$$

But since we know that

$$M_{\mathrm{t,rel}} = \frac{u_3 - w_\mathrm{t}}{a_3},$$

we can establish an error. We now iterate on $w_\mathrm{r}$ until the error meets a given tolerance.

To solve for the merging of two shocks of the same family, as illustrated in Figure 4.11b, we start by guessing the speed of the transmitted shock, $w_\mathrm{t}$:

$$w'_\mathrm{t} = (w_a + w_b)/2.$$

With $w'_\mathrm{t}$ and the Rankine–Hugoniot jumps we evaluate conditions in region 4. Then using (4.34) and $S_5 = S_1$, we evaluate the Riemann invariant $R_5 = 2a_5/(\gamma - 1) + u_5$, but since $R_5 = R_1$ we can evaluate an error which we reduce by iterating on the value of $w_\mathrm{t}$.

The shock interactions we have discussed are solved in the function `SHOCK_CROSS.M` of the program `Q1D.M`. First, the time at which the interaction takes place, $t^*$, is determined in the function `FIND_SHK_XING.M`, then the function `SHOCK_CROSS.M` is called from within the function `LAMBDA_Q1D.M` to compute the interaction.

Two more shock interactions are common when solving piston-driven flows. These are the interaction of a shock with a piston and the interaction of a shock with a constant pressure open end. These are illustrated in Figure 4.12. Both are very simple to solve. The first requires finding a reflected shock, $\Sigma_\mathrm{r}$, satisfying the boundary condition $u_3 = \mathrm{d}c/\mathrm{d}t$; the second requires finding an expansion such that $p_3 = p_{\mathrm{exit}}$, where $p_{\mathrm{exit}}$ is the given pressure at the open end of the tube at $t = t^*$. To solve the latter, we use the fact that $S_3 = S_1$ and $R_3 = R_1$.

These two interactions are computed in the function `SHOCK_WALL.M` of the program `Q1D.M`. The time at which the interactions take place is determined in the

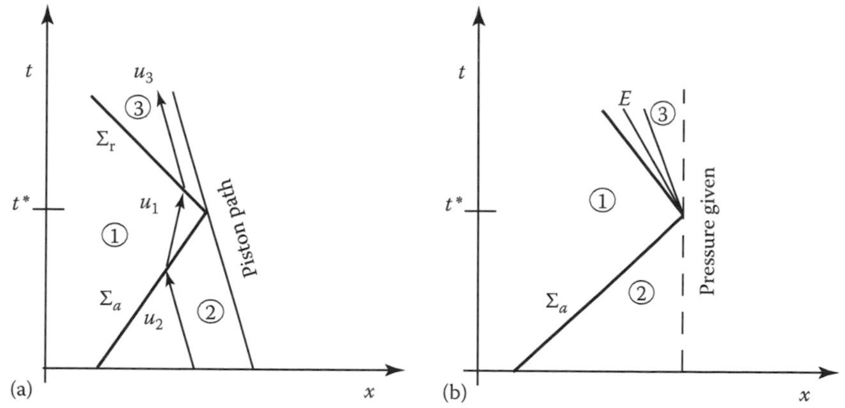


**FIGURE 4.12:** Illustration of a shock reflecting from (*a*), a piston; and (*b*), from a tube-end where the pressure is given.

function `FIND_WALL_XING.M`. The, user defined, input variable `bc_type(i)` determines the type of interaction required according to

`bc_type(1)` $= 0$, if the left end of the tube corresponds to a piston,

`bc_type(2)` $= 0$, if the right end of the tube corresponds to a piston,

`bc_type(1)` $= 1$, if the left end of the tube corresponds to an open end,

`bc_type(2)` $= 1$, if the right end of the tube corresponds to an open end.

### 4.3.2 Piston-Driven Flows: Numerical Examples

There are 18 default cases dealing with one-dimensional flows available in `Q1D.M`. These cases are described in Table 4.2. The first 13 cases are *unit problems* that check a specific aspect of the code, for example, the interaction of a shock and a contact discontinuity. The last 5 cases test the capabilities of the code to handle more complex problems. Figure 4.13 illustrates the resulting shock paths for case 15.

**TABLE 4.2:** Built-in one-dimensional test cases available in the code `Q1D.M`.

| Case | Description |
|---|---|
| 1 | Shock detection, right moving shock |
| 2 | Single shock moving right |
| 3 | Single shock moving left |
| 4 | Shock detection, left moving shock |
| 5 | Head-on collision of two shocks |
| 6 | Two right moving shocks coalescing |
| 7 | Two left moving shocks coalescing |
| 8 | Right moving shock interacting with a contact, resulting in a reflected shock |
| 8.5 | Right moving shock interacting with a contact, resulting in a reflected expansion |
| 9 | Left moving shock interacting with a contact, resulting in a reflected shock |
| 9.5 | Left moving shock interacting with a contact, resulting in a reflected expansion |
| 10 | Right moving shock interacting with a piston |
| 11 | Left moving shock interacting with a piston |
| 12 | Shock detection and reflection |
| 13 | Shock detection and reflection, mirror image of case 12 |
| 14 | Cases 12 and 13 combined |
| 15 | Asymmetric version of case 14 |
| 16 | Another asymmetric version of case 14 |

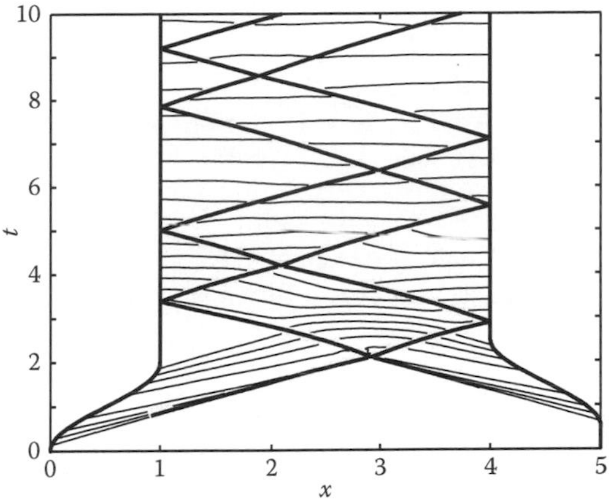


**FIGURE 4.13:** Heavy lines are traces of piston and shock paths for case 15 of `Q1D.M`. Thin lines show isobars.

This case corresponds of two pistons moving with a $t^3$ law, but the start of the right piston is delayed until $t = 0.5$. The calculation is terminated at $t = 10$. In addition to the paths of the pistons and the shocks (heavy lines), the figure shows lines of constant pressure (thin lines). The case illustrates shock detection, head-on shock collisions, and shock reflection of walls. The contact surfaces resulting from the head-on shock collisions were deemed too weak and were eliminated. In order to run one of these default cases, it is only necessary to input the case number.

`Q1D.M` is intended as a research code and requires some level of participation from the user to set up and run a nondefault case. Here is a checklist of things that need to be done:

1. In `Q1D.M`, create a new case number or modify an existing case number. A few parameters need to be defined here. These include the total time for the run, the number of mesh points, the Courant number, and a switch to detect or not detect shock formation. Other parameters might be needed depending on the complexity of the case, see examples in `Q1D.M`.

2. The piston paths, speeds, and accelerations need to be defined in the functions `PISTONLAW.M`, `PISTONLAWT.M`, `PISTONLAWTT.M`, `EXITBOUNDARY.M`, `EXITBOUNDARYT.M`, and `EXITBOUNDARYTT.M`.

3. In function `LAMBDA_Q1D.M`, arrays are stored for plotting lines of constant pressure. If this is desired, add the case number to line 74. Also add the case number to line 94 of function `PLOT_SOLUTIONPISTON.M`.

4. Boundary conditions are defined in functions `BCATNEQUAL1.M` and `BCATNEQUALNT.M`. These functions reside within function `LAMBDA_Q1D.M`. Follow the examples in these functions for computing wall boundaries, constant pressure boundaries, etc.

5. Initial conditions are defined in `INITIALPISTON.M`.

6. In the next section we discuss problems where the cross-sectional area of the channel changes as a function of both $x,t$. This information is provided in functions `AREA_F.M`, `AREAX_F.M`, and `AREAT_F.M`.

*A word of warning*. Chances are good that the code will run even when wrong or inconsistent inputs are defined, for example, inconsistent piston path and piston speed definition. Therefore, do not believe the results of this code (or any code, for that matter) until you have carefully checked the results!

---

## 4.4 Quasi-One-Dimensional Flows

Multidimensional problems where the rate of change of the flow variables in two of the space dimensions is small compared to the rate of change in the other space dimension can be studied by assuming that the dependent variables are only functions of one space dimension and time. As we shall see, with this assumption, only the continuity equation is modified. The momentum and energy equations, second and third of (4.1), remain unchanged.

### 4.4.1 Governing Equations

Consider a small volume element of a duct whose cross-sectional area, $A$, may be a function of $x$ and $t$. If we assume that the walls of the duct are impermeable, then the rate of change of the mass in an interval $\Delta x$ is equal to the balance of the incoming and outgoing mass flow. The mathematical expression of this statement of conservation of mass is

$$\int_x^{x+\Delta x} (\rho A)_t \mathrm{d}x = \rho u A - (\rho + \Delta\rho)(u + \Delta u)(A + \Delta A).$$

If we divide by $\Delta x$ and take the limit as $\Delta x \rightarrow 0$, we obtain the continuity equation for quasi-one-dimensional flows,

$$\rho_t + u\rho_x + \rho u_x = -\rho(uA_x + A_t)/A. \tag{4.36}$$

Similarly, the change in $x$-momentum within the interval $\Delta x$ has to balance the external force acting on the flow in the interval $\Delta x$. The mathematical expression is

$$\int_x^{x+\Delta x} (\rho u A)_t \mathrm{d}x + (\rho + \Delta\rho)(u + \Delta u)^2(A + \Delta A) - \rho u^2 A = pA - (p + \Delta p)(A + \Delta A) + \hat{p}\Delta A,$$

where $\hat{p}$ is a value of pressure between $p$ and $p + \Delta p$, and the term $\hat{p}\Delta A$ represents the force acting on the walls in the $x$-direction. If we now divide by $\Delta x$ and take the limit as $\Delta x \to 0$, we obtain the $x$-momentum equation in the form

$$u\rho_t + \rho u_t + \rho u\, A_t/A + u^2 \rho_x + 2\rho u u_x + \rho u^2\, A_x/A + p_x = 0.$$

If we multiply (4.36) by $u$ and subtract the resulting equation from the above momentum equation, we obtain the second equation of (4.1).

The derivation of the energy equation follows from a similar analysis. The end result is that the third equation of (4.1) remains unchanged.

For quasi-one-dimensional flows the system of equations equivalent to (4.1) is

$$\begin{aligned} P_t + uP_x + \gamma u_x &= -\gamma(uA_x + A_t)/A, \\ u_t + uu_x + \Theta P_x &= 0, \\ S_t + uS_x &= 0. \end{aligned} \tag{4.37}$$

The only change is the addition of the *source term* to the first equation. This source term accounts for area variations in space–time and does not contain derivatives of the dependent variable: *u, P, S*. The local properties of shock waves and contact surfaces remain exactly the same as for one-dimensional flows, as does the interaction of these discontinuities with each other. However, a new type of interaction involving a shock and an abrupt area change, to be discussed in the Section 4.4.3, is now possible.

A direct consequence of the fact that the source term does not contain derivatives of the dependent variables is that the characteristic slopes obtained from the first two equations of (4.37) are the same as for one-dimensional flows. The compatibility equations now read

$$aP' \pm \gamma u' = -\gamma a(uA_x + A_t)/A \text{ on } \lambda^{\pm}. \tag{4.38}$$

Equation 4.38 replaces Equation 4.6, and it should be clear that, even if the flow is isentropic, Equations 4.8 are no longer valid. It is also obvious that in quasi-one-dimensional flows the concept of simple waves is lost. Every perturbation interacts with the area change and waves are propagated in both directions. It is this kind of interaction that makes nontrivial steady states possible in quasi-one-dimensional flows and creates very interesting and complex flow patterns. Another feature that distinguishes quasi-one-dimensional flows from one-dimensional flows is that in the latter shocks could only form as a result of the action of pistons. In quasi-one-dimensional flows shocks can form as a result of waves interacting with area variations.

If we express the governing equations in terms of Riemann variables and computational coordinates (4.21) we obtain the following set of equations:

$$\begin{aligned} R_T + \Lambda^+(R_X - aS_X/\gamma(\gamma-1)) - aS_T/\gamma(\gamma-1) &= -a(uA_x + A_t)/A, \\ Q_T + \Lambda^-(Q_X - aS_X/\gamma(\gamma-1)) - aS_T/\gamma(\gamma-1) &= -a(uA_x + A_t)/A, \\ S_T + US_X &= 0. \end{aligned} \tag{4.39}$$

### 4.4.2 Steady Laval Nozzle

A variable area channel consisting of a smooth converging–diverging section connected to a large stagnation reservoir (plenum) at the left and discharging into another large reservoir with lower pressure at the right is known as a Laval nozzle after the Swedish inventor Gustaf de Laval. The steady flow in a Laval nozzle provides a convenient way of checking the results of a numerical code for quasi-one-dimensional flow, even in the presence of shocks, since it only involves the solution of transcendental equations.

If the flow is steady, (4.36) can be integrated; we find

$$m = \rho u A = \text{constant}. \tag{4.40}$$

It follows from the last of (4.37) that the entropy is constant (more precisely, it is piecewise constant). With constant entropy, the pressure, temperature, and speed of sound are given by

$$\begin{aligned} p &= \rho^{\gamma}, \\ \Theta &= \rho^{\gamma-1}, \\ a^2 &= \gamma\rho^{\gamma-1}. \end{aligned}$$

Using these relations in the second of (4.37), after setting $u_t = 0$, the equation can be integrated into the familiar form:

$$\frac{1}{2}u^2 + \frac{\gamma}{\gamma-1}\frac{p}{\rho} = \text{constant}. \tag{4.41}$$

Using (4.40) and (4.41), we can express all quantities as functions of the Mach number:

$$\begin{aligned} \frac{p}{p_o} &= \left(1+\frac{\gamma-1}{2}M^2\right)^{-\frac{\gamma}{\gamma-1}}, \\ \frac{\rho}{\rho_o} &= \left(1+\frac{\gamma-1}{2}M^2\right)^{-\frac{1}{\gamma-1}}, \\ \frac{u}{a_o} &= M\left(1+\frac{\gamma-1}{2}M^2\right)^{-\frac{1}{2}}, \\ \frac{A}{A*} &= \left(\frac{\gamma+1}{2}\right)^{\frac{\gamma+1}{2(\gamma-1)}} M\left(1+\frac{\gamma-1}{2}M^2\right)^{-\frac{\gamma+1}{2(\gamma-1)}}, \end{aligned} \tag{4.42}$$

where

$p_o$, $\rho_o$, and $a_o$ are the stagnation values of pressure, density, and speed of sound
$A*$ is the area where the flow velocity reaches the speed of sound

We refer to $A*$ as the *critical* area. It also follows that

$$(M^2-1)\frac{du}{u} = \frac{dA}{A}. \tag{4.43}$$

The last equation tells us that supersonic and subsonic flows have opposite behavior when the area changes and that a smooth transition from subsonic to supersonic flow, or vice versa, is only possible when $M^2 = 1$.

In the absence of shocks, (4.42) provide the steady solution for flow in a nozzle, of given area $A(x)$, as a function of $x$ for given stagnation conditions and exit pressure. The procedure is as follows. From the first of (4.42) one determines the exit Mach number. Next, since the exit area, $A_e$, is known, one determines $A_*$ from the last of (4.42). Now, the last of (4.42) is solved for $M$ at each value of $A(x)$. Once $M(x)$ is known, $p$, $\rho$, and $u$ follow from (4.42).

If the exit pressure is not much lower than the stagnation pressure, a steady subsonic flow, with pressure decreasing in the convergent section and increasing in the divergent section, develops in the nozzle. For this type of flow, we find that $A_* < A_{\mathrm{throat}}$, where $A_{\mathrm{throat}}$ is the *throat* area, the area where $\mathrm{d}A/\mathrm{d}x = 0$. At a certain value of the exit pressure, let us call it $p_e^*$, the flow becomes sonic at the throat. Using (4.42), we find that $p_e^*$ satisfies the following relation:

$$\frac{A_{\mathrm{throat}}}{A_e} = \left(\frac{\gamma+1}{2}\right)^{\frac{\gamma+1}{2(\gamma-1)}} \left(\frac{2}{\gamma-1}\right)^{\frac{1}{2}} \left(\frac{p_e^*}{p_o}\right)^{\frac{\gamma+1}{2\gamma}} \left[\left(\frac{p_e^*}{p_o}\right)^{-\frac{\gamma-1}{\gamma}} - 1\right]^{\frac{1}{2}}. \tag{4.44}$$

There are two values of $p_e^*$ that satisfy (4.44). We are interested in the highest one corresponding to subsonic exit Mach number, let this value be $p_{e,h}^*$ and let the other be $p_{e,l}^*$. With $p_e = p_{e,l}^*$, the flow accelerates in the converging section, until it becomes sonic at the throat, and then decelerates in the divergent section of the nozzle. For $p_{e,h}^* > p_e$, the mass flow is at its critical value

$$m_* = \rho_* u_* A_* = \left(\frac{2}{\gamma+1}\right)^{\frac{\gamma+1}{2(\gamma-1)}} \rho_o a_o A_{\mathrm{throat}},$$

and we say that the nozzle is *chocked*, meaning that the nozzle cannot sustain a higher flow rate. Under these conditions, the flow becomes supersonic in the divergent section immediately after the throat. Some distance from the throat, the supersonic flow is terminated by a shock and the flow becomes subsonic again. After the shock, the pressure rises until it reaches the value $p_e$ at the exit section. In the two regions of the nozzle, the one before the shock and the one after the shock, all of the above equations are valid. However, the stagnation values and the critical area are not the same in the two regions. The two regions are *connected* through the Rankine–Hugoniot relations which are valid at the shock, and which remain unchanged from their one-dimensional form. If we let $A_{*1}$ and $A_{*2}$ be the critical areas associated with the flow to the left and to the right of the shock, respectively, then it follows from (4.40) that

$$\frac{A_{*2}}{A_{*1}} = \frac{\rho_{o1}}{\rho_{o2}} = e^{[S]/(\gamma-1)}, \tag{4.45}$$

where $[\![S]\!]$ is the jump in entropy across the shock, hence, the second critical area is always larger than the first.

If we approach the shock from the left, we know that $A_* = A_{\text{throat}}$, hence, *if we know the area where the shock is located*, we can find the Mach number in front of the shock from the last equation of (4.42). With the Mach number in front of the shock and the Rankine–Hugoniot relations, we can find all of the flow variables after the shock. In particular, with the pressure, density, speed of sound, and Mach number, we can find new stagnation conditions, using (4.42), after the shock. These new stagnation conditions apply to the region to the right of the shock. A new critical throat area associated with the flow behind the shock follows from the last of (4.42). We use this last relation again, now evaluated at the exit plane, to find the exit Mach number. With the exit Mach number and the first of (4.42), we can find the exit pressure. If this exit pressure is not equal to $p_e$, then the area where the shock was located is not correct. This process can be repeated until the correct shock location is found.

As $p_e$ is reduced, the shock will move closer to the exit plane. The shock reaches the exit plane when the pressure in front of the shock is equal to $p^*_{e,1}$. The exit pressure for this condition follows from the Rankine–Hugoniot jump with the shock at the exit plane. Let this exit pressure be $\bar{p}_e$. For values of the exit pressure such that $\bar{p}_e > p_e > p^*_{e,1}$, the flow is compressed in the jet outside of the nozzle. When this condition exists, we say that the nozzle is *over-expanded*. When $p_e = p^*_{e,1}$, the flow leaves the nozzle smoothly and the nozzle is said to be *ideally expanded*. When $p_e < p^*_{e,1}$, the flow is expanded in the jet as it leaves the nozzle. When this condition exists, we say that the nozzle is *under-expanded*.

### 4.4.3 Interaction of a Shock with an Abrupt Area Change

The interaction of a shock wave with a channel of rapidly varying cross-sectional area is of interest in a number of practical problems, such as the passage of shocks through wire-mesh screens, the starting process in a supersonic tunnel, and flows associated with piston and jet engines. The problem can be simplified by modeling the area change as an abrupt (discontinuous) change and treating the flow with a self-similar inviscid model. The problem is solved by guessing a wave pattern and insuring that the pattern satisfies the governing equations for the given incident shock strength and area jump. Thus, the incident shock strength, measured by the shock Mach number, $M_i$, and the area jump ratio, $\alpha$, are a set of parameters that categorize the possible wave patterns.

Several investigators [158,184,193] have shown that for certain values of $M_i$, $\alpha$ corresponding to supersonic flow behind the incident shock and a certain range of area contraction, three wave patterns can satisfy all of the governing equations. However, it has been shown by [184] and [193] that if the problem is solved, not as a steady state problem, but as a time dependent problem, then only one solution occurs which coincides with the solution predicted by the *minimum entropy production principle* [107].

To find the analytic solution to this problem, we assume that the area change is over a very short distance which may be approximated by a jump in area. Let us also assume that the incident shock is moving from left to right and that the gas ahead of the shock is at rest. Let $x = 0$ be the location of the area jump, and let $t = 0$ be the time at which the incident shock strikes the area jump. Because there is no reference length, we expect the solution to be constant along rays originating at the origin. That is, the dependent variables are only functions of the ratio $x/t$. A typical wave pattern is shown in Figure 4.14. For the case illustrated, we have a transmitted shock, $\Sigma_t$, a contact surface, $C$, followed by an expansion, $E$, and a reflected shock, $\Sigma_r$. The flow is at rest in regions 1 and 2. The flow variables in these regions are

$$\begin{aligned} u_1 &= u_2 = 0, \\ p_1 &= p_2 = 1, \\ \rho_1 &= \rho_2 = 1, \\ a_1 &= a_2 = \sqrt{\gamma}, \\ S_1 &= S_2 = 0. \end{aligned}$$

The area ratio $\alpha$ is defined as

$$\alpha = A_l/A_r,$$

where
$A_l$ is the area to the left
$A_r$ is the area to the right

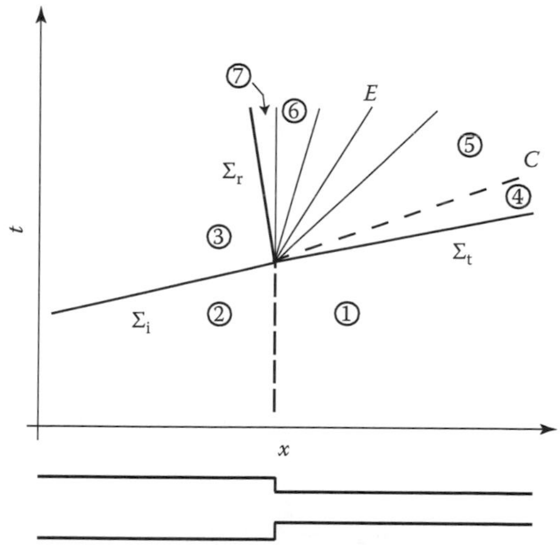


**FIGURE 4.14:** Typical wave diagram for the interaction of a shock with an abrupt area change.

The Mach number of the incident shock is

$$M_\mathrm{i} = w_\mathrm{i}/a_2.$$

The flow variables in region 3 are easily obtained from the Rankine–Hugoniot conditions. Across the contact surface, the following two conditions must be satisfied:

$$p_4 = p_5,$$
$$u_4 = u_5.$$

If the Mach number of the transmitted shock, $M_\mathrm{t}$, is known then the flow in region 4 could be determined by using the Rankine–Hugoniot jumps. This Mach number can be expressed in terms of the pressure ratio $p_\mathrm{r} = p_4/p_1 = p_5/p_1$:

$$M_\mathrm{t}^2 = \frac{(\gamma+1)p_\mathrm{r}/2 + (\gamma-1)/2}{\gamma}. \tag{4.46}$$

Therefore, with $p_5$ known, region 4 is completely defined. In general, the wave pattern between regions 3 and 5 will be different from that shown in Figure 4.14. The specific pattern will depend on the value of $M_\mathrm{i}$ and $\alpha$. We will show how the solution for this case is obtained, with the understanding that the procedure has to be modified for other wave patterns.

The conservation of mass relation provides a condition, last expression of (4.42), between regions 6 and 7, which we write as

$$\alpha = \frac{M_6}{M_7}\left(\frac{1+(\gamma-1)M_7^2/2}{1+(\gamma-1)M_6^2/2}\right)^{(\gamma+1)/2(\gamma-1)}.$$

Since, immediately to the right of the area jump the flow is sonic, $M_6 = 1$, it follows that $M_7$ is known. With region 3 known and $M_7$ known, we can obtain the solution in region 7. Now, since the flow from region 7 to region 6 is isentropic, we have that $S_6 = S_7$ and

$$a_6 = a_7\big((1+(\gamma-1)M_7^2/2)/(\gamma+1)/2\big)^{1/2}.$$

Also, since the flow is sonic in region 6, we have that $u_6 = a_6$. The density and pressure are given by

$$\begin{aligned}\rho_6 &= \exp\big((\ln(a_6^2/\gamma) - S_6)/(\gamma-1)\big),\\ p_6 &= \rho_6 a_6^2/\gamma.\end{aligned} \tag{4.47}$$

Across the expansion, the Riemann variable $Q$ is constant, $Q_5 = Q_6$. Using this relation, we can write $u_5$ in terms of the slope of the expansion tail, $\lambda_5^+ = u_5 + a_5$:

$$u_5 = a_6 + 2\lambda_5^+/(3-\gamma).$$

Therefore, if we guess $\lambda_5^+$, and since $S_5 = S_6$ and $a_5 = \lambda_5^+ - u_5$, the pressure and density in region 5 follow from (4.47) with an appropriate change of subscripts. We can now compute $M_t$ from (4.46) and can, therefore, find if $u_5 = u_4$. If this is not the case, we iterate on $\lambda_5^+$ until this condition is met.

The various wave patterns that can occur as a result of a shock interacting with an area jump are shown in Figures 4.15 and 4.16. Figure 4.15 is divided into four quadrants by the lines $\alpha = 1$ and $M_i = M_i^*$, where $M_i^*$ is the incident shock Mach number corresponding to $M_3 = 1$. For $\gamma = 1.4$, $M_i^* = 2.068$.

*Consider the first quadrant, $M_i \leq M_i^*$ and $\alpha \leq 1$.* For weak incident shocks, a weak rarefaction wave is reflected; pattern I$\beta$, Figure 4.16. The effect of the rarefaction is to accelerate the flow before it enters into the area divergence. However, because the flow remains subsonic, it is decelerated as it crosses into the bigger chamber. In general, the transmitted shock is weaker than the incident shock. As the strength of the incident shock increases, the rarefaction becomes stronger, eventually creating sonic conditions at the entrance to the divergence. The locus of such points corresponds to line $m$ of Figure 4.15. As the shock strength continues to increase, a standing shock develops just ahead of the area jump, $x = 0^-$; pattern $I\delta$, Figure 4.16. As the shock strength continues to increase, the standing shock moves to the other side of the area jump, $x = 0^+$. For these conditions, the flow in front of the shock goes through an isentropic expansion corresponding to the full area jump. The locus of points corresponding to these conditions map to line $n$ of Figure 4.15. The wave pattern changes to type I$\gamma$, Figure 4.16, with a further increase in shock strength. Now, the shock is swept downstream. As $M_i$ approaches $M_i^*$, the reflected expansion fan disappears.

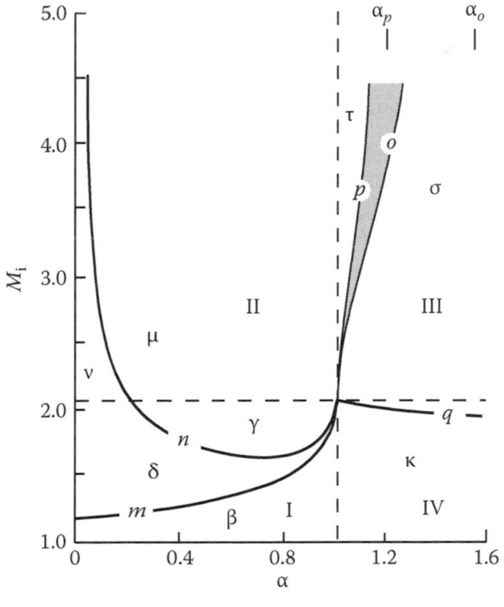


**FIGURE 4.15:** Parameter space $M_i$, $\alpha$ for shock interaction with an abrupt area change, $\gamma = 1.4$.

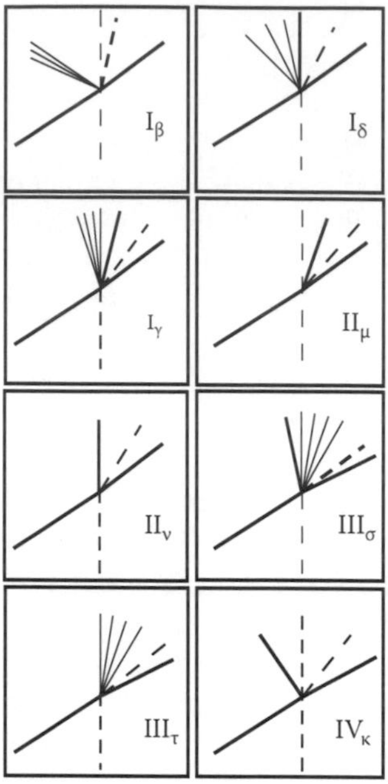


**FIGURE 4.16:** Wave patterns resulting from the interaction of a shock and abrupt area change.

*Consider the second quadrant*, $M_i \geq M_i^*$ *and* $\alpha \leq 1$. In this quadrant, the flow behind the incident shock is supersonic. For area ratios to the right of line $n$, the wave pattern is of type II$\mu$, Figure 4.16. In the region between the incident shock and the beginning of the area jump, $x=0^-$, the Mach number is bounded by 1.890 for $\gamma=1.4$. This Mach number limitation does not apply to the flow to the right of the area divergence. Here very high Mach numbers can be achieved by decreasing the area ratio, but keeping it to the right of line $n$. If the area ratio lies to the left of line $n$, the secondary shock becomes a standing shock, wave pattern II$\nu$, Figure 4.16. The wave pattern is similar to that of type I$\delta$, except that the flow behind the incident shock is supersonic and there is no reflected rarefaction wave.

*Consider the third quadrant*, $M_i \geq M_i^*$ *and* $\alpha \geq 1$. If the area ratio is large, wave pattern III$\sigma$ occurs, Figure 4.16. In this region, the reflected wave is a shock. The subsonic flow behind the reflected shock is accelerated to sonic conditions by the area convergence. The flow is then further accelerated by the rarefaction. In general, the transmitted shock is stronger than the incident shock. If we decrease the area ratio while holding $M_i$ fixed, we reach curve $o$ of Figure 4.15. The wave pattern remains of type III$\sigma$. If we continue to decrease the area ratio, we reach curve $p$. Now the reflected shock becomes a standing shock. This is the limiting case of pattern III$\sigma$. Curve $p$ consists of the locus of points for which the reflected shock is a standing shock. As $M_i \to \infty$, curve $p$ approaches

$$\alpha_p = 1/\left(\sqrt{\gamma-1}[\gamma(3-\gamma)/2]^{1/(\gamma-1)}\right).$$

For $\gamma = 1.4$, $\alpha_p = 1.193$. If the area ratio is just slightly greater than one, then we have pattern type III$\tau$, Figure 4.16. Under these conditions, the flow reaches the area jump at supersonic speed. The area contraction compresses the flow isentropically, but not enough to make the flow subsonic. Once within the small chamber, the flow is accelerated by the rarefaction wave. If we hold $M_\mathrm{i}$ fixed and increase the area ratio, we reach curve $p$. The wave pattern remains of type III$\tau$. If we further increase the area ratio, we reach curve $o$. Now the head of the expansion running downstream in the small chamber is sonic, but the pattern remains that of III$\tau$. Curve $o$ represents the locus of points for which the area ration produces sonic conditions after the area jump. As $M_\mathrm{i} \to \infty$, curve $o$ approaches

$$\alpha_o = \sqrt{\delta - 1}/(\gamma/2)^{1/(\gamma-1)}.$$

For $\gamma = 1.4$, $\alpha_o = 1.543$. The region between curves $o$ and $p$, the shaded region of Figure 4.15, is the *region of ambiguity* where wave patterns III$\tau$ and III$\sigma$ coexist. In addition, a third pattern with a standing shock within the area contraction and an expansion running downstream is also a valid solution of the self-similar model in this region. However, it is well known that a standing shock in a region where the area is contracting is unstable.

The entropy of an infinitesimal element of mass is $S\rho A\mathrm{d}x$. If we integrate between $x = -\infty$ and $x = \infty$ at a fixed time we get the total entropy in the channel. The entropy production in an interval of time $\Delta t$ is, therefore, given by

$$\chi = \int_{-\infty}^{\infty} [\rho(\Delta t)S(\Delta t) - \rho(0)S(0)]A\mathrm{d}x. \tag{4.48}$$

Equation 4.48 is easily integrated in closed form. For wave pattern III$\sigma$, Figure 4.14, we get

$$\chi/\Delta t = (\rho_3 S_3 - \rho_7 S_7)w_\mathrm{r} + \frac{S_5}{\alpha}\int_0^{\lambda_5^+} \rho\mathrm{d}\lambda + \frac{1}{\alpha}[\rho_5 S_5 a_5 + \rho_4 S_4(w_\mathrm{t} - u_4)]. \tag{4.49}$$

Across the expansion fan, the density is a function of $\lambda^+$ and is given by

$$\rho(\lambda^+) = \left[\left(\frac{\gamma - 1}{\gamma + 1}\right)^2 \frac{\exp(-S_5)}{\gamma}\right]^{1/(\gamma-1)} (Q_5 + \lambda^+)^{2/(\gamma-1)},$$

therefore, the integral appearing in (4.49) integrates to

$$\int_0^{\lambda_5^+} \rho\mathrm{d}\lambda = \frac{\gamma-1}{\gamma+1}\left[\left(\frac{\gamma - 1}{\gamma + 1}\right)^2 \frac{\exp(-S_5)}{\gamma}\right]^{1/(\gamma-1)} \times \left[(Q_5 + \lambda_5^+)^{(\gamma+1)/(\gamma-1)} - Q_5^{(\gamma+1)/(\gamma-1)}\right].$$

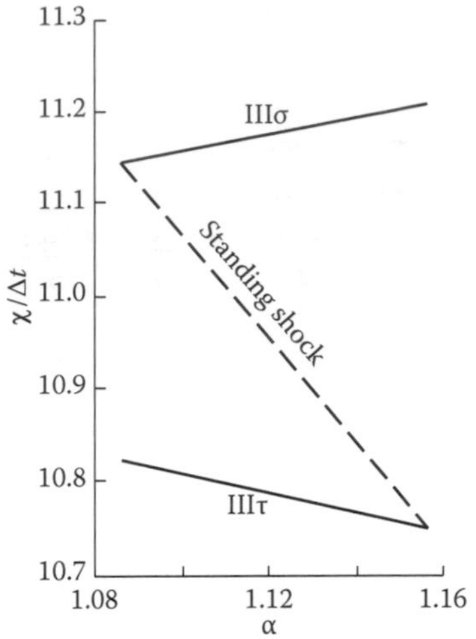


**FIGURE 4.17:** Entropy production in region of ambiguity, $M_\mathrm{i} = 3.5$, $\gamma = 1.4$.

Similar results are obtained for the other two wave patterns. The results are plotted in Figure 4.17. The *principle of minimum entropy production* states: when multiple stationary states satisfy the same governing equations and boundary conditions, the most stable stationary state is that that produces the minimum entropy. Therefore, we should expect to see wave pattern IIIτ σ if we solve the problem as a time dependent problem with the abrupt area jump replaced by a rapid continuous variation in area. This is indeed what happens.

If we consider the wave patterns along curves $o$ and $m$, Figures 4.15 and 4.16, we see that the patterns are very similar. Curves $o$ extends into the first quadrant as curve $m$. The same is true of curves $p$ and $n$.

Consider the fourth quadrant, $M_\mathrm{i} \leq M_\mathrm{i}^*$ and $\alpha \geq 1$. Here we find wave pattern IVκ, Figure 4.16. At low Mach numbers, the salient features are a reflected shock moving into the subsonic flow behind the incident shock and an isentropic acceleration of the flow entering the area contraction not sufficiently strong to generate supersonic flow in the small chamber. If we increase $M_\mathrm{i}$, we reach curve $q$. At this point the flow in the small chamber reaches sonic conditions. Curve $q$ is the locus of points separating patterns IIIσ and IVκ. If $\alpha \to \infty$, curve $q$ approaches a value of 1.718 for $\gamma = 1.4$.

The interaction of a shock with an abrupt area change should be treated as the interaction with a rapid continuous area variation in `Q1D.M`.

### 4.4.4 Quasi-One-Dimensional Flows: Numerical Examples

It will be easy to compile a book of quasi-one-dimensional flow examples. We will be frugal. There are a dozen built-in quasi-one-dimensional cases in `Q1D.M`. The cases are listed in Table 4.3. The first two, cases 17 and 18, show the interaction of a shock with a rapid area variation, see discussion in previous section. Case 19 is

**TABLE 4.3:** Built-in quasi-one-dimensional test cases available in the code `Q1D.M`.

| Case | Description |
|---|---|
| 17 | Shock moving right in variable area duct, contracting area duct, $M_i = 3.5$, $\alpha = 1.12$, see Section 4.4.3 |
| 18 | Shock moving right in variable area duct, expanding area duct, $M_i = \sqrt{2}$, $\alpha = 1/3$, see Section 4.4.3 |
| 19 | Laval nozzle with no shocks, $A(x) = x + 1/x$ |
| 20 | Laval nozzle with shock, $A(x) = x + 1/x$, $p_{\text{exit}} = 0.7735$ |
| 21 | Laval nozzle, nozzle start from quiescent conditions, $A(x) = x + 1/x$, $p_{\text{exit}} = 0.7735$ |
| 22 | Laval nozzle surge: explores the effect of exit plane "overpressure," $A(x) = x + 1/x$ |
| 23 | Oscillating Laval nozzle: $A(x,t) = x + 1/(x + \varepsilon^2 \sin^2(\omega t))$ |
| 24 | Oscillating Laval nozzle: $A(x,t) = x + \varepsilon \sin(\omega t) + 1/x$ |
| 25 | Oscillating Laval nozzle: $A(x,t) = x(1 + \varepsilon \sin(\omega t)) + 1/x$ |
| 26 | Oscillating Laval nozzle: $A(x,t) = x + 1/x + e^{-\delta(x-1)^2} \varepsilon \sin(\omega t)$ |
| 27 | Double throat nozzle, perturbation of exact initial solution with shock |
| 28 | Double throat nozzle, nozzle start from quiescent conditions |

to show that the code maintains (to within numerical accuracy) the exact solution to a steady Laval nozzle. Case 20 is similar to the previous case, except that now we have a steady solution with a shock. Case 21 is more interesting. The Laval nozzle starts with quiescent flow, and then the pressure at the exit plane is lowered abruptly from its stagnation value to the value corresponding to case 20. This simulates the sudden opening of a valve connecting the Laval nozzle to a lower pressure chamber. As the flow evolves, the throat chokes and eventually a shock forms and comes to rest at the location predicted by the steady solution.

Case 22 is a simulation of a phenomenon known as *compressor surge* [140]. Imagine the Laval nozzle to be an inlet followed by a compressor at the exit plane. We are interested on the effects on the nozzle flow created by a sudden pressure surge in the compressor. Let the area of the nozzle be given by

$$A(x) = x + 1/x, \quad 0 < x \leq 2, \tag{4.50}$$

with the compressor located at $x = 2$. At $t = 0$ we have in the nozzle a steady state solution with a shock at $x = 1.5$. This shock location corresponds to an exit pressure, $p_{\text{exit}}$, equal to 0.7735. For $t > 0$, the exit pressure is given by

$$p(2,t) = p_{\text{exit}} + \varepsilon \sin(t), \quad t < 1,$$
$$p(2,t) = p_{\text{exit}}, \quad t \geq 1,$$

where $\varepsilon$ is a parameter used to study the effect of the overpressure amplitude. Results for four values of $\varepsilon$ are shown in Figure 4.18. For $\varepsilon = 0.300$, panel (a) of Figure 4.18,

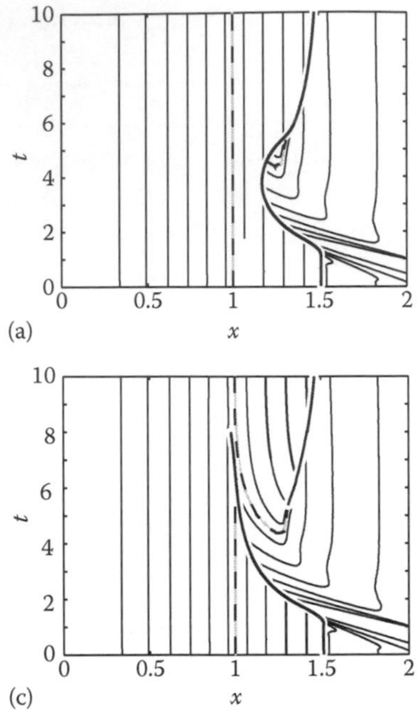

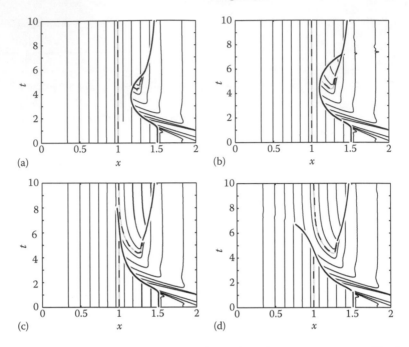

**FIGURE 4.18:** Case 22, Laval nozzle simulation of compressor surge. Shocks are shown as heavy black lines, sonic lines shown as dashed lines. Thin lines are constant pressure contours. Panel (a) corresponds to $\varepsilon = 0.300$, panel (b) corresponds to $\varepsilon = 0.325$, panel (c) corresponds to $\varepsilon = 0.340$, and panel (d) corresponds to $\varepsilon = 0.350$.

the pressure pulse deflects the *standing shock* from its steady state position, but eventually the shock returns to its original position. A slight increase in amplitude to $\varepsilon = 0.325$ sends the standing shock closer to the throat, but it begins to recover before reaching the throat, see panel (b). Sonic conditions develop behind the standing shock as it approaches the throat. The supersonic flow behind the standing shock is accompanied by compression waves which form a secondary shock. The standing shock and the secondary coalesce. The contact surface from this interaction is very weak and is eliminated. Eventually, the remaining shock returns to the steady state position. A further increase in amplitude, to 0.340, pushes the standing shock into the throat, panel (c). As in the previous case, supersonic flow develops behind the standing shock, and compression waves in this region form a secondary shock. However, this time the standing shock becomes too weak and is eliminated. The sonic line that had developed behind the standing shock moves to the throat and chokes the nozzle. The secondary shock slowly drifts to its steady state position. A further increase in amplitude to 0.350 pushes the standing shock further into the convergent section of the nozzle, panel (d). As the shock penetrates this subsonic region it weakens and is eliminated. The rest of the wave dynamics that ensues is similar to that in panel (c). The complexity and beauty introduced by the nonlinearity of the equations is captivating.

*Here is a second warning*. When dealing with nonlinear phenomena do not trust your intuition. Intuition is a result of a subconscious integration of past experiences that showed similar behavior. It sometimes works when the behavior is linear. It usually fails with nonlinear phenomena.

Cases 23 through 26 are studies of flows generated by forcing the nozzle wall to oscillate. For small wall deflections, asymptotic small perturbation methods [1,141] provide a good mathematical background for the understanding of the flow dynamics. For large deflections, these methods are less practical. In cases 23 through 26, we use a forcing function defined by $f(t) = \varepsilon \sin(\omega t)$, where $\omega = \pi/4$ and the amplitude $\varepsilon$ is a small free parameter. For cases 23a and case 23b the area is given by

$$A(x,t) = x + 1/(x + f^2(t)).$$

Case 23a corresponds to $\varepsilon^2 = 0.12$ and case 23b corresponds to $\varepsilon^2 = 0.20$. Figure 4.19 shows how $f(t)$ changes the area in time. At $t = 0$ the area is given by the solid line. The change is primarily a back and forth sliding of the shape along the $x$-axis.

The results for case 23a are shown in Figure 4.20. As shown in the figure, a repeating cycle is established after a short transient. The cycle repeats every four time units. During the beginning of the cycle, the standing shock moves toward the throat, then toward the end of the cycle it moves back into the divergent section. As it moves back, a supersonic pocket is formed behind the shock. This is accompanied by the coalescence of compression waves which form a secondary shock of the same family as the original shock. The two shocks soon coalesce, terminating the supersonic pocket, and creating a contact surface that moves toward the exit plane and exits the computational region. Note that the sonic line that originally sat at the throat has moved into the convergent section and is oscillating about some point where $A_x < 0$. (Equation 4.43 is only valid for steady flow).

For $\varepsilon^2 = 0.20$, case 23b, Figure 4.21, the behavior is quite different. Again, after a brief transient a flow cycle is established. The period is, as before, four time units.

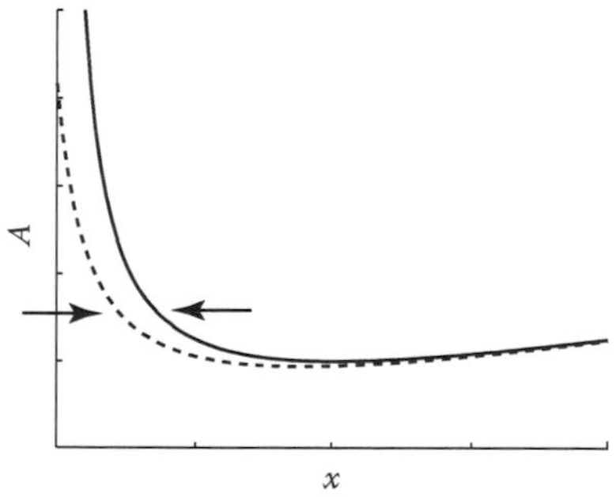


**FIGURE 4.19:** The figure indicates the direction of the nozzle wall oscillation. The solid line corresponds to the starting shape. The dashed line is the maximum deviation occurring at $t = 2$, 6, 10, etc., for $\varepsilon^2 = 0.12$. The oscillation is a horizontal displacement of the shape.

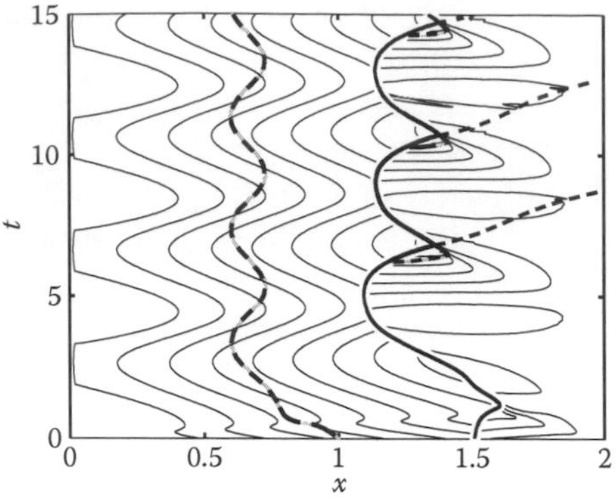


**FIGURE 4.20:** Case 23a, oscillating Laval nozzle, $\varepsilon^2 = 0.12$. Shocks are shown as heavy black lines, sonic lines shown as dashed lines. Dashed lines originating at shock interceptions and moving toward the nozzle exit correspond to contact lines. Thin lines are pressure contours.

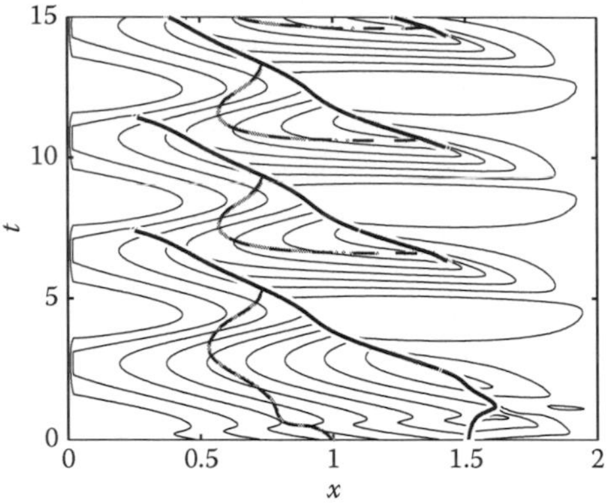


**FIGURE 4.21:** Case 23b, oscillating Laval nozzle, $\varepsilon^2 = 0.20$, shocks are shown as heavy black lines, sonic lines shown as dashed lines. Thin lines are pressure contours.

However, now the standing shock is pushed way into the convergent section of the nozzle where it weakens. Compression waves in the neighborhood of the original location of the standing shock coalesce and form a new shock which is in turn pushed into the convergent section. The pattern repeats.

Case 24 corresponds to $A(x, t) = x + 1/x + f(t)$ with $\varepsilon = 0.5$ and $\omega = \pi/4$. As shown in Figure 4.22, the primary effect of the oscillation is to displace the area shape up and down. The resulting flow field is shown in Figure 4.23. The flow characteristics are somewhat similar to that of case 23a. The period is now eight time units and the sonic line oscillates about the throat. As part of the cycle, the sonic line is pinched by the primary shock.

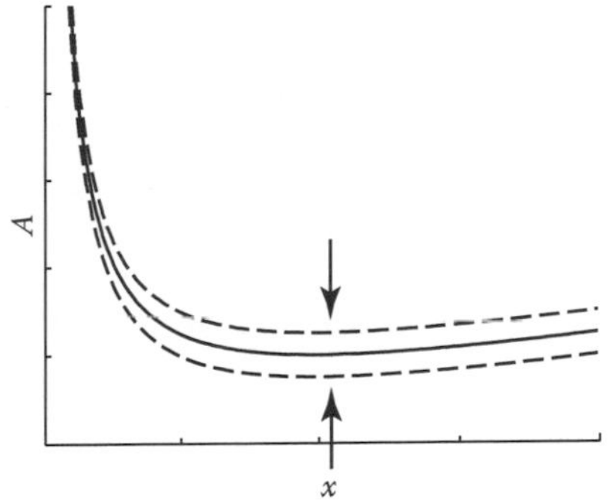


**FIGURE 4.22:** The figure indicates the direction of the nozzle wall oscillation. The solid line corresponds to the starting shape. The dashed lines are the maximum deviations occurring at $t = 2$, 6, 10, etc. The oscillation is a vertical displacement of the starting shape.

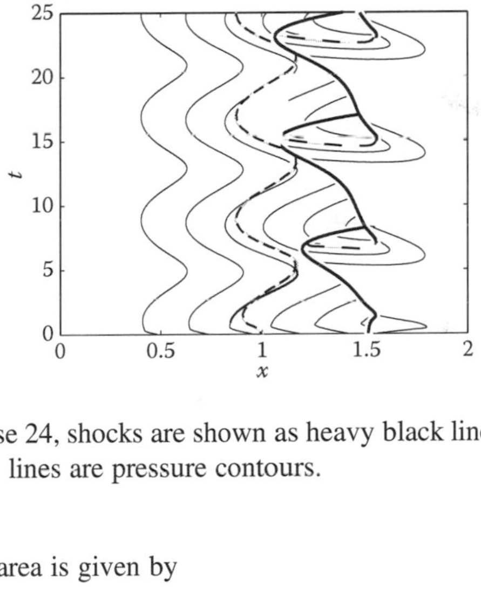


**FIGURE 4.23:** Case 24, shocks are shown as heavy black lines, sonic lines shown as dashed lines. Thin lines are pressure contours.

For case 25, the area is given by

$$A(x, t) = x(1 + f(t)) + 1/x,$$

with $\varepsilon = 0.10$ and $\omega = \pi/4$. The oscillation creates a flapping like motion with the maximum perturbations taking place at the exit plane, see Figure 4.24. The resulting flow field is shown in Figure 4.25. It is clear from the figure that subharmonic frequencies are present. The period of the main cycle is eight time units. A subharmonic with a period of 3/4 time units is imposed on the main cycle. The sonic line oscillates about the throat and the standing shock oscillates about its original position. The cycle is also characterized by the formation of two shocks which interact with the standing shock. The second of these two shocks forms at 3/4 time units from the formation of the first, but the appearance of the first is tied to the main cycle.

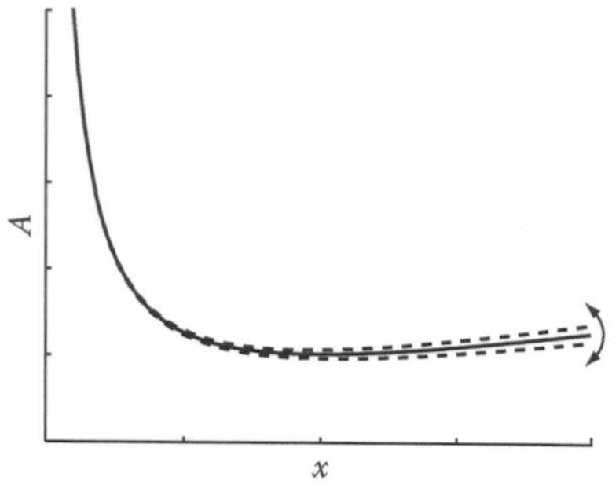


**FIGURE 4.24:** The figure indicates the direction of the nozzle wall oscillation. The solid line corresponds to the starting shape. The dashed lines are the maximum deviations occurring at $t = 2, 6, 10$, etc. The oscillation creates the effect of a flapping motion with maximum deviation at the exit plane.

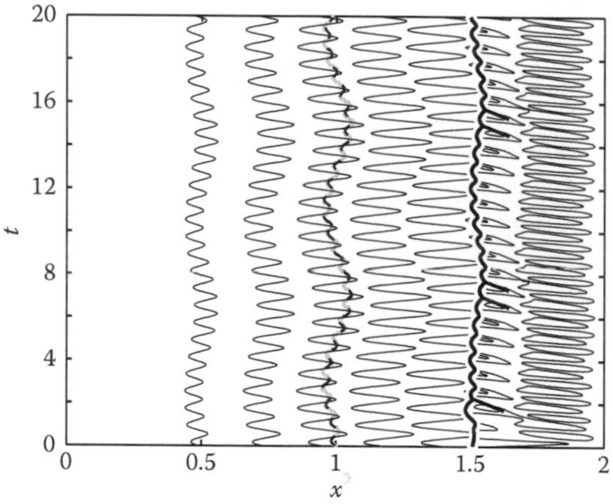


**FIGURE 4.25:** Case 25, shocks are shown as heavy black lines, sonic lines shown as dashed lines. Thin lines are pressure contours.

The final case in this series, case 26, corresponds to an area defined by

$$A(x,t) = x + 1/x + e^{-30(x-1)^2} f(t),$$

with $\varepsilon = 0.20$ and $\omega = \pi/4$. As shown in Figure 4.26, this corresponds to a pulse concentrated at the throat. The resulting flow field is shown in Figure 4.27. The pulse makes the original standing shock oscillate just ahead of its steady state location. It also pushes the sonic line into the convergent section; a secondary shock is then formed in the neighborhood of the throat and is pushed into the convergent section where it weakens. The cycle repeats as a dance involving a tight coupling of the sonic line and the secondary shock.

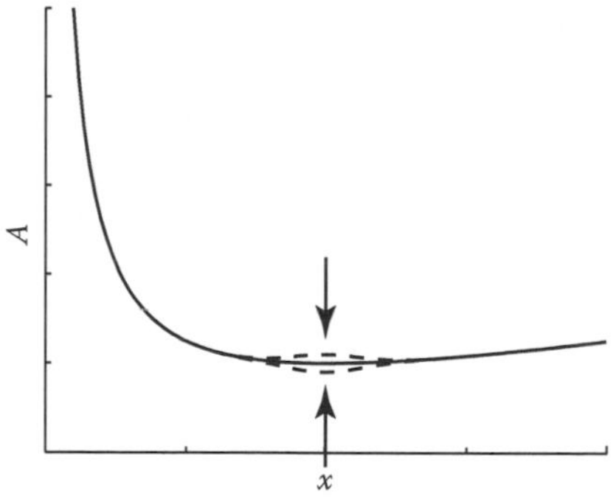


**FIGURE 4.26:** The figure indicates the direction of the nozzle wall oscillation. The solid line corresponds to the starting shape. The dashed lines are the maximum deviations occurring at $t = 2, 6, 10$, etc. The oscillation is a vertical displacement of the starting shape concentrated at $x = 1$ and quickly decaying for $|x - 1| > 0$.

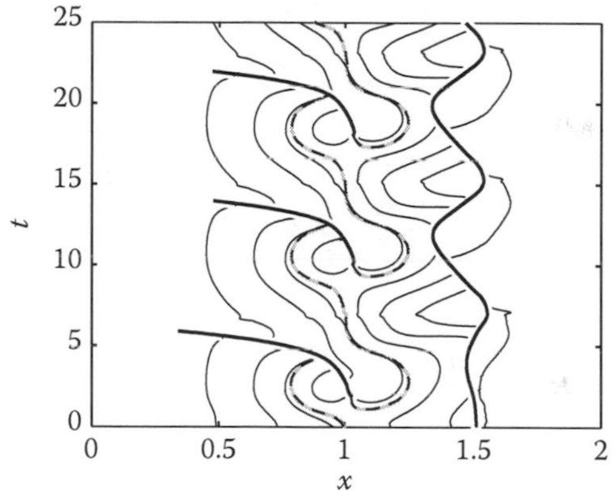


**FIGURE 4.27:** Case 26, shocks are shown as heavy black lines, sonic lines shown as dashed lines. Thin lines are pressure contours.

The last two cases study the stability of shocks in dual-throat nozzles [130,192,197]. For case 27, the dual-throat area is defined by

$$A(x) = x + \frac{1}{x}, \quad x \leq 2,$$

$$A(x) = 13 - \frac{45x}{4} + 3x^2, \quad 2 < x < 2.7964,$$

$$A(x) = 5, \quad 2.7964 \leq x \leq 4.2124,$$

$$A(x) = x - 3.92 + \frac{1}{x - 4}, \quad x > 4.2124.$$

The initial conditions for this case correspond to a steady flow with a standing shock at $x = 1.5$. This is consistent with an exit pressure of 0.79005 at $x = 6$. The first throat occurs at $x = 1$, with an area equal to 2. The resulting throat area to choke the flow after the shock is 2.0591. The actual area of the second throat is 2.0800, hence the flow remains subsonic at the second throat. At $t = 0$, a pressure pulse is sent into the nozzle from the exit plane. The pressure pulse is defined by

$$p(6,t) = p_{\text{exit}} + \varepsilon \sin(t), \quad 0 < t \leq 1,$$

$$p(6,t) = p_{\text{exit}}, \quad t > 1,$$

where

$\varepsilon = 0.20$

$p_{\text{exit}} = 0.79005$

The pressure pulse evolves into a shock that traverses the nozzle and interacts with the standing shock, see Figure 4.28. Note that even before the shock that was generated by the pressure pulse reaches the standing shock, the standing shock had already started to move. As the flow field evolves, the second throat chokes, and a new shock forms after the second throat. The shock resulting from the interaction of the original standing shock and the shock from the pressure pulse settles at a new position closer to the first throat.

Since it appears that we have two solutions that satisfy the steady state equations, we must ask which is consistent with the minimum entropy production principle. The entropy production, measured at the exit plane, is given by $\rho SA$, but, since the

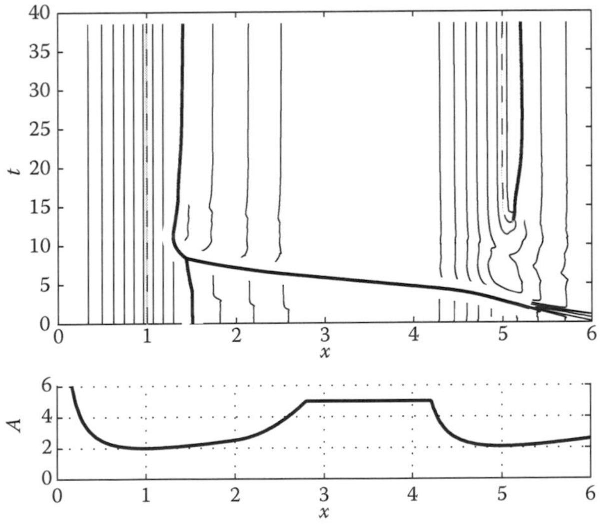


**FIGURE 4.28:** Case 27, corresponding to a dual-throat nozzle. Initial conditions are perturbed by a pressure pulse originating at the exit plane. Heavy solid lines correspond to shock paths, dashed lines correspond to sonic lines, and thin lines are constant pressure lines.

area is the same for both solutions, we omit the area. For the steady flow with a single shock at $t = 0$, we have $\rho S = .0098$ measured at the exit plane. For the near-steady flow with two shocks at $t = 39$, the entropy production is $\rho S = 0.0056$, measured at the exit plane. Thus, the stationary solution with two shocks is the stable solution and is consistent with the minimum entropy production principle.

Case 28 corresponds also to a dual-throat nozzle. In this case the nozzle is defined by

$$A(x) = x + \frac{1}{x}, \quad x \leq 1,$$
$$A(x) = 2 + 8[(x - 1)(x - 2)]^2, \quad 1 < x < 2,$$
$$A(x) = x - 1 + \frac{1}{x - 1}, \quad x \geq 2.$$

This nozzle geometry has throats at $x = 1$ and $x = 2$ and at both locations the area is equal to 2. The initial conditions correspond to a gas at rest.

If we assume the flow to be steady and if we have a shock in the divergent section after the first throat, then from (4.45) it follows that the flow would become supersonic before reaching the second throat. Thus, it would appear that in this case the flow cannot be steady. It is, however, consistent with the steady solution to have a shock in the divergent section after the second throat with both throats choked. Is this configuration stable? Is there another stable solution? To answer these questions, we simulate a nozzle start by setting the exit pressure to 0.7735 at $t = 0$ and holding it at that value for the rest of the simulation. This value of the exit pressure is consistent with a steady shock at $x = 2.5$. The resulting flow field is shown in Figure 4.29. At about $t \approx 20$, both throats choke. A shock forms near the first throat, but is pushed into the first divergent section where it becomes weak and is eliminated. A steady shock forms near $x = 2.5$. A second shock forms in the divergent region near the first throat. This shock oscillates near the first throat. During the oscillation, as the shock comes close to the throat, the sonic line at the throat disappears, and then it reappears as the shock moves away from the throat.

This result raises some questions. Are these self-excited oscillations stable, or will they eventually damp out? After running the code to $t = 150$, the answer to this question appears to be that the oscillations do not damp out. Is the entropy production of this case smaller than that of a steady solution with a standing shock satisfying the same exit pressure (i.e., with a shock at $x = 2.5$)? As shown in Figure 4.30, the answer to the latter is no. Is this an instance where the minimum entropy production principle fails? Perhaps the more fundamental question is: Can we apply the principle to rule on the stability of a stationary field vs. a nonstationary field? We let the reader ponder on that question. If at $t = 50$ we lower the exit pressure to 0.72 and let the flow evolve, we find that the shock near the second throat moves closer to the exit plane without disturbing the flow between the two throats, as required by the domain of dependence rule. If the oscillating solution found between the two throats depends on how the nozzle is started, then this means that there are an infinite number of solutions for cases with the same exit pressure. This is not the case. The oscillating solution that we find, for $t$ large, between the two throats is independent of the exit pressure or how the exit pressure is imposed, as shown in Figure 4.31. In this

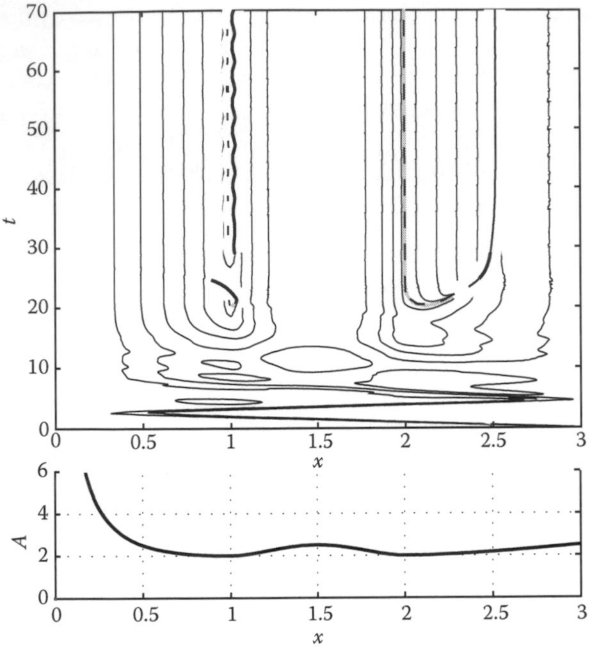


**FIGURE 4.29:** Case 28, simulation of nozzle start for a dual-throat nozzle. Heavy solid lines correspond to shock paths, dashed lines correspond to sonic lines, and thin lines are constant pressure lines.

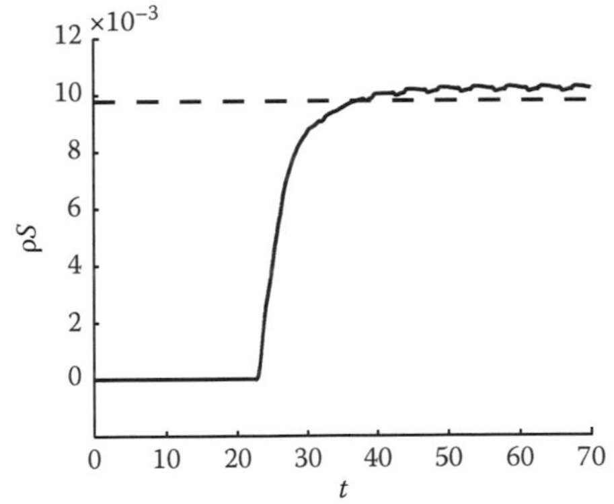


**FIGURE 4.30:** Entropy production measured at the exit plane for the oscillating flow, solid line, and for a steady flow with a shock at $x = 2.5$, dashed line.

figure we have plotted the location of the oscillating shock vs. time for case 28 (exit pressure 0.7735) and for a similar case where the exit pressure is set to 0.72 at $t = 0$. The paths of the oscillating shock are the same to within numerical accuracy. These problems with dual throats that have the same area are very sensitive to conditions in

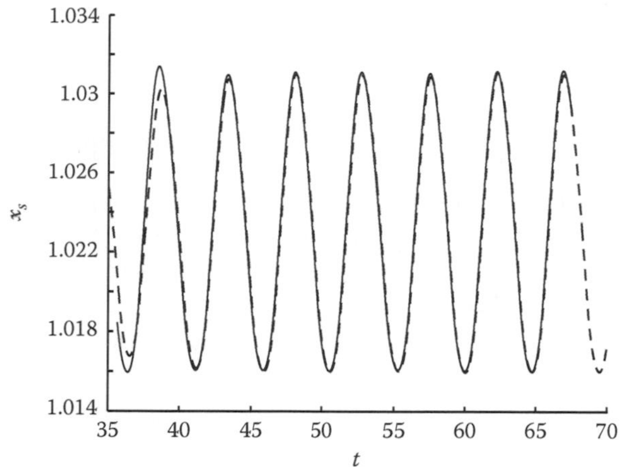


**FIGURE 4.31:** Shock position for a case with exit pressure equal to 0.72, solid line, and for a case with exit pressure equal to 0.7735, dashed line. The solid line curve has been shifted in time to line up with the dashed curve.

the neighborhood of the throats and small discretization errors can lead to large global changes (see problem 4.7 below).

Wave diagrams, such as Figure 4.28 are very helpful tools for visualizing the flow field. Another useful tool is a *movie-like* depiction of the shock motion in the nozzle. Such a tool is the code `MOVIE_MAKER.M`. To create a movie, first run `Q1D.M` with the input parameter `movie` set to 1. This will create a data file called `TEMPORARY.M` which is read by `MOVIE_MAKER.M`. When `Q1D.M` runs with the input parameter `movie` set to 1, most outputs are suppressed in order to make the run more efficient. Not all cases are good candidates for a movie. `MOVIE_MAKER.M` is set up to make movies for the following default cases: Case 1, cases 14 through 18, and cases 21 through 28. A typical film clip is shown in Figure 4.32.

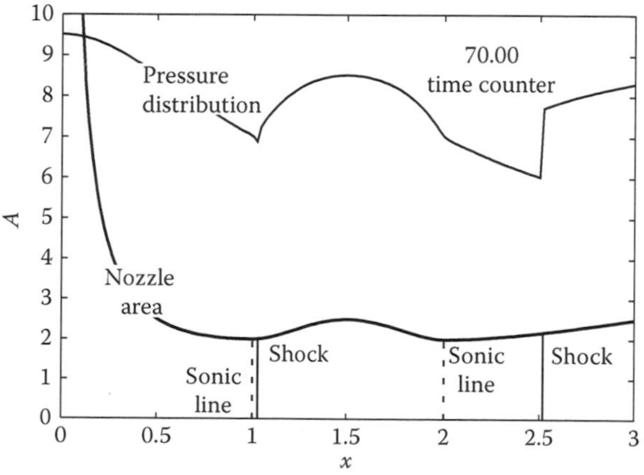


**FIGURE 4.32:** Typical information provided in a film clip.

## Problems

**4.1** Find the stability constrain of the λ-scheme for the linear equation $u_t + \lambda u_x = 0$ using von Neumann's method.

**4.2** Find the stability constrain of the λ-G scheme for the linear equation $u_t + \lambda u_x = 0$ using von Neumann's method.

**4.3** Derive the conservation of energy equation for quasi-one-dimensional flow. Then show that $S_t + uS_x = 0$ is valid.

**4.4** Let ψ be the mass flow escaping through a porous wall per unit length. Derive the continuity equation for a quasi-one-dimensional flow with porous walls.

**4.5** With reference to case 26, discussed above, suppose we center the pulse at $x_l$, i.e.,

$$A(x,t) = x + \frac{1}{x} + e^{-30(x-x_l)^2} f(t),$$

where $f(t) = \varepsilon \sin(\omega t)$, $\omega = \pi/4$, and $\varepsilon = 0.20$, and suppose we place a microphone at the throat ($x = 1$) and record the pressure from $t = 0$ to $t = 25$. What will be the maximum pressure recorded as a function of the pulse location $x_l$?
The answer is show in Figure 4.33, The maximum pressure is shown as a heavy solid line. There are four peaks, labeled *a*, *b*, *c*, *d* in the figure. The pressure at $x = 1$, averaged over $t = 0$ to $t = 25$, is shown as the dashed line. If the pulse is placed to the right of point *e*, its perturbation doesn't reach the throat. (See CHAP_4/Q1D/ prob.4.5 in the companion CD).
Repeat this exercise, but find the maximum pressure anywhere along $0.8 \leq x \leq 1.2$.

**4.6** Study the shock oscillation of case 28 in the frequency domain. (See CHAP_4/Q1D/ Problem 4.6 in the companion CD).

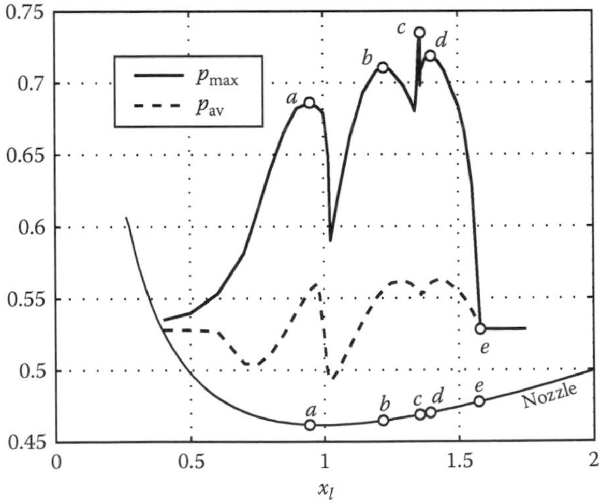


**FIGURE 4.33:** Maximum pressure and average pressure measured at $x = 1$ as the pulse location $x_l$ is varied.

**4.7** Consider the following three dual-throat nozzles:
Case (a)

$$A(x) = \frac{1}{x} - \frac{1}{x-3} + 1.16667e^{-2.8(x-1.5)^2} + 0.00240269,$$
$$0 < x \leq 0.8289,$$
$$A(x) = 2 + 2.479101[(x-0.8289)(x-2.1708)]^2,$$
$$0.8289 < x < 2.1708,$$
$$A(x) = \delta(x-0.1708) + \frac{1}{x-1.1708} + 1 - \delta,$$
$$2.1708 \leq x \leq 3,$$

Case (b)

$$A(x) = \frac{1}{x} - \frac{1}{x-3} + 1.16667e^{-2.8(x-1.5)^2} + 0.00240269,$$
$$0 < x < 2.1708,$$
$$A(x) = \delta(x-0.1708) + \frac{1}{x-1.1708} + 1 - \delta,$$
$$2.1708 \leq x \leq 3,$$

and Case (c)

$$A(x) = \frac{1}{x} - \frac{1}{x-3} + 1.16667e^{-2.8(x-1.5)^2} + 0.00240269,$$
$$0 < x < 0.8289,$$
$$A(x) = 2 + 2.479101(x-0.8289)(2.1708-x),$$
$$0.8289 \leq x < 2.1708,$$
$$A(x) = \delta(x-0.1708) + \frac{1}{x-1.1708} + 1 - \delta,$$
$$2.1708 \leq x \leq 3,$$

where $\delta$ is a free parameter. For $\delta = 1$ and $\delta = 2$, compare the nozzle areas for all three cases. Then compute the flow for each nozzle starting from a gas at rest. Let the exit plane be at $x = 3$, try both $\delta = 1$ and $\delta = 2$, and let the pressure at the exit plane be given by

$$p(3,t) = \frac{(1+p_{\text{exit}})}{2} + \frac{(1-p_{\text{exit}})}{2}\cos(\pi t),\ 0 < t \leq 1,$$
$$p(3,t) = p_{\text{exit}}, \quad t > 1,$$

where $p_{\text{exit}} = 0.7735$. Use 150 mesh intervals and run the calculation to $t = 70$.

**4.8** A simple model to study the coupling of an elastic nozzle wall with the flow in the nozzle can be achieved by assuming that the wall deformations are small and the motion is governed by the linear beam equation

$$mA_{tt} + DA_{xxxx} = \Delta p$$

where

$m$ is the linear mass of the wall
$D$ is the bending stiffness
$\Delta p$ is the local pressure difference across the wall

Write the quasi-one-dimensional equations that couple to the beam equation and study the small perturbations of the resulting system of equations. (See [105]).

# Chapter 5

# Euler Equations, Two-Dimensional Problems

I started working on the blunt body problem because I did not like any of the results published in journals and reports. What counted was the idea of putting aside, once and for all, the myth of the steady equations and of replacing the search for a "solution" with the description of an evolution which, in that case, happened to converge to a steady solution.

Gino Moretti [153].

## 5.1 The Blunt Body Problem

Blunt body flows came under intensive study during the mid-1950s to mid-1960s because of the interest at that time in the United States and the Soviet Union to develop atmospheric reentry vehicles. The heating rate on a spherical body is inversely proportional to the square root of the radius of curvature, so blunt bodies with large nose radii were used to mitigate the intensive heating occurring at reentry speeds. In the mid-1960s, Rusanov [186] in the Soviet Union and Moretti [154] in the United States independently developed time asymptotic methods to obtain high-quality solutions for these types of flows. The problem is now considered a classic in computational fluid dynamics and continues to draw the attention of researchers who use it as a way to bench mark new methods. Unfortunately, it is not uncommon to see in the current literature solutions whose qualities are questionable.

### 5.1.1 Cylindrical and Spherical Coordinates

The geometry of the body lends itself to treatment by either cylindrical or spherical coordinates.

For cylindrical coordinates, let $\theta$ be the polar angle measured from the $x$-axis as shown in Figure 5.1. In terms of the notation introduced in Section 3.4, for cylindrical coordinates we have

$$\varsigma = r,\ \eta = \theta,\ \xi = z,$$

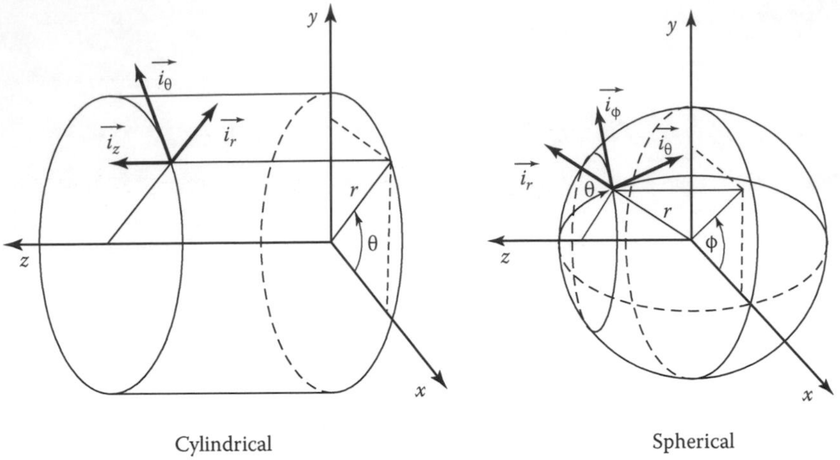


**FIGURE 5.1:** Cylindrical and spherical coordinates.

where $r$, $\theta$, and $z$ are defined by

$$x = r\cos\,\theta,\ y = r\sin\,\theta,\ z = z,$$

or, equivalently

$$r = \sqrt{x^2 + y^2},\ \theta = \arctan\,(y/x),\ z = z.$$

The metric coefficients are (see Section 3.4)

$$h_r = 1,\ h_\theta = 1/r,\ h_z = 1.$$

The gradient, divergence, and curl operators are, respectively,

$$\begin{aligned} \nabla\phi &= \vec{i}_r\frac{\partial\phi}{\partial r} + \vec{i}_\theta\frac{1}{r}\frac{\partial\phi}{\partial\theta} + \vec{i}_z\frac{\partial\phi}{\partial z}, \\ \nabla\boldsymbol{v} &= \frac{1}{r}\frac{\partial ru}{\partial r} + \frac{1}{r}\frac{\partial v}{\partial\theta} + \frac{\partial w}{\partial z}, \quad \text{and} \\ \nabla\times\boldsymbol{v} &= \vec{i}_r\left(\frac{1}{r}\frac{\partial w}{\partial\theta} - \frac{\partial v}{\partial z}\right) + \vec{i}_\theta\left(\frac{\partial u}{\partial z} - \frac{\partial w}{\partial r}\right) + \vec{i}_z\frac{1}{r}\left(\frac{\partial rv}{\partial r} - \frac{\partial u}{\partial\theta}\right), \end{aligned} \tag{5.1}$$

where $\boldsymbol{v} = u\vec{i}_r + v\vec{i}_\theta + w\vec{i}_z$.

For spherical coordinates, define $\theta$ to be the azimuthal angle measured from the $z$-axis, and define $\phi$ to be the polar angle in the $x$–$y$ plane, measured from the $x$-axis, as shown in Figure 5.1, then following the notation of Section 3.4, for spherical coordinates we have

$$\varsigma = r,\ \eta = \theta,\ \xi = \phi,$$

where $r$, $\phi$, and $\theta$ are defined by

$$x = r\cos\phi\sin\theta,\ y = r\sin\phi\sin\theta,\ z = r\cos\theta,$$

or, equivalently

$$r = \sqrt{x^2 + y^2 + z^2},\ \theta = a\tan\left(\sqrt{x^2 + y^2}/z\right),\ \phi = a\tan(y/x).$$

The metric coefficients are

$$h_r = 1,\ h_\theta = 1/r,\ h_\phi = 1/r\sin\theta.$$

The gradient, divergence, and curl operators are, respectively,

$$\begin{aligned}
\nabla\phi &= \vec{i}_r\frac{\partial\phi}{\partial r} + \vec{i}_\theta\frac{1}{r}\frac{\partial\phi}{\partial\theta} + \vec{i}_\phi\frac{1}{r\sin\theta}\frac{\partial\phi}{\partial\phi},\\
\nabla\boldsymbol{v} &= \frac{1}{r^2}\frac{\partial r^2 u}{\partial r} + \frac{1}{r\sin\theta}\frac{\partial\sin\theta v}{\partial\theta} + \frac{1}{r\sin\theta}\frac{\partial w}{\partial\phi},\quad\text{and}\\
\nabla\times\boldsymbol{v} &= \vec{i}_r\frac{1}{r\sin\theta}\left(\frac{\partial\sin\theta w}{\partial\theta} - \frac{\partial v}{\partial\phi}\right) + \vec{i}_\theta\frac{1}{r}\left(\frac{1}{\sin\theta}\frac{\partial u}{\partial\phi} - \frac{\partial rw}{\partial r}\right)\\
&\quad + \vec{i}_\phi\frac{1}{r}\left(\frac{\partial rv}{\partial r} - \frac{\partial u}{\partial\theta}\right),
\end{aligned}\tag{5.2}$$

where $\boldsymbol{v} = u\vec{i}_r + v\vec{i}_\theta + w\vec{i}_\phi$.

### 5.1.2 Euler Equations in Cylindrical and Spherical Coordinates

With the vector relations (5.1) and (5.2) we can write the Euler equations in cylindrical and spherical coordinates:

$$\begin{aligned}
&P_t + uP_r + \frac{v}{r}P_\theta + \gamma\left(u_r + \frac{v_\theta}{r} + \vartheta\right) = 0,\\
&u_t + uu_r + \frac{v}{r}u_\theta - \frac{v^2}{r} + \Theta P_r = 0,\\
&v_t + uv_r + \frac{v}{r}v_\theta + \frac{uv}{r} + \frac{\Theta}{r}P_\theta = 0,\\
&S_t + uS_r + \frac{v}{r}S_\theta = 0,
\end{aligned}\tag{5.3}$$

where

$$\vartheta = \begin{cases} u/r, & \text{for cylindrical}\\ (2u + v\cot\theta)/r, & \text{for spherical.}\end{cases}$$

On the symmetry line we must take the limit:

$$\lim_{\theta\to\pi} v\cot\theta = v_\theta.$$

### 5.1.3 Computational Coordinates

We want to transform the governing equations from the physical frame ($r$, $\theta$, $t$) to a computational frame ($Z$, $Y$, $T$). We have two reasons for doing this. The first is that we want to have a uniform mesh spacing to simplify the numerical approximation of derivatives. The second is that we want to make the bow shock and the wall coincide with constant computational surfaces; the $Z=1$ surface for the bow shock, and the $Z=0$ surface for the wall. The transformation we will use is defined by

$$\begin{aligned} T &= t, & t &= T, \\ Z &= \frac{r - b(\theta)}{c(\theta, t) - b(\theta)}, & r &= b(\theta) + (c(\theta, t) - b(\theta))Z, \\ Y &= \pi - \theta, & \theta &= \pi - Y. \end{aligned} \tag{5.4}$$

In (5.4), $b(\theta)$ is the wall radius which is a function of $\theta$ and is assumed to be stationary, and $c(\theta, t)$ is the bow shock radius which is a function of both $\theta$ and $t$. We will call the component of the bow shock velocity along a radial line $w$, i.e., $\partial c/\partial t = w$. The first of (5.4) can be misleading. Do not take $T=t$ to mean that $T$ and $t$ are the same. Penrose [166, p. 189] says that Nick Woodhouse calls this "the second fundamental confusion of calculus." The two are far from being the same as we can see by looking at the derivatives:

$$\begin{aligned} \frac{\partial}{\partial t} &= \frac{\partial}{\partial T} + Z_t \frac{\partial}{\partial Z}, \\ \frac{\partial}{\partial r} &= Z_r \frac{\partial}{\partial Z}, \\ \frac{\partial}{\partial \theta} &= Z_\theta \frac{\partial}{\partial Z} + Y_\theta \frac{\partial}{\partial Y}. \end{aligned}$$

The vectors tangent to $t$ and $T$ point in different directions as long as the bow shock is not stationary.

In the computational plane, the governing equations are

$$\begin{aligned} & P_T + UP_Z + VP_Y + \gamma(Z_r u_Z + (Z_\theta v_Z + Y_\theta v_Y)/r + \vartheta) = 0, \\ & u_T + Uu_Z + Vu_Y - \frac{v^2}{r} + \Theta Z_r P_Z = 0, \\ & v_T + Uv_Z + Vv_Y + \frac{uv}{r} + \frac{\Theta}{r}(Z_\theta P_Z + Y_\theta P_Y) = 0, \\ & S_T + US_Z + VS_Y = 0, \end{aligned} \tag{5.5}$$

where the contravariant velocity components are defined by

$$\begin{aligned} U &= Z_t + uZ_r + \frac{v}{r} Z_\theta, \\ V &= \frac{v}{r} Y_\theta. \end{aligned}$$

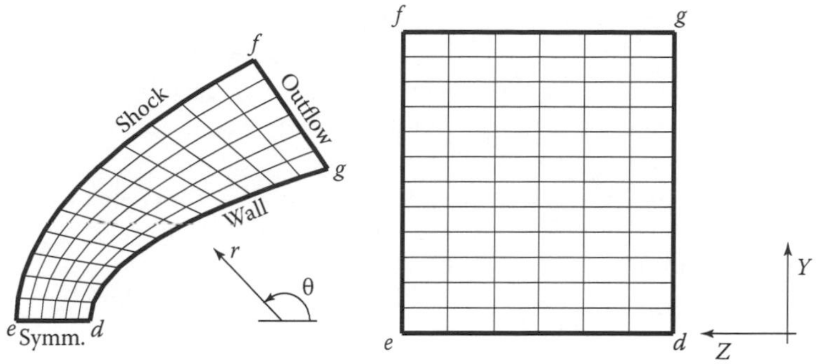


**FIGURE 5.2:** The physical plane is shown on the left panel; the computational plane is shown on the right panel.

Figure 5.2 illustrates the transformation of the physical plane to the computational plane. The symmetry line $\overline{ed}$, corresponding to $\theta = \pi$, is mapped to the line $Y = 0$. The bow shock $\overline{ef}$, corresponding to $r = c(\theta, t)$, is mapped to the line $Z = 1$. The outflow boundary $\overline{fg}$, corresponding to $\theta = \theta_{\max}$, is mapped to the line $Y = \pi - \theta_{\max}$. Finally, the wall $\overline{dg}$, corresponding to $r = b(\theta)$, is mapped to the line $Z = 0$.

### 5.1.4 Initial Conditions

The definition of initial conditions for the blunt body problem is a nontrivial exercise. We need to come up with a reasonable bow shock shape as well as providing reasonable initial values to the flow variables over the entire shock layer. Fortunately, the bow shock shape is not very sensitive to the blunt body shape. This is one reason why the inverse problem is so poorly behaved, see Section 1.3. Therefore, we start by assuming that the bow shock can be approximated by a parabolic profile,

$$c(\theta, 0) = \left(\cos\,\theta + \sqrt{\cos^2\theta + 4\mu c(\pi, 0)}\right)\Big/ 2\mu, \tag{5.6}$$

which we can adapt to a specific body shape and free stream Mach number by means of two free parameters: $c(\pi, 0)$ and $\mu$. The first parameter is the radius of the bow shock on the symmetry line which we write as

$$c(\pi, 0) = b(\pi) + \Delta,$$

where $\Delta$ is the bow shock standoff distance. We created an empirical correlation for $\Delta$ in terms of the free stream Mach number, the body shape and the flow type, i.e., plane or axisymmetric, initially from [145] and subsequently from many runs of the `BLUNT.M` code. The parameter $\mu$ is related to the bow shock curvature, but we use it as a free parameter to control the shape of the bow shock downstream of the shock sonic point. It is also empirically correlated from known bow shock shapes from previous calculations. With the bow shock shape approximated by (5.6), we can use the Rankine–Hugoniot conditions to obtain the flow field on the high pressure side of the shock.

The stagnation conditions at the nose of the blunt body are known exactly, since we know that the streamline that wets the surface is the streamline on the symmetry line. Since the entropy on the surface is known (it is the entropy carried by the streamline on the symmetry line), we also know exactly the flow conditions at the wall sonic point. What we do not know is the location of the sonic point, $\theta^*$, on the wall. This is the third empirical parameter. It is also correlated from known solutions. With $\theta^*$ known, we approximate the Mach number on the surface of the blunt body by the relation

$$M_b^2 = (Y/Y^*)^{2.22\beta},$$

where

$Y^* = \pi - \theta^*$

$\beta$ is the ellipticity, if the body has an elliptical cross section, or is one otherwise

With the stagnation conditions and $M_b$ known on the surface, the flow field on the surface is defined.

The rest of the shock layer is initialized by interpolating the flow variables between the bow shock and the wall along $Y=$constant lines. A linear interpolation is used for $P$ and $S$, and a quadratic interpolation is used for the two velocity components.

### 5.1.5 Outflow Boundary

The outflow boundary, line $fg$ of Figure 5.3, should lie within the supersonic region of the shock layer. If this is the case, then perturbations from this boundary do

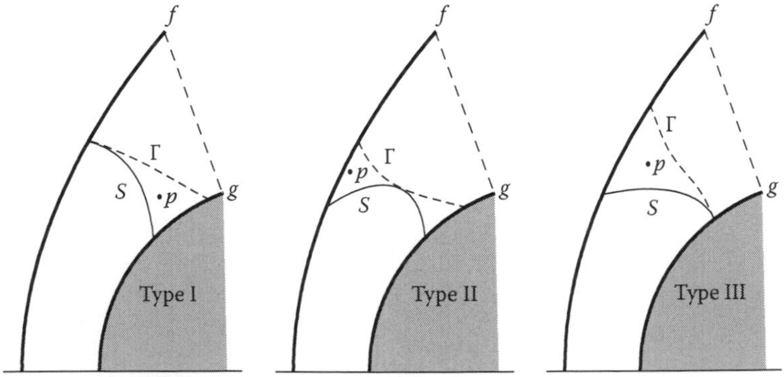


**FIGURE 5.3:** Types of transonic regions.

not propagate back into the shock layer, and a simple linear extrapolation of flow variables or a computation with one sided differences would suffice for this line. What must be avoided is having the outflow boundary cut through the transonic region. The transonic region is that region of the shock layer bounded on one side by the sonic line, line $S$ of Figure 5.3, and on the other by the *limiting characteristic*, line $\Gamma$ of Figure 5.3. The limiting characteristic is the characteristic whose locus of points has only one point of contact with the sonic line. The point of contact might be somewhere along the sonic line or where the sonic line meets the shock or the wall. The reason we must avoid locating the outflow boundary within the transonic region is that any perturbation created at some point $p$ within the transonic region eventually propagates to the sonic line and, therefore, affects the rest of the shock layer.

Studies of the limiting characteristic occurring on blunt body flows can be found in [99,187,204]. The nature of this line depends on the character of the flow, i.e., whether it is plane or axisymmetric, the free stream Mach number, $\gamma$, and the shape of the body. The three types of limiting characteristics arising from elliptical cylinders and ellipsoids are shown in Figure 5.3. For plane flows, only types I and II occur. Type I occurs for plane flow if the free stream Mach number is less than $\sqrt{(\gamma^2+4\gamma+5)/(\gamma+3)}$,* otherwise type II occurs. For axisymmetric flow, type I has not been observed in numerical or experimental studies and, although not proven, it is believed that it cannot occur in axisymmetric flows [187]. The occurrence of type II or III for axisymmetric flow depends on the angle between the velocity vector and the normal to the sonic line at the body sonic point. If this angle is less than zero, the limiting characteristic is of type II, otherwise it is of type III. There is no a priori analytical way of establishing this condition.

### 5.1.6 Shock Computation

We present two methods for the computation of the bow shock wave. The first, which we call the *standard method*, is an extension of the method we used for one-dimensional flows. The second method, which we call *nuovo method*, because it is new and was inspired by work by Paciorri and Bonfiglioli [163], has only been tested with blunt body flows. In order to develop the standard method, we must first find the characteristic surfaces associated with two-dimensional unsteady flows.

#### 5.1.6.1 Characteristic Conoid

As in Section 4.1.1, we want to look for a linear combination of the continuity and momentum equations to find total derivatives of the flow variables along certain directions. In one-dimensional unsteady flows, we found these directions along three lines: the particle path, and the characteristics with slopes $\lambda^{\pm}$. The extension of this concept to two-dimensional unsteady flows leads us to search for a *characteristic surface* on which the derivatives of the flow variables combine into a total derivative. To facilitate the analysis, we write the equations of motion in a Cartesian $(x, y)$ frame

* This function takes the value 1.69 for $\gamma = 1.4$.

with the velocity vector defined by $\mathbf{v} = \bar{u}\vec{i}_x + \bar{v}\vec{i}_y$. Of course, the results of the analysis are valid, after suitable transformation, in any other frame of reference.

The continuity and momentum equations are

$$\begin{aligned} P_t + \bar{u}P_x + \bar{v}P_y + \gamma(\bar{u}_x + \bar{v}_y) &= 0, \\ \bar{u}_t + \overline{uu}_x + \overline{vu}_y + \Theta P_x &= 0, \\ \bar{v}_t + \overline{uv}_x + \overline{vv}_y + \Theta P_y &= 0. \end{aligned} \tag{5.7}$$

As in Section 4.1.1, we multiply (5.7) by three arbitrary multipliers $\mu_1$, $\mu_2$, and $\mu_3$, and add them to obtain

$$\begin{aligned} &[\mu_1 P_t + (\mu_1\bar{u} + \mu_2\Theta)P_x + (\mu_1\bar{v} + \mu_3\Theta)P_y] \\ &+ [\mu_1\bar{u}_t + (\mu_1\gamma + \mu_2\bar{u})\bar{u}_x + \mu_2\overline{vu}_y] \\ &+ [\mu_1\bar{v}_t + \mu_3\overline{uv}_x + (\mu_1\gamma + \mu_3\bar{v})\bar{v}_y] = 0. \end{aligned} \tag{5.8}$$

For one-dimensional flows, we looked for a slope $\lambda$ to define the characteristic line. The extension of this concept to two dimensions requires us to look for a vector $\vec{n} = n_t\vec{i}_t + n_x\vec{i}_x + n_y\vec{i}_y$ normal to the characteristic surface. Consider the first bracket of (5.8). This term represents the dot product of $\nabla P$ with a vector $\vec{\tau} = \mu_1\vec{i}_t + (\mu_1\bar{u} + \mu_2\Theta)\vec{i}_x + (\mu_1\bar{v} + \mu_3\Theta)\vec{i}_y + (\mu_1\bar{v} + \mu_3\Theta)\vec{i}_y$. This product is the derivative of $P$ in the direction of $\vec{\tau}$. If $\vec{\tau}$ has to lie on the characteristic surface, then it must be normal to $\vec{n}$, hence $\vec{n} \cdot \vec{\tau} = 0$:

$$n_t\mu_1 + n_x(\mu_1\bar{u} + \mu_2\Theta) + n_y(\mu_1\bar{v} + \mu_3\Theta) = 0. \tag{5.9}$$

With (5.9) and the two other conditions from the two other brackets of (5.8), we can write the following three equations for the three unknown multipliers $\mu_1$, $\mu_2$, and $\mu_3$.

$$\begin{aligned} &(n_t + \bar{u}n_x + \bar{v}n_y)\mu_1 + \Theta n_x\mu_2 + \Theta n_y\mu_3 = 0, \\ &\gamma n_x\mu_1 + (n_t + \bar{u}n_x + \bar{v}n_y)\mu_2 = 0, \\ &\gamma n_y\mu_1 + (n_t + \bar{u}n_x + \bar{v}n_y)\mu_3 = 0. \end{aligned} \tag{5.10}$$

The equations (5.10) have a nontrivial solution only if the determinant of the coefficients vanishes. This is the case, if either of the following conditions is satisfied:

$$n_t + \bar{u}n_x + \bar{v}n_y = 0, \tag{5.11}$$

$$(n_t + \bar{u}n_x + \bar{v}n_y)^2 = a^2\left(n_x^2 + n_y^2\right). \tag{5.12}$$

A vector tangent to the particle path is defined by $\mathbf{v}_\mathrm{p} = \mathbf{v} + \vec{i}_t$. Therefore, Equation 5.11 represents the dot product of $\mathbf{v}_\mathrm{p}$ with $\vec{n}$. That is to say, the normal to a possible characteristic surface is normal to the particle path in the $(x, y, t)$ space. Hence, Equation 5.11 defines surfaces which are loci of particle paths.

Equation 5.12 defines surfaces which are envelopes of quadrics in terms of the direction cosines of the surface's normal at each point. If we expand (5.12), it takes the form

$$\sum_{ij} \alpha_{ij}\chi_i\chi_j = 0, \quad i,j = 1,2,3, \tag{5.13}$$

where the $\chi_i$'s represent the direction cosines and the coefficient matrix $\alpha_{ij}$ is given by

$$|\alpha_{ij}| = \begin{vmatrix} \bar{u}^2 - a^2 & \overline{uv} & \bar{u} \\ \overline{uv} & \bar{v}^2 - a^2 & \bar{v} \\ \bar{u} & \bar{v} & 1 \end{vmatrix}. \tag{5.14}$$

Equation 5.13 may also be written in the form

$$\chi^{\mathrm{T}}|\alpha_{ij}|\chi = 0, \tag{5.15}$$

where $\chi^{\mathrm{T}} = [n_x \quad n_y \quad n_t]$. If a quadric is defined by its direction cosines, then its equation in terms of its Cartesian components is given by

$$\mathrm{X}^{\mathrm{T}}|\mathrm{A}_{ij}|\mathrm{X} = 0, \tag{5.16}$$

where $\mathrm{X}^{\mathrm{T}} = [x \quad y \quad t]$ the coefficients $\mathrm{A}_{ij}$ are the signed complementary minors of the determinant* (5.14):

$$|\mathrm{A}_{ij}| = \begin{vmatrix} 1 & 0 & -\bar{u} \\ 0 & 1 & -\bar{v} \\ -\bar{u} & -\bar{v} & \bar{v}^2 + \bar{u}^2 - a^2 \end{vmatrix}.$$

If we expand (5.16) we find the envelope of characteristic surfaces through the origin of the reference frame:

$$\mathbf{F} = (x - \bar{u}t)^2 + (y - \bar{v}t)^2 - a^2t^2 = 0. \tag{5.17}$$

This equation is a cone with apex at the origin of the frame of reference. On any $t =$ constant plane, the cross section of the cone is a circle with radius $at$, centered at $x = \bar{u}t$, $y = \bar{v}t$. The surface of the cone carries perturbations originating at the origin. These perturbations travel in all directions at the speed of sound relative to the gas particles. In addition the perturbations are advected with the particle velocity. In general, the path followed by the center of the circle is not a straight line, and the radius of the circle varies from one time level to another. However, let us simplify the picture by assuming that both the particle velocity and the speed of sound are

* The $\alpha_{ij}$'s can be obtained from the $\mathrm{A}_{ij}$'s in a similar manner.

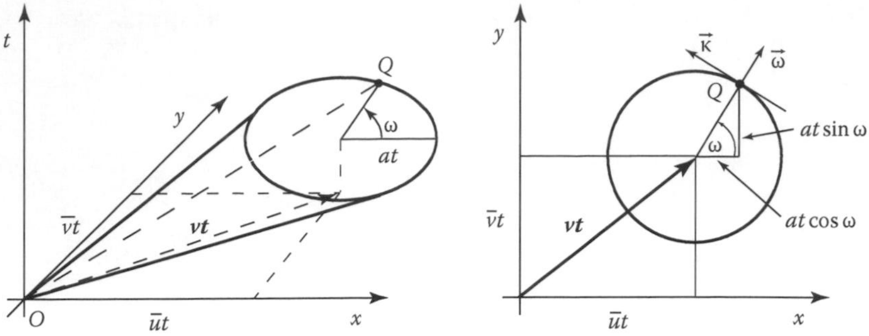


**FIGURE 5.4:** Characteristic conoid.

constant. The cone, then, is as shown in Figure 5.4; a series of expanding circles in the $(x, y)$ plane as time increases.

The cone is known as the *Mach cone* and it represents the *range of influence* of its vertex, point $O$ in Figure 5.4. If we extend the cone backward in time, we have the reciprocal concept of *domain of dependence*. The base of the backward cone is now the domain of dependence of its vertex.

The unknowns in (5.10) are proportional to the signed complementary minors of any row in the determinant of its coefficients:

$$\begin{aligned}\mu_1 &= (n_t + \bar{u}n_x + \bar{v}n_y)^2,\\ \mu_2 &= -\gamma n_x(n_t + \bar{u}n_x + \bar{v}n_y),\\ \mu_3 &= -\gamma n_y(n_t + \bar{u}n_x + \bar{v}n_y),\end{aligned} \tag{5.18}$$

and the components of the normal, $\vec{n}$, to the cone are proportional to the components of $\nabla\mathbf{F}$ of (5.17):

$$-\frac{n_t}{\bar{u}(x-\bar{u}t)+\bar{v}(y-\bar{v}t)+a^2t} = \frac{n_x}{x-\bar{u}t} = \frac{n_y}{y-\bar{v}t}. \tag{5.19}$$

Let $Q$ represent any point on the surface of the cone, and let us define its position by the angle, $\omega$, measured as shown in Figure 5.4. On a $t=$constant plane, the coordinates of $Q$ are

$$\begin{aligned}x-\bar{u}t &= at\cos\omega,\\ y-\bar{v}t &= at\sin\omega.\end{aligned} \tag{5.20}$$

With (5.20), we can rewrite (5.19) in the form

$$-\frac{n_t}{\bar{u}\cos\omega+\bar{v}\sin\omega+a} = \frac{n_x}{\cos\omega} = \frac{n_y}{\sin\omega}.$$

It follows that $n_t + \bar{u}n_x + \bar{v}n_y$ is proportional to $-\bar{u}\cos\omega - \bar{v}\sin\omega - a + \bar{u}\cos\omega + \bar{v}\sin\omega$, which reduces to $-a$. We can, therefore, rewrite (5.18) as

$$\mu_1 = a,$$
$$\mu_2 = \gamma \cos\omega, \tag{5.21}$$
$$\mu_3 = \gamma \sin\omega.$$

With (5.21) and (5.8), we can write the compatibility equation valid on the characteristic surface:

$$\begin{aligned} &a[P_t + (\bar{u} + a\cos\omega)P_x + (\bar{v} + a\sin\omega)P_y] \\ &\quad + \gamma\cos\omega[\bar{u}_t + (\bar{u} + a\cos\omega)\bar{u}_x + (\bar{v} + a\sin\omega)\bar{u}_y] \\ &\quad + \gamma\sin\omega[\bar{v}_t + (\bar{u} + a\cos\omega)\bar{v}_x + (\bar{v} + a\sin\omega)\bar{v}_y] \\ &\quad - \gamma a\sin\omega(-\bar{u}_x\sin\omega + \bar{u}_y\cos\omega) + \gamma a\cos\omega(-\bar{v}_x\sin\omega + \bar{v}_y\cos\omega) = 0. \end{aligned} \tag{5.22}$$

Each square bracket of (5.22) represents a total differential along a generator of the cone passing through the point $Q$. Each of the last two terms represents partial derivatives in a direction tangent to the cross section of the cone at point $Q$, see Figure 5.4. If we introduce the vectors

$$\vec{\lambda} = (\bar{u} + a\cos\omega)\vec{i}_x + (\bar{v} + a\sin\omega)\vec{i}_y,$$
$$\vec{\kappa} = -\sin\omega\vec{i}_x + \cos\omega\vec{i}_y,$$

we can rewrite (5.22) in a more compact, more expressive form:

$$\begin{aligned} &a[P_t + \vec{\lambda}\cdot\nabla P] + \gamma\cos\omega[\bar{u}_t + \vec{\lambda}\cdot\nabla\bar{u}] + \gamma\sin\omega[\bar{v}_t + \vec{\lambda}\cdot\nabla\bar{v}] \\ &\quad = \gamma a[\sin\omega(\vec{\kappa}\cdot\nabla\bar{u}) - \cos\omega(\vec{\kappa}\cdot\nabla\bar{v})]. \end{aligned} \tag{5.23}$$

Now note that the angle $\omega$ is an independent parameter; therefore, we can bring the direction cosines depending on $\omega$ inside the differentials and write

$$\begin{aligned} &a[P_t + \vec{\lambda}\cdot\nabla P] + \gamma[\partial(\cos\omega\bar{u} + \sin\omega\bar{v})/\partial t + \vec{\lambda}\cdot\nabla(\cos\omega\bar{u} + \sin\omega\bar{v})] \\ &\quad = \gamma a[\vec{\kappa}\cdot\nabla(\sin\omega\bar{u} - \cos\omega\bar{v})]. \end{aligned} \tag{5.24}$$

Let us define the vector $\vec{\omega} = \cos\omega\vec{i}_x + \sin\omega\vec{i}_y$, see Figure 5.4, and the two velocity components $\tilde{u} = \vec{\omega}\cdot\boldsymbol{v} = \cos\omega\bar{u} + \sin\omega\bar{v}$ and $\tilde{v} = \vec{\kappa}\cdot\boldsymbol{v} = -(\sin\omega\bar{u} - \cos\omega\bar{v})$, then (5.24) takes the form

$$a[P_t + \vec{\lambda}\cdot\nabla P] + \gamma[\tilde{u}_t + \vec{\lambda}\cdot\nabla\tilde{u}] = -\gamma a[\vec{\kappa}\cdot\nabla\tilde{v}].$$

Finally, introduce the notation

$$( )' = \partial( )/\partial t + \vec{\lambda} \cdot \nabla( ),$$
$$( )_\kappa = \vec{\kappa} \cdot \nabla( ),\tag{5.25}$$

to obtain

$$aP' + \gamma\tilde{u}' = -\gamma a\tilde{v}_\kappa.\tag{5.26}$$

Compare (5.26) with (4.38). These are essentially the same equations. The right-hand side of (4.38) is, of course, an approximation to the right-hand side of (5.26) which expresses exactly the contribution from terms perpendicular to $\vec{\omega}$. Equation 5.26 is valid along any cone generator $OQ$ (called a *bicharacteristic*).

#### 5.1.6.2 Standard Shock Computation

Assuming that the supersonic side of a shock is known, and assuming that we also know the shape of the shock, the Rankine–Hugoniot jumps provide sufficient conditions to evaluate all the flow variables on the subsonic side. However, to complete the analysis of the shock one additional condition is needed to evaluate the shock speed. The additional condition can be satisfied with Equation 5.26 evaluated on the subsonic side of the shock. Consider the upper panel of Figure 5.5. We want to evaluate the shock speed at point $O$ at time $t + \Delta t$. The domain of dependence of point $O$ is defined by the Mach cone with vertex at $O$. Since point $O$ is on the subsonic side of the shock, the base of the cone at time $t$ has its center, $C$, on the supersonic side as shown in Figure 5.5. The radius of the circle, at the base of the cone, is $a\Delta t$, where $a$ is the speed of sound on the subsonic side. The domain of dependence of point $O$ at time $t$ consists of the region bounded on one side by the circular arc $\widehat{BAD}$ and the shock on the other.* Now draw a fixed Cartesian frame of reference $(\eta, \xi)$ at point $O$, such that its axes are normal and tangent to the shock at $O$ in the plane $t + \Delta t =$ constant. Let $\tilde{u}$ and $\tilde{v}$ be the velocity components in the $(\eta, \xi)$ plane. We want to write (5.26) along a generator of the cone on the subsonic side of the shock to determine the speed of the shock at $O$. To this end, draw a line $CA$ parallel to $\eta$ and choose the angle $\omega$ as indicated in Figure 5.5. Any generator $QO$ can be used to solve the problem. We are free to choose the most convenient one. Since the Rankine–Hugoniot jumps are evaluated on a plane normal to the shock at $O$, it is convenient to take $QO$ parallel to $\eta$, i.e., $\omega = 0$. Now that we have chosen $Q = A$, let us draw the range of influence of point $A$. The range of influence of point $A$ at time $t + \Delta t$ is the circle drawn on the lower panel of Figure 5.5. The dashed line $AO$ is the generator of the cone (bicharacteristic) with vertex at $A$ along which we want to integrate (5.26). $AO$ is defined by $\omega = \pi$ and the vector $\vec{\tau}$, at $O$, points in a direction opposite to $\eta$.

* Of course the Mach cone $OBDO$ does not extend into the supersonic region, but it is convenient to draw it this way to explain how (5.26) is applied.

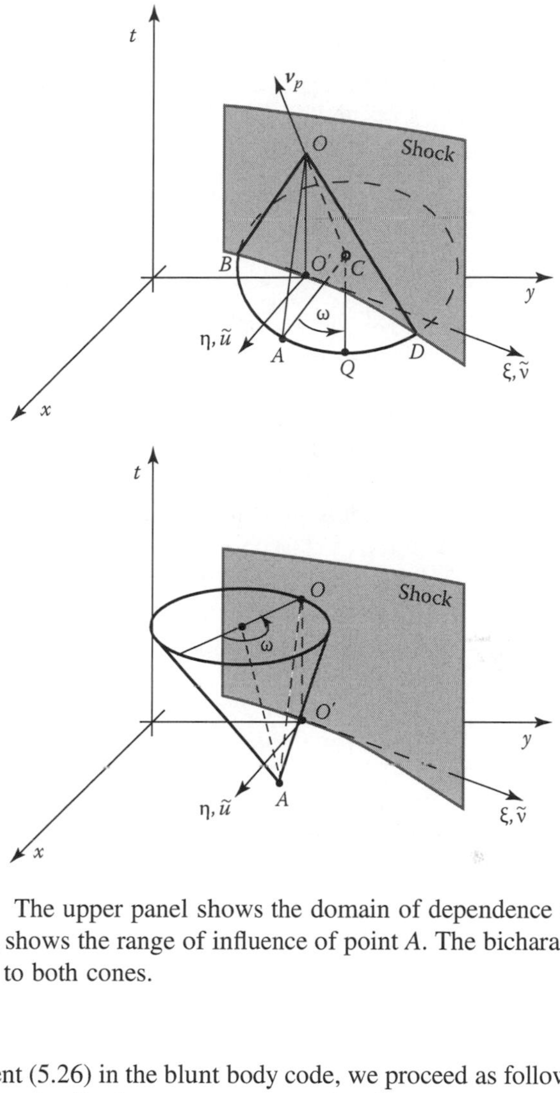


**FIGURE 5.5:** The upper panel shows the domain of dependence of point $O$, and the lower panel shows the range of influence of point $A$. The bicharacteristic labeled $OA$ is common to both cones.

To implement (5.26) in the blunt body code, we proceed as follows. Take a fixed frame of reference ($\eta$, $\xi$) whose axes are normal and tangent to the shock at the point of interest, with $\eta$ pointing into the shock layer. Let $\vec{\eta}$ be the unit normal to the shock pointing into the shock layer: $\vec{\eta} = \eta_r \vec{i}_r + \eta_\theta \vec{i}_\theta$. It follows that the velocity component normal to the shock is $\tilde{u} = \boldsymbol{v} \cdot \vec{\eta}$. Multiply the $u$-momentum equation in (5.5) by $\eta_r$ and similarly multiply the $v$-momentum equation by $\eta_\theta$. Add these two equations to obtain an equation for $\partial\tilde{u}/\partial T$. Now multiply the continuity equation in (5.5) by $a/\gamma$ and add the resulting equation to the equation for $\partial\tilde{u}/\partial T$. We have now effectively written (5.26) in terms of our computational coordinates. The next step is to replace $\partial\tilde{u}/\partial T$ and $\partial P/\partial T$ with the Rankine–Hugoniot jumps as was done in Section 4.3, Equations 4.31, to obtain an equation for the shock acceleration: $w_T$. The shock acceleration is then used to evaluate the shock speed and the shock radius. Two details are different from Section 4.3. First, for the blunt body problem, the flow conditions on the supersonic side of the shock are constant, so all the time derivatives of supersonic side variables in (4.31) vanish. Second, in the blunt body code it was convenient to define the shock velocity in the radial direction, therefore, the component of the shock velocity entering in the Rankine–Hugoniot jumps is $w\eta_r$.

The details of this calculation can be found in the function `BLUNT_MAC_1.M` of the `BLUNT.M` code.

#### 5.1.6.3 Nuovo Shock Computation

The new shock-fitting method takes a slightly different point of view from the standard method. Suppose we compute a time derivative of a flow variable at a shock point on the subsonic side using Equation 5.26 and accept this derivative as a *given value*. If we could write a Rankine–Hugoniot jump for this derivative, then we would have the additional condition needed to evaluate $w_T$. However, Equation 5.26 expresses a balance between the primitive variables $P$ and $\tilde{u}$, and there is no obvious reason to select one over the other. It is apparent that we should take into account the combined effect of $P$ and $\tilde{u}$ in the calculation of $w_T$. We can accomplish this if we rewrite (5.26) in terms of a Riemann variable. We will do this in Section 5.1.8 when we derive the λ-scheme for two space dimensions. For now, let us just state that the appropriate Riemann variable for the bow shock calculation is

$$Q_{\mathrm{h}} = \frac{2a_{\mathrm{h}}}{\gamma - 1} - \tilde{u}_{\mathrm{h}}, \tag{5.27}$$

where the subscript refers to values on the high pressure side of the shock and $\tilde{u}_{\mathrm{h}} = v_{\mathrm{h}} \cdot \vec{\eta}$. It follows from the Rankine–Hugoniot jumps, see Section 6.5.4, that $Q_{\mathrm{h}}$ is related to the Mach number relative to the shock,

$$M_{\mathrm{l,rel}} = \frac{\tilde{u}_{\mathrm{l}} - w\eta_r}{a_{\mathrm{l}}}, \tag{5.28}$$

by the relation

$$g(M_{\mathrm{l,rel}}) = \frac{\gamma + 1}{2a_{\mathrm{l}}}(Q_{\mathrm{h}} - \tilde{u}_{\mathrm{l}}), \tag{5.29}$$

where

$$g(M_{\mathrm{l,rel}}) = \frac{1}{\gamma - 1}\sqrt{\left(2\gamma M_{\mathrm{l,rel}}^2 - (\gamma - 1)\right)\left((\gamma - 1)M_{\mathrm{l,re}}^2 + 2\right)\Big/ M_{\mathrm{l,rel}}^2} \\ + \left(M_{\mathrm{l,rel}}^2 - 1\right)/M_{\mathrm{l,rel}},$$

and the l subscript refers to values on the low pressure side of the bow shock (the free stream). Equation 5.29 is the same as Equation 6.51. The equation for the shock acceleration follows by differentiating (5.29) with respect to $T$:

$$w_T = \frac{1}{\eta_r}\left(\left(1+\frac{\gamma+1}{2g'}\right)\tilde{u}_{1,T} - \frac{\gamma+1}{2g'}Q_{h,T} - w\eta_{r,T}\right), \tag{5.30}$$

where $g' = \partial g/\partial M_{1,\text{rel}}$. Let $\sigma = c_\theta = -c_Y$, then

$$\begin{aligned} \eta_r &= -1\Big/\sqrt{1+(\sigma/c)^2}, \\ \eta_\theta &= -\sigma\eta_r/c, \\ \eta_{\theta,T} &= \eta_r^3(w_Y + \sigma w/c)/c, \\ \eta_{r,T} &= \sigma\eta_{\theta,T}/c, \text{ and} \\ \tilde{u}_{1,T} &= u_\infty(\cos\theta\eta_{r,T} - \sin\theta\eta_{\theta,T}). \end{aligned} \tag{5.31}$$

The evaluation of $\eta_{\theta,T}$ involves the sum of two small quantities, $w_Y$ and $\sigma w/c$, that vanish in the steady state. This calculation is prone to odd–even oscillations which can be eliminated by doing a weighted average of $\eta_{\theta,T}$.

It follows that if we know $Q_{h,T}$, we can evaluate $w_T$. Ideally, $Q_{h,T}$ should be evaluated with an upwind scheme written in terms of Riemann variables, such as the $\lambda$-scheme of Section 5.1.8. Would it work if $Q_{h,T}$ is evaluated by other means? Surprisingly, it works rather well even when $Q_{h,T}$ is evaluated by a central differences scheme. The function `BLUNT_MAC_2.M` does the shock computation using (5.30) with $Q_{h,T}$ evaluated from

$$Q_{h,T} = \frac{a}{\gamma(\gamma-1)}[(\gamma-1)P_T + S_T] - [u_T\eta_r + v_T\eta_\theta + u\eta_{r,T} + v\eta_{\theta,T}], \tag{5.32}$$

with the derivatives of the primitive variables ($P_T$, $S_T$, $u_T$, and $v_T$) evaluated with the MacCormack scheme. The results obtained by this method are equivalent to those obtained by the standard shock-fitting method of Section 5.1.6.2.

Let us summarize the new shock calculation algorithm. Assume that we are at time level $t$, and the flow field, as well as, the shock geometry and shock velocity is known. The pseudocode to compute the shock is

1. Evaluate $P_T$, $u_T$, $v_T$, and $S_T$ at the shock using (5.5) with the *background* numerical scheme, preferably an upwind scheme.

2. Evaluate $\eta_r$, $\eta_\theta$, $\eta_{r,T}$, $\eta_{\theta,T}$, and $\tilde{u}_{1,T}$ from (5.31).

3. Evaluate $Q_{h,T}$ from (5.32).

4. Evaluate $w_T$ from (5.30).

5. Update the shock speed and shock location:

$$\begin{aligned} w^{t+\Delta t} &= w^t + w_T^t\Delta t, \\ c^{t+\Delta t} &= c^t + w^t\Delta t + \tfrac{1}{2}w_T^t\Delta t^2. \end{aligned}$$

6. Repeat 1 through 5 for all shock points.
7. Using the new shock location, $c^{t+\Delta t}$, evaluate $\eta_r$, $\eta_\theta$, again using (5.31). These are used in the next step to evaluate the free stream Mach number component normal to the shock.
8. Using the Rankine–Hugoniot jumps, Section 3.6.4, evaluate $P$, $u$, $v$, and $S$ at the shock at $t + \Delta t$.

### 5.1.7 Wall Computation

At the wall, we take a fixed frame of reference ($\eta$, $\xi$) whose axes are normal and tangent to the wall at the point of interest. Let $\vec{\eta}$ be the unit normal pointing to the wall, $\vec{\eta} = \eta_r \vec{i}_r + \eta_\theta \vec{i}_\theta$, and $\vec{\tau}$ the unit tangent, $\vec{\tau} = \eta_\theta \vec{i}_r - \eta_r \vec{i}_\theta$. Let $\tilde{u}$ and $\tilde{v}$ be the velocity components in the $\vec{\eta}$, $\vec{\tau}$ directions, respectively. The boundary condition at the wall is $\tilde{u} = 0$, which implies that $\tilde{u}_T = 0$. Combining the two momentum equations in (5.5) into a momentum equation in the tangential direction and simplifying terms, we find

$$
\begin{aligned}
&P_T + VP_Y + \gamma(Z_r\tilde{u}_Z/n_r + Y_\theta v_Y/r + \vartheta) = 0,\\
&\tilde{v}_T + V\tilde{v}_Y + \Theta Y_\theta n_r P_Y/r = 0, \qquad\qquad (5.33)\\
&S_T + VS_Y = 0.
\end{aligned}
$$

From the second of (5.33) we have

$$
\begin{aligned}
v_T &= -\tilde{v}_T n_r,\\
u_T &= \tilde{v}_T n_\theta.
\end{aligned} \qquad (5.34)
$$

The first and third of (5.33) together with (5.34) are integrated with the same scheme used for the rest of the field.

### 5.1.8 The Lambda Scheme in Two Space Dimensions

The extension of the $\lambda$-scheme to two space dimensions was carried out by Zannetti and Colasurdo in [241]. In that work, the assumption was made that the flow was isentropic and they ended up solving a system of equations equivalent to (5.26), but with the space gradients defined in terms of Riemann variables, not unlike the equation set (4.24). Here we derive the $\lambda$-scheme starting from (5.5) with the goal of obtaining an equation set equivalent to (4.29). The first step, see Section 4.2, is to replace the pressure in (5.5) with the speed of sound, to obtain

$$
\begin{aligned}
& a_T + Ua_Z + Va_Y + \frac{\gamma-1}{2} a\left( Z_r u_Z + \frac{Z_\theta}{r} v_Z + \frac{Y_\theta}{r} v_Y + \vartheta \right) = 0, \\
& u_T + Uu_Z + Vu_Y + \frac{2}{\gamma-1} a Z_r a_Z - \frac{a^2}{\gamma(\gamma-1)} Z_r S_Z - \frac{v^2}{r} = 0, \\
& v_T + Uv_Z + Vv_Y + \frac{2}{\gamma-1} a \frac{Z_\theta}{r} a_Z + \frac{2}{\gamma-1} a \frac{Y_\theta}{r} a_Y + \\
& \quad - \frac{a^2}{\gamma(\gamma-1)} \frac{Z_\theta}{r} S_Z - \frac{a^2}{\gamma(\gamma-1)} \frac{Y_\theta}{r} S_Y + \frac{vu}{r} = 0, \\
& S_T + US_Z + VS_Y = 0.
\end{aligned}
\tag{5.35}
$$

Now we form a linear combination of the first three equations, as we did in Section 5.1.6.1, and choose four bicharacteristic directions, namely $\pm\vec{i}_r$ and $\pm\vec{i}_\theta$, to find the following set of compatibility conditions:

$$
\begin{aligned}
& Q^{(r)'} - \frac{a}{\gamma(\gamma-1)} S' = H^-, \text{ on } \vec{\Lambda}^-_{(r)}, \\
& R^{(r)'} - \frac{a}{\gamma(\gamma-1)} S' = H^+, \text{ on } \vec{\Lambda}^+_{(r)}, \\
& Q^{(\theta)'} - \frac{a}{\gamma(\gamma-1)} S' = K^+, \text{ on } \vec{\Lambda}^-_{(\theta)}, \\
& Q^{(0)'} - \frac{a}{\gamma(\gamma-1)} S' = K^-, \text{ on } \vec{\Lambda}^+_{(\theta)}, \\
& S' = 0, \text{ on } \vec{\Lambda}_{(S)}.
\end{aligned}
\tag{5.36}
$$

Here, the symbol $(\ )'$ has the same meaning as in (5.25) with $\vec{\lambda}$ replaced by the bicharacteristic directions $\vec{\Lambda}$, defined below. This set is the extension of (4.29) to two space dimensions. But unlike (4.29), we now have five equations for the four unknowns *a, u, v*, and *S*. This, however, does not represent a significant problem, since it is rather easy and obvious how to define the unknowns from *Q* and *R*. The Riemann variables *Q* and *R* are defined by

$$
\begin{aligned}
& Q^{(r)} = \frac{2a}{\gamma-1} - u, \quad R^{(r)} = \frac{2a}{\gamma-1} + u, \\
& Q^{(\theta)} = \frac{2a}{\gamma-1} - v, \quad R^{(\theta)} = \frac{2a}{\gamma-1} + v,
\end{aligned}
\tag{5.37}
$$

the bicharacteristic directions are

$$
\begin{aligned}
& \vec{\Lambda}^\pm_{(r)} = \vec{i}_T + (U \pm aZ_r)\vec{i}_X + V\vec{i}_Y, \\
& \vec{\Lambda}^\pm_{(\theta)} = \vec{i}_T + \left( U \pm a\frac{Z_\theta}{r} \right)\vec{i}_X + \left( V \pm a\frac{Y_\theta}{r} \right)\vec{i}_Y, \\
& \vec{\Lambda}_{(S)} = \vec{i}_T + U\vec{i}_X + V\vec{i}_Y,
\end{aligned}
\tag{5.38}
$$

and the right-hand sides of (5.36) are

$$
\begin{aligned}
H^{\pm} &= -\left(a(Z_\theta v_Z + Y_\theta v_Y + \vartheta) \pm v^2\right)/r, \\
K^{\pm} &= -(aZ_r u_Z + (a\vartheta \pm uv)/r)).
\end{aligned}
\tag{5.39}
$$

The original unknowns are obtained from the Riemann variables from

$$
\begin{aligned}
a &= \frac{\gamma-1}{8}(Q^{(r)} + R^{(r)} + Q^{(\theta)} + R^{(\theta)}), \\
u &= \frac{1}{2}(R^{(r)} - Q^{(r)}),\ v = \frac{1}{2}(R^{(\theta)} - Q^{(\theta)}).
\end{aligned}
\tag{5.40}
$$

If we expand (5.36), we find

$$
\begin{aligned}
& Q_T^{(r)} + (U - aZ_r)\left[Q_Z^{(r)} - \frac{aS_Z}{\gamma(\gamma-1)}\right] + V\left[Q_Y^{(r)} - \frac{aS_Y}{\gamma(\gamma-1)}\right] \\
& = \frac{aS_T}{\gamma(\gamma-1)} + H^-, \text{ on } \Lambda^-_{(r)}, \\
& R_T^{(r)} + (U + aZ_r)\left[R_Z^{(r)} - \frac{aS_Z}{\gamma(\gamma-1)}\right] + V\left[R_Y^{(r)} - \frac{aS_Y}{\gamma(\gamma-1)}\right] \\
& = \frac{aS_T}{\gamma(\gamma-1)} + H^+, \text{ on } \Lambda^+_{(r)}, \\
& Q_T^{(\theta)} + \left(U - a\frac{Z_\theta}{r}\right)\left[Q_Z^{(\theta)} - \frac{aS_Z}{\gamma(\gamma-1)}\right] + \left(V - a\frac{Y_\theta}{r}\right)\left[Q_Y^{(\theta)} - \frac{aS_Y}{\gamma(\gamma-1)}\right] \\
& = \frac{aS_T}{\gamma(\gamma-1)} + K^+, \text{ on } \Lambda^-_{(\theta)}, \\
& R_T^{(\theta)} + \left(U + a\frac{Z_\theta}{r}\right)\left[R_Z^{(\theta)} - \frac{aS_Z}{\gamma(\gamma-1)}\right] + \left(V + a\frac{Y_\theta}{r}\right)\left[R_Y^{(\theta)} - \frac{aS_Y}{\gamma(\gamma-1)}\right] \\
& = \frac{aS_T}{\gamma(\gamma-1)} + K^-, \text{ on } \Lambda^+_{(\theta)}, \\
& S_T + U[S_Z] + V[S_Y] = 0, \text{ on } \Lambda_{(s)}.
\end{aligned}
\tag{5.41}
$$

Pandolfi [164] gives a geometrical interpretation of the $\lambda$-scheme as follows. Consider the mesh point $(n,m)$ at time $t+\Delta t$, see Figure 5.6. Although, the mesh point $(n,m)$ is fixed in the $(Z,Y)$ plane, it moves an amount equal to $wZ_n\Delta t$ in the physical plane, because of the motion of the bow shock. The domain of dependence of this point is the Mach cone with vertex at $(n,m)$ at time $t+\Delta t$, with its base at time $t$. This base is shown in the figure as the circle with center at 0 and radius $a\Delta t$. The *CFL* condition ensures that the circle is contained within the parallelogram formed by the mesh lines $Z_{n+1}$, $Z_{n-1}$, $Y_{m+1}$, and $Y_{m-1}$. On the surface of the Mach cone, we have selected four bicharacteristics, originating at points 1, 4, 2, and 3 of the figure, to propagate signals to point $(n,m)$. These points are located at the interception of the

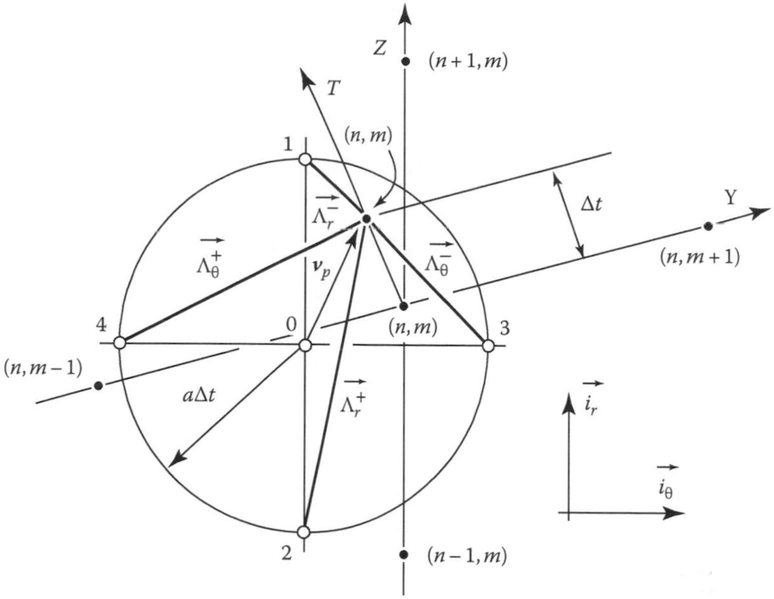


**FIGURE 5.6:** Geometrical interpretation of the λ-scheme.

circle with the two straight lines through 0 and parallel to the unit vectors $\vec{i}_r$ and $\vec{i}_\theta$. The slopes of the bicharacteristics in the volume $(Z,Y,T)$ are defined by the first two equations of (5.38). Each of these four bicharacteristics carries perturbations of its associated Riemann variable. In addition, originating from the center of the circle, the particle path carries a perturbation associated with the entropy. The slope of the particle path in the volume $(Z,Y,T)$ is defined by the last of (5.38). These five signals converge on point $(n,m)$ at time $t+\Delta t$ to provide new values for the Riemann variables and the entropy. These values are unscrambled by (5.40) to obtain new values for the primitive variables $a$, $u$, $v$.

In order to be consistent with the domain of dependence of each of these five signals, we discretize each term of (5.41) in a square bracket according to the sign of the coefficient in front of it, as was done for (4.29). Also as for (4.29), the time derivative of the entropy appearing on the first four equations of (5.41) is obtained from the last equation of (5.41). Finally, the derivatives appearing in the terms $H^\pm$ and $K^\pm$ are evaluated by central differences.

### 5.1.9 Blunt Body Flow Results

We present results for blunt body flows corresponding to elliptical cylinders, ellipsoids, and a paraboloid. All results are for $\gamma = 1.4$.

Let the ellipse/ellipsoid cross section be defined by

$$(x/a)^2 + (y/b)^2 = 1, \tag{5.42}$$

and let $b=1$. The *ellipticity* (the deviation from a circle) is defined by $e=b/a=1/a$. Because we chose $b=1$, the radius of curvature at $\theta=\pi$, $r_b$, is equal to the ellipticity. In polar coordinates, the radius of the body is defined by

$$b(\theta)=\frac{-e^2x_0\cos\theta+\sqrt{e^2\cos^2\theta+\left(1-e^2x_0^2\right)\sin^2\theta}}{e^2\cos^2\theta+\sin^2\theta},$$

where $x_0$ is an offset for the origin of the polar coordinates $(r,\theta)$:

$$x=r\cos\theta+x_0$$
$$y=r\sin\theta$$

This offset provides more control over the placement of the outflow boundary. Results are presented for ellipticity values of [0.5,1,1.5]. The body shapes corresponding to these values are shown in Figure 5.7.

In Figure 5.8 the bow shock and sonic line shapes for spherical and circular blunt bodies ($e=1$) are compared for several free stream Mach numbers. It is typical for the axisymmetric cases to have the bow shock closer to the body than for the equivalent plane cases. This is a consequence of the additional relief the flow has around the sides of the axisymmetric bodies. As a result, the axisymmetric bodies are easier to compute. Table 5.1 compares the results of our calculations with those presented by Rusanov in [187] for the spherical blunt body. The first two rows compare the bow shock standoff distance, while the last two rows compare the radius of curvature of the bow shock at $\theta=\pi$. The latter, is a much harder quantity to evaluate accurately.

Figure 5.9 compares the bow shock standoff distance for the three ellipsoids, and Figure 5.10 makes the same comparison for the three elliptical cylinders. Note that the results show a nice trend when plotted against $\left(M_\infty^2-1\right)^{-1}$. For the axisymmetric cases, the rate of growth of the standoff distance is decreasing as $M_\infty\to 1$, while the opposite is true for the plane cases. On the other hand, the standoff distance becomes insensitive to the free stream Mach number as $M_\infty\to\infty$. Figures 5.11 and 5.12 show the behavior of the body sonic angle for elliptical cylinders and ellipsoids, respectively, as a function of $M_\infty^{-2}$.

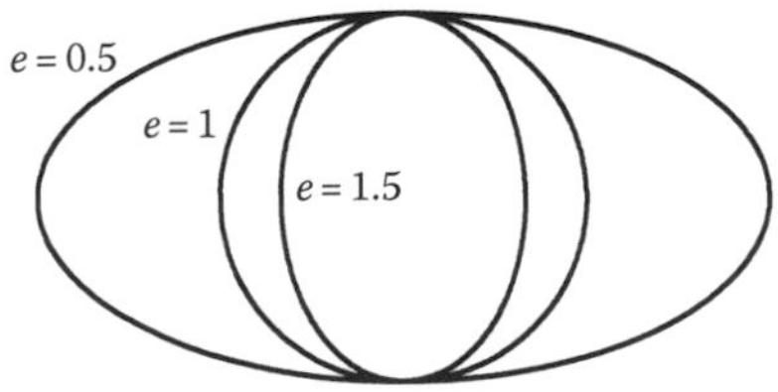


**FIGURE 5.7:** Elliptical cross sections for various values of the ellipticity parameter.

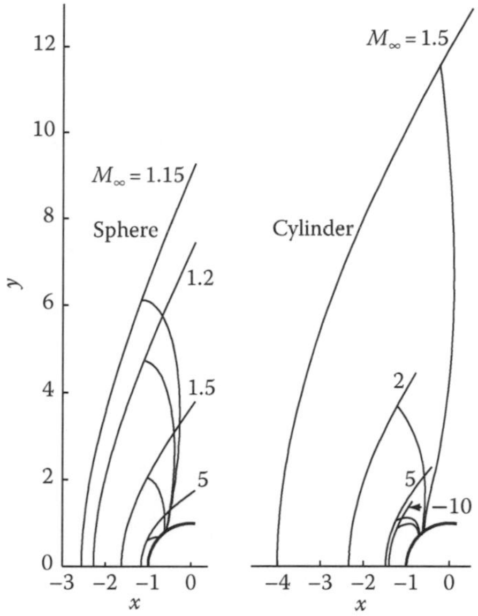


**FIGURE 5.8:** Shock and sonic line shapes for a sphere and a circular cylinder at various free stream Mach numbers.

**TABLE 5.1:** Shock standoff distance and shock radius of curvature at symmetry line for spherical blunt body from Rusanov [187] and present calculations.

| $M_\infty$ | 1.5 | 2.0 | 4.0 | 6.0 | 8.0 | $\infty$ |
|---|---|---|---|---|---|---|
| $\Delta/r_b$ [187] | 0.616 | 0.350 | 0.175 | 0.149 | 0.140 | 0.129 |
| $\Delta/r_b$ | 0.616 | 0.349 | 0.175 | 0.149 | 0.140 | 0.129 |
| $1/r_b\kappa$ [187] | 3.169 | 2.052 | 1.442 | 1.356 | 1.328 | 1.292 |
| $1/r_b\kappa$ | 3.148 | 2.040 | 1.438 | 1.353 | 1.325 | 1.293 |

Parabolic and paraboloid blunt bodies are defined by

$$x = y^2/2r_b,$$

where $r_b$ is the blunt nose radius of curvature which we define as unity. In polar coordinates, the parabolic cross section is defined by

$$b(\theta) = \frac{\cos\theta + \sqrt{\cos^2\theta - 2(2 + x_0)\sin^2\theta}}{\sin^2\theta},$$

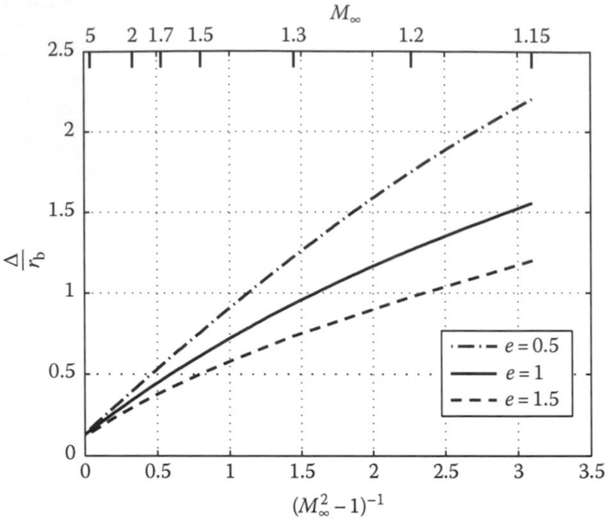


**FIGURE 5.9:** Bow shock standoff distance plotted vs. $(M_\infty^2 - 1)^{-1}$ for ellipsoid bodies of revolution.

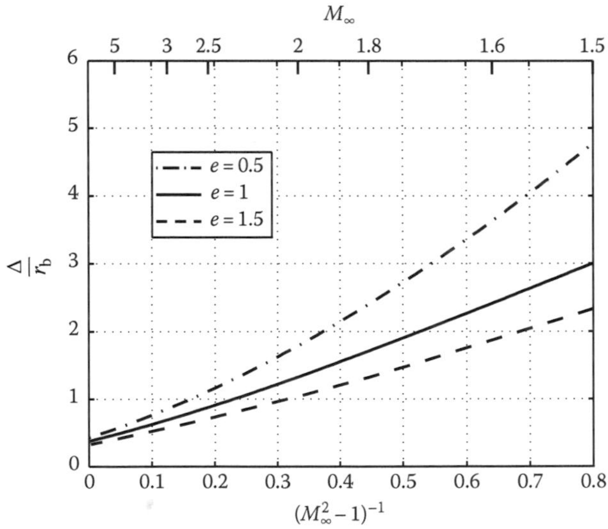


**FIGURE 5.10:** Bow shock standoff distance plotted vs. $(M_\infty^2 - 1)^{-1}$ for elliptical cylinders.

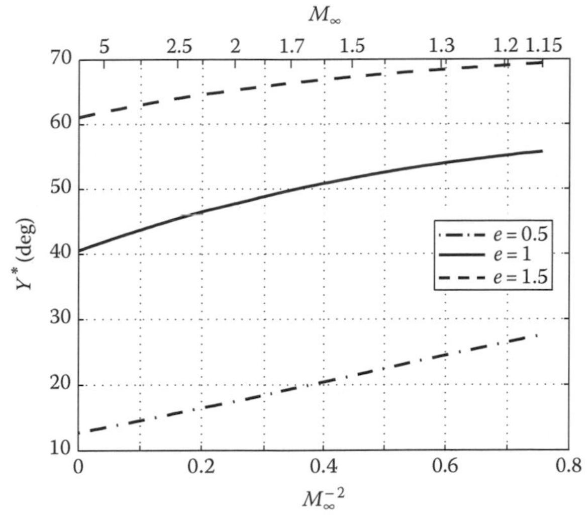


**FIGURE 5.11:** Body sonic angle in degrees plotted vs. $M_\infty^{-2}$ for an ellipsoid body of revolution.

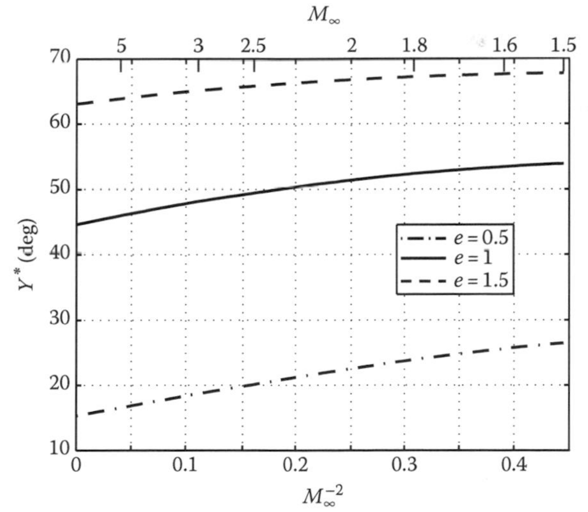


**FIGURE 5.12:** Body sonic angle in degrees plotted vs. $M_\infty^{-2}$ for an elliptical cylinder.

**TABLE 5.2:** Shock standoff distance and shock radius of curvature at symmetry line for a paraboloid blunt body from Rusanov [187] and present calculations.

| $M_\infty$ | **1.5** | **2.0** | **4.0** | **6.0** | **8.0** | $\infty$ |
|---|---|---|---|---|---|---|
| $\Delta/r_b$ [187] | 0.863 | 0.432 | 0.196 | 0.163 | 0.153 | 0.139 |
| $\Delta/r_b$ | 0.863 | 0.430 | 0.195 | 0.163 | 0.153 | 0.136 |
| $1/r_b\kappa$ [187] | 4.598 | 2.540 | 1.628 | 1.495 | 1.461 | 1.410 |
| $1/r_b\kappa$ | 4.718 | 2.599 | 1.638 | 1.513 | 1.470 | 1.415 |

which places the nose, $b(\pi)$, at $x = -2$, and the origin of the $(r, \theta)$ frame at $y = 0, x = x_0$.

Table 5.2 shows a comparison with the results of Rusanov [187] for a paraboloid blunt body. Figure 5.13 shows the behavior of the bow shock standoff distance for a paraboloid blunt body as a function of $\left(M_\infty^2 - 1\right)^{-1}$. Figure 5.14 shows the behavior of the location of the sonic line on the body surface as a function of $M_\infty^{-2}$. Unlike the elliptical cylinders and ellipsoids, the rate of growth of the sonic angle on the wall continues to increase as $M_\infty \to 1$.

To within the numerical accuracy of the results, the behavior of the standoff distance is linear and is given by

$$\Delta/r_b = 0.8998\chi + 0.1359,$$

where $\chi = \left(M_\infty^2 - 1\right)^{-1}$. The same results show that the radius of curvature of the bow shock, $r_s/r_b = 1/r_b\kappa$, at the symmetry line is given by

$$r_s/r_b = 1.22\chi^2 + 3.156\chi + 1.415,$$

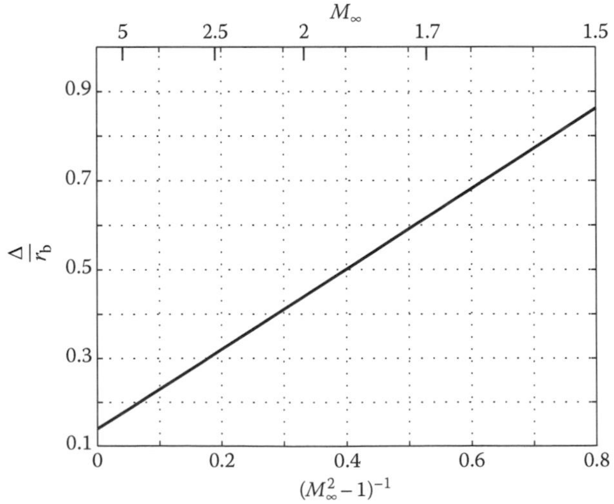


**FIGURE 5.13:** Bow shock standoff distance for paraboloid blunt body.

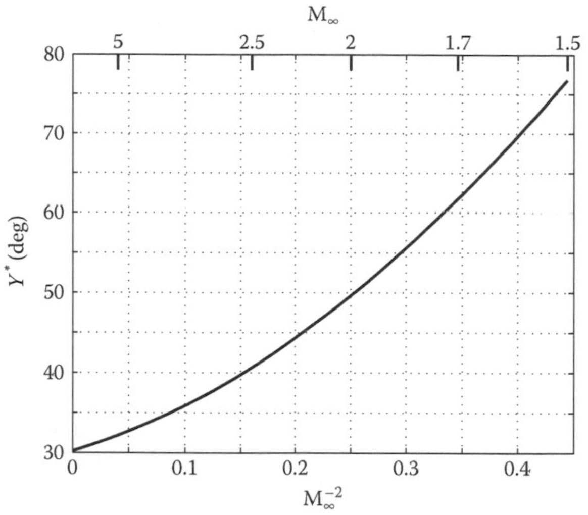


**FIGURE 5.14:** Body sonic angle for a paraboloid blunt body.

to within numerical accuracy. Assuming that near the symmetry line the bow shock has a parabolic profile, we find the following equation for the bow shock shape created by a paraboloid blunt body:

$$x_s = y^2/(2r_s/r_b) - (\Delta/r_b + 2). \tag{5.43}$$

When (5.43) is compared to bow shock shapes from numerical simulations, we find that (5.43) is accurate up to where the sonic line meets the bow shock.

Table 5.3 shows the results of calculations with five grid sizes for a paraboloid blunt body for $M_\infty = 2$ using both shock-fitting methods. The first column lists the grid size. $N$ is the number of intervals in the radial direction and $M$ is the number of intervals in the circumferential direction. Listed on the table are the bow shock radius of curvature, the bow shock standoff distance, the percent error of the standoff distance, and the body sonic angle. The table shows that accurate results are obtained even with a grid as small as $10 \times 20$. Figure 5.15 shows the Mach number contours for the $10 \times 20$ grid and for the $80 \times 160$ grid. The Mach contours for the $20 \times 40$ and $40 \times 80$ grids are visually identical to those of the $80 \times 160$ and are not shown.

### 5.1.10 Running the Blunt Body Code

The input parameters used to run the blunt body code are listed in Table 5.4, and are entered in the main program `BLUNT.M`. Parameters 2 through 5 are used to name data files that are used to restart a previous calculation either as a continuation run with the same grid or with a finer grid. If the finer grid option is used, the grid

**TABLE 5.3:** Results for a paraboloid at $M_\infty = 2$ and five different grid sizes with the outflow boundary at $Y_{\max} = 70°$.

| $N \times M$ | $1/r_b\kappa$ | $\Delta/r_b$ | % Err $\Delta/r_b$ | $Y^*$ |
|---|---|---|---|---|
| 5 × 10 {st} | 3.1579 | 0.4314 | 0.37 | 52.00 |
| 5 × 10 {nu} | 3.1945 | 0.4335 | 0.86 | 51.93 |
| 10 × 20 {st} | 2.8555 | 0.4304 | 0.14 | 50.32 |
| 10 × 20 {nu} | 2.8749 | 0.4310 | 0.28 | 50.33 |
| 20 × 40 {st} | 2.6760 | 0.4300 | 0.00 | 49.88 |
| 20 × 40 {nu} | 2.6834 | 0.4302 | 0.00 | 49.88 |
| 40 × 80 {st} | 2.6127 | 0.4298 | 0.00 | 49.75 |
| 40 × 80 {nu} | 2.6140 | 0.4298 | 0.00 | 49.75 |
| 80 × 160 {st} | 2.5991 | 0.4298 | 0.00 | 49.73 |
| 80 × 160 {nu} | 2.5985 | 0.4298 | 0.00 | 49.73 |

*Notes:* $N$ is the number of intervals in the radial direction and $M$ is the number of intervals in the circumferential direction. Listed are the shock radius of curvature at the symmetry line, the shock standoff distance, the percent error of the standoff distance based on the 80 × 160 solution, and the body sonic angle in degrees. The standard shock-fitting method is labeled {st}, and the nuovo shock-fitting method is labeled {nu}.

interval will be halved in both directions. Input parameters 6 and 7 control the number of intervals used to define the mesh. The code produces high-quality results with few mesh points, so be stingy with these inputs. The most important consideration about your choice of interval size is to try to maintain a reasonable mesh aspect ratio, a ratio between 1 and 2, as this has a strong influence on the solution. Input

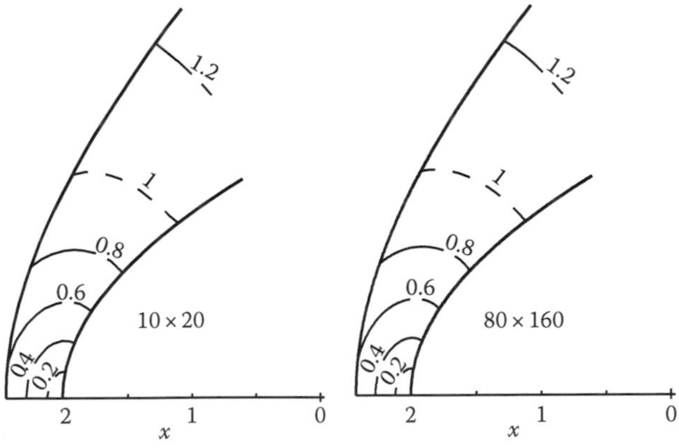


**FIGURE 5.15:** Mach number contours for $M_\infty = 2$ and a paraboloid blunt body. The left panel corresponds to a calculation with 10 radial intervals and 20 circumferential intervals; the right panel is for 80 radial intervals and 160 circumferential intervals.

**TABLE 5.4:** Input parameters for *BLUNT.M*.

| | | |
|---|---|---|
| 1 | `runnum` | Run number, to keep tract of runs |
| 2 | `loadrunint` | Name of file previously saved, this file is interpolated to finer mesh |
| 3 | `loadruncon` | Name of file previously saved, this file is used to continue calculation on same grid |
| 4 | `saverunana` | Name of file to save solution for analysis |
| 5 | `saveruncon` | Name of file to save solution for continuation |
| 6 | `na` | Number of intervals in radial direction |
| 7 | `ma` | Number of intervals in circumferential direction |
| 8 | `ka` | Number of time steps |
| 9 | `method` | 1 for standard shock-fitting<br>2 for nuovo shock-fitting |
| 10 | `stab` | CFL number (0.75) |
| 11 | `ja` | 0 no movie, otherwise every other ja frame is plotted |
| 12 | `blunt_movie` | 1 for movie of shock/sonic line shape<br>2 movie of shock speed |
| 13 | `jb` | 0 do not read previous solution<br>1 interpolate previous solution to finer mesh<br>2 continue with solution on same mesh |
| 14 | `jc` | 0 do not save solution, 1 save solution |
| 15 | `lb` | partial output printed every lb steps |
| 16 | `le` | 1 for elliptical $x$-section,<br>2 for parabolic $x$-section<br>3 for super-circular $x$-section |
| 17 | `ld` | 1 for plane flow<br>2 for axisymmetric flow |
| 18 | `ell` | ellipticity, for le = 1 |
| 19 | `supcir` | supcir >=2, exponent for super-circle or super-sphere, supcir =2 for circular $x$-section, can be a noninteger |
| 20 | `themax` | Outflow boundary angle, measured from $\theta = \pi$ |
| 21 | `x0` | Origin shift for $(r, \theta)$ frame for le = 1, 2 |
| 22 | `angle` | Angle (deg) for wedge or cone following blunt nose, used only with le = 1 |
| 23 | `mach` | Free stream Mach number |
| 24 | `gamma` | Isentropic exponent |

parameter 9 controls the method used for fitting the bow shock, see Sections 5.1.6.2 and 5.1.6.3. Input parameters 11 and 12 are for playing a movie of the evolution of the bow shock and sonic line (`blunt_movie` = 1) or the evolution of the shock speed $w$ (`blunt_movie` = 2). The movie displays every other `ja` frames as the calculation is made. The movie option is a very useful tool for debugging. Most other outputs are suppressed when this option is used. Input parameter 15 controls the output of parameters that monitor the progression toward a steady state. These parameters are the step counter, the time, the shock standoff distance, the *range*, defined as $\frac{1}{u_\infty}\sqrt{\sum_m^M w_m^2/M}$, the shock speed at $\theta = \pi$, the absolute value of the maximum shock speed, the relative error in the stagnation pressure, and the relative error in the body sonic pressure. A typical value for `lb` is 100. Input parameter 16 selects a body cross sectional shape from those that are currently available in the code. These shapes are defined by $b(\theta)$ and $b_\theta(\theta)$ in the function `BLUNT_GEOM.M`. In addition to the elliptical and parabolic cross sections, a super-circular cross section defined by

$$b(\theta) = (|\sin\,\theta|^k + |\cos\,\theta|^k)^{-\frac{1}{k}},$$
$$b_\theta(\theta) = -b^{1+k}\sin\,\theta\cos\,\theta(|\sin\,\theta|^{k-2} - |\cos\,\theta|^{k-2}),$$

is available with `le` = 3. The super-circular cross section can take noninteger values of the exponent $k$ (input parameter 19) and it results in flat-nosed bodies as shown in Figure 5.16. Neither the parabolic nor the super-circular cross sections have been tested for plane flows. We leave this as an exercise for the reader; some fine-tuning of initial conditions might be required for best performance. Input parameter 18 is used to define the ellipticity of elliptical cross sections. When running a super-circular cross section `ell` should be set to 1, since the initial conditions for a circular cross section are used with the super-circular cross section. Input parameter 22 is the angle, in degrees, of a cone or wedge following an elliptical blunt nose. The code figures out the point of tangency. The cone or wedge geometry will not come into play if `angle` is set to a large negative value.

Computations of shock layers with low free stream Mach numbers, Mach numbers less than 1.2, are handled with a *low Mach number strategy* which approaches the low free stream Mach number from above. The computation is

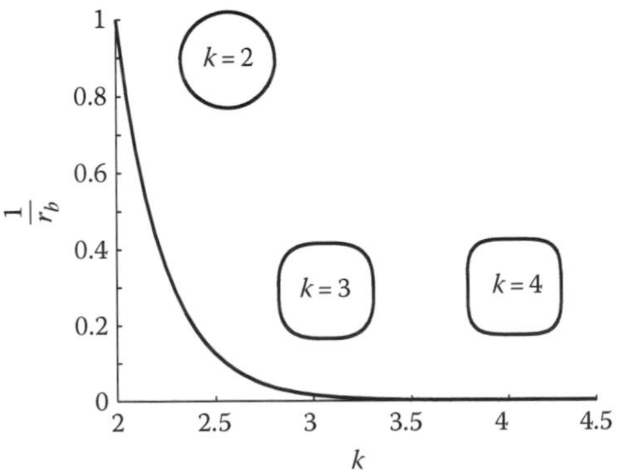


**FIGURE 5.16:** Nose curvature as a function of the exponent $k$.

started at a higher free Mach number than the desired value (the *target Mach number*) and is gradually reduced until the desired value is reached. The logic for this process is carried out in function BLUNT_RUN.M. When running a low free stream Mach number, use input parameters 2 through 5 to save the solution for a continuation run.

## 5.2 External Conical Corners

Conical flows are characterized by having constant flow properties along rays emanating from a common vertex. For this to be the case, the flow should be steady, inviscid, and supersonic. A distinguishing feature of conical flows is the absence of a reference length. The first study of conical flows was done by Busemann in 1929 [24]. In this section we will consider only conical flows representative of corners of the type shown in Figure 5.17. The figure shows two planes, shaded dark gray, forming an angle $\Omega + \pi/2$. If $\Omega = 0$, the two planes meet at a right angle, while if $\Omega = \pi/2$ the planes lie on the same plane. Lying on these planes are two wedges whose leading edges are swept $\Lambda$ degrees from the $x$-axis. The point where the leading edges meet is the vertex of the conical field. The wedge angles turn the flow by $\delta$ degrees in the $y$–$z$ plane. The surfaces of these wedges are labeled $\Sigma_1$ and $\Sigma_2$ in the figure. Where these two surfaces meet a corner is formed. This corner lies on a plane $\phi_c$ degrees from the surface $\Sigma_2$, see Figure 5.17. We will assume that $\delta_1 = \delta_2$ and $\Lambda_1 = \Lambda_2$, such that the configuration is symmetric with respect to the plane $\phi_c$ = constant. Furthermore, we will consider only cases where the leading

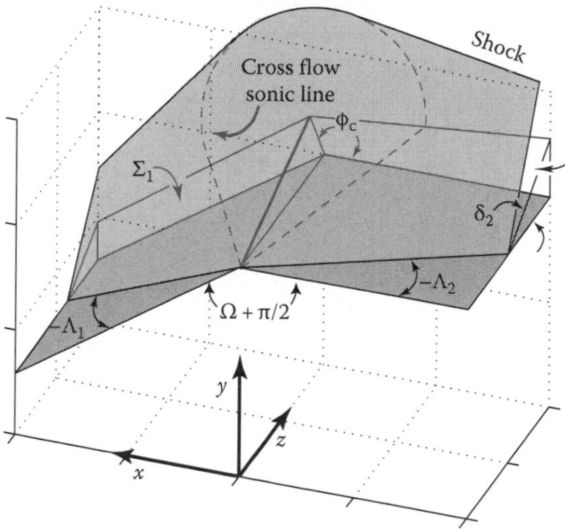


**FIGURE 5.17:** External corner geometry.

edges are swept forward as shown in the figure. If the leading edges are swept back, a cross flow shock forms and the disturbed region is larger than the region bounded by the Mach conoid. The location and shape of this cross flow shock would have to be obtained as part of the solution. It is also assumed that a supersonic flow directed along the $z$-axis impinges on this configuration generating a shock wave that is attached to the wedge leading edges, as shown in the figure. Away from the corner, the flow field between the shock and the wedge surface is two dimensional. This flow can be easily defined by solving the Rankine–Hugoniot conditions such that the flow downstream of the shock is tangent to the wedge surface. However, in the neighborhood of the corner, inside a region bounded by the cross flow sonic lines and a curved shock, the flow field is unknown. The subject of this section is the study of this flow field. Although, this flow has been the subject of several studies [9,11,189,191,195], and is fairly well understood, we hope to be able to add a few new nuances.

In the study of conical flows it is convenient to take advantage of the conical self-similarity by taking the projection of the flow field onto the surface of the unit sphere centered at the vertex. On this surface, the governing equations exhibit mixed-type behavior, i.e., elliptic-hyperbolic, not unlike the blunt body problem of the previous section. The projection of the streamlines onto the surface of the sphere in the region of interest, between the cross flow sonic lines and the curved shock, in general, appear to be directed toward the corner. The exact details of the topology of these cross flow streamlines are the subject of a large body of work. We begin by reviewing this work, as well as other aspects of the structure of this flow field, before we discuss the numerical method used to solve this problem.

### 5.2.1 Governing Equations

The steady Euler equations written in spherical coordinates are

$$\begin{aligned}
v\rho_\theta + \frac{w}{\sin\theta}\rho_\phi + \rho\left(v_\theta + \frac{w_\phi}{\sin\theta} + 2u + v\text{cotan }\theta\right) &= 0,\\
vu_\theta + \frac{w}{\sin\theta}u_\phi - (v^2 + w^2) &= 0,\\
vv_\theta + \frac{w}{\sin\theta}v_\phi + \frac{1}{\rho}p_\theta + uv - w^2\text{cotan }\theta &= 0,\\
vw_\theta + \frac{w}{\sin\theta}w_\phi + \frac{1}{\rho}\frac{p_\phi}{\sin\theta} + uw - vw\text{cotan }\theta &= 0,\\
vS_\theta + \frac{w}{\sin\theta}S_\phi &= 0,
\end{aligned} \tag{5.44}$$

where $r$ is the distance from the conical vertex, and since we are writing the equations on the unit sphere $r$ is set to one, $\theta$ is the polar angle and $\phi$ is the azimuthal angle. The velocity components are $u$, $v$, and $w$ along the $r$, $\theta$, and $\phi$ directions, respectively, see Figure 5.18, and as usual $\rho$, $p$, and $S$ are the density, pressure, and entropy.

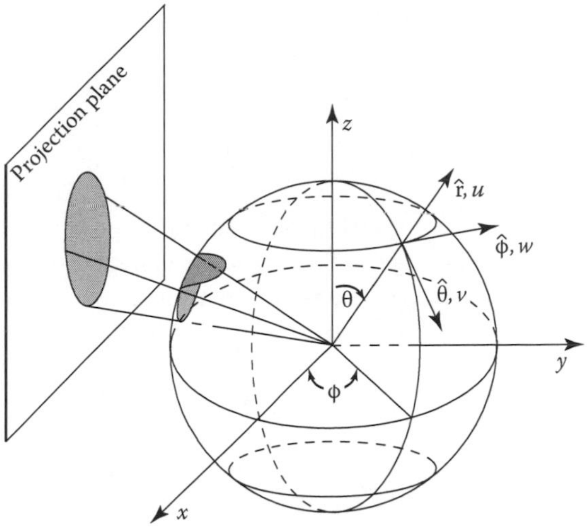


**FIGURE 5.18:** Spherical coordinates and projection plane.

If we introduce $q$ and $\beta$ as the magnitude of the cross flow velocity component and its inclination, respectively,

$$
\begin{aligned}
q^2 &= v^2 + w^2, \\
\tan\beta &= v/w,
\end{aligned}
\tag{5.45}
$$

Equations 5.44 can be written as

$$
\begin{aligned}
&\left(1 - \frac{a^2}{q^2}\right)\frac{1}{\gamma p} p_s + \beta_n + \frac{u}{q} + \sin\beta \,\text{cotan}\,\theta = 0, \\
&u_s = q, \\
&q_s + \frac{a^2}{\gamma q^2 p} p_s + u = 0, \\
&\beta_s + \frac{a^2}{\gamma q^2 p} p_n - \cos\beta \,\text{cotan}\,\theta = 0, \\
&S_s = 0,
\end{aligned}
\tag{5.46}
$$

where $a$ is the speed of sound and $s$ and $n$ are coordinates locally tangent and normal to the cross flow streamlines, such that

$$
\begin{aligned}
\frac{\partial}{\partial s} &= \sin\beta\frac{\partial}{\partial\theta} + \frac{\cos\beta}{\sin\theta}\frac{\partial}{\partial\phi}, \\
\frac{\partial}{\partial n} &= \cos\beta\frac{\partial}{\partial\theta} - \frac{\sin\beta}{\sin\theta}\frac{\partial}{\partial\phi}.
\end{aligned}
\tag{5.47}
$$

If we eliminate the pressure from the first equation of (5.46) and using the third equation of (5.46) the resulting equation is

$$\frac{1}{q}\left(1-\frac{q^2}{a^2}\right)q_s+\beta_n+\frac{u}{q}\left(2-\frac{q^2}{a^2}\right)+\sin\beta\,\text{cotan}\,\theta=0. \tag{5.48}$$

Equation 5.48 is elliptic in regions of the unit sphere where $q^2 < a^2$ and hyperbolic in regions where $q^2 > a^2$. The locus of points on the unit sphere where $q^2 = a^2$ form a boundary between these regions and is known as the cross flow sonic line.

The radial component of vorticity on the surface of the unit sphere is given by

$$\omega_r = q_n - q\beta_s + q\cos\beta\,\text{cotan}\,\theta. \tag{5.49}$$

If the flow is irrotational, then $\omega_r = 0$, and the fourth equation of (5.46) may be replaced by the following equation:

$$q_n + \frac{a^2}{\gamma q p}p_n = 0. \tag{5.50}$$

In hyperbolic regions we have at each point two real characteristic directions defined by

$$\frac{dn}{ds} = \pm\tan\mu, \tag{5.51}$$

where $\mu$ is the cross flow Mach angle defined in terms of the cross flow Mach number, $M = q/a$, as

$$\sin\mu = 1/M. \tag{5.52}$$

Along these curves, disturbances are propagated satisfying the compatibility condition obtained from the first and fourth of (5.46):

$$\frac{d\beta}{dn} \pm \frac{a^2}{q^2}\frac{\text{cotan}\,\mu}{\gamma p}\frac{dp}{dn} = \pm\text{cotan}\,\mu\cos\beta\,\text{cotan}\,\theta - \left(\frac{u}{q}+\sin\beta\,\text{cotan}\,\theta\right). \tag{5.53}$$

### 5.2.2 Flow Properties in the Neighborhood of the Sonic Line

For *two-dimensional irrotational* flows, the flow conditions along the sonic line are constant. The situation is quite different for conical flows. First let us derive the conservation of total enthalpy. If the flow is steady and $\boldsymbol{v}$ is the velocity vector, it follows from (3.40) that

$$\nabla\cdot(\rho H\boldsymbol{v}) = 0.$$

Expanding the divergence, we obtain

$$H\nabla(\rho \boldsymbol{v}) + \rho \boldsymbol{v} \cdot \nabla H = 0,$$

but from (3.36), in the steady state, $\nabla(\rho \boldsymbol{v}) = 0$, therefore

$$\boldsymbol{v} \cdot \nabla H = 0. \tag{5.54}$$

Equation 5.54 asserts that for an inviscid, steady flow the total enthalpy is conserved along streamlines. The total enthalpy might change from streamline to streamline, but if the flow is uniform at infinity, then the total enthalpy is the same everywhere. If this is the case, we say the flow is *homenergic*. Since the total enthalpy is conserved across a steady shock, a homenergic flow remains homenergic even in the presence of steady shocks. From this we can infer that a flow need not be isentropic or irrotational to be homenergic. Bernoulli's equation for a homenergic flow follows from (5.54):

$$\begin{aligned} \frac{a^2}{\gamma - 1} + \frac{u^2}{2} + \frac{q^2}{2} &= \frac{q_{\max}^2}{2}, \\ &= \gamma\left(\frac{M_\infty^2}{2} + \frac{1}{\gamma - 1}\right). \end{aligned} \tag{5.55}$$

From (5.55) it follows that along the cross flow sonic line

$$\frac{\mathrm{d}q}{\mathrm{d}l} = -\left(\frac{\gamma - 1}{\gamma + 1}\right)\frac{u^*}{q^*}\frac{\mathrm{d}u}{\mathrm{d}l}, \tag{5.56}$$

where

$\mathrm{d}l$ is an infinitesimal element of length along the sonic line

superscript * denotes critical conditions

The rate of change of entropy in a direction normal to a streamline is given by [190]

$$\frac{1}{\gamma}\frac{a^2}{\gamma - 1}\frac{\partial S}{\partial n} = -u\frac{\partial u}{\partial n} - q\omega_r, \tag{5.57}$$

and therefore, if the flow is irrotational,

$$\frac{\partial u}{\partial n} = 0. \tag{5.58}$$

Now, since

$$\frac{\mathrm{d}u}{\mathrm{d}l} = \frac{\partial u}{\partial n}\frac{\partial n}{\partial l} + \frac{\partial u}{\partial s}\frac{\partial s}{\partial l},$$

it follows that

$$\frac{du}{dl} = q^* \frac{ds}{dl},$$

which, with reference to Figure 5.19, can be written in the form

$$\frac{du}{dl} = q^* \cos \alpha,$$

and therefore

$$\frac{dq}{dl} = -\left(\frac{\gamma-1}{\gamma+1}\right) u^* \cos \alpha. \tag{5.59}$$

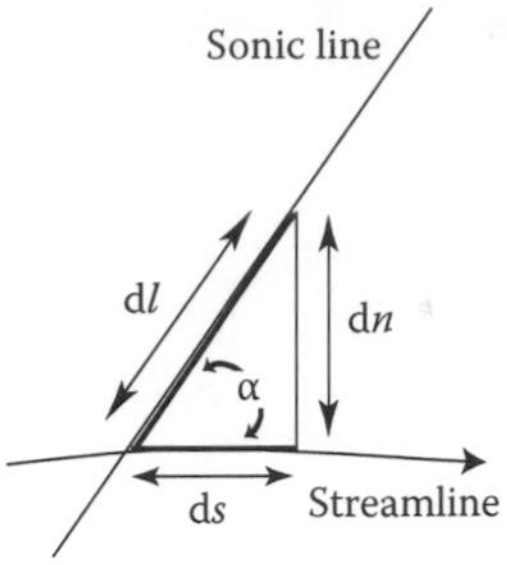


**FIGURE 5.19:** Inclination of sonic line to streamline.

Equation 5.59 was first introduced by Ferri in [75] in a slightly different form. The equation shows that the cross flow velocity increases along the sonic line if the angle that the cross flow velocity makes with the sonic line is less than $\pi/2$. From this relation it is easy to obtain the behavior of other flow quantities. For example, the change in the speed of sound, radial component of Mach number, and the pressure are given, respectively, by

$$\frac{da}{dl} = -\left(\frac{\gamma-1}{\gamma+1}\right) u^* \cos \alpha,$$

$$\frac{d(u/a)}{dl} = \left(1 + \frac{\gamma-1}{\gamma+1}\left(\frac{u^*}{a^*}\right)^2\right) \cos \alpha,$$

$$\frac{1}{p^*}\frac{dp}{dl} = -\frac{2\gamma}{\gamma+1}\left(\frac{u^*}{a^*}\right) \cos \alpha.$$

If we assume that in the neighborhood of the sonic line the flow is irrotational and the flow properties are analytic functions of $s$ and $n$, then a Taylor series expansion about the sonic line gives

$$\begin{aligned}
q &= a^* + q_n^* n + q_s^* s + \dots,\\
u &= u^* + a^* s + \dots,\\
a &= a^* - \frac{\gamma-1}{2} q_n^* n - \frac{\gamma-1}{2}(u^* + q_s^*) s + \dots,\\
\beta &= \beta^* + \left(\cos \beta^* \operatorname{cotan} \theta^* + \frac{1}{a^*} q_n^*\right) s - \left(\frac{u^*}{a^*} + \sin \beta^* \operatorname{cotan} \theta^*\right) n + \dots,\\
p &= p^*\left(1 - \gamma\left(\frac{u^*}{a^*}\right)\left(1 + \frac{1}{u^*} q_s^*\right) s - \frac{\gamma}{a^*} q_n^* n + \dots\right).
\end{aligned} \tag{5.60}$$

Introducing the above Taylor expansion into the definition of the cross flow sonic line, we obtain the expression for the sonic line slope:

$$\frac{dn}{ds} \equiv \tan\alpha = \frac{-\left(\frac{\gamma-1}{\gamma+1} + \frac{1}{u^*}q_s^*\right)}{\frac{1}{u^*}q_n^*}. \tag{5.61}$$

Similarly the slope of the isobar at the sonic line is given by

$$\frac{dn}{ds} \equiv \frac{-p_s^*}{p_n^*} = -\frac{1+\frac{1}{u^*}q_s^*}{\frac{1}{u^*}q_n^*}.$$

The angle made by the cross flow Mach lines and the meridional plane $\phi$, at some point along the sonic line is given by $\varepsilon = \beta \pm \mu$. A measure of the curvature of the Mach lines, $d\varepsilon/dn$, can be obtained from the compatibility condition, Equation 5.53. To first order we find

$$\frac{d\varepsilon}{dn} \approx \pm\frac{1}{a^*}\frac{\gamma+1}{2}\frac{1}{\cos\mu^*}\frac{dq}{dn}\bigg|^*.$$

At a *singular point* of the sonic line, the sonic line becomes tangent to the Mach line and a weak singularity, i.e., a jump in first derivatives, may occur across the sonic line. From Equation 5.61, it follows that at such a point

$$\frac{\partial q}{\partial n}\bigg|^* = 0,$$

but since

$$\frac{dq}{dn} = \frac{\partial q}{\partial n} + \frac{\partial q}{\partial s}\frac{ds}{dn} = \frac{\partial q}{\partial n} + \frac{\partial q}{\partial s}\frac{1}{\tan\alpha},$$

it follows that $dq/dn = 0$ and the curvature of the Mach line is finite. At a *non-singular point* the curvature is infinite.

To obtain the shapes of the Mach lines near a sonic point we expand the coordinates $s$ and $n$ in a power series:

$$s = \sum_{i=1}^{\infty} b_i \nu^i, \qquad n = \sum_{i=1}^{\infty} c_i \nu^i, \tag{5.62}$$

where $\nu = \pi/2 - \mu$. First we consider a nonsingular sonic point. Since for this case both the slope and curvature of the Mach lines are infinite, we set $b_1 = b_2 = 0$. The remaining coefficients are found by expanding both the Mach line slope

$$\frac{dn}{ds} = \pm\text{cotan}\,\nu,$$

and the strip condition satisfied on a Mach line

$$\frac{dq}{d\nu} = q_n^* \frac{dn}{d\nu} + q_s^* \frac{ds}{d\nu},$$

with the aid of (5.55), (5.60), and (5.62), and equating coefficients of equal powers. After some manipulation, we find

$$s = \pm\frac{2}{3}\frac{a^*}{\gamma+1}\frac{1}{q_n^*}\nu^3 + \cdots,$$
$$n = \frac{a^*}{\gamma+1}\frac{1}{q_n^*}\nu^2 \mp \frac{2}{3}\frac{a^*}{\gamma+1}\frac{1}{q_n^{*2}}\left(\frac{\gamma-1}{\gamma+1}u^* + q_s^*\right)\nu^3 + \cdots.$$

The leading terms are the same as in two-dimensional flow [100], but the higher order terms are different. As in two-dimensional flow, near a sonic point each Mach line behaves like one branch of the semi-cubical parabola

$$n^3 = \frac{9}{4}\frac{a^*}{\gamma+1}\frac{1}{q_n^*}s^2 \pm \ldots,$$

with a cusp at the sonic point. Now consider a singular sonic point. Since the curvature of the Mach lines is finite we must retain the coefficient $b_2$. Then proceeding as before, we find the leading terms of the expansion:

$$s = \frac{a^*}{(\gamma-1)u^* + (\gamma+1)q_s^*}\nu^2 + \ldots$$
$$n = \pm\frac{2a^*}{(\gamma-1)u^* + (\gamma+1)q_s^*}\nu + \ldots.$$

Therefore, near a singular sonic point the Mach lines behave like a branch of the parabola

$$n^2 = \frac{4a^*}{(\gamma-1)u^* + (\gamma+1)q_s^*}s \pm \ldots,$$

which, unlike the two-dimensional case, defines only two real Mach lines at the sonic point.

### 5.2.3 Jump Conditions at a Singular Sonic Point

The external corner configuration we will study has a cross flow sonic line that is singular. Here we will explore in more detail the flow behavior in the neighborhood of this line. In the linearized theory of conical flows the pressure downstream of the sonic line behaves like

$$p \sim \pm\sqrt{s},$$

so that the pressure gradient becomes unbounded as $s \to 0$. Lighthill [126] has shown that the linear theory is the first term in a series that diverges near the sonic line and that the true behavior of the pressure gradient is

$$\left.\frac{1}{p^*}\frac{\partial p}{\partial s}\right|^* \sim -\frac{2\gamma}{\gamma+1}\frac{M_\infty^2-1}{M_\infty^2}, \quad s>0,$$
$$\left.\frac{1}{p^*}\frac{\partial p}{\partial s}\right|^* = 0, \quad s<0, \tag{5.63}$$

assuming that the pressure decreases downstream of the sonic line, and

$$\left.\frac{1}{p^*}\frac{\partial p}{\partial s}\right|^* \sim \frac{\gamma}{\gamma+1}\frac{M_\infty^2-1}{M_\infty^2}, \quad s>0,$$
$$\left.\frac{1}{p^*}\frac{\partial p}{\partial s}\right|^* = 0, \quad s<0,$$

if the pressure increases; where $M_\infty$ is the free stream Mach number. Note that although the pressure gradient is finite, the jump in value makes the first derivative singular.

The constraints on the first derivative jumps may be obtained by examining Equations 5.46. Let us denote the jump by $[\![f]\!] = f_+ - f_-$ where the $+$ and $-$ subscripts indicated downstream and upstream values, respectively. We find,

$$[\![S_n]\!] = [\![S_s]\!] = [\![u_n]\!] = [\![u_s]\!] = [\![\beta_n]\!] = [\![\beta_s]\!] = [\![p_n]\!] = [\![q_n]\!] = 0,$$

and

$$[\![q_s]\!] = -\frac{a^*}{\gamma p^*}[\![p_s]\!] = -\frac{2}{\gamma-1}[\![a_s]\!].$$

Consider now the second of (5.46). Differentiating this equation with respect to $s$ and using the third of (5.46):

$$u_{ss} + u = -\frac{a^2}{\gamma p q}p_s. \tag{5.64}$$

If we assume that near the sonic point the right-hand side of (5.64) behaves like a delta function $h(s)$ such that

$$h(s) = \begin{cases} h_+, & s>0, \\ h_-, & s<0, \end{cases}$$

and $u$ satisfies the boundary conditions:

$$u(0) = u^*,$$
$$u_s(0) = a^*,$$

we find,

$$\frac{u}{u^*} = \left(1 - \frac{h}{u^*}\right)\cos s + \frac{a^*}{u^*}\sin s + \frac{h}{u^*},$$
$$\frac{q}{u^*} = -\left(1 - \frac{h}{u^*}\right)\sin s + \frac{a^*}{u^*}\cos s.$$

The speed of sound follows from Equation 5.55 and, assuming isentropic flow, the pressure is given by

$$p = \left(\frac{a^2}{\gamma}\right)^{-\frac{\gamma-1}{\gamma}}. \tag{5.65}$$

For the external corner problem, Roe [180] has found the delta function at the point on the wall where the cross flow sonic line originates:

$$h(s) = \begin{cases} \dfrac{u^*}{\gamma+1}, & s > 0, \\ 0, & s < 0. \end{cases} \tag{5.66}$$

Roe's derivation begins by using the first of (5.46) to write

$$\frac{1}{p}p_s = -\frac{\gamma q^2\left(\beta_n + \dfrac{u}{q} + \sin\beta \operatorname{cotan}\theta\right)}{(q^2 - a^2)}.$$

When this expression is evaluated at the point where the sonic line meets the wall, both numerator and denominator vanish. To continue, we apply l'Hôspitals rule and use (5.49), assuming that the flow is irrotational, to evaluate $\beta_{ns}$. The result is

$$\frac{1}{p^*}p_s^* = -\frac{\gamma}{\gamma+1}\frac{u^*}{a^*}, \tag{5.67}$$

which leads to (5.66). Lighthill's result (5.63) is an approximation to (5.67) for weak shocks.

### 5.2.4 Flow Behavior at a Non-Stagnating Corner Point

If a pressure difference exists across a corner point,* flow will spill over from the high pressure side to the low pressure side of the corner. Let the axis of the spherical coordinates lie along the axial corner such that the plane $\phi = 0$ lies on the surface of the wall $\Sigma_2$ and $\theta = 0$ corresponds to the corner. To obtain a set of equations valid

* The corner point is not on a symmetry plane.

at the corner, multiply (5.44) by $\theta$ and take the limit as $\theta \to 0$. The resulting equations are

$$
\begin{aligned}
& w\rho_\phi + \rho(w_\phi + v) = 0, \\
& wu_\phi = 0, \\
& \rho w(v_\phi - w) = 0, \\
& \rho w(w_\phi + v) + p_\phi = 0, \\
& wS_\phi = 0.
\end{aligned}
\tag{5.68}
$$

Using the definition of entropy to eliminate the density in (5.68), we find

$$
(w^2 - a^2)(w_\phi + v) = 0. \tag{5.69}
$$

Therefore, two solutions are possible. One corresponding to $w^2 - a^2 = 0$ is the conical analog of a Prandtl–Meyer expansion, the other corresponding to $w_\phi + v = 0$ is valid at a cross flow stagnation point when $\theta = 0$.

Integration of the third equation of (5.68) results in the following equations that define the flow variables at the corner as a function of an angle $\upsilon$ measured from the perpendicular to the oncoming cross flow direction:

$$
\begin{aligned}
v &= \sqrt{q_{\max}^2 - u^{*2}} \sin\left(\sqrt{\frac{\gamma - 1}{\gamma + 1}}\upsilon\right), \\
w &= \sqrt{\frac{\gamma - 1}{\gamma + 1}}\sqrt{q_{\max}^2 - u^{*2}} \cos\left(\sqrt{\frac{\gamma - 1}{\gamma + 1}}\upsilon\right), \\
p &= p^*\left[\cos\left(\sqrt{\frac{\gamma - 1}{\gamma + 1}}\upsilon\right)\right]^{\frac{2\gamma}{\gamma-1}}, \\
\rho &= \rho^*\left[\cos\left(\sqrt{\frac{\gamma - 1}{\gamma + 1}}\upsilon\right)\right]^{\frac{2}{\gamma-1}}, \\
S &= S^*, \\
u &= u^*.
\end{aligned}
$$

The usual Prandtl–Meyer relation between the flow inclination and the Mach number is given by

$$
\beta = -\sqrt{\frac{\gamma - 1}{\gamma + 1}}\, a\tan\left(\sqrt{\frac{\gamma - 1}{\gamma + 1}}\sqrt{M^2 - 1}\right) + a\tan\left(\sqrt{M^2 - 1}\right),
$$

where $M$ is the cross flow Mach number.

### 5.2.5 Flow Properties in the Neighborhood of a Cross-Flow Stagnation Point

If the flow is steady, the entropy is constant along a cross flow streamline. Hence on a cross flow streamline

$$dS = \frac{\partial S}{\partial \varphi} + \frac{\partial S}{\partial \theta}\frac{d\theta}{d\phi}\bigg|_{sl} = 0,$$

where $\frac{d\theta}{d\phi}\Big|_{sl}$ denotes the slope of the streamline. Therefore, with the last of (5.44) we write the equation for the cross flow streamline as

$$\frac{d\theta}{d\phi}\bigg|_{sl} = \frac{v \sin \theta}{w}. \tag{5.70}$$

Equation 5.70 provides the direction field of the streamlines at every point on the unit sphere except at points where both $v$ and $w$ vanish. These points are singular points of (5.70) and correspond to cross flow stagnation points. Assume that a singular point occurs at $(\theta_0, \phi_0)$ and introduce new coordinates

$$\vartheta = \theta - \theta_0,$$
$$\psi = (\phi - \phi_0) \sin \theta_0,$$

so that we may write (5.70) in the form

$$\frac{d\vartheta}{d\psi} = \frac{v(\vartheta, \psi)}{w(\vartheta, \psi)}. \tag{5.71}$$

If along a streamline, we consider $\vartheta$ and $\psi$ to be functions of a parameter $t$, we can write (5.71) in the more convenient form

$$\begin{aligned}\frac{d\vartheta}{dt} &= v(\vartheta, \psi),\\ \frac{d\psi}{dt} &= w(\vartheta, \psi).\end{aligned} \tag{5.72}$$

If there are no other singular points near the origin, and $v$ and $w$ are analytic functions of $\vartheta$ and $\psi$ near the origin, we can expand $v$ and $w$ in a Taylor series*:

* The appearance of other singularities, in the real plane or in the complex plane, reduces the radius of convergence of the Taylor series. Since we are interested in the bifurcation of the flow pattern, other singularities will be present, and, as we will see in Section 5.2.8, $v$ and $w$ are not necessarily analytic at the origin.

$$v = v_\vartheta \vartheta + v_\psi \psi + V(\vartheta, \psi),$$
$$w = w_\vartheta \vartheta + w_\psi \psi + W(\vartheta, \psi), \tag{5.73}$$

where it is understood that the derivatives are evaluated at the origin and $V$ and $W$ are at least second order in $\vartheta$ and $\psi$. For sufficiently small $\vartheta$ and $\psi$, (5.73) can be approximated by

$$v = v_\vartheta \vartheta + v_\psi \psi,$$
$$w = w_\vartheta \vartheta + w_\psi \psi.$$

Thus, we have replaced the nonlinear system (5.72) by the simpler linear system

$$\frac{d\vartheta}{dt} = v_\vartheta \vartheta + v_\psi \psi,$$
$$\frac{d\psi}{dt} = w_\vartheta \vartheta + w_\psi \psi. \tag{5.74}$$

The solution of (5.74) is of the form

$$\vartheta = C_1 e^{\Gamma t}, \quad \psi = C_2 e^{\Gamma t}. \tag{5.75}$$

Substituting (5.75) into (5.74), we find a solution if $\Gamma$ satisfies the characteristic equation

$$\Gamma^2 + h\Gamma + g = 0, \tag{5.76}$$

where the trace,* $h$, and the Jacobian, $g$, are $h = -(v_\vartheta + w_\psi)$, $g = v_\vartheta w_\psi - v_\psi w_\vartheta$. The solutions of (5.76) (eigenvalues) are

$$\Gamma_{1,2} = -\frac{1}{2}h \pm \frac{1}{2}\sqrt{h^2 - 4g}.$$

In terms of the eigenvalues, $\Gamma_{1,2}$, we have $h = -(\Gamma_1 + \Gamma_2)$ and $g = \Gamma_1\Gamma_2$. The type of solutions we find depend on the value of $h$ and the discriminant $\Delta = h^2 - 4g = (v_\vartheta - w_\psi)^2 + 4v_\psi w_\vartheta = (\Gamma_1 - \Gamma_2)^2$. By constructing solutions for various values of $h$ and $\Delta$, we classify the singularities as shown in Table 5.5. The stable node and saddle singularities are illustrated in Figure 5.20.

Before proceeding any further, let us return to the $\theta$, $\phi$ coordinate system. In this system we have

$$h = -\left(v_\theta + \frac{w_\phi}{\sin \theta_0}\right), \tag{5.77}$$

* Sum of diagonal terms.

**TABLE 5.5:** Classification of singularities according to $\Delta$ and $h$.

| $\Delta, g, h$ | Singularity Type | Stable If |
|---|---|---|
| $\Delta = 0$ | Node | $h > 0$ |
| $\Delta > 0,\ g > 0$ | Node | $h > 0$ |
| $\Delta > 0,\ g < 0$ | Saddle | Always |
| $\Delta < 0,\ h = 0$ | Center | Always |
| $\Delta < 0,\ h \neq 0$ | Spiral | $h > 0$ |

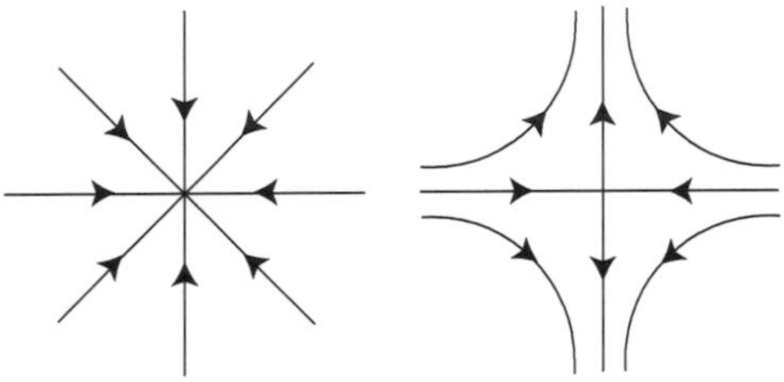


**FIGURE 5.20:** Illustration of stable node and saddle singularities.

$$g = \frac{v_\theta w_\phi - v_\phi w_\theta}{\sin\,\theta_0}, \tag{5.78}$$

$$\Delta = \left(v_\theta - \frac{w_\phi}{\sin\,\theta_0}\right)^2 + 4w_\theta \frac{v_\phi}{\sin\,\theta_0}, \tag{5.79}$$

and we recall that all derivatives are evaluated at the stagnation point.

In order to see the conditions imposed by the governing equations (5.44) on the above quantities, let us first rewrite (5.44) with the density replaced by the log of the pressure, $P$:

$$\begin{aligned}
vP_\theta + \frac{w}{\sin\,\theta}P_\phi + \gamma\left(v_\theta + \frac{w_\phi}{\sin\,\theta} + 2u + v\text{cotan}\,\theta\right) &= 0,\\
vu_\theta + \frac{w}{\sin\,\theta}u_\phi - (v^2 + w^2) &= 0,\\
vv_\theta + \frac{w}{\sin\,\theta}v_\phi + \frac{a^2}{\gamma}P_\theta + uv - w^2\text{cotan}\,\theta &= 0,\\
vw_\theta + \frac{w}{\sin\,\theta}w_\phi + \frac{a^2}{\gamma}\frac{P_\phi}{\sin\,\theta} + uw - vw\text{cotan}\,\theta &= 0,\\
vS_\theta + \frac{w}{\sin\,\theta}S_\phi &= 0.
\end{aligned} \tag{5.80}$$

Evaluating (5.80) at the stagnation point, we find

$$\begin{aligned} v_\theta + \frac{w_\phi}{\sin\theta_0} + 2u_0 &= 0, \\ P_\theta &= 0, \\ P_\phi &= 0. \end{aligned} \tag{5.81}$$

We obtain three additional relations as follows. The first one is obtained by differentiating the third equation of (5.80) with respect to $\phi$ and the fourth with respect to $\theta$, then eliminating the mixed pressure derivative. The two others are obtained by taking derivatives with respect to $\phi$ and $\theta$ of the second equation of (5.80). The resulting equations are evaluated at the stagnation point to yield

$$\begin{aligned} \left(\frac{v_\phi}{\sin\theta_0} - w_\theta\right)\left(v_\theta + \frac{w_\phi}{\sin\theta_0} + u_0\right) &= 0, \\ \left(\frac{v_\theta w_\phi - v_\phi w_\theta}{\sin\theta_0}\right)u_\phi &= 0, \\ \left(\frac{v_\theta w_\phi - v_\phi w_\theta}{\sin\theta_0}\right)u_\theta &= 0. \end{aligned} \tag{5.82}$$

From the first of (5.81) and the first of (5.82), we conclude that at a stagnation point

$$w_\theta = \frac{v_\phi}{\sin\theta_0}. \tag{5.83}$$

The term in parenthesis on the last two equations of (5.82) is the Jacobian, which by assumption is not zero, therefore $u_\theta = u_\phi = 0$ at a stagnation point. Using the first of (5.81) and (5.83), we can write $g, h$, and $\Delta$ as

$$\begin{aligned} h &= 2u_0, \\ g &= -\left(w_\theta^2 + (v_\theta + u_0)^2 - u_0^2\right), \\ \Delta &= \left(v_\theta - \frac{w_\phi}{\sin\theta_0}\right)^2 + 4w_\theta^2. \end{aligned} \tag{5.84}$$

We conclude that $h > 0$ and $\Delta \geq 0$, and therefore, only stable nodal ($g > 0$) and saddle ($g < 0$) points occur in these flows. It is now clear that the nature of the singularity is governed by the Jacobian, which we rewrite as

$$\bar{g} = \frac{g}{u_0^2} = 1 - (\xi + 1)^2 - \eta^2, \tag{5.85}$$

where

$$\xi = v_\theta / u_0$$

$$\eta = w_\theta / u_0$$

In the $\xi, \eta$ phase plane $\bar{g} = 0$ is a circle with unit radius, centered at $\xi = -1, \eta = 0$. Nodal singularities lie inside the circle and saddle singularities lie outside. Along the circle, the nature of the singularity is not defined by the linear system (5.74). To study the nature of the singularity along $\bar{g} = 0$ we have to include higher order terms in the analysis.

Let us consider again the external corner shown in Figure 5.17. Let the axis of the spherical coordinates lie along the axial corner such that the plane $\phi = 0$ lies on the surface of the wall $\Sigma_2$. The symmetry plane would also correspond to a $\phi$ = constant plane. It follows that $\eta = 0$ on these two planes, since $w = 0$ along these two planes. The Jacobian simplifies to

$$\bar{g} = 1 - (\xi + 1)^2. \tag{5.86}$$

$\bar{g}(\xi)$ is the parabola shown in Figure 5.21. We now make the following important observation: the Jacobian is an intrinsic property of the singularity. Its value is the same whether we measure $\xi$ along the wall or along the symmetry plane. This implies that $\xi$ along the wall, $\xi_w$, is connected to $\xi$ along the symmetry plane, $\xi_s$, by the relation

$$2 + \xi_w + \xi_s = 0. \tag{5.87}$$

We say that a flow pattern bifurcates when small changes in some flow parameter cause a topological structural change in the flow pattern. An example of bifurcation is when a node changes to a saddle and two nodes. The bifurcation occurs when the Jacobian vanishes. Therefore, if for this bifurcation $\xi_w = 0$, then we must also have $\xi_s = -2$. The same is true if we measure $\xi$ along any $\phi$-plane. In that case, Equation 5.85 would define $\eta$ such that $\bar{g}$ is constant. The case $\xi_w = \xi_s$ occurs only when the Jacobian is a maximum. This is the case when a two-dimensional uniform flow is viewed in spherical coordinates. For this case $u = \sqrt{\gamma}M_\infty \cos\theta$, $v = -\sqrt{\gamma}M_\infty \sin\theta$, and $w = 0$, hence at $\theta = 0$ we have $u_0 = \sqrt{\gamma}M_\infty$, $v_\theta/u_0 = -1$, $w_\theta/u_0 = 0$, and $\bar{g} = 1$.

### 5.2.6 Higher Order Singularities

When the Jacobian vanishes, the structure of the flow goes through a bifurcation. In order to understand the nature of the flow when this takes place we must include

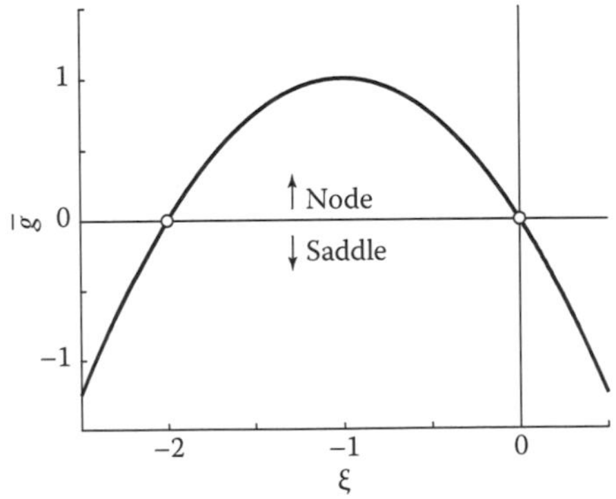


**FIGURE 5.21:** Jacobian as a function of $\xi$.

the higher order terms $V_2$ and $W_2$ in the analysis. The mathematical analysis of this problem is treated by Andronov et al. [3]. Bakker et al. [10] and Bakker [5] have studied the behavior of higher order singularities of conical flows assuming potential flow. Since the problems we are interested in are strongly rotational, we will avoid, whenever possible, making that assumption.

Let us assume that we are investigating the flow about an external symmetrical corner. Let $\theta_0, \phi_0$ be the coordinates of the corner point, and let the flow conditions be such that at the corner stagnation point the Jacobian vanishes and $\xi_s = v_\theta/u_0 = -2$. Let the spherical coordinates have its center at the vertex and let the $\phi = 0$ corresponds to the $y =$ constant plane on which the compression wedge $\delta_2$ lies, see Figure 5.17. The symmetry plane, $\phi_c$, corresponds to the plane $\phi_0 = \phi_c$. This choice of $\phi = 0$ is made to ensure that the angle $\theta_0$ is different from zero. At the corner we have $v_\vartheta = -2u_0$, $v_\psi = 0$, $w_\vartheta = 0$, and $w_\psi = 0$, therefore, the expansion of 5.73 leads to

$$\begin{aligned}\frac{d\vartheta}{dt} &= -2u_0\vartheta + v_{\vartheta\vartheta}\frac{\vartheta^2}{2} + v_{\vartheta\psi}\vartheta\psi + v_{\psi\psi}\frac{\psi^2}{2}\\ &\quad + v_{\vartheta\vartheta\vartheta}\frac{\vartheta^3}{6} + v_{\vartheta\vartheta\psi}\frac{\vartheta^2\psi}{2} + v_{\vartheta\psi\psi}\frac{\vartheta\psi^2}{2} + v_{\psi\psi\psi}\frac{\psi^3}{6} + \dots,\\ \frac{d\psi}{dt} &= w_{\vartheta\vartheta}\frac{\vartheta^2}{2} + w_{\vartheta\psi}\vartheta\psi + w_{\psi\psi}\frac{\psi^2}{2}\\ &\quad + w_{\vartheta\vartheta\vartheta}\frac{\vartheta^3}{6} + w_{\vartheta\vartheta\psi}\frac{\vartheta^2\psi}{2} + w_{\vartheta\psi\psi}\frac{\vartheta\psi^2}{2} + w_{\psi\psi\psi}\frac{\psi^3}{6} + \dots.\end{aligned} \tag{5.88}$$

To determine the type of higher order singularly that is compatible with (5.88) we need to apply Theorem 65 of [3]. To do this, we first must transform (5.88) to its *canonical* form:

$$\begin{aligned}\frac{d\vartheta}{dt} &= \vartheta + \bar{V}_2(\vartheta, \psi),\\ \frac{d\psi}{dt} &= \bar{W}_2(\vartheta, \psi).\end{aligned} \tag{5.89}$$

This is easily accomplished by the transformation $\bar{t} = -2u_0 t$. Therefore we rewrite (5.88) as

$$\begin{aligned}\frac{d\vartheta}{d\bar{t}} &= \vartheta - v_{\vartheta\vartheta}\frac{\vartheta^2}{4u_0} - v_{\vartheta\psi}\frac{\vartheta\psi}{2u_0} + v_{\psi\psi}\frac{\psi^2}{4u_0}\\ &\quad - v_{\vartheta\vartheta\vartheta}\frac{\vartheta^3}{12u_0} - v_{\vartheta\vartheta\psi}\frac{\vartheta^2\psi}{4u_0} - v_{\vartheta\psi\psi}\frac{\vartheta\psi^2}{4u_0} - v_{\psi\psi\psi}\frac{\psi^3}{12u_0} + \dots,\\ \frac{d\psi}{d\bar{t}} &= -w_{\vartheta\vartheta}\frac{\vartheta^2}{4u_0} - w_{\vartheta\psi}\frac{\vartheta\psi}{2u_0} - w_{\psi\psi}\frac{\psi^2}{4u_0}\\ &\quad - w_{\vartheta\vartheta\vartheta}\frac{\vartheta^3}{12u_0} - w_{\vartheta\vartheta\psi}\frac{\vartheta^2\psi}{4u_0} - w_{\vartheta\psi\psi}\frac{\vartheta\psi^2}{4u_0} - w_{\psi\psi\psi}\frac{\psi^3}{12u_0} + \dots.\end{aligned}$$

We now want to find the function $\vartheta = f(\psi)$ that solves the equation $\vartheta + \bar{V}_2(\vartheta, \psi) = 0$. The form of $\bar{V}_2$ suggests we try

$$f(\psi) = c_1\psi + c_2\psi^2 + c_3\psi^3 + \dots. \tag{5.90}$$

Substituting (5.90) into $\vartheta + \bar{V}_2(\vartheta, \psi) = 0$ we find

$$c_1 = 0, \quad c_2 = \frac{v_{\psi\psi}}{4u_0}, \quad f(\psi) = \frac{v_{\psi\psi}}{4u_0}\psi^2 + O(\psi^3).$$

Therefore we have

$$\begin{aligned} \vartheta &= \frac{v_{\psi\psi}}{4u_0}\psi^2 + O(\psi^3), \\ \frac{\mathrm{d}\psi}{\mathrm{d}\bar{t}} &= -w_{\psi\psi}\frac{\psi^2}{4u_0} - \left(w_{\vartheta\psi}\frac{v_{\psi\psi}}{8u_0^2} + \frac{w_{\psi\psi\psi}}{12u_0}\right)\psi^3 + O(\psi^4). \end{aligned} \tag{5.91}$$

Theorem 65 of [3] classifies the singularities corresponding to (5.91) as follows. Let **P** (0, 0) be an isolated equilibrium state of system (5.89) where $\bar{V}_2$ and $\bar{W}_2$ are analytic functions in some small neighborhood around **P**(0, 0).* Let $\vartheta = f(\psi)$ be a solution of the equation $\vartheta + \bar{V}_2(\vartheta, \psi) = 0^{\dagger}$ in the neighborhood of **P**(0, 0), and assume that the series expansion of the function $h(\psi) = \bar{W}_2(f(\psi), \psi)$ has the form $h(\psi) = \Delta_m\psi^m + \dots$, where $\Delta_m$ is the first nonvanishing coefficient and $m \geq 2$. Then,

1. If $m$ is odd and $\Delta_m > 0$, **P**(0, 0) is a *topological node.*
2. If $m$ is odd and $\Delta_m < 0$, **P**(0, 0) is a *topological saddle.*
3. If $m$ is even, **P**(0, 0) is a *saddle node.*

The streamline patterns corresponding to these higher order singularities are illustrated in Figure 5.22. To complete the analysis we must be able to evaluate the

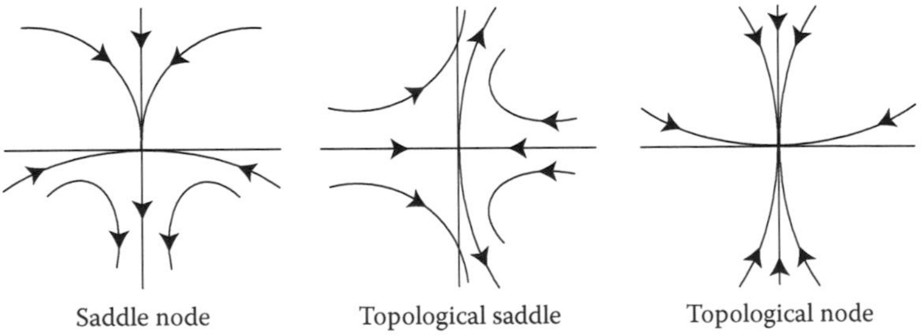


**FIGURE 5.22:** Streamline patterns for higher order singularities.

* The flows we are interested in do not necessarily satisfy this condition.
† By the implicit function theorem there is only one solution.

derivatives in (5.91). Since $w$ is antisymmetric about $\phi = \phi_c$, $w_{\phi\phi} = 0$. Now, differentiating the first of (5.80) with respect to $\theta$ and evaluating the result at the stagnation point yields

$$\frac{w_{\phi\theta}}{\sin\,\theta_0} = 2u_0 \mathrm{cotan}\;\theta_0 - v_{\theta\theta}. \tag{5.92}$$

Therefore,

$$\begin{aligned} \vartheta &= \frac{v_{\psi\psi}}{4u_0}\psi^2 + O(\psi^3), \\ \frac{d\psi}{d\bar{t}} &= -\left((2u_0 \mathrm{cotan}\;\theta_0 - v_{\vartheta\vartheta})\frac{v_{\psi\psi}}{8u_0^2} + \frac{w_{\psi\psi\psi}}{12u_0}\right)\psi^3 + O(\psi^4). \end{aligned} \tag{5.93}$$

Therefore, we have $m = 3$ and

$$\Delta_m = -\left((2u_0 \mathrm{cotan}\;\theta_0 - v_{\vartheta\vartheta})\frac{v_{\psi\psi}}{8u_0^2} + \frac{w_{\psi\psi\psi}}{12u_0}\right). \tag{5.94}$$

The singularity is either a topological saddle if $\Delta_m < 0$ or a topological node if $\Delta_m > 0$. Our numerical results show that the bifurcation occurs via a topological node ($\Delta_m > 0$) and it results in one stable saddle point and two stable nodal points as shown in Figure 5.23. We will discuss this again when we examine the numerical results.

### 5.2.7 Bifurcation of Higher Order Singularities

The vanishing of the Jacobian opens the door for the flow pattern to bifurcate. The type of higher order singularity that exists when the Jacobian vanishes determines the resulting flow pattern after bifurcation. In [7] Bakker analyzed the bifurcation of the saddle node, topological node, and topological saddle singularity.

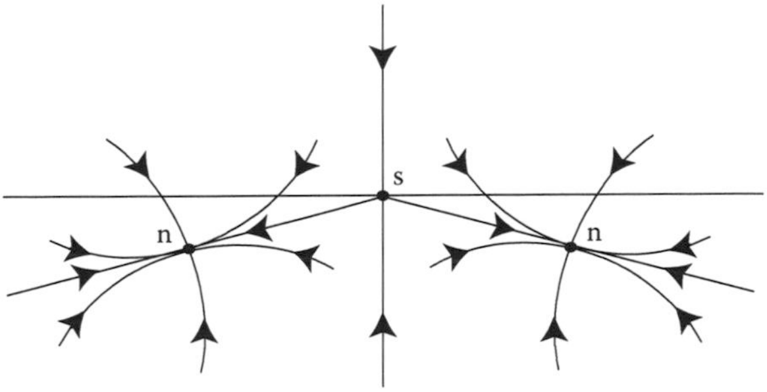


**FIGURE 5.23:** Bifurcation of topological node into two stable nodes (n) and one stable saddle (s).

In what follows, we use the same method used by Bakker in [7] to study the saddle node bifurcation without assuming that the flow satisfies a potential function.

We begin by perturbing the streamline patterns around $g = 0$. The perturbation comes about due to a change in some input parameter; this could be a change in the free stream flow direction or a change in the corner geometry. Let $\tau$ represent this parameter. We consider flow properties to be functions of this parameter: $v = v(\vartheta, \psi, \tau), w = w(\vartheta, \psi, \tau)$, and $g = g(\tau)w = w(\vartheta, \psi, \tau)g = g(\tau)$. When the Jacobian vanishes, we say that $\tau$ attains a critical value $\tau_0$. Therefore, near $\tau_0$ we have $v_\vartheta = v_\vartheta|_0 + \left.\dfrac{\partial v_\vartheta}{\partial \tau}\right|_0 (\tau - \tau_0)$, where the notation $f|_0$ indicates the value of $f$ at $g(\tau_0) = 0$. Since the location of the singular points changes after bifurcation, we expect the coordinates of these points to depend on the perturbation parameter. Hence, new coordinates $\vartheta(\tau), \psi(\tau)$ are introduced into the equation describing the streamline patterns. The unknown coefficients in the definition of the new coordinates are determined by equating terms of the same order in the perturbation parameter. The locations of the singular points are identified and the streamline patterns in the neighborhood of these new stagnation points are determined.

We apply the method to the bifurcation of the saddle node singularity, corresponding to $w_{\psi\psi} \neq 0$. The Jacobian vanishes when $v_\vartheta = -2u_0$. To simplify the notation let $\varepsilon = \left.\dfrac{\partial v_\vartheta}{\partial \tau}\right|_0 (\tau - \tau_0)$. We assume that the solution in a small neighborhood of the stagnation point at $g=0$ is analytic in $\tau$ and that $|\varepsilon| \ll 1$. We write $v_\vartheta = -2u_0 + \varepsilon$, and from (5.81) it follows that $w_\psi = -\varepsilon$. With these assumptions, Equations 5.72 are

$$\begin{aligned}
\frac{d\vartheta}{dt} &= (-2u_0 + \varepsilon)\vartheta + v_{\vartheta\vartheta}\frac{\vartheta^2}{2} + v_{\vartheta\psi}\vartheta\psi + v_{\psi\psi}\frac{\psi^2}{2} \\
&\quad + v_{\vartheta\vartheta\vartheta}\frac{\vartheta^3}{6} + v_{\vartheta\vartheta\psi}\frac{\vartheta^2\psi}{2} + v_{\vartheta\psi\psi}\frac{\vartheta\psi^2}{2} + v_{\psi\psi\psi}\frac{\psi^3}{6} + \dots, \\
\frac{d\psi}{dt} &= -\varepsilon\psi + w_{\vartheta\vartheta}\frac{\vartheta^2}{2} + w_{\vartheta\psi}\vartheta\psi + w_{\psi\psi}\frac{\psi^2}{2} \\
&\quad + w_{\vartheta\vartheta\vartheta}\frac{\vartheta^3}{6} + w_{\vartheta\vartheta\psi}\frac{\vartheta^2\psi}{2} + w_{\vartheta\psi\psi}\frac{\vartheta\psi^2}{2} + w_{\psi\psi\psi}\frac{\psi^3}{6} + \dots.
\end{aligned} \tag{5.95}$$

Let us introduce new coordinates $\vartheta_i, \psi_i$ such that in small neighborhood of the new stagnation points we have

$$\begin{aligned}
\vartheta_i &= \alpha_{1,i}\varepsilon + \alpha_{2,i}\varepsilon^2 + O(\varepsilon^3), \\
\psi_i &= \beta_{1,i}\varepsilon + \beta_{2,i}\varepsilon^2 + O(\varepsilon^3).
\end{aligned} \tag{5.96}$$

We note that if $\varepsilon \to 0$ then $(\vartheta_i, \psi_i) \to (0, 0)$ and we recover the saddle node singularity. To evaluate the coefficients in (5.96), we introduce (5.96) into (5.95) and equate terms of like powers in $\varepsilon$. We find

$$
\begin{aligned}
&-2u_0\alpha_{1,i} = 0,\\
&\alpha_{1,i} - 2u_0\alpha_{2,i} + \frac{v_{\vartheta\vartheta}}{2}\alpha_{1,i}^2 + v_{\vartheta\psi}\alpha_{1,i}\beta_{1,i} + \frac{v_{\psi\psi}}{2}\beta_{1,i}^2 = 0,\\
&-\beta_{1,i} + \frac{w_{\vartheta\vartheta}}{2}\alpha_{1,i}^2 + w_{\vartheta\psi}\alpha_{1,i}\beta_{1,i} + \frac{w_{\psi\psi}}{2}\beta_{1,i}^2 = 0,\\
&\alpha_{2,i} + v_{\vartheta\vartheta}\alpha_{1,i}\alpha_{2,i} + v_{\vartheta\psi}(\alpha_{1,i}\beta_{2,i} + \alpha_{2,i}\beta_{1,i}) + v_{\psi\psi}\beta_{1,i}\beta_{2,i}\\
&+ \frac{v_{\vartheta\vartheta\vartheta}}{6}\alpha_{1,i}^3 + \frac{v_{\vartheta\vartheta\psi}}{2}\alpha_{1,i}^2\beta_{1,i} + \frac{v_{\vartheta\psi\psi}}{2}\alpha_{1,i}\beta_{1,i}^2 + \frac{v_{\psi\psi\psi}}{6}\beta_{1,i}^3 = 0,\\
&-\beta_{2,i} + w_{\vartheta\vartheta}\alpha_{1,i}\alpha_{2,i} + w_{\vartheta\psi}(\alpha_{1,i}\beta_{2,i} + \alpha_{2,i}\beta_{1,i}) + w_{\psi\psi}\beta_{1,i}\beta_{2,i}\\
&+ \frac{w_{\vartheta\vartheta\vartheta}}{6}\alpha_{1,i}^3 + \frac{w_{\vartheta\vartheta\psi}}{2}\alpha_{1,i}^2\beta_{1,i} + \frac{w_{\vartheta\psi\psi}}{2}\alpha_{1,i}\beta_{1,i}^2 + \frac{w_{\psi\psi\psi}}{6}\beta_{1,i}^3 = 0.
\end{aligned}
\tag{5.97}
$$

From (5.97) we find the following three stagnation points:

$$
P_1 \to \begin{cases} \alpha_{1,1} = 0,\\ \alpha_{2,1} = 0,\\ \beta_{1,1} = 0,\\ \beta_{2,1} = 0. \end{cases} \tag{5.98}
$$

$$
P_2 \to \begin{cases} \alpha_{1,2} = 0,\\ \alpha_{2,2} = \dfrac{v_{\psi\psi}}{u_0 w_{\psi\psi}^2},\\ \beta_{1,2} = \dfrac{2}{w_{\psi\psi}},\\ \beta_{2,2} = -\left(\frac{2}{3}v_{\psi\psi\psi} + \dfrac{v_{\psi\psi}}{u_0}\left(v_{\vartheta\psi} + \frac{1}{2}w_{\psi\psi}\right)\right)\Big/\left(v_{\psi\psi}w_{\psi\psi}^2\right). \end{cases} \tag{5.99}
$$

$$
P_3 \to \begin{cases} \alpha_{1,3} = 0,\\ \alpha_{2,3} = \dfrac{v_{\psi\psi}}{u_0 w_{\psi\psi}^2},\\ \beta_{1,3} = \dfrac{2}{w_{\psi\psi}},\\ \beta_{2,3} = -2\left(\frac{2}{3}w_{\psi\psi\psi} + \dfrac{v_{\psi\psi}w_{\vartheta\psi}}{u_0}\right)\Big/w_{\psi\psi}^3. \end{cases} \tag{5.100}
$$

If

$$
\frac{\mathrm{d}x}{\mathrm{d}t} = R(x, y), \quad \frac{\mathrm{d}y}{\mathrm{d}t} = Q(x, y)
$$

represent the equations for the streamlines for a singular point $x_0, y_0$ not at the origin, we can rewrite the equations at the origin by means of the transformation $\hat{x} = x - x_0, \hat{y} = y - y_0$ as

$$
\frac{\mathrm{d}\hat{x}}{\mathrm{d}t} = R_x\hat{x} + R_y\hat{y}, \quad \frac{\mathrm{d}\hat{y}}{\mathrm{d}t} = Q_x\hat{x} + Q_y\hat{y}.
$$

Therefore, from (5.95), retaining only first order terms, and letting $\hat{\vartheta} = \vartheta - \vartheta_i, \hat{\psi} = \psi - \psi_i$, we find

$$\begin{aligned}\frac{d\hat{\vartheta}}{dt} &= (-2u_0 + \varepsilon + v_{\vartheta\vartheta}\vartheta_i + v_{\vartheta\psi}\psi_i)\hat{\vartheta} + (v_{\vartheta\psi}\vartheta_i + v_{\psi\psi}\psi_i)\hat{\psi},\\ \frac{d\hat{\psi}}{dt} &= (w_{\vartheta\vartheta}\vartheta_i + w_{\vartheta\psi}\psi_i)\hat{\vartheta} + (-\varepsilon + w_{\vartheta\psi}\vartheta_i + w_{\psi\psi}\psi_i)\hat{\psi}.\end{aligned} \tag{5.101}$$

The types of solutions corresponding to (5.101) depend on the following parameters:

$$\begin{aligned}\bar{g} &= (2 - \bar{\varepsilon})\bar{\varepsilon} + (\bar{\varepsilon}(\bar{w}_{\vartheta\psi} - \bar{v}_{\vartheta\vartheta}) - 2\bar{w}_{\vartheta\psi})\vartheta_i\\ &\quad + (\bar{\varepsilon}(\bar{w}_{\psi\psi} - \bar{v}_{\vartheta\psi}) - 2\bar{w}_{\psi\psi})\psi_i + (\bar{v}_{\vartheta\vartheta}\bar{w}_{\vartheta\psi} - \bar{v}_{\vartheta\psi}\bar{w}_{\vartheta\vartheta})\vartheta_i^2\\ &\quad + (\bar{v}_{\vartheta\vartheta}\bar{w}_{\psi\psi} - \bar{v}_{\psi\psi}\bar{w}_{\vartheta\vartheta})\vartheta_i\psi_i + (\bar{v}_{\vartheta\psi}\bar{w}_{\psi\psi} - \bar{v}_{\psi\psi}\bar{w}_{\vartheta\psi})\psi_i^2,\\ \bar{h} &= 2 + (\bar{v}_{\vartheta\vartheta} + \bar{w}_{\vartheta\psi})\vartheta_i + (\bar{v}_{\vartheta\psi} + \bar{w}_{\psi\psi})\psi_i,\\ \bar{\Delta} &= \bar{h}^2 - 4\bar{g},\end{aligned}$$

where we have introduced the following normalization: $\bar{g} = g/u_0^2$, $\bar{h} = h/u_0$, $\bar{\Delta} = \Delta/u_0^2$, $\bar{v} = v/u_0$, and $\bar{w} = w/u_0$. From the first of (5.80) it follows that $\bar{v}_{\vartheta\psi} + \bar{w}_{\psi\psi} = 0$. Therefore,

$$\bar{h} = 2 + (\bar{v}_{\vartheta\vartheta} + \bar{w}_{\vartheta\psi})\vartheta_i.$$

For $P_1$, we have $(\vartheta_1, \psi_1) \to (0, 0)$ and $\bar{g} = 2\bar{\varepsilon} + O(\bar{\varepsilon}^2)$, $\bar{h} = 2$ and $\bar{\Delta} = 4(1 - 2\bar{\varepsilon}) + O(\bar{\varepsilon}^2)$. Since $|\bar{\varepsilon}| \ll 1$, we conclude that for $\bar{\varepsilon} < 0$ the $P_1$ singularity is a saddle, and for $\bar{\varepsilon} > 0$ the $P_1$ singularity is a node. For both $P_{2,3}$ we have

$$\begin{aligned}\bar{h} &= 2 + O(\bar{\varepsilon}^2),\\ \bar{g} &= -2\bar{\varepsilon} + O(\bar{\varepsilon}^2),\\ \bar{\Delta} &= 4(1 + 2\bar{\varepsilon}) + O(\bar{\varepsilon}^2).\end{aligned}$$

Therefore $P_{2,3}$ behave like the inverse of $P_1$ with changes in $\bar{\varepsilon}$. To first order in $\bar{\varepsilon}$, the $P_2$ and $P_3$ singularities are indistinguishable. The effect of $\bar{\varepsilon}$ on the streamline pattern is shown in Figure 5.24.

The bifurcation of the topological node and topological saddle involve two perturbation parameters. The details of these bifurcations can be found in [7].

### 5.2.8 Flow Model near the Corner Cross Flow Stagnation Point

To further investigate the behavior of the flow near the corner stagnation point, we now assume that the flow is isentropic and can be described by a conical potential function $rF(\theta, \phi)$. In reality, at the corner the flow can be highly rotational and the results we obtain from this analysis should be considered only of qualitative value.

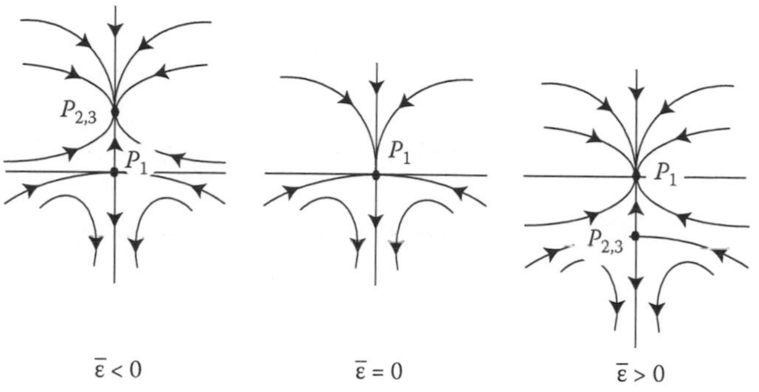


**FIGURE 5.24:** Sketch of saddle node bifurcation as a function of an $\bar{\varepsilon}$ perturbation, $w_{\psi\psi} \neq 0$.

Combining the continuity equation with Bernoulli's equation, the $u$-momentum equation and the isentropic relation, we find

$$\left(1 - \frac{v^2}{a^2}\right)v_\theta + \left(1 - \frac{w^2}{a^2}\right)\frac{w_\phi}{\sin\,\theta} + \left(2 - \frac{q^2}{a^2}\right)u - \frac{vw}{a^2}\left(w_\theta + \frac{v_\phi}{\sin\,\theta}\right) + v\text{cotan}\,\theta = 0. \tag{5.102}$$

In the vicinity of the stagnation point, we assume that $v^2$, $w^2$, and $vw$ can be neglected compared to $a^2$. Neglecting these terms, and aligning the $z$-axis with the corner, we find from (5.102) that the conical potential on the unit sphere satisfies the equation

$$F_{\theta\theta} + \frac{1}{\theta}F_\theta + \frac{1}{\theta^2}F_{\phi\phi} + 2F = 0, \tag{5.103}$$

with the velocity components given by

$$u = F, \quad v = F_\theta, \quad w = F_\phi/\theta.$$

The boundary conditions are

$$u = u_0 \text{ at } \theta = 0,$$
$$w = 0 \text{ at } \phi = 0, \quad \text{and } \phi = \hat{\phi}_c,$$

where $\hat{\phi}_c$ is the angle between the wedge surface $\Sigma_2$ and the symmetry plane $\hat{\phi}_c = \phi_c - a\tan\,(\tan\,\Lambda_2 \tan\,\delta_2)$. The solution is easily obtained by separation of variables. It is

$$F(\theta, \phi) = \sum_n^\infty C_n \cos\,(2nl\phi)B_k(\sqrt{2}\theta), \tag{5.104}$$

where

$l = \pi/2\hat{\phi}_c$
$k = 2nl$
$n = 0, 1, 2, 3, \ldots$
$C_0 = u_0$

All other $C_n$'s are undetermined constants. $B_k$ is a Bessel function of the first kind of order $k$. The series expansion of the Bessel function is defined in terms of a series in gamma functions. The gamma function is related to the factorial by $\Gamma(n+1) = n!$. The expansion is

$$B_k(x) = \sum_{m=0}^{\infty} \frac{(-1)^m}{m!\Gamma(m+k+1)} \left(\frac{x}{2}\right)^{2m+k},$$

and therefore the leading terms in the expansion of the velocity components about $\theta = 0$ are

$$\begin{aligned} u &= u_0\left(1 - \frac{1}{2}\theta^2\right) + \frac{C_1}{(2l)!} \cos(2l\phi)\left(\frac{\theta}{\sqrt{2}}\right)^{2l} + \ldots, \\ v &= -u_0\theta + \frac{\sqrt{2}lC_1}{(2l)!} \cos(2l\phi)\left(\frac{\theta}{\sqrt{2}}\right)^{2l-1} + \ldots, \text{ and} \\ w &= -\frac{\sqrt{2}lC_1}{(2l)!} \sin(2l\phi)\left(\frac{\theta}{\sqrt{2}}\right)^{2l-1} + \ldots. \end{aligned} \tag{5.105}$$

The speed of sound follows from Bernoulli's relation (5.55) and the pressure and density follow from the isentropic condition (5.65).

The character of the singularity follows from

$$\frac{v_\theta}{u_0} = -1 + C_1\frac{(2l-1)l}{u_0(2l)!} \cos(2l\phi)\left(\frac{\theta}{\sqrt{2}}\right)^{2(l-1)} + \ldots, \tag{5.106}$$

in the limit of $\theta \to 0$. The nature of the flow at the corner is, thus, strongly dependent on the value of the angle $\hat{\phi}_c = \pi/2l$ and the value of $C_1$. For internal corners, $\infty > l > 1$, we have a nodal singularity at the corner independent of the value of $C_1$, since $v_\theta/u_0 = -1$. For a flat plate or circular cone, $l = 1$, the cross flow stagnation point is a node or saddle depending on the value of $C_1$. By matching the local solution with the linear solution for a slender cone at angle of attack [195], the value of $C_1$ is found to be

$$C_1 = 2u_0(2\alpha/\delta - 1), \tag{5.107}$$

where

$\alpha$ is the angle of attack
$\delta$ is the cone half angle, see Problem 5.6

With the aid of (5.107), we find that bifurcation takes place when $\alpha/\delta > 1$. For external corners, $\frac{1}{2} < l < 1$, the singularity at the corner is a saddle of streamlines with a weak singularity ($C^1$) in $v$ unless $C_1$ vanishes. In the range $\frac{1}{2} < l < \frac{3}{4}$, there is also a weak singularity in the pressure, again, unless $C_1$ vanishes. If the pressure is singular, the singularity appears as a cusp in the pressure distribution.

Finally, we show that (5.106) satisfies (5.87):

$$2 + \frac{v_\theta(0)}{u_0} + \frac{v_\theta(\hat{\phi}_c)}{u_0} = C_1 \frac{(2l-1)l}{u_0(2l)!}\left(\frac{\theta}{\sqrt{2}}\right)^{2(l-1)}\left(1 + \cos\,(2l\hat{\phi}_c)\right),$$

but $1 + \cos\,(2l\hat{\phi}_c) = 0$, since $2l\hat{\phi}_c = \pi$. This is true for all odd-order approximations of (5.104).

### 5.2.9 Generalized Coordinate Transformation

We want to solve (5.80) in the elliptical region bounded by the cross flow sonic line, the curved shock, the symmetry plane $\phi = \phi_c$, and the wedge surface $\Sigma_2$. The elliptic nature of the equations presents a problem which we overcome by introducing a *pseudo-time* coordinate into the governing equations by adding a $P_t$, $u_t$, $v_t$, $w_t$, and $S_t$ term to the corresponding equations of (5.80). The conical assumption implicit in (5.80) is only consistent with steady flows, thus this pseudo-time is a numerical artifact introduced to make the equations hyperbolic. As in the case of the blunt body, we want to carry out the numerical integration in a computational plane that maps the shock wave to a computational boundary and allows for a uniformly spaced mesh. Although not strictly necessary for this problem, we will present the coordinate mapping in a generalized way.

Let $(Y, X, T)$ be the coordinates in the computational space. Then, introduce the mapping

$$\begin{aligned} f(\theta, \phi, t, Y, X, T) &= Y - \frac{\theta - \theta_w(X)}{\theta_s(X, t) - \theta_w(X)} = 0, \\ g(\theta, \phi, t, Y, X, T) &= X - \frac{\phi_c - \phi}{\phi_c - \phi_{sl}(Y)} = 0, \text{ and} \\ h(\theta, \phi, t, Y, X, T) &= T - t = 0, \end{aligned} \tag{5.108}$$

where

$\theta_w$ defines the $\Sigma_2$ wedge surface
$\phi_{sl}$ defines the cross flow sonic line
$\theta_s$ defines the shock surface

The wedge surface is given by

$$\theta_w(\phi) = a\tan\left[\frac{\tan\,\delta_2}{\sin\,\phi - \tan\,\Lambda_2 \tan\,\delta_2 \cos\,\phi}\right]. \tag{5.109}$$

The definition of the cross flow sonic line follows from the condition that $a^2 - q^2 = 0$:

$$\begin{aligned} a_2^2 &- (\bar{u}_2 \sin \phi_{sl} + \bar{v}_2 \cos \phi_{sl})^2 \\ &- [(\bar{v}_2 \sin \phi_{sl} - \bar{u}_2 \cos \phi_{sl}) \cos \theta - \bar{w}_2 \sin \theta]^2 = 0, \end{aligned} \tag{5.110}$$

where $\bar{u}_2$, $\bar{v}_2$, $\bar{w}_2$ are the velocity components in the $x$, $y$, $z$ directions, respectively, in the region behind the two-dimensional shock and outboard of the cross flow sonic line, and $a_2$ is the speed of sound in the same region. The symmetry plane is given in terms of $\Omega$ by $\phi_c = \frac{1}{2}(3\pi/2 - \Omega)$, and, of course, the shock surface, $\phi_s$, remains to be determined. The mapping is illustrated in Figure 5.25. The so-called physical plane in the illustration is already a mapping from the physical plane to the surface of the unit sphere to a projection from the surface of the unit sphere to a plane.

The Jacobian of the transformation defined by (5.108) is given by

$$J = \begin{bmatrix} f_Y & f_X & f_T \\ g_Y & g_X & g_T \\ h_Y & h_X & h_T \end{bmatrix}.$$

From (5.108) it follows that

$$\begin{aligned} f_Y &= 1, & f_X &= (\theta_{wX} + Y(\theta_{sX} - \theta_{wX}))/(\theta_s - \theta_w), & f_T &= 0, \\ g_Y &= -X\phi_{slY}/(\phi_c - \phi_{sl}), & g_X &= 1, & g_T &= 0, \\ h_Y &= 0, & h_X &= 0, & h_T &= 1, \end{aligned}$$

and, therefore,

$$J = 1 - f_X g_Y.$$

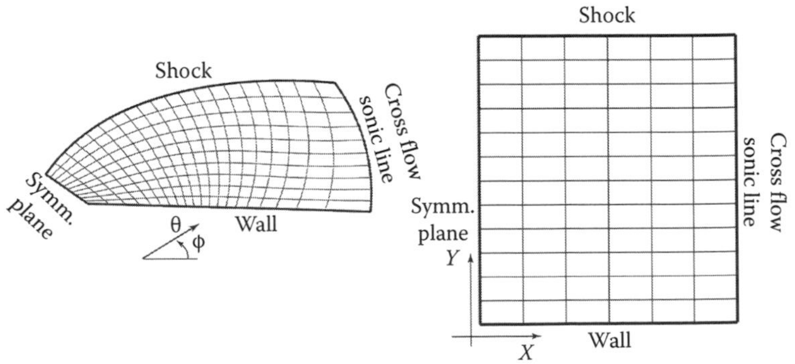


**FIGURE 5.25:** Mapping from "physical" plane, top panel, to computational plane, bottom panel.

The partial derivatives $X_\theta$, $X_\phi, \ldots$ are defined by

$$X_\theta = -\frac{1}{J}\begin{bmatrix} f_Y & f_\theta & f_T \\ g_Y & g_\theta & g_T \\ h_Y & h_\theta & h_T \end{bmatrix},\ X_\phi = -\frac{1}{J}\begin{bmatrix} f_Y & f_\phi & f_T \\ g_Y & g_\phi & g_T \\ h_Y & h_\phi & h_T \end{bmatrix},$$

with similar expressions for the other derivatives. The resulting expressions are

$$\begin{aligned} X_\theta &= f_\theta g_Y/J, & X_\phi &= -g_\phi/J, & X_t &= f_t g_Y/J, \\ Y_\theta &= -f_\theta/J, & Y_\phi &= g_\phi f_X/J, & Y_t &= -f_t/J, \\ T_\theta &= 0, & T_\phi &= 0, & T_t &= 1, \end{aligned}$$

where

$$\begin{aligned} f_\theta &= -1/(\theta_s - \theta_w), & f_\phi &= 0, & f_t &= -Yw_s f_\theta, \\ g_\theta &= 0, & g_\phi &= 1/(\phi_c - \phi_{sl}), & g_t &= 0, \\ h_\theta &= 0, & h_\phi &= 0, & h_t &= -1, \end{aligned}$$

and $w_s = \partial\theta_s/\partial t$.

The equations of motion are mapped to the computational space using the chain rule

$$\frac{\partial}{\partial\theta} = \frac{\partial}{\partial Y}Y_\theta + \frac{\partial}{\partial X}X_\theta + \frac{\partial}{\partial T}T_\theta,$$

with similar expressions for $\partial/\partial\phi$ and $\partial/\partial t$. The equations are

$$\begin{aligned} P_T + UP_X + VP_Y + \gamma\left(Y_\theta v_Y + X_\theta v_X + \frac{w_X X_\phi + w_Y Y_\phi}{\sin\theta} + 2u + v\text{cotan}\,\theta\right) &= 0, \\ v_T + Uv_X + Vv_Y + \Theta(Y_\theta P_Y + X_\theta P_X) + uv - w^2\text{cotan}\,\theta &= 0, \\ w_T + Uw_X + Vw_Y + \frac{\Theta}{\sin\theta}(Y_\phi P_Y + X_\phi P_X) + uw - vw\text{cotan}\,\theta &= 0, \\ S_T + US_X + VS_Y &= 0. \end{aligned} \tag{5.111}$$

The contravariant velocity components are

$$\begin{aligned} U &= vX_\theta + \frac{wX_\phi}{\sin\theta} + X_t, \\ V &= vY_\theta + \frac{wY_\phi}{\sin\theta} + Y_t. \end{aligned}$$

The $u$-momentum equation is replaced by Bernoulli's equation (5.55). The expressions for $\phi_{slY}$ and $\theta_{wX}$ follow from

$$\frac{\partial\phi_{sl}}{\partial\theta} = \frac{\partial\phi_{sl}}{\partial Y}Y_\theta\bigg|_X, \tag{5.112}$$

and

$$\frac{\partial \theta_w}{\partial \phi} = \frac{\partial \theta_w}{\partial X} X_\phi \bigg|_Y. \tag{5.113}$$

In (5.112), the derivative is taken along $X =$ constant line, and, therefore, the derivative $f_X$ appearing in the Jacobian in $Y_\theta$ vanishes. Similarly, in (5.113) the derivative $g_Y$ appearing in the Jacobian in $X_\phi$ vanishes.

### 5.2.10 Initial Conditions

The location of the cross flow sonic line is defined by (5.110). Along this line, we specify the values corresponding to the two-dimensional flow region. The reason we use the cross flow sonic line as a boundary is that, although the flow variables are continuous across this line, some of their first derivatives are not continuous and we want to avoid differentiation across this line.

The initial shock wave shape is defined by

$$\theta_s(\phi, 0) = c_1(\phi - \phi_{sl})^3 + c_2(\phi - \phi_{sl})^2 + \frac{\partial \theta_s}{\partial \phi}\bigg|_{sl} (\phi - \phi_{sl}) + \theta_s(\phi_{sl}, 0).$$

That is, we match the slope and location of the two-dimensional shock where it meets the cross flow sonic line and we are left with two free parameters, $c_1$ and $c_2$. These two parameters are evaluated by satisfying the following two boundary conditions on the symmetry plane:

$$\theta_s(\phi_c, 0) = \theta_{s0},$$
$$\theta_{s\phi}(\phi_c, 0) = 0.$$

The initial shock standoff distance, $\theta_{s0}$, is obtained from a correlation of previous runs. Knowing the shock shape, the solution along the shock is obtained by applying the Rankine–Hugoniot jumps.

At the corner point, corresponding to $\theta_w(\phi_c)$, the cross flow velocity components are zero. We also assume that the corner point is a node and prescribe the value of the entropy to be the same as the value at the shock on the symmetry line. The pressure at the corner, $p_0$, is defined from a correlation of previous runs. With the flow field defined at the corner and at the shock, the flow along the symmetry line is defined by linearly interpolating all variables between the shock and the corner point.

At other points along the wall, the flow is defined by

$$P(\theta_w, \phi_w, 0) = \log\left[p_0 + (p_2 - p_0)(1 - \sin(\pi\sqrt{(1 - X)}/2))\right],$$
$$S(\theta_w, \phi_w, 0) = S_2,$$
$$v(\theta_w, \phi_w, 0) = q_2 X\sqrt{X} n_\theta,$$
$$w(\theta_w, \phi_w, 0) = q_2 X\sqrt{X} n_\phi,$$

where

the subscript 2 refers to values in the two-dimensional region $q_2 = \sqrt{v_2^2 + w_2^2}$

$n_\theta$ and $n_\phi$ are the direction cosines of the unit tangent to the wall

The radial velocity component follows from Bernoulli's equation.

The rest of the flow field is assumed to be a linear function of $Y$ at constant $X$.

### 5.2.11 Shock Computation

The shock computation is based on a simplified method that takes into account only the pressure variation on the high pressure side of the shock to find the shock acceleration. The methods discussed in Sections 5.1.6.2 and 5.1.6.3 are more consistent approaches, since they simultaneously account for the variations in the pressure and velocity fields. However, for this problem which is not intended to be time accurate, this simplified approach was found to work well. The change in pressure at the shock is connected to the shock acceleration through the Rankine–Hugoniot jump:

$$P_T = 2\gamma M_{\infty\mathrm{rel}} M_{\infty\mathrm{rel}T} / \left(\gamma M_{\infty\mathrm{rel}}^2 - (\gamma - 1)/2\right). \tag{5.114}$$

The free stream Mach number relative to the shock is given by

$$M_{\infty\mathrm{rel}} = (\vec{V}_\infty - w_\mathrm{s}\vec{Y}) \cdot \vec{n}_\mathrm{s} / \sqrt{\gamma},$$

where

$\vec{V}_\infty$ is the free stream velocity vector

the shock velocity is defined by $\vec{w}_\mathrm{s} = w_\mathrm{s}\vec{Y}$

$\vec{Y}$ and $\vec{n}_\mathrm{s}$ are the unit vectors tangent to $X =$ constant and normal to the shock, respectively

The shock acceleration, $w_{sT}$ follows from (5.114):

$$\frac{\partial w_\mathrm{s}}{\partial T} = \frac{1}{\vec{Y} \cdot \vec{n}_s} \left[ \frac{\partial \vec{V}_\infty}{\partial T} \cdot \vec{n}_\mathrm{s} + \vec{V}_\infty \cdot \frac{\partial \vec{n}_\mathrm{s}}{\partial T} - w_\mathrm{s} \frac{\partial (\vec{Y} \cdot \vec{n}_\mathrm{s})}{\partial T} - \frac{\sqrt{\gamma}\left(M_{\infty\mathrm{rel}}^2 + \dfrac{1-\gamma}{2\gamma}\right) P_T}{2M_{\infty\mathrm{rel}}} \right]. \tag{5.115}$$

The time derivatives of the direction cosines in (5.115) are defined in terms of $w_s$ and $\partial w_\mathrm{s}/\partial\phi$ as in Sections 5.1.6.2 and 5.1.6.3, and $P_T$ is evaluated from the first of (5.111).

### 5.2.12 Wall Computation

At the wall, the second and third equations in (5.111) are combined into a cross flow momentum equation

$$q_T + U(e_\theta v_X + e_\phi w_X) + \Theta\left(e_\theta X_\theta + \frac{e_\phi X_\phi}{\sin\theta_w}\right)P_X + uq = 0, \tag{5.116}$$

where $e_\theta$ and $e_\phi$ are the direction cosines of the unit tangent to the wall. Once (5.116) is integrated, the cross flow velocity components are obtained from

$$v = e_\theta q,$$
$$w = e_\phi q.$$

The pressure is obtained from a modified form of the first of (5.111). Namely,

$$P_t + qP_s + \gamma(q_s + q\beta_n + 2u + v\mathrm{cotan}\,\theta) = 0, \tag{5.117}$$

which is obtained by using the following coordinate transformations:

$$\frac{\partial}{\partial\theta} = \sin\beta\frac{\partial}{\partial s} + \cos\beta\frac{\partial}{\partial n},$$
$$\frac{1}{\sin\theta_w}\frac{\partial}{\partial\phi} = \cos\beta\frac{\partial}{\partial s} - \sin\beta\frac{\partial}{\partial n},$$

where
$s$ and $n$ are coordinates tangent and normal to the cross flow streamlines
$\beta$ is the cross flow slope

$$\tan\beta = \frac{v}{w} = \frac{1}{\sin\theta_w}\frac{\partial\theta_w}{\partial\phi}$$

Recasting (5.117) in terms of computational coordinates, we find

$$P_T + UP_X + \gamma\left[\left(e_\theta X_\theta + \frac{e_\phi X_\phi}{\sin\theta_w}\right)q_X + q\beta_n + 2u + v\mathrm{cotan}\,\theta\right] = 0,$$

and

$$q\beta_n = \frac{q}{1+\tan^2\beta}\frac{\partial\tan\beta}{\partial X} + \left(e_\phi Y_\theta - \frac{e_\theta Y_\phi}{\sin\theta_w}\right)(e_\phi v_Y - e_\theta w_Y).$$

The entropy is evaluated by upwinding according to the sign of the contravariant velocity component $U$. At the corner, because it is a cross flow stagnation point, the entropy equation degenerates to $S_t = 0$, providing no information on the value of

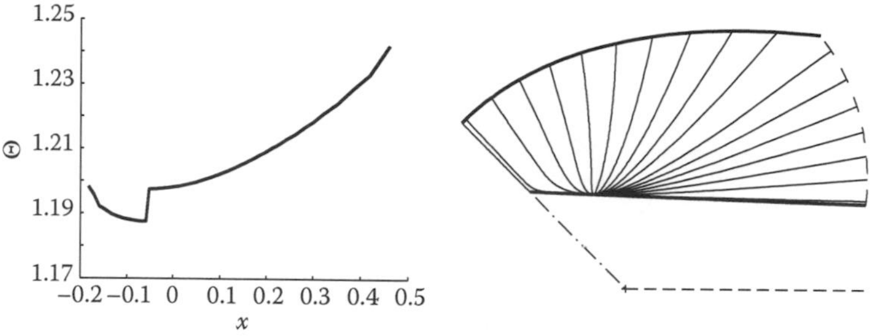


**FIGURE 5.26:** The left panel show the wall temperature for an external corner flow with the following parameters: $M_\infty = 3$, $\Lambda_2 = -10°$, $\delta_2 = 10°$, $\Omega = 0°$. The node singularity lies on the wall at a distance of 0.124 from the corner, as shown on the right panel depicting the cross flow streamlines.

entropy at the corner. To resolve this problem, values of entropy are assigned to the corner depending on the type of singularity occurring there. If the corner is a nodal point, then two values of entropy are defined at the corner (and, consequently, two values of temperature and radial velocity component). One value is obtained by extrapolating from the neighboring point along the wall, the other from the neighboring point along the symmetry plane. If the corner is a saddle point, then only one value of entropy is assigned to the corner. This value is determined depending on whether the streamline reaching the corner comes from the wall or the symmetry plane.

The technique outlined here allows us to capture the jump in entropy, temperature, and radial velocity component associated with a node singularity as illustrated in Figure 5.26 for a case where the node singularity has moved from the corner to the wall.

## 5.2.13 External Corner Flow Results

We present results for external corner flows representing three geometrical configurations, a triangular cross section delta wing, a sharp external corner configuration, and a rounded corner *wing-tip* configuration. The first set of results is for the delta wing illustrated in Figure 5.27.

The delta wing configuration is obtained by setting $\Omega = 90°$, see Figure 5.17, such that the two compression wedges lie in the same plane. For this configuration we want to investigate the bifurcation pattern as a function of the wedge compression angle $\delta_2$. To this end, we fix the free stream Mach number and the leading edge sweep angle at: $M_\infty = 3$, $\Lambda_2 = -30°$. The results are summarized in the bifurcation diagram of Figure 5.28. The bifurcation diagram shows the computed value of the parameter $\xi$ along the wall and the symmetry plane. For $\delta_2 = 0$, the free stream flow remains unperturbed and, as discussed at the end of Section 5.2.5, $\xi_w = \xi_s = -1$. As the wedge angle is increased, the value of $\xi_w$ approaches zero, while the value of $\xi_s$ approaches $-2$. Since $\xi_w$ and $\xi_s$ are related through (5.87), we only need to know one in order to create the bifurcation diagram. However, the values of $\xi_w$ and $\xi_s$ plotted in Figure 5.28 are computed independently. Any deviation from the condition

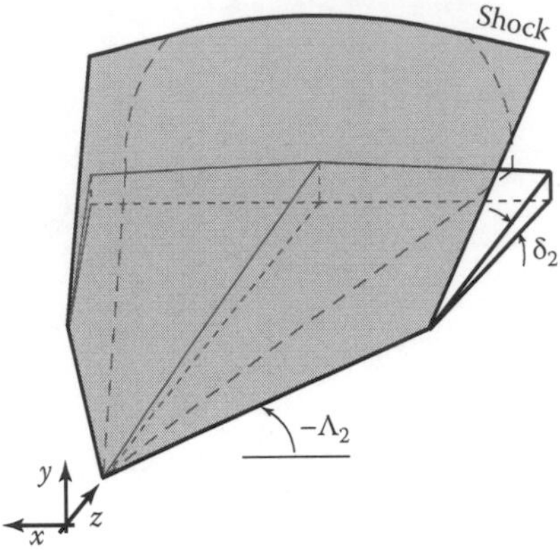


**FIGURE 5.27:** Delta wing configuration, $\Omega = 90°$.

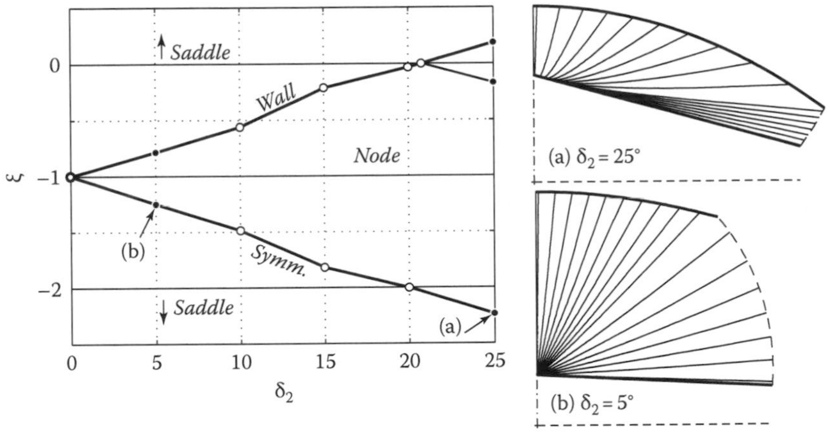


**FIGURE 5.28:** Bifurcation diagram for delta wing configuration, $M_\infty = 3$, $\Lambda_2 = -30°$, $\delta_2$ in degrees.

$2 + \xi_w + \xi_s = 0$ is an indication of the error in the numerical computation of these parameters. At $\delta_2 \approx 20°$, the pattern bifurcates with the node singularity moving from the corner to the wall proper. At the corner, the stagnation point becomes a saddle singularity. The condition (5.87) remains valid for the saddle singularity. We will discuss this further in the next example. The free stream Mach cone angle is $\mu_\infty \approx 19.5°$, thus it seems reasonable to speculate that the bifurcation takes place when $\delta/\mu_\infty > 1$. Beyond $\delta_2 > 27.50°$, the shock wave detaches from the leading edge and the flow ceases to be conical. The right panel of Figure 5.28 shows the cross flow streamlines for $\delta_2 = 5°$ and $25°$.

Let us now study the bifurcation that takes place with the more general external corner problem represented by Figure 5.17. This configuration is related to the

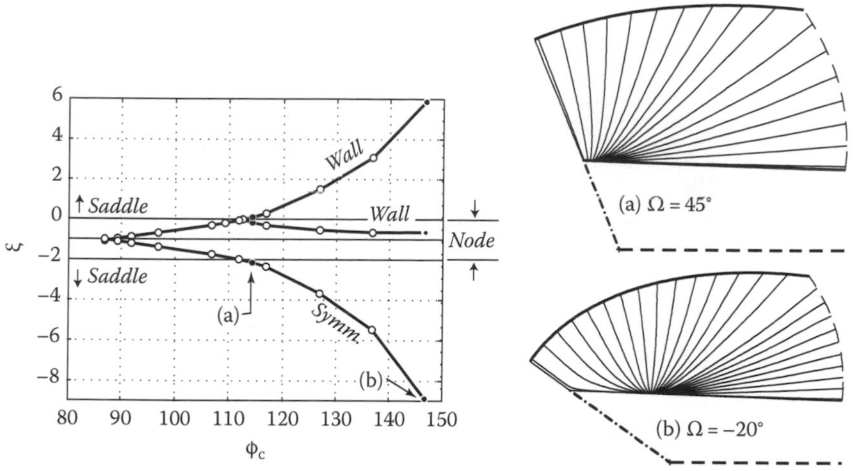


**FIGURE 5.29:** Bifurcation diagram for sharp corner as a function of the symmetry plane angle $\phi_c$. Calculations are for $M_\infty = 3$, $\Lambda_2 = -10°$, $\delta_2 = 10°$. Cross flow streamlines are shown for (a) $\Omega = 45°$ and (b) $\Omega = -20°$.

so-called *caret-wing* shape. For this study, we fix the following parameters: $M_\infty = 3$, $\delta_2 = 10°$, and $\Lambda_2 = -10°$, leaving $\Omega$ as a free parameter. The relation between $\Omega$ and the symmetry plane angle $\phi_c$, see Figure 5.17, is $\phi_c = (3\pi/2 - \Omega)/2$. The bifurcation diagram for this case is shown in Figure 5.29. The two roots, $\xi_w$ and $\xi_s$, are equal at $\phi_c \approx 85°$. At this angle the corner is slightly convex. As the angle $\phi_c$ increases ($\Omega$ decreases), we see the same pattern we saw with the delta wing bifurcation diagram. Bifurcation now takes place at $\phi_c \approx 115°$ (corresponding to $\Omega \approx 40°$). Figure 5.30 shows that the error in satisfying (5.87) is well behaved prior to bifurcation, but increases rapidly after bifurcation. There are several reasons that could account for this. But the main reason, we believe, is the increasingly singular behavior

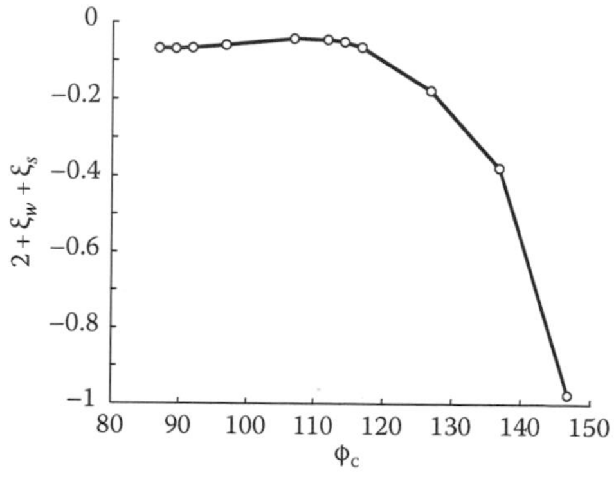


**FIGURE 5.30:** Error in meeting condition (5.87) for external corner problem.

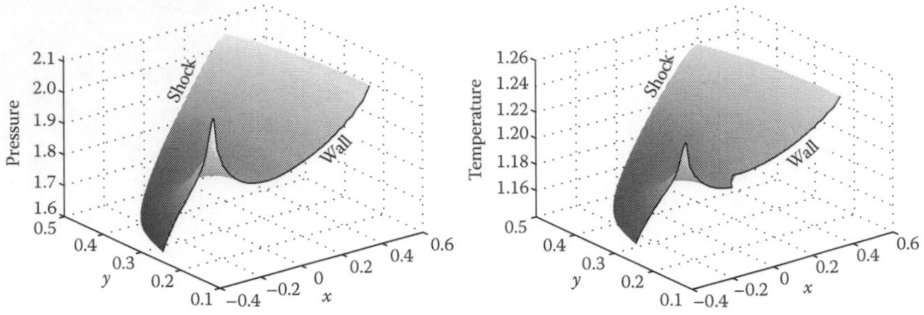


**FIGURE 5.31:** Carpet plot of pressure and temperature for case (b) of Figure 5.29.

of $v_\theta$ at the corner as the angle $\phi_c$ increases, see 5.2.8. The singular behavior at the corner is shown in the carpet plots of Figure 5.31 for case (b) of Figure 5.29.

The last case we want to study is the bifurcation problem associated with the *wing-tip* configuration shown in Figure 5.32. This configuration is created by setting $\Omega = -90°$ such that the two wedges lie back-to-back. The corner for this configuration is rounded such that the radius of the corner corresponds to a cone half angle equal to $\delta_2$. For this configuration, the free stream velocity can have a side slip inclination in the $x$–$z$ plane measured by the angle $\mu$ as shown in the figure. We want to study the bifurcation problem associated with variations of the side slip angle. To this end, we fix the following parameters: $M_\infty = 2$, $\Lambda_2 = -10°$, and $\delta_2 = 10°$.

The bifurcation diagram is shown in Figure 5.33. The behavior of $\xi$ at the corner, along the wall and along the symmetry plane, is linear with $\mu$. At $\mu \approx 3°$, $\xi = -1$. An increase in $\mu$ pushes the singularity toward the symmetry plane. Bifurcation takes place at $\mu \approx 9°$. For $\mu > 9°$, the node "lifts off" and moves to the symmetry plane, leaving a saddle singularity at the corner. The cross flow streamline pattern for this type of flow is shown on the right panel of Figure 5.33, case (b). If $\mu$ is decreased,

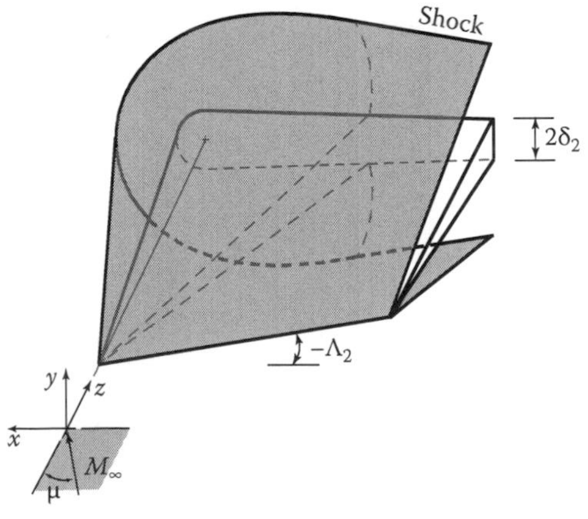


**FIGURE 5.32:** *Wing-tip* configuration.

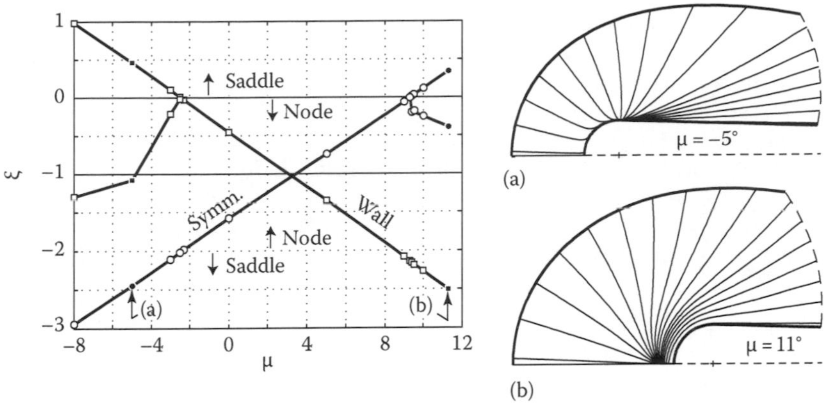


**FIGURE 5.33:** Bifurcation diagram for *wing-tip* configuration. Calculations are for $M_\infty = 2$, $\Lambda_2 = -10°$, $\delta_2 = 10°$. Cross flow streamlines are shown for (a) $\mu = -5°$ and (b) $\mu = 11°$.

the corner singularity is pushed toward the wall with bifurcation occurring at $\mu \approx -2°$. For $\mu < -2°$, the nodal singularity lies on the wall and a saddle singularity is left at the corner. The cross flow streamline pattern for this type of flow is shown on the right panel of Figure 5.33, case (a). Note that condition (5.87) is reasonably well satisfied throughout the range of side slip angles tested. The reason for this is that for this case $\hat{\phi}_c = \pi/2$, and the flow is well behaved at the corner.

Finally, we compare in Figure 5.34 the theoretical value of $P_\theta$, at the interception of the cross flow sonic line and the wall, predicted by (5.67) with its computed value

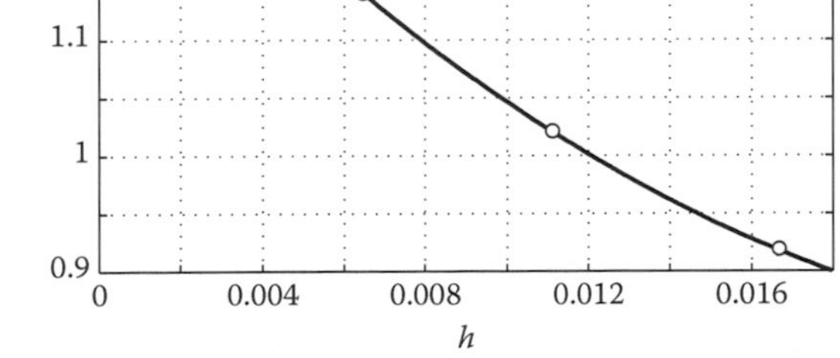

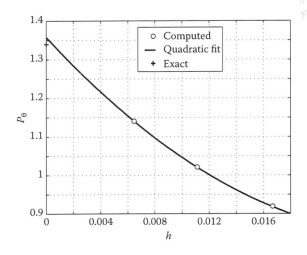

**FIGURE 5.34:** $P_\theta$ evaluated at the interception of the cross flow sonic line and the wall compared to the theoretical value.

in the limit of vanishing mesh spacing. The case corresponds to $M_\infty = 3$, $\Lambda_2 = -10°$, $\delta_2 = 10°$, and $\Omega = 0°$. The three mesh sizes shown correspond to (60,40), (90,40), and (120,40).

### 5.2.14 Running the External Corner Flow Code

The external corner flow code is named `CORNER_MU.M`, because it is the version that can compute *wing-tip* configurations with side slip. The flows associated with these corner flows typically have strong gradients, both near the cross flow sonic line and near the cross flow stagnation points. For this reason, these computations require more mesh points than are required for blunt body flows. A typical external corner flow requires a mesh with 60 intervals along the wall and 40 intervals between the wall and the shock. The input parameters for `CORNER_MU.M` are listed in Table 5.6. The first five parameters do not need any further comments. The parameter labeled `iload` is used to read a previous solution for continuing a calculation. If `iload`=1, the program loads the file named `cor_save`. At the end of a computation, the program will save the solution to a file named

**TABLE 5.6:** Input parameters for *CORNER_MU.M*.

| | Input Parameter | Description |
|---|---|---|
| 1 | `id` | Run number, to keep tract of runs |
| 2 | `kout` | Number of time steps between partial output |
| 3 | `klast` | Total number of time steps |
| 4 | `na` | Number of intervals along wall |
| 5 | `ma` | Number of intervals between wall and shock |
| 6 | `iload` | 0 does not load previous solution<br>1 loads previous solution for continuation run |
| 7 | `icut` | 0 delta wing, triangular *x*-section<br>1 truncated delta wing<br>2 circular arc rounded corner<br>3 sharp corner, general case<br>4 *wing-tip* configuration, rounded corner, omega = −90° |
| 8 | `alam2` | Leading edge sweep angle in degrees |
| 9 | `del2` | Compression wedge angle in degrees |
| 10 | `omega` | Angle defining corner opening, Figure 5.17 |
| 11 | `mu` | Free stream side slip angle in degrees, Figure 5.32 |
| 12 | `ach` | Free stream Mach number |
| 13 | `gamma` | Isentropic exponent |
| 14 | `car0` | Circular arc radius, used with icut = 2 |
| 15 | `cax0` | *x* location of circular arc center for icut = 4 |
| 16 | `istream` | 0 no streamline plot<br>1 streamline plot |

`cor_save1`. Therefore, to continue a run it is just a matter of changing the name of `cor_save1` to `cor_save`. The parameter `icut` is used to select the type of configuration to be computed. If `icut` = `1`, the configuration is that of a truncated delta wing. The wing is truncated at the cross flow sonic point and the surface $\Sigma_2$ stays flat from this point to the symmetry plane. Note that for a delta wing configuration, $\Omega$ should be set to $\pi/2$. If `icut` = `2`, the corner is rounded with a circular arc defined by input parameter 14. The other parameters do not need further comments, except for `istream`. This parameter is used to plot the cross flow stream lines. The locus of cross flow streamlines is computed in the computational plane using a fourth-order Runge–Kutta scheme to integrate the streamline slopes. The locus is then mapped to the physical plane. The integration uses the MATLAB function `griddata` at every step of the Runge–Kutta integration to interpolate the solution along the streamline path. This is a very slow process (about 3 min per streamline on a PC laptop computer) and should be used only for final plots and when working with a saved solution.

The partial output generated by `kout` consists of the following parameters: the step counter, the pseudo-time, the average shock speed, the corner pressure, the minimum pressure along the wall, the location of the minimum pressure, the shock standoff distance, the angle $\theta_s(\phi_c)$, the pressure at the corner divided by the pressure in two-dimensional region, the shock angle $\theta_s$ at the symmetry line divided by the shock angle $\theta_s$ at the cross flow sonic line, the pressure at the shock at the symmetry line, and the CPU time in seconds.

## 5.3 Supersonic Flow over Elliptical Wings

If the flow is steady, inviscid, and supersonic, the equations of motion are hyperbolic in the general direction of the velocity vector. Under these conditions, the flow can be evaluated by a space marching technique not unlike the time integration method used in the previous two sections. Thus, although the flow involves three space coordinates, the computational effort is the same as that of a two-dimensional time-dependent problem. In this section, we will describe techniques that are applicable to the calculation of somewhat general airframe cross sections, but will only apply these techniques to the computation of the flow over elliptical wing cross sections.

### 5.3.1 Governing Equations

We assume that the free stream is uniform, that the airframe of interest has a plane of symmetry, and that the free stream velocity vector lies in this plane. We introduce a Cartesian frame of reference $(x, y, z)$ with corresponding unit vectors $(\vec{i}_x, \vec{i}_y, \vec{i}_z)$ such that the plane of symmetry corresponds to the $x = 0$ plane and the axial coordinate $z$ lines up with the axis of the airframe. The free stream is characterized by the free stream velocity vector defined by

$$\vec{V}_\infty = V_\infty(\sin\alpha\vec{i}_y + \cos\alpha\vec{i}_z),$$

where the angle of attack $\alpha$ is the angle between $\vec{V}_\infty$ and $\vec{i}_z$. In terms of our usual nondimensionalization, the magnitude of the free stream velocity is related to the free stream Mach number by $V_\infty = \sqrt{\gamma}M_\infty$. If we let the velocity vector be defined by $v = u\vec{i}_x + v\vec{i}_y + w\vec{i}_z$, then the equations of motion are

$$\begin{aligned} wP_z + uP_x + vP_y + \gamma(w_z + u_x + v_y) &= 0, \\ ww_z + uw_x + vw_y + \Theta P_z &= 0, \\ wu_z + uu_x + vu_y + \Theta P_x &= 0, \\ wv_z + uv_x + vv_y + \Theta P_y &= 0, \\ wS_z + uS_x + vS_y &= 0. \end{aligned} \tag{5.118}$$

If we take the first and second equations of (5.118) and solve them for $P_z$ and $w_z$, we obtain the following system

$$\begin{aligned} P_z &= -\frac{1}{(1 - a^2/w^2)}\left[\sigma P_x + \eta P_y + \frac{\gamma}{w}(u_x + v_y - (\sigma w_x + \eta w_y))\right], \\ w_z &= -\frac{1}{(1 - a^2/w^2)}\left[\sigma w_x + \eta w_y - \frac{a^2}{\gamma w}\left(\sigma P_x + \eta P_y + \frac{\gamma}{w}(u_x + v_y)\right)\right], \\ u_z &= -\left[\sigma u_x + \eta u_y + \frac{a^2}{\gamma w}P_x\right], \\ v_z &= -\left[\sigma v_x + \eta v_y + \frac{a^2}{\gamma w}P_y\right], \\ S_z &= -[\sigma S_x + \eta S_y], \end{aligned} \tag{5.119}$$

where, for convenience, we have introduced the streamline slopes $\sigma = u/w, \eta = v/w$. The system (5.119) is hyperbolic in the $z$ direction if $w^2 > a^2$. Thus, even if the flow is supersonic, we might not be able to march in the $z$ direction if it is not properly aligned with the flow direction. In two-dimensional time-dependent problems we found that the intersection of the Mach cone (the envelope of characteristic surfaces) with a $t =$ constant plane was a circle with the particle path going through the center of the circle and the vertex of the cone. Now, the circle corresponds to the interception of the Mach cone with a plane perpendicular to the streamline that goes through the vertex of the cone. The interception of the Mach cone with a $z =$ constant plane is in general elliptical in shape.

Since the flow is steady, inviscid, and the free stream is uniform, the total enthalpy is conserved, see Section 5.2.2. This is expressed by Bernoulli's equation:

$$\frac{1}{2}(w^2 + u^2 + v^2) + \frac{a^2}{\gamma - 1} = \gamma\left(\frac{1}{2}M_\infty^2 + \frac{1}{\gamma - 1}\right). \tag{5.120}$$

Bernoulli's equation can be used to replace the $w$-momentum equation in (5.119).

## 5.3.2 Conformal Mapping

The conformal mapping of an airframe cross section from the physical plane to a circle or near circle in the map plane provides a convenient way of generating computational grids that are either orthogonal or close to orthogonal. The idea has been successfully exploited in [136,149,151] with airframe cross sections of various levels of complexity. Today, this approach is not a match to the advanced, sophisticated grid generation methods that are commercially available, but it still useful as a quick, efficient method for generating grids in a research environment. The goal then is to transform a given airframe cross section from the physical space $(x, y, z)$ to a circle or near circle space $(r, \theta, \mathfrak{z})$, see Figure 5.35. If necessary, the transformation from a near circle to an exact circle can be achieved efficiently by implementing the Theodorsen–Garrick mapping [219] with fast Fourier transforms.

Let $G = x + iy$, $\varsigma = re^{i\theta} = \vartheta + i\omega$, and consider the following sequence of conformal mappings:

$$
\begin{array}{ccc}
G = x + iy, & W_J = W_J(W_{J-1}, \mathfrak{z}), & \varsigma = re^{i\theta}, \\
G = G(W_J, \mathfrak{z}), & W_{J-1} = W_{J-1}(W_{J-2}, \mathfrak{z}), & \\
 & \cdot & \\
 & W_2 = W_2(W_1, \mathfrak{z}), & \cdot \\
 & W_1 = W_1(\varsigma, \mathfrak{z}), & \\
\text{physical plane} & \text{intermediate planes} & \text{near circle plane}
\end{array}
\tag{5.121}
$$

In addition to the coordinate transformation represented by (5.121) between $G$ and $\varsigma$, we define $\mathfrak{z} = z$. The sequence of maps represented by $W_J$ through $W_1$ can be very simple or very complex depending on the complexity of the airframe cross section. A constrain on the definition of the $W_i$'s is that they be easily invertible in order to

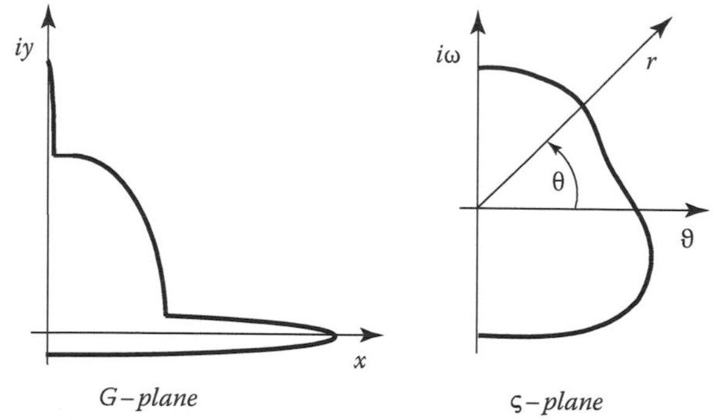


**FIGURE 5.35:** Physical plane cross section transformed to a near circular cross section in the $\varsigma$-plane.

apply the inverse transform from the near circle plane to the physical plane. An example of such a sequence is

$$\begin{aligned}
G &= W_5 + ic_5(z),\\
W_5 &= W_4 + c_4(z)/W_4,\\
W_4 &= W_3 + ic_3(z),\\
W_3 &= W_2 + c_2(z)/W_2^2,\\
W_2 &= W_1 + ic_1(z),\\
W_1 &= \varsigma + c_0(z)/\varsigma.
\end{aligned} \tag{5.122}$$

This sequence was used in [136] to map F-14 and Space Shuttle-like cross sections to near circles. In (5.122), the $c_i$'s are functions of the axial coordinate $z$. Their definition is such as to place the singularities of the mappings inside the airframe cross section, and depends on the specific geometry of the airframe.

Another important sequence consists of repeated applications of the Kármán–Trefftz mapping:

$$\frac{W_{j+1} - \delta_j d_j}{W_{j+1} + \delta_j d_j} = \left(\frac{W_j - h_j}{W_j + \bar{h}_j}\right)^{\delta_j},$$

where

$h_j$ is the location, in the $W_j$ plane, of the "hinge-point", the point in the cross section where a corner occurs
$\bar{h}_j$ is the complex conjugate of $h_j$
$d_j = \mathrm{Re}(h_j)$
$\delta_j = \pi/(2\pi - \phi)$ and $\phi$ is the internal corner angle at the hinge-point

This mapping was used by Moretti [151,153] to design the Southern Mbwanga Air Force (SMAF) airplane.

In order to apply the mappings represented by (5.121), we must be able to transform the governing equations from the physical plane to the near circle plane. This is accomplished as follows. By the chain rule, we have

$$\frac{\partial G}{\partial \varsigma} = \frac{\partial G}{\partial W_J}\left(\prod_{j=2}^{J}\frac{\partial W_j}{\partial W_{j-1}}\right)\frac{\partial W_1}{\partial \varsigma},$$

and its inverse is simply $\varsigma_G = 1/G_\varsigma$. If we define the Cartesian coordinates in the near circle plane by

$$\vartheta = r\cos\theta, \quad \omega = r\sin\theta,$$

then their derivatives with respect to $x$, $y$ are

$$\begin{aligned}
\vartheta_x &= \mathrm{Re}(\varsigma_G),\\
\omega_x &= \mathrm{Im}(\varsigma_G),
\end{aligned} \tag{5.123}$$

and, using the Cauchy–Riemann conditions,

$$\begin{aligned} \vartheta_y &= -\omega_x, \\ \omega_y &= \vartheta_x. \end{aligned} \tag{5.124}$$

From (5.123) and (5.124), we find

$$\begin{aligned} r_x &= (\vartheta\vartheta_x + \omega\omega_x)/r, \\ \theta_x &= (\vartheta\omega_x - \omega\vartheta_x)/r^2, \\ \mathfrak{z}_x &= 0, \\ r_y &= (\vartheta\vartheta_y + \omega\omega_y)/r, \\ \theta_y &= (\vartheta\omega_y - \omega\vartheta_y)/r^2, \\ \mathfrak{z}_y &= 0. \end{aligned} \tag{5.125}$$

It remains for us to find the derivatives with respect to the axial coordinate $z$. These follow from

$$\frac{\partial G}{\partial \mathfrak{z}} = \frac{\partial G}{\partial W_J}\left[\frac{\partial \varsigma}{\partial \mathfrak{z}}\prod_{j=2}^{J}\frac{\partial W_j}{\partial W_{j-1}} + \sum_{k=0}^{J-1}\left(\left.\frac{\partial W_{J-k}}{\partial \mathfrak{z}}\right|_{W_{J-(k+1)}}\prod_{j=0}^{k}\frac{\partial W_{J-j}}{\partial W_{J-(j+1)}}\right)\right] + \left.\frac{\partial G}{\partial \mathfrak{z}}\right|_{W_J}. \tag{5.126}$$

Therefore,

$$\begin{aligned} x_{\mathfrak{z}} &= \mathrm{Re}(G_{\mathfrak{z}}), \\ y_{\mathfrak{z}} &= \mathrm{Im}(G_{\mathfrak{z}}), \\ z_{\mathfrak{z}} &= 1. \end{aligned}$$

Since $r$, $\theta$ are independent of $\mathcal{Z}$, we can write

$$\begin{aligned} r_{\mathfrak{z}} &= r_x x_{\mathfrak{z}} + r_y y_{\mathfrak{z}} + r_z = 0, \\ \theta_{\mathfrak{z}} &= \theta_x x_{\mathfrak{z}} + \theta_y y_{\mathfrak{z}} + \theta_z = 0, \end{aligned} \tag{5.127}$$

and solving (5.127) for $r_z$, $\theta_z$, we find

$$\begin{aligned} r_z &= -(r_y y_{\mathfrak{z}} + r_x x_{\mathfrak{z}}), \\ \theta_z &= -(\theta_y y_{\mathfrak{z}} + \theta_x x_{\mathfrak{z}}), \\ \mathfrak{z}_z &= 1. \end{aligned} \tag{5.128}$$

The transformation of derivatives from the physical plane to the near circle is then

$$\begin{aligned}
\frac{\partial}{\partial x} &= r_x \frac{\partial}{\partial r} + \theta_x \frac{\partial}{\partial \theta}, \\
\frac{\partial}{\partial y} &= r_y \frac{\partial}{\partial r} + \theta_y \frac{\partial}{\partial \theta}, \\
\frac{\partial}{\partial z} &= r_z \frac{\partial}{\partial r} + \theta_z \frac{\partial}{\partial \theta} + \frac{\partial}{\partial \mathfrak{z}}.
\end{aligned} \tag{5.129}$$

### 5.3.3 Computational Coordinates

We want to further map the near circle plane to a computational plane where the mesh is equally spaced and the body surface and bow shock surface map to coordinate surfaces. This mapping is no different than the one introduced in the Sections 5.2.1 and 5.2.2. As before, let (*X*, *Y*, *Z*) be the computational coordinates and let $b(\theta, \mathfrak{z})$ and $c(\theta, \mathfrak{z})$ represent the body surface and the bow shock surface, respectively. The computational coordinates are defined by

$$\begin{aligned}
X &= \frac{r - b(\theta, \mathfrak{z})}{c(\theta, \mathfrak{z}) - b(\theta, \mathfrak{z})}, \\
Y &= \theta/\pi + \frac{1}{2}, \\
Z &= \mathfrak{z}.
\end{aligned} \tag{5.130}$$

The transformation defined by (5.130) maps the body surface to the plane $X = 0$ and the bow shock surface to the plane $X = 1$, while the symmetry planes at $\theta = \pm\pi/2$ are mapped to the planes $Y = 0$ and $Y = 1$.

The transformation of the governing equations from the near circle plane to the computational plane follows from

$$\begin{aligned}
\frac{\partial}{\partial r} &= X_r \frac{\partial}{\partial X}, \\
\frac{\partial}{\partial \theta} &= X_\theta \frac{\partial}{\partial X} + Y_\theta \frac{\partial}{\partial Y}, \\
\frac{\partial}{\partial \mathfrak{z}} &= X_{\mathfrak{z}} \frac{\partial}{\partial X} + \frac{\partial}{\partial Z},
\end{aligned}$$

where the derivatives $X_r$, $X_\theta$, etc., are obtained from (5.130). The transformation of derivatives from the physical plane to the computational plane is given by

$$\frac{\partial}{\partial x} = X_x \frac{\partial}{\partial X} + Y_x \frac{\partial}{\partial Y},$$
$$\frac{\partial}{\partial y} = X_y \frac{\partial}{\partial X} + Y_y \frac{\partial}{\partial Y},$$
$$\frac{\partial}{\partial z} = X_z \frac{\partial}{\partial X} + Y_z \frac{\partial}{\partial Y} + \frac{\partial}{\partial Z},$$

where

$$\begin{aligned}
X_x &= r_x X_r + \theta_x X_\theta \\
Y_x &= \theta_x Y_\theta \\
X_y &= r_y X_r + \theta_y X_\theta \\
Y_y &= \theta_y Y_\theta \\
X_z &= r_z X_r + \theta_z X_\theta + X_z \\
Y_z &= \theta_z Y_\theta
\end{aligned}$$

The governing equations in the computational plane are given by

$$\begin{aligned}
P_Z &= -\left[\hat{U}P_X + \hat{V}P_Y + \frac{\gamma}{w}(\Delta - (\bar{U}w_X + \bar{V}w_Y))\right], \\
w_Z &= -\left[\hat{U}w_X + \hat{V}w_Y - \frac{a^2}{\gamma w}\left(\bar{U}P_X + \bar{V}P_Y + \frac{\gamma}{w}\Delta\right)\right], \\
u_Z &= -\left[Uu_X + Vu_Y + \frac{a^2}{\gamma w}(X_x P_X + Y_x P_Y)\right], \\
v_Z &= -\left[Uv_X + Vv_Y + \frac{a^2}{\gamma w}(X_y P_X + Y_y P_Y)\right], \\
S_Z &= -[US_X + VS_Y],
\end{aligned} \tag{5.131}$$

where

$$\Delta = (X_x u_X + Y_x u_Y + X_y v_X + Y_y v_Y)/(1 - a^2/w^2),$$

and

$$\begin{aligned}
\bar{U} &= (\sigma X_x + \eta X_y)/(1 - a^2/w^2), \\
\bar{V} &= (\sigma Y_x + \eta Y_y)/(1 - a^2/w^2), \\
U &= \sigma X_x + \eta X_y - X_z, \\
V &= \sigma Y_x + \eta Y_y - Y_z, \\
\hat{U} &= \bar{U} - X_z, \\
\hat{V} &= \bar{V} - Y_z.
\end{aligned}$$

### 5.3.4 Body Surface Computation

At a rigid wall the boundary condition is $\boldsymbol{v} \cdot \vec{n} = 0$, where $\vec{n}$ is the unit vector normal to the surface. To compute the pressure on the surface, the rigid wall boundary condition is combined with the compatibility condition reaching the surface along a characteristic in the $X$, $Z$ plane from the flow field. To find this compatibility condition we have to first perform some manipulations with the equations in (5.131) as follows.

We begin by moving all $X$ and $Z$ derivatives in the first of (5.131) to the left side of the equal to sign, the resulting equation, after some minor simplifications, is

$$P_Z + \hat{U}P_X + \frac{\gamma}{w}\frac{1}{(1 - a^2/w^2)}(X_x\sigma_X + X_y\eta_X) = R_P, \tag{5.132}$$

where the right-hand side term, $R_P$, contains all the $Y$ derivative terms. Now subtract the second equation in (5.131), after multiplying it by $u$, from the third equation, after multiplying it by $w$, to obtain

$$w^2\sigma_Z + \frac{a^2}{\gamma}(X_x + \bar{U}\sigma)P_X + w^2\left(X_z + \frac{X_x\sigma}{1 - a^2/w^2} + X_y\eta\right)\sigma_X$$
$$+ \frac{a^2\sigma}{1 - a^2/w^2}X_y\eta_X = wR_u - uR_w, \tag{5.133}$$

where the terms $R_w$ and $R_u$ contain all the $Y$ derivatives of the second and third of (5.131), respectively. A similar operation with the fourth and second equations of (5.131) results in

$$w^2\eta_Z + \frac{a^2}{\gamma}(X_y + \bar{U}\eta)P_X + w^2\left(X_z + X_x\sigma + \frac{X_y\eta}{1 - a^2/w^2}\right)\eta_X$$
$$+ \frac{a^2\eta}{1 - a^2/w^2}X_x\sigma_X = wR_v - vR_w, \tag{5.134}$$

where the term $R_v$ contains all the $Y$ derivatives of the fourth equation of (5.131). Now, we multiply (5.132), (5.133), and (5.134) by arbitrary multipliers $\mu_1$, $\mu_2$, and $\mu_3$, respectively, and add the three equations. We then follow the procedure of Section 5.1.6.1 and find the characteristic slopes:

$$\lambda^{\pm} = (a\bar{U}/w)^2[1 \pm \sqrt{1 + \Upsilon}], \tag{5.135}$$

where

$$\Upsilon = \frac{X_z^2 + X_x^2 + X_y^2}{(a\bar{U}/w)^2(1 - a^2/w^2)}.$$

The characteristic reaching the wall has the $\lambda^-$ slope. The compatibility equation on this characteristic is

$$\left[\lambda^- + \frac{a^2}{w^2}\frac{X_z}{1-a^2/w^2}\right](P_Z + \lambda^- P_X) + \frac{\gamma}{1-a^2/w^2}(X_x\sigma_Z + X_y\eta_Z)$$
$$+ \frac{\gamma\lambda^-}{1-a^2/w^2}(X_x\sigma_X + X_y\eta_X) = \frac{R}{\lambda^-}, \tag{5.136}$$

where

$$R = \left[\lambda^- + \frac{a^2}{w^2}\frac{X_z}{1-a^2/w^2}\right]R_P + \frac{\gamma\lambda^-}{w(1-a^2/w^2)}[X_x(R_u - \sigma R_w) + X_y(R_v - \eta R_w)].$$

We want to solve (5.136) for $P_Z$ after eliminating its dependence on $X_z$, $\sigma_Z$, and $\eta_Z$. This can be accomplished by invoking the boundary condition as follows. The wall corresponds to the surface $X=0$, therefore, the boundary condition can be written as

$$\left(\frac{\boldsymbol{v}}{w}\right)\cdot\nabla X = X_x\sigma + X_y\eta + X_z = 0. \tag{5.137}$$

This relation is used to eliminate $X_z$ from (5.135) and (5.136). Since (5.137) holds along the surface, we can differentiate it with respect to $Z$ to obtain

$$X_x\sigma_Z + X_y\eta_Z = -(X_{xZ}\sigma + X_{yZ}\eta + X_{zZ}). \tag{5.138}$$

Equation 5.138 is used to replace $X_x\sigma_Z + X_y\eta_Z$ in Equation 5.136 with the right hand side of Equation 5.138 which is evaluated in terms of second derivatives of the surface shape. The resulting equation for $P_Z$ is integrated to find the new value of the pressure on the surface.

The entropy at the surface is evaluated by upwinding the $S_Y$ derivative. The coefficient of $S_X$ vanishes, since it is the contravariant velocity component.

To evaluate the velocity at the surface, the $u$- and $v$-momentum equations are integrated using the interior point scheme. The values thus obtained do not necessarily satisfy the surface boundary condition. Let these temporary values of $u$ and $v$ be labeled $\bar{u}$ and $\bar{v}$. With $\bar{u}$ and $\bar{v}$, we find also a temporary value of $w$ from (5.120) and similarly label it $\bar{w}$. If $\vec{\tau} = \tau_x\vec{i}_x + \tau_y\vec{i}_y + \tau_z\vec{i}_z$ and $\vec{\kappa} = \kappa_x\vec{i}_x + \kappa_y\vec{i}_y + \kappa_z\vec{i}_z$ are unit tangents to the surface, such that

$$\vec{\tau} = \vec{n}\times\vec{i}_z,$$
$$\vec{\kappa} = \vec{\tau}\times\vec{n},$$

then we enforce the boundary condition at the surface by defining the velocity vector as $\boldsymbol{v} = \bar{q}\vec{\tau} + \bar{w}\vec{\kappa}$, where $\bar{q} = (\bar{u}\vec{i}_x + \bar{v}\vec{i}_y)\cdot\vec{\tau}$. From $\boldsymbol{v}$ we find $u = \boldsymbol{v}\cdot\vec{i}_x$, $v = \boldsymbol{v}\cdot\vec{i}_y$, and $w = \boldsymbol{v}\cdot\vec{i}_z$.

### 5.3.5 Bow Shock Computation

The bow shock shape is updated as in the standard method described in Section 5.1.6.2, for a different approach see [54]. We want to replace $P_Z$, $\sigma_Z$, and $\eta_Z$ with expressions for the second derivative of the bow shock radius, $c_{ZZ}$. This derivative is then used to update the bow shock radius in the usual way.

The surface of the bow shock is defined by

$$F = r - c(\theta, \mathfrak{z}).$$

Therefore, the unit normal to the bow shock is $\vec{n} = n_x\vec{i}_x + n_y\vec{i}_y + n_z\vec{i}_z$, where

$$n_x = F_x/\upsilon, \quad n_y = F_y/\upsilon, \quad n_z = F_z/\upsilon,$$
$$\upsilon = \sqrt{F_x^2 + F_y^2 + F_z^2},$$

and

$$F_x = r_x - c_\theta\theta_x,$$
$$F_y = r_y - c_\theta\theta_y,$$
$$F_z = r_z - c_\theta\theta_z - c_{\mathfrak{z}}.$$

Now, we apply the Rankine–Hugoniot conditions. The free stream velocity component normal to the bow shock is

$$\tilde{u}_\infty = \vec{V}_\infty \cdot \vec{n} = V_\infty(\cos\alpha n_z + \sin\alpha n_y).$$

The logarithm of the pressure behind the bow shock is given by

$$P = \ln\left(\frac{2}{\gamma - 1}\right) + \ln\left(\tilde{u}_\infty^2 - \frac{\gamma - 1}{2}\right), \tag{5.139}$$

(remember that $P_\infty = 0$), and the normal velocity component behind the bow shock is given by

$$\tilde{u} = \frac{\gamma - 1}{\gamma + 1}\tilde{u}_\infty + \frac{2\gamma}{\gamma - 1}\frac{1}{\tilde{u}_\infty}. \tag{5.140}$$

Therefore, the velocity vector behind the bow shock is

$$\boldsymbol{v} = \vec{V}_\infty + (\tilde{u} - \tilde{u}_\infty) \cdot \vec{n}. \tag{5.141}$$

Since (5.139) and (5.140) are only functions of $\tilde{u}_\infty$, it follows that

$$P_Z = \frac{\partial P}{\partial \tilde{u}_\infty}\tilde{u}_{\infty Z},$$
$$\tilde{u}_Z = \frac{\partial \tilde{u}}{\partial \tilde{u}_\infty}\tilde{u}_{\infty Z},$$

where

$$\frac{\partial P}{\partial \tilde{u}_\infty} = \frac{2\tilde{u}_\infty}{\tilde{u}_\infty^2 - \frac{\gamma-1}{2}}, \quad \frac{\partial \tilde{u}}{\partial \tilde{u}_\infty} = \frac{\gamma-1}{\gamma+1} - \frac{2\gamma}{\gamma+1}\frac{1}{\tilde{u}_\infty^2},$$

and

$$\tilde{u}_{\infty Z} = V_\infty(\cos \alpha n_{zZ} + \sin \alpha n_{yZ}).$$

The derivatives of the direction cosines are

$$\begin{aligned} n_{xZ} &= (F_{xZ}\upsilon - F_x[F_xF_{xZ} + F_yF_{yZ} + F_zF_{zZ}])/\upsilon^2, \\ n_{yZ} &= (F_{yZ}\upsilon - F_y[F_xF_{xZ} + F_yF_{yZ} + F_zF_{zZ}])/\upsilon^2, \\ n_{zZ} &= (F_{zZ}\upsilon - F_z[F_xF_{xZ} + F_yF_{yZ} + F_zF_{zZ}])/\upsilon^2. \end{aligned} \tag{5.142}$$

We are looking for $c_{ZZ}$; this term comes from the $F_{zZ}$ derivative in (5.142). Let us rewrite (5.142) to bring this term out:

$$\begin{aligned} n_{xZ} &= [\Omega + F_xF_zc_{ZZ}]/\upsilon^2, \\ n_{yZ} &= [\Phi + F_yF_zc_{ZZ}]/\upsilon^2, \\ n_{zZ} &= \left[\Psi + \left(F_z^2 - \upsilon\right)c_{ZZ}\right]/\upsilon^2, \end{aligned}$$

where

$$\begin{aligned} \Omega &= F_{xZ}\upsilon - F_x[F_xF_{xZ} + F_yF_{yZ} + F_z\beta] \\ \Phi &= F_{yZ}\upsilon - F_y[F_xF_{xZ} + F_yF_{yZ} + F_z\beta] \\ \Psi &= -F_z[F_xF_{xZ} + F_yF_{yZ}] - \left[F_z^2 - \upsilon\right]\beta \\ \beta &= r_{zZ} - c_{\theta Z}\theta_z - c_\theta\theta_{zZ} \end{aligned}$$

The dependence of $\tilde{u}_{\infty Z}$ on $c_{ZZ}$ is

$$\tilde{u}_{\infty Z} = V_\infty\left[\cos \alpha\Psi + \sin \alpha\Phi + \left(F_yF_z \sin \alpha + \left(F_z^2 - \upsilon\right)\cos \alpha\right)c_{ZZ}\right]/\upsilon^2.$$

The velocity components $u$, $v$, and $w$ behind the bow shock can be found from (5.141). To find $\sigma_Z = (u_Z - \sigma w_Z)/w$ and $\eta_Z = (v_Z - \eta w_Z)/w$, we need the expressions for $u_Z$, $v_Z$, and $w_Z$. These are given by

$$\begin{aligned}
u_Z &= -\frac{2}{\gamma+1}\left(1+\frac{\gamma}{\tilde{u}_\infty^2}\right)n_x\tilde{u}_{\infty Z} + (\tilde{u}-\tilde{u}_\infty)n_{xZ}, \\
v_Z &= -\frac{2}{\gamma+1}\left(1+\frac{\gamma}{\tilde{u}_\infty^2}\right)n_y\tilde{u}_{\infty Z} + (\tilde{u}-\tilde{u}_\infty)n_{yZ}, \\
w_Z &= -\frac{2}{\gamma+1}\left(1+\frac{\gamma}{\tilde{u}_\infty^2}\right)n_z\tilde{u}_{\infty Z} + (\tilde{u}-\tilde{u}_\infty)n_{zZ}.
\end{aligned} \tag{5.143}$$

We have all the ingredients needed for the recipe of Section 5.1.6.2. The derivatives $P_Z$, $\sigma_Z$, and $\eta_Z$ appearing in the compatibility equation reaching the bow shock on the $\lambda^+$ characteristic, similar to (5.136), are written in terms of $c_{ZZ}$. The resulting equation is solved for $c_{ZZ}$ which is used to update the bow shock position. The details of the implementation can be found in the function `SW_SHOCK.M` of the `SUPERWING.M` program.

### 5.3.6 Second Derivatives of the Mappings

The update at the body boundary and at the bow shock requires second derivatives of the mappings. These derivatives are obtained as follows:

$$\begin{aligned}
G_{\varsigma\varsigma} &= G_\varsigma\left(\frac{\partial \ln(W_{1\varsigma})}{\partial\varsigma} + \sum_{j=2}^{J}\frac{\partial \ln\left(W_{jW_{j-1}}\right)}{\partial W_{j-1}} + \frac{\partial \ln(G_{W_J})}{\partial W_J}\right), \\
\varsigma_{GG} &= -\frac{G_{\varsigma\varsigma}}{G_\varsigma^3}.
\end{aligned}$$

The second derivatives of the Cartesian coordinates $\vartheta$ and $\omega$ are

$$\begin{aligned}
\vartheta_{xx} &= \mathrm{Re}(\varsigma_{GG}), \\
\omega_{xx} &= \mathrm{Im}(\varsigma_{GG}), \\
\vartheta_{yy} &= -\vartheta_{xx}, \\
\vartheta_{yx} &= -\omega_{yy}, \\
&\text{etc.}
\end{aligned}$$

The second derivatives of the polar coordinates are

$$\begin{aligned}
r_{xr} &= (\vartheta_r\vartheta_x + \vartheta\vartheta_{xr} + \omega_r\omega_x + \omega\omega_{xr})/r - r_x/r, \\
\theta_{xr} &= (\vartheta_r\omega_x + \vartheta\omega_{xr} - \omega_r\vartheta_x - \omega\vartheta_{xr})/r^2 - 2\theta_x/r, \\
&\text{etc.}
\end{aligned}$$

The derivatives with respect to $\mathfrak{z}$ are

$$G_{\varsigma\mathfrak{z}} = G_{\varsigma}\frac{\partial}{\partial\mathfrak{z}}\left(\ln\left(W_{1\varsigma}\right) + \sum_{j=2}^{J}\ln\left(W_{jW_{j-1}}\right) + \ln\left(G_{W_J}\right)\right),$$

$$\varsigma_{G\mathfrak{z}} = -\left(G_{\varsigma\mathfrak{z}} + G_{\varsigma\varsigma}\varsigma_G G_{\mathfrak{z}}\right)\Big/G_{\varsigma}^2,$$

$$\vartheta_{x\mathfrak{z}} = \mathrm{Re}\left(\varsigma_{G\mathfrak{z}}\right),$$

$$\omega_{x\mathfrak{z}} = \mathrm{Im}\left(\varsigma_{G\mathfrak{z}}\right),$$

etc.

The derivatives $r_{x\mathfrak{z}}$, $\theta_{x\mathfrak{z}}$, etc. follow from (5.125), and the derivatives $r_{z\mathfrak{z}}$, $\theta_{z\mathfrak{z}}$ follow from (5.128) and (5.126).

### 5.3.7 Results for Flow over Elliptical Cross Sections

The supersonic marching code is named `SUPERWING.M`. In this code only a simple Joukowsky mapping is used to transform elliptical cross sections to circles. The mapping is defined by

$$G = h\left(\varsigma + \frac{1}{\varsigma}\right),$$

where $h$ is a real number. If A and B, functions of $z$, are the semi-major and semi-minor axes, respectively, of an elliptical cross section, then

$$h(z) = \tfrac{1}{2}\sqrt{A(z)^2 - B(z)^2}.$$

The elliptical cross section in the physical plane is transformed to a circle with center at the origin of the $\varsigma$-plane and a radius, $r_0$, defined by,

$$b(\mathfrak{z}) = r_0 = \frac{1}{2}(\mathrm{A(z)} + \mathrm{B(z)}).$$

In Sections 5.3.7.1 through 5.3.7.3 we discuss results obtained with the `SUPERWING.M` code for the following configurations: a circular cone, an ogive of revolution, and an elliptical wing.

#### 5.3.7.1 Circular Cone

The supersonic flow over a cone at zero angle of attack appears at first glance as a very simple flow field to compute. However, under some conditions, it can present some challenges. The first entry in Jones' cone tables [110] corresponds to a 5° cone at Mach number 2 and zero incidence. It is the only entry at this Mach number. The problem is governed by the Taylor–Maccoll equations [218] which are

$$\begin{aligned}
\frac{d\bar{u}}{d\bar{\theta}} &= \bar{v}, \\
\frac{d\bar{v}}{d\bar{\theta}} &= -\bar{u} + (\bar{u} + \bar{v}/\tan(\bar{\theta}))/(\bar{v}^2/a^2 - 1), \\
a^2 &= (\gamma - 1)\left(H_{st} - \frac{1}{2}(\bar{u}^2 + \bar{v}^2)\right),
\end{aligned} \tag{5.144}$$

where $\bar{u}$ and $\bar{v}$ are the velocity components, in the $(x,y)$ plane, in the directions of the polar coordinates $(\bar{\rho}, \bar{\theta})$, respectively. The total enthalpy is $H_{st}$ and $a$ is the speed of sound. These equations are solved by guessing a shock angle, applying the Rankine–Hugoniot jumps, and integrating (5.144) from the shock to the cone surface. If $\bar{v} = 0$ at the cone surface, then the problem is solved. Otherwise, a new guess for the shock angle is made and the process is repeated until the surface boundary condition is satisfied. This process is implemented in the function `SW_INITIAL.M` using a fourth-order accurate Runge–Kutta scheme. The second equation of (5.144) is ill behaved when $\bar{v}$ approaches $a$, and in general $\bar{v}_{\bar{\theta}}$ exhibits a boundary-layer-like behavior at the shock that gets more pronounced as the free stream Mach number decreases, as shown in Figure 5.36.

To see how `SUPERWING.M` handles this problem, we start with initial conditions at some plane $z = z^*$ corresponding to the Taylor–Maccoll solution and march downstream to some $z_{last}$ station. Since the problem is conical, the solution at $z_{last}$ can be scaled back to $z^*$ and the process can be repeated until convergence is achieved. Figure 5.37 compares the pressure field computed with `SUPERWING.M` to the values from Jones' tables [110] for $M_\infty = 2$ and a 5° half cone angle. There are small

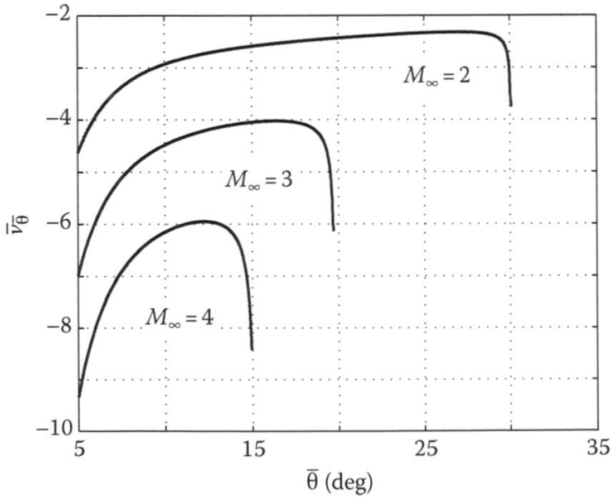


**FIGURE 5.36:** Boundary layer like behavior of $\bar{v}_{\bar{\theta}}$ at the bow shock, cone half angle is 5°.

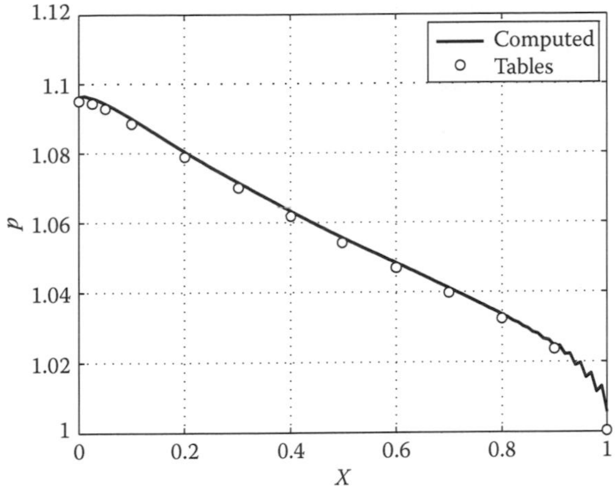


**FIGURE 5.37:** Computed pressure distribution across the shock layer compared to values from [110].

wiggles at the shock (at $X = 1$) which are a result of a sharp pressure gradient there. If a small amount of artificial dissipation (in the form of a second derivative of pressure acting only near the shock, see function `SW_POINTS.M`) is added to the continuity equation and $w$-momentum equation, we obtain the results shown in Figure 5.38. In general, we prefer to treat this ailment by introducing a stretching in the mappings to resolve the high gradients in the layer near the shock. However, this is an instance

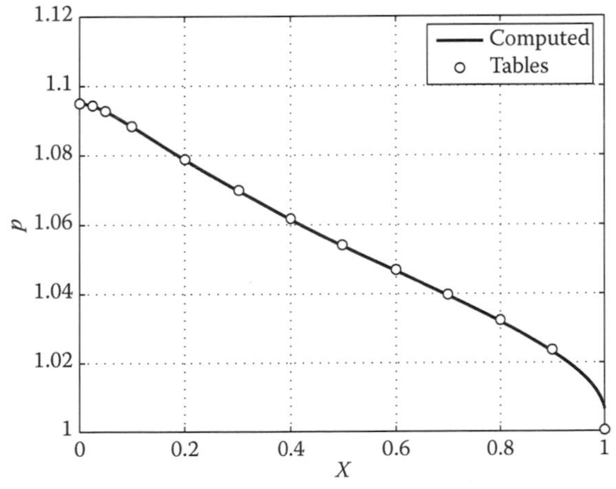


**FIGURE 5.38:** Computed pressure distribution across the shock layer, with added dissipation, compared to values from [110].

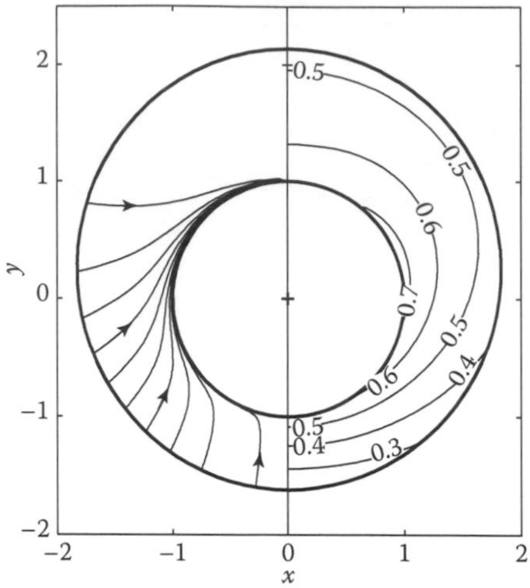


**FIGURE 5.39:** Results for a 10° half angle cone at $M_\infty = 4$ and 5° angle of attack. The left panel shows the computed cross flow streamlines, and the right panel shows the cross flow iso-Mach lines.

where a small amount of artificial dissipation has solved the problem without smearing out the rest of the solution.

As a second example, Figure 5.39 shows the computed flow field over a 10° half angle cone at Mach 4° and 5° angle of attack. The figure shows the cross flow streamlines on the left panel and the cross flow iso-Mach lines in the right panel. The solution was obtained with 40 intervals around the body and 40 intervals between the cone and the bow shock. The cross flow streamlines are actually lines of constant entropy. Clearly, an entropy layer builds up on the lee-side of the cone surface where a node singularity resides. A comparison of the computed surface pressure distribution with the tabulated values in [110] is shown in Figure 5.40.

#### 5.3.7.2 Ogive

Here we present results for the ogive of revolution shown in Figure 5.41. The geometry for this configuration consists of a 10° half angle nose cone followed by a circular cylinder. The transition between the nose cone and the cylinder is achieved with a cubic polynomial. The geometry for this configuration is one of the default geometries in `SUPERWING.M`, and the details can be found in the function `SW_GEOM.M`. The initial conditions, imposed at $z=0$, correspond to the 10° cone solution shown in Figures 5.39 and 5.40. The computed pressure distribution is shown in Figure 5.42. In this figure, we show the isobars on the windward plane, the ogive surface, and the leeward plane. The ogive surface is *unwrapped* in the plot by plotting the isobars as a function, $\xi$, of the distance from the leading edge (the interception of the ogive surface with the $y=0$ plane).

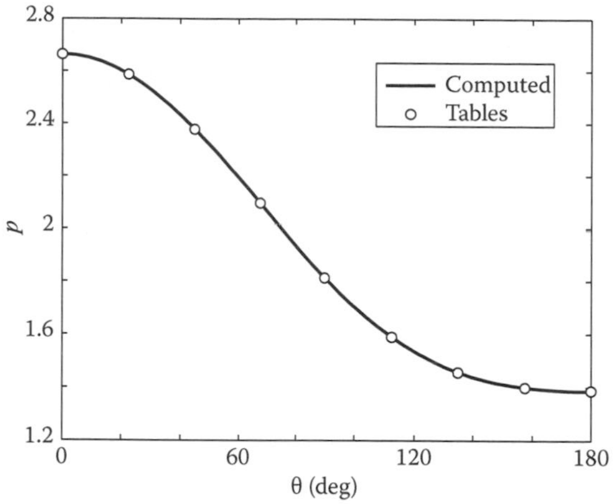


**FIGURE 5.40:** Computed surface pressure distribution for a 10° half angle cone at $M_\infty = 4$ and 5° angle of attack compared to tabulated values [110].

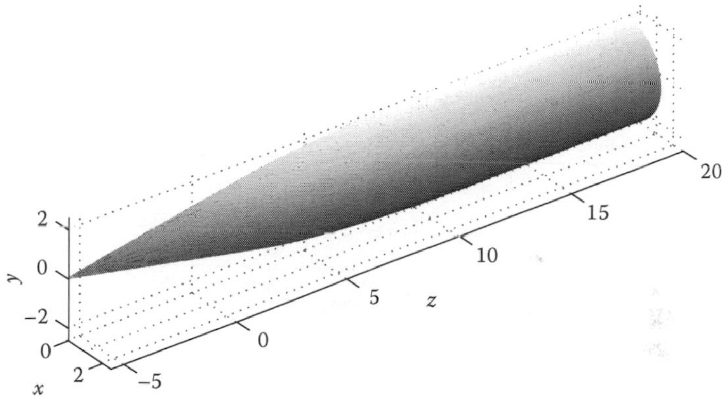


**FIGURE 5.41:** Ogive of revolution consisting of a 10° half angle cone followed by a circular cylinder.

### 5.3.7.3 Elliptical Wing

The final set of results is for the elliptical wing depicted in Figure 5.43. The ellipse has an ellipticity equal to $1/2$. The ellipse grows at a constant rate (i.e., is conical) until $z = 0$. The 20° leading edge (major axis of the ellipse) grows at a constant rate, the semi-minor axis levels off to a maximum height of 2 units. The transition is achieved with a cubic fit. This geometry is one of the default configurations in `SUPERWING.M` and the geometrical details can be found in the function `SW_GEOM.M`. The starting plane solution, at $z = 0$, is shown in Figure 5.44. The left panel of this figure shows the cross flow streamlines. The streamline that wets the

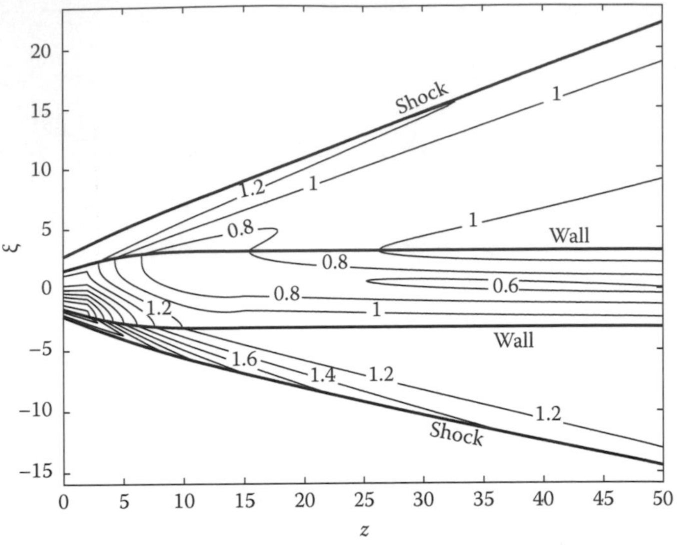


**FIGURE 5.42:** Computed isobars on windward plane, ogive surface and leeward plane at $M_\infty = 4$ and 5° angle of attack.

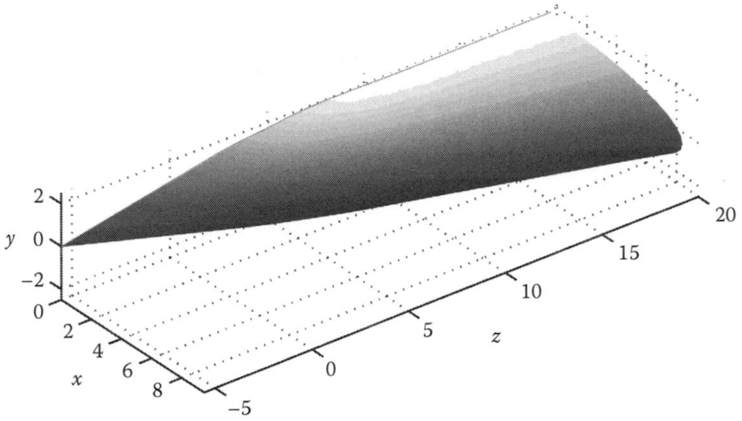


**FIGURE 5.43:** Elliptical wing geometry.

surface is indicated by the arrow at about 80° from the windward plane. There are vortical singularities forming on both the windward and leeward planes. The right panel shows the cross flow iso-Mach lines. For this case the free stream Mach number is 2 and the angle of attack is 5°.

The computed isobars on the windward plane, the wing surface, and the leeward plane are shown in Figure 5.45. As before, the wing surface is unwrapped in the plot. Notice the high pressure gradient near the leading edge as the cross flow expands as it turns around the leading edge. Again, a 40 × 40 mesh was used for this calculation.

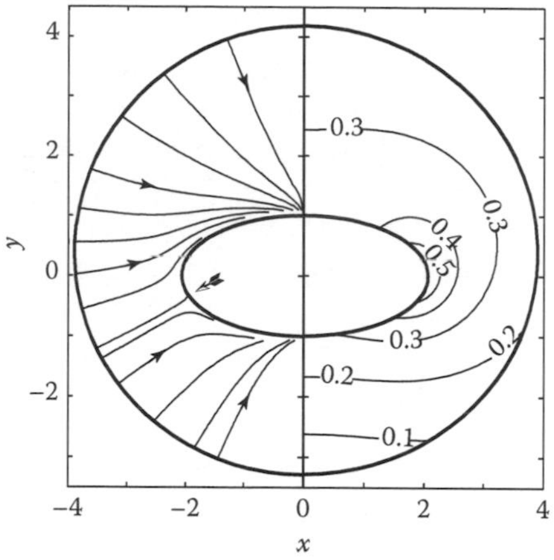


**FIGURE 5.44:** Initial conditions used at $z = 0$.

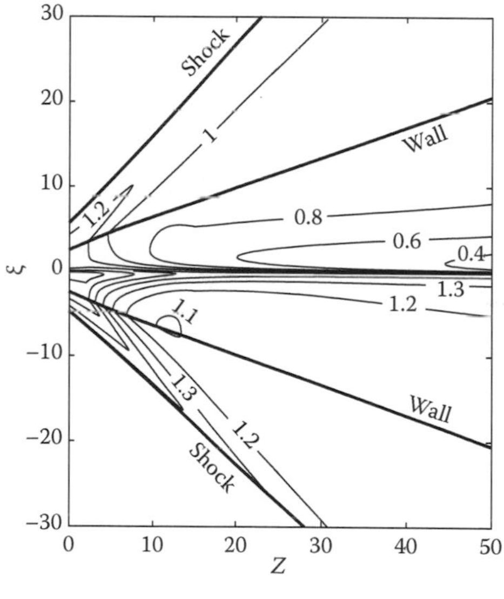


**FIGURE 5.45:** Computed isobars for elliptical wing configuration.

### 5.3.8 Running the Supersonic Marching Code

The first step in setting up a run of the `SUPERWING.M` code is the definition of the geometry parameters A($z$) and B($z$) that control the shape of the elliptical cross section. This is done in function `SW_GEOM.M`. There are three default geometry configurations in this function: (1) an ogive, (2) an elliptical wing, and (3) simple circular cone. These three examples should serve as guidance for the definition of other geometries. Take note that `SUPERWING.M` is designed to handle only the bow shock. If the geometry is such that it creates a compression of the flow that results in an internal shock, the results will not be very accurate and the calculation might fail.

If the nose of the configuration consists of a circular cone and the flow is at zero incidence, the Taylor–Maccoll solution build into the code is sufficient to generate a starting plane. If the flow is at some angle of attack or the nose geometry has an elliptical cross section, then a starting plane solution must be developed first. This process will consist of running the conical initial nose shape until it converges. The best way to accomplish this is to run the conical nose for some axial distance and save the file. Then when the file is reloaded, to continue the calculation of the conical solution, the flow field will be rescaled to the original starting plane dimensions. This is done in the function `SW_SCALE.M`. A few iterations in this way should provide a converged starting plane.

The calculation of interior points is done in the function `SW_POINTS.M`. This function integrates the five governing equations (5.131). A very similar function, named `SW_POINTS_W.M`, does not integrate the $w$-momentum equation; instead, Bernoulli's equation is used to find $w$. The results obtained with `SW_POINTS_W.M`,

**TABLE 5.7:** Input parameters for the *SUPERWING.M* code.

| | Input Parameter | Description |
|---|---|---|
| 1 | `nrun` | Run number, to keep tract of runs |
| 2 | `na` | Number of intervals along wall, from −90° to 90° |
| 3 | `ma` | Number of intervals between wall and shock |
| 4 | `ka` | Number of steps in $z$ direction, this is a safety factor |
| 5 | `ja` | Output frequency |
| 6 | `loadx` | 0 does not load previous solution<br>1 loads previous solution for continuation run, use only with conical solution |
| 7 | `iadd` | 0 no mesh points are added<br>1 doubles mesh in both directions |
| 8 | `gindex` | 1 ogive cylinder geometry<br>2 elliptical wing geometry<br>3 circular cone geometry |
| 9 | `Ivisc` | 0 no dissipation added<br>1 dissipation added to continuity equation and $w$-momentum equation |
| 10 | `mach` | Free stream Mach number |
| 11 | `gamma` | Isentropic exponent |
| 12 | `cone` | Cone half angle in degrees |
| 15 | `dist` | $z$ distance to end calculation, initial cone plane at $1/\tan(\text{cone})$ |
| 16 | `attack` | Angle of attack in degrees |
| 17 | `guess_shk_ang` | −1 uses built-in guess for cone shock angle, otherwise user inputs guess in degrees |

in general, are not as good as those obtained with SW_POINTS.M. To use SW_POINTS_W.M, you have to replace the call in SUPERWING.M to SW_POINTS.M with a call to SW_POINTS_W.M.

The input parameters of SUPERWING.M are listed in Table 5.7. For the most part, they are the same input parameters we have used in other codes. The parameter loadx is used to load a previous solution to continue a calculation. But, keep in mind that this is only intended for developing a starting conical solution, and that the loaded solution will be rescaled to the original dimensions of the starting plane. The parameter, Ivisc is used to add artificial dissipation to the continuity and $w$-momentum equations. It is intended for low Mach number cases only. Use it sparingly! The last parameter, guess_shk_ang, is used for guessing the shockangle in the Taylor–Maccoll solution in the function SW_INITIAL.M. If it is set to $-1$, the code defaults to a build in correlation. If not set to $-1$, then it should be set to a user provided shock angle guess. For most cases, the build in guess works fine.

The output consists of various plots of the flow field.

## Problems

**5.1** Build a MATLAB function to evaluate the angle between the sonic line and the velocity vector at the surface of a blunt body. Use this function to show that, for $\gamma = 1.4$ and ellipsoid bodies, the angle is acute if

$$\begin{aligned} M_\infty &< 2, \quad \text{for } e = 1/2, \\ M_\infty &< 3.15, \quad \text{for } e = 1, \\ M_\infty &< 4.2, \quad \text{for } e = 3/2, \end{aligned}$$

and obtuse otherwise [204].

**5.2** Replace the MacCormack scheme with the $\lambda$-scheme in the BLUNT.M code. Compare the results from this scheme to those of the MacCormack scheme for the cases listed in Tables 5.1 and 5.2.

**5.3** Build a MATLAB function to draw the limiting characteristic in blunt body flows. Use this function to show type I, type II, and type III transonic regions.

**5.4** Modify the blunt body code so that it can compute nonsymmetric blunt bodies and flows at angle of attack. There is a supposition that the streamline wetting the surface of a blunt body is the streamline crossing the shock at its strongest point (where the free stream direction is normal to the bow shock). Study this issue with the modified code.

**5.5** Evaluate the bow shock curvature using (3.120) and compare it to the curvature calculated by BLUNT.M. Note that in the limit of $\theta \to \pi$, the vorticity and the tangential velocity component vanish, and l'Hôspital's rule needs to be applied to find the value of the curvature. The results shown in Figure 5.46 correspond to a first-order calculation of vorticity at the shock. Can you improve the results with a more accurate approximation for the vorticity?

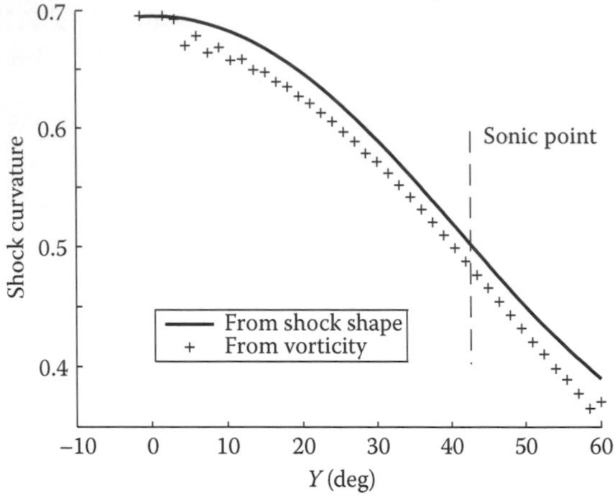


**FIGURE 5.46:** Bow shock curvature for a sphere at $M_\infty = 4$ evaluated from the shock shape and from Equation 3.120.

**5.6** When a circular cone is at a small angle of incidence to an incoming supersonic free stream, a nodal singularity forms on the leeward side of the cone as shown in Figure 5.47a. At a larger angle of incidence, the node "lift-off" from the surface, as shown in Figure 5.47b, and leaves in its place a saddle singularity. The potential slender body solution* is given by [8]

$$\frac{v}{u_0} = -\frac{\delta}{\hat{\theta}^2}(\hat{\theta}^2 - 1)(\hat{\theta} - \hat{\alpha}\cos\phi) + O(\varepsilon^2 \log \varepsilon),$$

$$\frac{w}{u_0} = -\frac{\hat{\alpha}\delta}{\hat{\theta}^2}(1 + \hat{\theta}^2)\sin\phi + O(\varepsilon^2 \log \varepsilon),$$

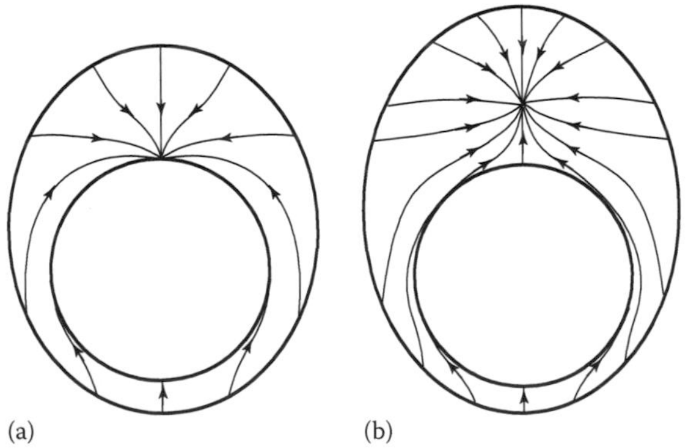


**FIGURE 5.47:** Cross flow streamline pattern for a circular cone at angle of attack.

* The MATLAB code CONE.M compares the potential and slender cone solutions.

where $\delta$ is the cone half angle, $\hat{\theta} = \theta/\delta$, $\hat{\alpha} = \alpha/\delta$, $\alpha$ is the angle of incidence, $\varepsilon = \sqrt{M_\infty^2 - 1}\tan\theta$, and $\phi$ is measured from the leeward plane. (a) Find the coordinates of the stagnation points. (b) Find the coefficient $C_1$ of (5.105) by matching $v_\theta/u_0$. Study the bifurcation associated with this problem and find the condition for "lift-off."

**5.7** Linear theory for supersonic flow past a thin delta wing with a wedge compression angle equal to $\delta$ and whose leading edge coincides with the free stream Mach cone (i.e., $\Lambda_2 = -a\sin 1/M_\infty$, $\delta \ll \pi + 2\Lambda_2$) is given in terms of the potential function $\Phi$ by [36, p. 457]

$$\Phi = \sqrt{\gamma}M_\infty(z + \delta\phi(x, y, z; M_\infty) + \dots), \tag{5.145}$$

where

$$\phi = -\frac{1}{\pi\beta}\sqrt{z^2 - \beta^2(y^2 + x^2)} + \frac{y}{\pi}\text{acos}\frac{\beta y}{\sqrt{z^2 - \beta^2 x^2}},$$

and $\beta = \sqrt{M_\infty^2 - 1}$. (a) Use (5.145) to find the velocity components in spherical coordinates. (b) Match the global solution to the local solution at the corner (5.105) and find the coefficient $C_1$. (c) Study the behavior (find $\partial\xi_{s,w}/\partial\delta$, $\partial\xi_{s,w}/\partial M_\infty$) of the corner singularity as a function of $\delta$ and $M_\infty$.

**5.8** For steady, supersonic flow, Bernoulli's equation (5.120) is valid along the shock. Write Bernoulli's equation in terms of the velocity components behind the shock and find an expression for the second derivative of the bow shock radius with respect to $z$. Use this expression to write a MATLAB function to update the bow shock radius.

**5.9** The "Butler" wing is an elliptical wing defined by

$$A(z) = cz,$$
$$B(z) = \begin{cases} cz, \ 0 \le z \le 0.2, \\ cz\left(1 - \left(\dfrac{z - 0.2}{0.8z}\right)^4\right), \ 0.2 \le z \le 1, \end{cases} \tag{5.146}$$

where A and B are the semi-major and semi-minor axis, respectively, of the elliptical cross section. The value of $c$ in (5.146) is 0.29814. Study this flow field at $M_\infty = 3.5$ and zero incidence and compare to the results in [25] and [209].

**5.10** Study the flow over an elliptical wing with semi-major and semi-minor axis defined by

$$A(z) = (1 - z/28)z\tan(20°),$$
$$B(z) = \frac{1}{2}A(z),$$

respectively, at a free stream Mach number equal to 3.95 and zero incidence, and compare your results to those in [221].

# Chapter 6

# *Floating Shock-Fitting with Unstructured Grids*

> Another effort in shock-fitting may therefore be warranted using floating methods of imposing jump relations to individual cells rather than treating the discontinuities as a boundary surface. Such a procedure could enrich a shock captured solution...
>
> Joseph L. Steger [210]

> I feel very excited and optimistic about these methods. If bad physics has achieved so much in the past, what might not good physics achieve in the future?
>
> Philip L. Roe [183]

## 6.1 Introduction

In the previous chapter we studied the implementation of boundary shock-fitting for several problems. It is possible, see [136], to extend this technique to more complex problems. However, as the number of shocks increases, the method becomes increasingly complicated. Floating shock-fitting was developed to circumvent this problem. As shown in Chapter 4 and [188], floating shock-fitting works well for one-dimensional problems. For two- and three-dimensional problems, many of the operations (many of them being just simple bookkeeping) that are required to implement floating shock-fitting remained cumbersome within the confines of structured grids. Thus, although several demonstrations of the method were made by Moretti [147,152] and other collaborators, the method has not been widely used. With the current level of maturity of unstructured grid methods, it now appears that floating shock-fitting is a viable technique. The pioneering work in floating shock-fitting with unstructured grids is part of a collaborative research effort originating at the University of Rome, La Sapienza, and the University of Basilicata, particularly in the work of Aldo Bonfiglioli and Renato Paciorri [163].

This chapter is different from the rest of this book. In the previous chapters, I have described matured shock-fitting techniques and provided working codes to demonstrate how these techniques work. Floating shock-fitting with unstructured grids is still in its infancy and much research is still needed before we can consider it a matured method. Also, the implementation of floating shock-fitting with unstructured grids requires considerable infrastructure (the entire infrastructure associated with an unstructured grid solver, including the codes needed for grid generation). This is well beyond the scope of this book. However, it is my opinion that the future of shock-fitting, as well as the future of unstructured grid solvers for high-speed flows, lies in this effort. Thus, in this chapter I will describe how floating shock-fitting can be implemented in an unstructured grid code, but will not provide a code to demonstrate these techniques. The work described follows closely the work of Bonfiglioli and Paciorri, and I am indebted to them for many private communications describing their method and many of the results included in this chapter.

---

## 6.2 Unstructured Grids: Preliminaries

In 1941, in a lecture to the American Mathematical Society concerning the computation of the stiffness of a rectangular plate with a rectangular hole, Richard Courant [39] discussed one of the first uses of a triangular mesh with the unknowns represented by piecewise continuous functions. Courant observed that "these results show in themselves and by comparison that the generalized method of triangular nets seems to have advantages. It was applied with similar success to the case of a square with four holes, and it is obviously adaptable to any type of domain . . . " Courant's formulation, constructed as a calculus of variation problem, was a precursor of what we call today the finite element method. The finite element method was firmly established in the early 1960s and quickly gained wide acceptance in the field of solid mechanics. For fluid mechanics, it was not until the mid 1980s that unstructured grid techniques, either as finite elements [101] (variational formulation) or finite volumes [139] (integral formulation), were seriously pursued. The push in this direction was driven by a need to solve problems of great geometrical complexity, e.g., full aircraft configurations, where the application of traditional structured or block-structured grids had become awkward, and by the availability of very fast, large memory computers. Twenty years later, unstructured grid solvers are widely used in computational fluid dynamics. However, as of this writing, several challenges remain. The generation of high quality unstructured grids remains a challenge. The treatment of shock waves, particularly for high Mach number flows, has proven difficult [26,85]. The formulation of efficient and robust unstructured algorithm remains a work in progress.

The unstructured grid method can be broken up into three interrelated parts: (1) the mesh, (2) the solver, and (3) the data structure. The mesh can be subdivided into the type of mesh and the method used for generating the mesh. The mesh types are simple meshes (simplicial) made solely of triangles in two dimensions and tetrahedral in three dimensions, or mixed element meshes which combine triangles and quadrilaterals in two dimensions and tetrahedral, prisms, pyramids, and hexahedra in three dimensions. The most common simplicial mesh generation methods are the advancing-front technique and the Delaunay triangulation method.

The solver can be subdivided into a finite-volume solver or a finite-element solver depending on how the governing equations are formulated, i.e., from an integral or variational representation. In the finite-volume method, the governing equations are represented as volume integrals that surround each node point. From (3.41), after integrating over the volume $\Omega$, we have

$$\int_{\Omega} \frac{\partial U}{\partial t} \mathrm{d}\Omega + \int_{\Omega} \nabla \cdot \mathbf{F}(U) \mathrm{d}\Omega = 0. \tag{6.1}$$

Now using Gauss' theorem and taking the time derivative outside the volume integral,

$$\frac{\partial}{\partial t} \int_{\Omega} U \mathrm{d}\Omega + \oint_{\Gamma} \mathbf{F}(U) \cdot \mathrm{d}\vec{\Gamma} = 0, \tag{6.2}$$

where

$\Gamma$ is the surface enclosing the volume $\Omega$

$\mathrm{d}\vec{\Gamma} = \mathrm{d}\Gamma \vec{\eta}$, where $\vec{\eta}$ is the unit normal to the surface, pointing outward

Thus, the integral representing a divergence term is replaced by surface integral over the surface of the control volume. The surface integrals represent fluxes entering and leaving the control volume. Within the finite-volume representation, we must also distinguish between methods that locate the unknowns at the vertex of the cells and those that locate them at the centroid of the cells. Figure 6.1 illustrates the two methods of storing unknowns in two dimensions. The dashed lines represent a control area for the vertex-based method, while the solid line represents a control area for the centroid-based method. Note that fluxes entering one control area are in turn leaving the adjacent control area. Thus, the total flux exchange across internal control area boundaries is zero, which is a restatement of the conservation property. Considering the differences in control areas of the two methods, it follows that centroid-based and vertex-based schemes use very different discretization stencils. For the same grid, the vertex-based scheme evaluates fewer unknowns using a larger stencil than the centroid-based scheme. Nonetheless, a recent study based on Poisson's equation indicates that both methods produce results with comparable accuracy when solving for the same number of unknowns [59]. We can further characterize the solver as a central or upwind method. For a central method, the fluxes are estimated at the surface of the control volume by averaging the fluxes at the surface of the control volume. For second-order schemes this results in a compact stencil. For upwind methods the fluxes at the surface of the control volume are estimated by taking into account the direction of the wave propagation. This could result in a larger stencil. A modification of the finite-volume vertex-based method is to apply the finite-volume method to triangles to evaluate the residual. This is followed by a distribution of the residuals, according to some rule, to vertices on appropriate control volumes. This is the so-called fluctuation-splitting method, which results in a compact stencil and is discussed in more detail below.

In order to evaluate (6.2), we must have access to data related to the triangle faces, $F_i$, edges, $E_j$, and vertices, $V_k$. The *data structure* refers to how we store this data in memory and how we link and reference the data. Figure 6.2 shows two

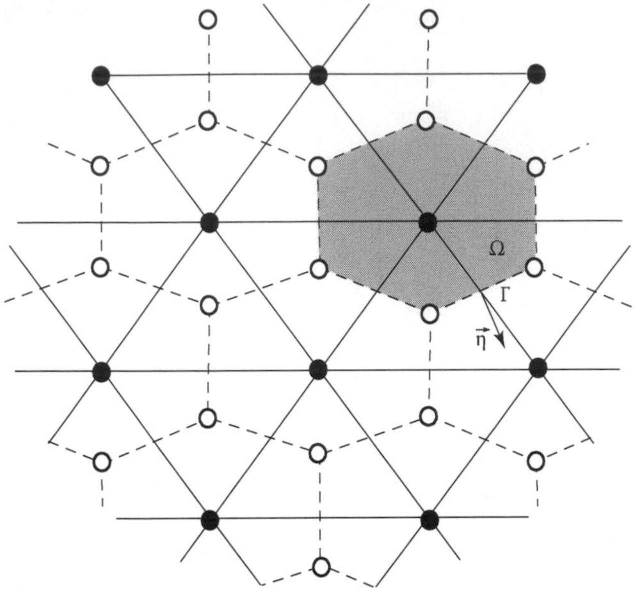


**FIGURE 6.1:** Illustration of a triangular mesh with unknowns located at the vertex (solid symbols), and a triangular mesh with unknowns located at the centroid (open symbols). For a vertex-scheme, $\Omega$ is the area of the shaded control area, $\Gamma$ is the boundary of the control area, and $\vec{\eta}$ is its unit normal.

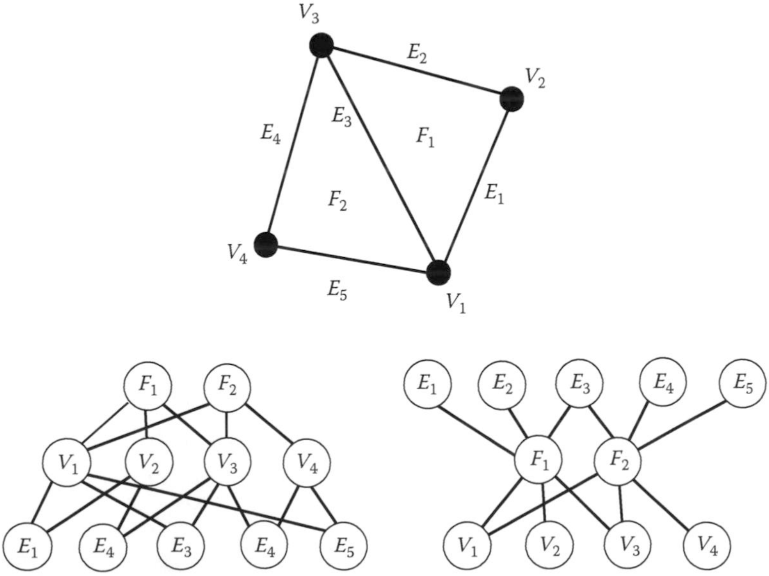


**FIGURE 6.2:** Illustration of two ways of linking data in an unstructured database.

triangles. The diagram on the left illustrates a face-based data structure. The face points to its vertices, which in turn points to the associated edges. The diagram on the right illustrates an edge-based data structure. Here, an edge points to the neighboring faces, which in turn point to their associated vertices. Another typical edge data structure uses a quadruple dataset associated with each edge, which refers to its two vertices and two adjacent faces. The data structure has to be designed with care to satisfy the needs of a particular solver, the type of search operations that need to be performed, and the efficient insertion and deletion of cells. All of these will be critical elements for the smooth implementation of a floating shock-fitting algorithm. We will not discuss the details of the data structure further. The interested reader can consult the articles by Barth [12] and Morgan [156].

### 6.2.1 Delaunay Triangulation

There are several ways that a given set of points in a plane can be triangulated, see Figure 6.3. The Russian mathematician Boris Delone [57] (spelled Delaunay in French) developed a triangulation technique that maximizes the minimum angle of all triangles in the triangulation, and, therefore, results in triangular grids that are of good quality for numerical calculations. A triangulation of a set of points P in the plane is a Delaunay triangulation if no point in P is inside the circumcircle of any triangle in the triangulation. For any set of points P in the plane, the Delaunay triangulation is unique if no four points in P are cocircular (e.g., the vertices of a rectangle). For a set of points that lie in a straight line there is no Delaunay triangulation. For any triangulation in the plane, if there are $k$ vertices on the convex hull* of a set of points P with $n$ points, then the number of triangles generated is $2n$-2-$k$ and the number of edges is $3n$-3-$k$. On average, each vertex in a Delaunay triangulation has six neighboring triangles. If we connect the centers of the circumcircles formed by a Delaunay triangulation to their nearest neighbors, we create a

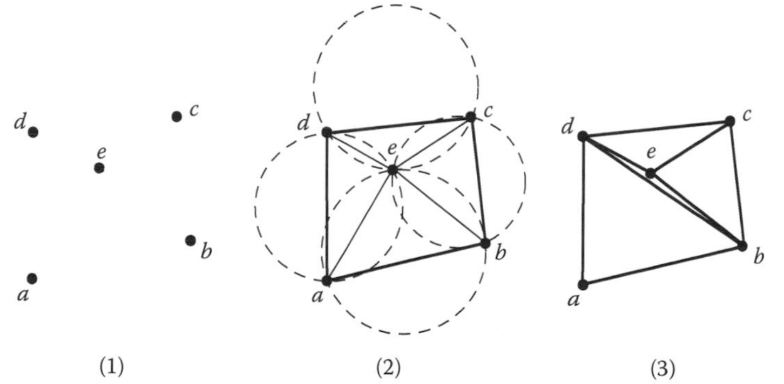


**FIGURE 6.3:** Illustration of two triangulations for a given set of points. The set of points (1) is Delaunay triangulated in (2), but not in (3).

* The convex hull of a set of points P is the smallest convex polygon surrounding the points in P. If we stretch a rubber band over all the points in P, the convex hull is the shape taken by the rubber band.

Voronoi tessellation of the plane. Likewise, if one draws a line between any two points whose Voronoi domains (cells) touch, a Delaunay triangulation is obtained. The Voronoi tessellation is the dual graph of the Delaunay triangulation. Despite its nice properties, a Delaunay triangulation is not useful unless it also conforms to user-specified boundaries, e.g., the surface of an airfoil, or the surface of a shock wave. A modified version of the Delaunay triangulation that includes user-defined edges in the triangulation is known as a constrained Delaunay triangulation. In 1987, Chew [33] developed an algorithm with the following properties: (1) the user-specified edges are included in the triangulation, and (2) it is as close as possible to the Delaunay triangulation. Shewchuk [203] has developed a mesh generator called *Triangle*, which implements Chew's constrained Delaunay triangulation using *Delaunay refinement*. The Delaunay refinement algorithm operates by performing a Delaunay or constrained Delaunay triangulation on a subset of vertices. Once this triangulation is done, the vertices are refined by inserting carefully placed vertices until the mesh meets constraints on triangle quality and size. The code is publicly available on the Web at http://www.cs.cmu.edu/quake/triangle.html.

### 6.2.2 Some Mathematical Relations

The median of a triangle is the line connecting one of its vertices to the midpoint of the opposite side. The three medians of any triangle intercept at the centroid of the triangle. If $(x_i, y_i), i = 1, 2, 3$, are the coordinates of the three vertices of a triangle, then the centroid is given by $(\bar{x}, \bar{y})$ where $\bar{x} = \frac{1}{3}\sum_{i=1}^{3} x_i$ and $\bar{y} = \frac{1}{3}\sum_{i=1}^{3} y_i$. A median bisects the area of a triangle, and the three medians divide the triangle into six triangles of equal area. The circle passing through the points $(x_i, y_i)$ has its center at

$$\begin{aligned} x_c &= -\delta/2\alpha, \\ y_c &= -\varepsilon/2\alpha, \end{aligned} \tag{6.3}$$

and radius

$$r_c = \sqrt{(\delta^2 + \varepsilon^2)/4\alpha^2 - \beta/\alpha}, \tag{6.4}$$

where

$$\alpha = \begin{vmatrix} x_1 & y_1 & 1 \\ x_2 & y_2 & 1 \\ x_3 & y_3 & 1 \end{vmatrix}, \quad \delta = -\begin{vmatrix} x_1^2 + y_1^2 & y_1 & 1 \\ x_2^2 + y_2^2 & y_2 & 1 \\ x_3^2 + y_3^2 & y_3 & 1 \end{vmatrix},$$

$$\varepsilon = \begin{vmatrix} x_1^2 + y_1^2 & x_1 & 1 \\ x_2^2 + y_2^2 & x_2 & 1 \\ x_3^2 + y_3^2 & x_3 & 1 \end{vmatrix}, \quad \beta = -\begin{vmatrix} x_1^2 + y_1^2 & x_1 & y_1 \\ x_2^2 + y_2^2 & x_2 & y_2 \\ x_3^2 + y_3^2 & x_3 & y_3 \end{vmatrix}.$$

A linear interpolation polynomial for some function $f(x, y)$ over the surface of the triangle is given by the expression

$$f(x, y) = c_1 + c_2 x + c_3 y. \tag{6.5}$$

The function (6.5) is $C^0$-continuous across adjacent triangles. If we let $f_i = f(x_i, y_i)$, then the coefficients $c_j$ are given by

$$C = \boldsymbol{B}^{-1} F,$$

where $C = [c_1, c_2, c_3]^{\mathrm{T}}$, $F = [f_1, f_2, f_3]^{\mathrm{T}}$, and

$$\boldsymbol{B} = \begin{bmatrix} 1 & x_1 & y_1 \\ 1 & x_2 & y_2 \\ 1 & x_3 & y_3 \end{bmatrix}.$$

The inverse of $\boldsymbol{B}$ exists if the three vertices of the triangle do not lie on a straight line. In that case, the solution is given by

$$\boldsymbol{B}^{-1} = \frac{1}{2A} \begin{bmatrix} x_2 y_3 - x_3 y_2 & x_3 y_1 - x_1 y_3 & x_1 y_2 - x_2 y_1 \\ y_2 - y_3 & y_3 - y_1 & y_1 - y_2 \\ x_3 - x_2 & x_1 - x_3 & x_2 - x_1 \end{bmatrix},$$

where

$$A = \tfrac{1}{2} \left| (x_3 - x_1)(y_2 - y_1) - (x_2 - x_1)(y_3 - y_1) \right| \tag{6.6}$$

is the area of the triangle.

Gauss' theorem can be used to define an average gradient of a scalar function $f(x, y)$ on some control volume $\Omega$ with surface boundary $\Gamma$:

$$\int_{\Omega} \nabla f \, \mathrm{d}\Omega = \oint_{\Gamma} f \, \mathrm{d}\vec{\Gamma}.$$

The average gradients in the $x$ and $y$ directions, respectively, are given by

$$\overline{\frac{\partial f}{\partial x}} = \frac{1}{\Omega} \int_{\Omega} \frac{\partial f}{\partial x} \mathrm{d}\Omega = \frac{1}{\Omega} \oint_{\Gamma} f \vec{i}_x \cdot \mathrm{d}\vec{\Gamma} \tag{6.7}$$

and

$$\overline{\frac{\partial f}{\partial y}} = \frac{1}{\Omega} \int_{\Omega} \frac{\partial f}{\partial y} \mathrm{d}\Omega = \frac{1}{\Omega} \oint_{\Gamma} f \vec{i}_y \cdot \mathrm{d}\vec{\Gamma}, \tag{6.8}$$

where $\vec{i}_x$ and $\vec{i}_y$ are the unit normals in the $x$ and $y$ directions, respectively. Let us apply (6.7) to the triangle defined by vertices $(x_i, y_i)$, $(i = 1,2,3,4)$, where, for notational convenience, we let $x_4 = x_1$ and $y_4 = y_1$. The length of the edge between vertices $i+1$ and $i$ is given by

$$\mathrm{d}l_{i+1,i} = \sqrt{(x_{i+1} - x_i)^2 + (y_{i+1} - y_i)^2}, \quad i = 1,2,3, \tag{6.9}$$

and the outward pointing unit normal to the edge is given by

$$\vec{\eta}_{i+1,i} = \frac{(y_{i+1} - y_i)\vec{i}_x - (x_{i+1} - x_i)\vec{i}_y}{\mathrm{d}l_{i+1,i}}, \quad i = 1,2,3. \tag{6.10}$$

Note that $\sum_{i=1}^{3} \vec{\eta}_{i+1,i} = 0$; this is true for any closed figure in the plane. The product $\vec{i}_x \cdot \vec{\eta}_{i+1,i} \mathrm{d}l_{i+1,i} = y_{i+1} - y_i$, and hence

$$\begin{aligned}\overline{\frac{\partial f}{\partial x}} &\approx \frac{1}{A} \sum_{\text{edges}} f \vec{i}_x \cdot \mathrm{d}\vec{l} = \frac{1}{A} \sum_{i=1}^{3} \frac{f_{i+1} + f_i}{2} (y_{i+1} - y_i), \\ &\approx \frac{1}{2A} [f_1(y_2 - y_3) + f_2(y_3 - y_1) + f_3(y_1 - y_2)],\end{aligned}$$

which is the same as the coefficient $c_2$ in the linear interpolation (6.5).

The relations (6.7) and (6.8) can be used recursively to define averaged higher derivatives. For example, let $f = g_x$, then from (6.7)

$$\overline{\frac{\partial^2 g}{\partial x^2}} = \frac{1}{\Omega} \oint_{\Gamma} g_x \vec{i}_x \cdot \mathrm{d}\vec{\Gamma}. \tag{6.11}$$

Evaluation of (6.11) requires knowing $g_x$ along the boundary $\Gamma$. If $g_x$ along $\Gamma$ was evaluated from the values of $g$ at cell vertices, a rather large stencil will be used to evaluate (6.11).

---

## 6.3 Unstructured Grid Solver

The floating shock-fitting algorithm developed by Paciorri and Bonfiglioli [163] is only weakly coupled to the unstructured grid solver. In theory, the unstructured grid solver they use could be replaced by another unstructured grid solver, but not any other unstructured grid solver. There are some qualities of the solver they use that are essential to the shock-fitting algorithm as currently implemented. Among them are (1) the solver is a cell-vertex algorithm; (2) the solver uses only triangular cells; (3) the algorithm is capable of adding and subtracting cells *on the fly*;

(4) a constrained Delaunay triangulation is used to conform the mesh to the shock surface; (5) to obtain better results at the shock, the algorithm is upwind based.

The unstructured grid solver used by Paciorri and Bonfiglioli was developed by Bonfiglioli [18]. We will review the algorithm in this section. However, because we are perhaps more interested in unsteady problems than steady problems, inviscid than viscous, some aspects of this algorithm, dealing with preconditioning the equations to accelerate convergence to the steady state and the treatment of viscous terms, will not be discussed.

### 6.3.1 Fluctuation Splitting for the Advection Equation

The extension of upwind schemes and Riemann solvers to multidimensional problems in the late 1970s and early 1980s was done by applying one-dimensional upwinding (and one-dimensional Riemann solvers) along mesh directions, irrespective of how the mesh was oriented with respect to the signal propagation directions. In the mid-1980s, around the same time that unstructured grid techniques started to gain acceptance in computational fluid dynamics, there was a realization that a multidimensional reinterpretation of the upwinding concept was needed. Among the key players in this effort were Philip Roe, David Sidilkover, Bram van Leer, and Herman Deconinck.* Bonfiglioli's flow solver is based on Roe's multidimensional upwinding method, also known as *fluctuation splitting* [182].

Consider the tessellation depicted in Figure 6.4. The figure shows seven triangles ($T_k$, $k = 1, \ldots, 7$) with a common vertex labeled 1. The dashed lines form the dual control area mesh created by connecting the triangle centroid to the midpoint of its edges. The control area for vertex 1 is the shaded region in the figure. Let us denote this area by $\Omega_1$ and its boundary by $\Gamma_1$. The area of triangle $T_k$, $A_k$, is given by (6.6). The area formed by $\Omega_1 \cap A_k = \frac{1}{3}A_k$, and, hence, the control area $\Omega_1 = \frac{1}{3}\sum_{k=1}^{7} A_k$. Let us consider how to solve the two-dimensional advection equation

$$\frac{\partial u}{\partial t} + \nabla \cdot \vec{F}(u) = 0,$$

or equivalently

$$\frac{\partial u}{\partial t} + \vec{\Lambda} \cdot \nabla u = 0, \tag{6.12}$$

where $\vec{\Lambda} = \lambda_x \vec{i}_x + \vec{\lambda}_y \vec{i}_y$, ($\lambda_x = \partial(\vec{i}_x \cdot \vec{F})/\partial u$, $\lambda_y = \partial(\vec{i}_y \cdot \vec{F})/\partial u$), is the local flow direction. The vector field $\vec{\Lambda}$ is shown in Figure 6.4 as the thin lines with arrowheads. Let the coordinates of the vertices of triangle $T_k$ be defined by $(x_i^k, y_i^k)$, $(i = 1, 2, 3, 4)$, where, for notational convenience we let $x_4^k = x_1^k$ and $y_4^1 = y_1^1$, and the vertices are

---

* Of course there were other important contributors, but these, in my opinion, led the effort. It is not a coincidence that most of them were associated with ICASE. The institute recognized the importance of the problem and nourished the effort [104].

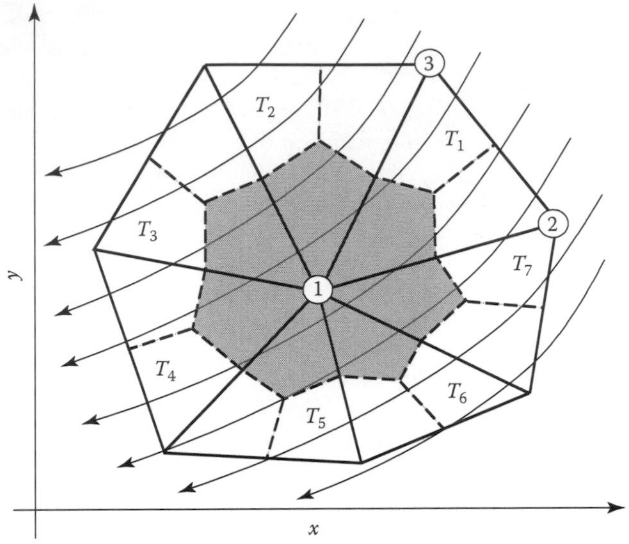


**FIGURE 6.4:** Tessellation of the plane into triangles (solid lines). The control area is the dual mesh (dashed lines) connecting a centroid to the midpoint the triangle edges. The shaded area is the control area for vertex 1.

labeled counterclockwise. Let $u(x, y)$ be defined at the vertices and be a linear function, (6.5), within the triangle.

If we integrate (6.12) over the area of a triangle, and use Gauss' theorem, we obtain

$$\begin{aligned} \int_A \frac{\partial u}{\partial t} \mathrm{d}A &= -\int_A \vec{\Lambda} \cdot \nabla u \mathrm{d}A, \\ &= \oint_l u\vec{\Lambda} \cdot \vec{n} \mathrm{d}l, \end{aligned} \tag{6.13}$$

where

$l$ is the boundary of the triangle,

$\vec{n}$ is the inward pointing unit normal to the triangle boundary ($\vec{n} = -\vec{\eta}$, defined by (6.10)).

Note that the contour integral in (6.13) is not preceded by a minus sign because we are using the inward normal. The left hand side of (6.13) represents the change in $u$ over the triangle. We refer to it as the *fluctuation* and denote it by $\phi_A$,

$$\phi_A = \oint_l u\vec{\Lambda} \cdot \vec{n} \mathrm{d}l.$$

Carrying out the integration, we find

$$\phi_A \approx \tfrac{1}{2}\bar{\vec{\Lambda}} \cdot \sum_{i=1}^{3} (u_i + u_{i+1}) dl_{i+1,i}\, \vec{n}_{i+1,i}, \tag{6.14}$$

where
$\bar{\vec{\Lambda}}$ is the average $\vec{\Lambda}$ over the triangle such that discrete conservation is recovered*,
$dl_{i+1,i}$ is defined by (6.9).

Since $\sum_{i=1}^{3} \vec{n}_{i+1,i} = 0$, we can write (6.14) as

$$\phi_A = -\sum_{i=1}^{3} \kappa_i u_i, \tag{6.15}$$

where

$$\kappa_i = \tfrac{1}{2}\bar{\vec{\Lambda}} \cdot (dl\vec{n})_{m+1,m}, \; i = 1,2,3 \quad \text{and} \quad 2m = -4 + 11i - 3i^2. \tag{6.16}$$

The term $dl\vec{n}$ is the *scaled* normal. Note that $\sum \kappa_i = 0$, therefore, not all the $\kappa_i$'s are positive. Negative $\kappa_i$'s are a consequence of the vector $\bar{\vec{\Lambda}}$ being directed outward with respect to one or two of the triangle's edge.

To obtain the updated value of $u$, the fluctuation $\phi_A$ is distributed among vertices of the triangle by the following expression

$$u_i^{t+\Delta t} = u_i^t + \frac{\Delta t}{\Omega_i} \sum_{\forall T_k} \alpha_i^k \phi_A^k, \tag{6.17}$$

where the summation is taken over all the triangles in $\Omega_i$ with common vertex $i$. The $\alpha_i$'s in (6.17) are yet to be determined, they are the degrees of freedom in the *scheme design space*, but to enforce conservation we have the constrain that for each triangle $\sum \alpha_i = 1$. The need for this constrain is evident by considering the weighted sum of all updates over all $N$ vertices:

$$\begin{aligned}\sum_{i=1}^{N} \Omega_i u_i^{t+\Delta t} &= \sum_{i=1}^{N} \Omega_i u_i^t + \Delta t \sum_{i=1}^{N} \sum_{\forall T_k} \alpha_i^k \phi_A^k, \\ &= \sum_{i=1}^{N} \Omega_i u_i^t + \Delta t \sum_{\forall T_k} \sum_{i=1}^{N} \alpha_i^k \phi_A^k.\end{aligned}$$

Now if $\sum_{i=1}^{3} \alpha_i = 1$ over all triangles, then fluxes across internal boundaries cancel, and all that is left is the integral around the outermost boundary $L$: $\sum_{\forall T_k} \phi_A^k = \oint_L u\vec{\Lambda} \cdot \vec{n} dL$.

---

* $\bar{\lambda}_x = \frac{1}{A} \int_A \partial(\vec{i}_x \cdot \vec{F})/\partial u dA$ and similarly for $\bar{\lambda}_y$.

### 6.3.2 Central Differences

If we distribute the fluctuation equally among the vertices, $\alpha_i = \frac{1}{3}$, the result is equivalent to a central difference scheme (the Galerkin scheme in finite elements). This scheme ignores the domain of dependence of the advection equation and is unstable. The scheme can be improved by adding some dissipation (which is the backdoor way of upwinding), as in a Lax–Wendroff (LW) scheme:

$$\int_{\Omega_i} \left(u_i^{t+\Delta t} - u_i^t\right) \mathrm{d}\Omega = \int_{\Omega_i} \left(\frac{\partial u}{\partial t}\Delta t + \frac{\partial^2 u}{\partial t^2}\frac{\Delta t^2}{2}\right) \mathrm{d}\Omega.$$

The first term of the right-hand side integral gives

$$\int_{\Omega_i} \frac{\partial u}{\partial t}\Delta t \mathrm{d}\Omega = \frac{1}{3}\Delta t \phi_{\mathrm{A}}.$$

For the second term, since $\dfrac{\partial u}{\partial t} = \phi_{\mathrm{A}}/A$, we find the contribution from a triangle to vertex $i$:

$$\int_{\Omega_i} \frac{\partial^2 u}{\partial t^2}\frac{\Delta t^2}{2} \mathrm{d}\Omega = \frac{\Delta t^2}{2}\oint_l \frac{\partial u}{\partial t}\vec{\lambda}\cdot\vec{n}\mathrm{d}l = \frac{\kappa_i \Delta t}{2A}\Delta t \phi_{\mathrm{A}}.$$

Summing the two parts, we find

$$\alpha_i = \frac{1}{3} + \frac{\kappa_i \Delta t}{2A}.$$

### 6.3.3 Linearity Preservation Property

A desirable property of a numerical scheme is to be able to preserve a steady state solution if one exits. That is, we want $\alpha_i \phi_{\mathrm{A}} \to 0$, when $\vec{\Lambda}\cdot\nabla u \to 0$. Since our basis function $u_i$ is linear in $(x, y)$, the linearity preservation property requires $\alpha_i \phi_{\mathrm{A}} \to 0$ when the exact steady state solution is linear in $(x, y)$. Since $\vec{\Lambda}\cdot\nabla u \to 0$ implies $\phi_{\mathrm{A}} \to 0$, linearity preservation requires that the $\alpha_i$'s be bounded. For a linear scheme, this is equivalent to requiring that the $\alpha_i$'s be independent of $u$. On a triangulated Cartesian grid with diagonals constructed uniformly (a *regular* triangular grid), a scheme that is linearity preserving (LP) is at least second-order accurate.

Consider the flow direction illustrated in Figure 6.4. For triangle $T_1$, the flow enters through the edge opposite vertex 1. For triangle $T_4$, the flow enters through the two edges meeting at vertex 1. Let us call the first case, a triangle with one inflow edge, a "Type I" triangle, and the later, a triangle with two inflow edges, a "Type II" triangle. In the design of upwind schemes, a "Type I" triangle corresponds to only one $\kappa_i > 0$. For example, in the case of $T_1$, we have $\kappa_1 > 0$, $\kappa_2 < 0$, and $\kappa_3 < 0$.

For "Type I" triangles, the distribution coefficients are $\alpha_i = 1$, $\alpha_j = 0$, and $\alpha_k = 0$, i.e., the full fluctuation is sent to vertex $i$. It follows that for "Type I" triangles the $\alpha_i$'s are bounded and LP is satisfied.

### 6.3.4 Positivity Property

In Section 2.3.10, we discussed the requirement for a scheme to be monotonicity preserving (no new extrema are created) and Godunov's theorem stating that if an advection scheme, based on linear discretization, preserves monotonicity it cannot be better than first-order accurate. To extend this condition to triangular meshes, let us write the solution at vertex $i$ as a function of the values at surrounding nodes

$$u_i^{t+\Delta t} = \sum_{j=1}^{N} c_{i,j} u_j^t. \tag{6.18}$$

A linear scheme of the form (6.18) is said to be *positive* (P) if all the coefficients, $c_{i,j}$, $j = 1, \ldots, N$, are nonnegative. Godunov's theorem when extended to multidimensional advection schemes on triangular meshes states that a *linear scheme cannot be linearity preserving and positive*. The upwind update discussed in Section 6.3.3 for a "Type I" triangle is both P (with $\Delta t$ small, $\kappa_1 \Delta t \leq \Omega_1$) and LP. However, no linear scheme can be both P and LP for "Type II" triangles, which must be present in order to satisfy zero divergence [55]. Thus, to satisfy both LP and P, we must look for schemes where the $\alpha_i$'s depend on the $u_i$'s.

### 6.3.5 Low Diffusion Schemes

Two linear upwind schemes that satisfy the LP property, but not P, are the low diffusion schemes A and B. Consider a "Type II" triangle, as in Figure 6.5. The vector $\vec{\Lambda}$ is drawn from vertex 1 and it cuts edge 3–2 at 4. In the low diffusion

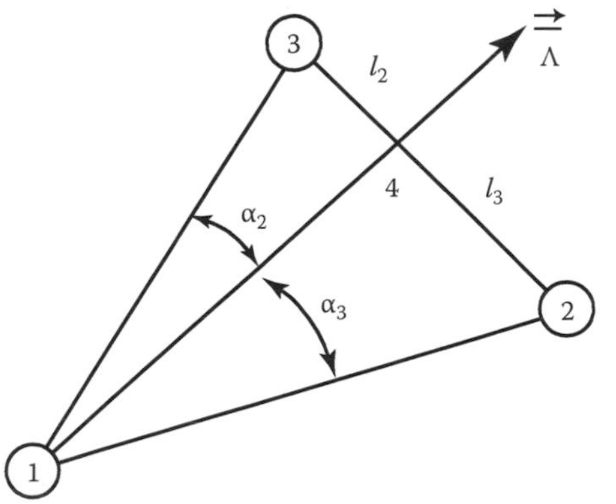


**FIGURE 6.5:** Low Diffusion schemes A and B, "Type II" triangle.

scheme A (LDA), the fluctuation to vertices 2 and 3 is weighted by a fraction of the area of the triangle:

$$\alpha_2 = \frac{A_{143}}{A} = \frac{l_2}{l_2 + l_3} = -\frac{\kappa_2}{\kappa_1},$$

$$\alpha_3 = \frac{A_{142}}{A} = \frac{l_3}{l_2 + l_3} = -\frac{\kappa_3}{\kappa_1},$$

where $A$ is the total area of the triangle. Making use of $\sum \kappa_i = 0$, the update is given by

$$u_1^{t+\Delta t} = u_1^t + \oplus,$$

$$u_2^{t+\Delta t} = u_2^t - \frac{\Delta t}{\Omega_2}[\kappa_2(u_2 - u_1) + \kappa_3(u_3 - u_1)]\frac{\kappa_2}{\kappa_2 + \kappa_3} + \oplus,$$

$$u_3^{t+\Delta t} = u_3^t - \frac{\Delta t}{\Omega_3}[\kappa_2(u_2 - u_1) + \kappa_3(u_3 - u_1)]\frac{\kappa_3}{\kappa_2 + \kappa_3} + \oplus,$$

where the symbol $\oplus$ is there to remind us that there are other contributions coming from neighboring triangles. Note that $\alpha_1$ is defined by $\sum \alpha_i = 1$.

The low diffusion scheme B is similarly defined by

$$\alpha_2 = \frac{\sin\alpha_2 \cos\alpha_3}{\sin(\alpha_2 + \alpha_3)},$$

$$\alpha_3 = \frac{\cos\alpha_2 \sin\alpha_3}{\sin(\alpha_2 + \alpha_3)}.$$

### 6.3.6 The N Scheme

The N scheme is an upwind linear scheme that is P preserving. It was labeled the N scheme by Sidilkover [205], because of its narrow stencil. The geometric interpretation of the scheme is as follows. We begin by splitting the advection vector into components parallel to the edges originating from the upstream vertex, see Figure 6.5,

$$\vec{\vec{\Lambda}} = \vec{\lambda}_{1,3} + \vec{\lambda}_{2,1},$$

hence, $\vec{\lambda}_{1,3} \cdot \vec{n}_{1,3} = \vec{\lambda}_{2,1} \cdot \vec{n}_{2,1} = 0$. Since the fluctuation $\phi_A$ is a linear function of $\vec{\vec{\Lambda}}$ we can decompose $\phi_A$ such that,

$$\phi_A\left(\vec{\vec{\Lambda}}\right) = \phi_2\left(\vec{\lambda}_{2,1}\right) + \phi_3\left(\vec{\lambda}_{1,3}\right).$$

Therefore,

$$\phi_2 = \oint_l u\vec{\lambda}_{2,1}\vec{n}\mathrm{d}l = -\kappa_2(u_2 - u_1),$$

and similarly,

$$\phi_3 = -\kappa_3(u_3 - u_1).$$

The update is given by

$$\begin{aligned} u_1^{t+\Delta t} &= u_1^t + \oplus, \\ u_2^{t+\Delta t} &= u_2^t - \frac{\Delta t}{\Omega_2}(u_2 - u_1)\kappa_2 + \oplus, \\ u_3^{t+\Delta t} &= u_3^t - \frac{\Delta t}{\Omega_3}(u_3 - u_1)\kappa_3 + \oplus. \end{aligned} \tag{6.19}$$

The scheme does not satisfy the LP property, since if $\phi_A \to 0$ we have $\phi_2 = -\phi_3$, but not necessarily $\phi_2 = \phi_3 = 0$. As written in (6.19) the scheme is linear, since the update depends only on the geometry of the triangle and the advection vector. However, if we look at the $\alpha_i$'s,

$$\begin{aligned} \alpha_1 &= 0, \\ \alpha_2 &= \frac{(u_2 - u_1)\kappa_2}{(u_2 - u_1)\kappa_2 + (u_3 - u_1)\kappa_3}, \\ \alpha_3 &= \frac{(u_3 - u_1)\kappa_3}{(u_2 - u_1)\kappa_2 + (u_3 - u_1)\kappa_3}. \end{aligned}$$

We find that the $\alpha_i$'s depend on $u$. The scheme satisfies P if $\Delta t$ satisfies the constrain

$$\Delta t \leq \min\left(\frac{\Omega_2}{\kappa_2}, \frac{\Omega_3}{\kappa_3}\right).$$

Among the linear P schemes of the form (6.17), the N scheme has the largest time step and the smallest cross diffusion.

### 6.3.7 The Positive Streamwise Invariance Scheme

The N scheme has some nice properties, but is not LP preserving. As already discussed, to satisfy both LP and P we have to look for a nonlinear scheme. A standard way of creating a nonlinear scheme is by the introduction of a *limiter*, see Section 2.3.11. Let us consider a limited fluctuation of the form

$$\begin{aligned} \hat{\phi}_2 &= \phi_2 - \psi(\phi_2, -\phi_3), \\ \hat{\phi}_3 &= \phi_3 + \psi(\phi_2, -\phi_3). \end{aligned}$$

The total fluctuation, $\phi_A = \hat{\phi}_2 + \hat{\phi}_3$, remains unchanged, and, therefore, conservation is not affected. The properties we want from the limiter are (1) the limiter is symmetric with respect to the fluctuations, $\psi(\phi_2, -\phi_3) = \psi(-\phi_3, \phi_2)$; and (2) the limiter recovers the LP property. A limiter meeting these requirements and retaining the same constrain on $\Delta t$ is the *MinMod* limiter defined by

$$\psi_{\mathrm{MM}}(a,b) = \mathrm{sgn}(a)\max[0,\ \min(|a|,\mathrm{sgn}(a)b)].$$

The N scheme with the *MinMod* limiter is a nonlinear upwind scheme that satisfies the P and LP properties. It is identical to the positive stream invariant scheme introduced in [215].

---

## 6.4 Application to Euler Equations

### 6.4.1 Change of Variables

Bonfigioli's code works with the quasilinear form of the Euler equations and the so-called symmetrizing variables:

$$\partial \boldsymbol{Q} = (\partial p/\rho a, \partial \breve{u}, \partial \breve{v}, \partial p - a^2 \partial \rho)^{\mathrm{T}},$$

where $\partial p - a^2\partial\rho = p\partial S$ and $\breve{u}$ and $\breve{v}$ are the components of the velocity vector in a local streamwise coordinate system $(\xi, \eta)$. In Section 2.4, we discussed how to introduce mappings and perform variable transformations. We apply the same principles here.

The divergence form (Section 3.3.1) is given by

$$\frac{\partial \boldsymbol{U}}{\partial t} + \nabla \cdot \mathbf{F}(\boldsymbol{U}) = 0, \tag{6.20}$$

where the vector of unknowns, $\boldsymbol{U} = [\rho \quad \rho u \quad \rho v \quad \rho E]^{\mathrm{T}}$, is written in terms of the Cartesian velocity components $u$ and $v$ in the $x$ and $y$ directions, respectively. The matrix $\mathbf{F}$ may be decomposed along any direction. Let $\vec{m} = m_x\vec{i}_x + m_y\vec{i}_y$ be an arbitrary unit vector, and let $\vec{u} = u\vec{i}_x + v\vec{i}_y$ be the velocity vector, then:

$$\mathbf{F} = \begin{bmatrix} \rho u_{\vec{m}} \\ \rho u u_{\vec{m}} + p m_x \\ \rho v u_{\vec{m}} + p m_y \\ \rho H u_{\vec{m}} \end{bmatrix}, \tag{6.21}$$

where $u_{\vec{m}} = \vec{u}\cdot\vec{m}$. The Jacobian matrix $\mathbf{A}_{\vec{m}} = \partial \mathbf{F}/\partial \boldsymbol{U}$ is given by

$$\mathbf{A}_{\vec{m}} = \begin{bmatrix} 0 & m_x & m_y & 0 \\ \frac{1}{2}(\gamma-1)q^2 m_x - u u_{\vec{m}} & u_{\vec{m}} - (\gamma-2)u m_x & u m_y - (\gamma-1) v m_x & (\gamma-1)m_x \\ \frac{1}{2}(\gamma-1)q^2 m_y - v u_{\vec{m}} & v m_x - (\gamma-1) u m_y & u_{\vec{m}} - (\gamma-2) v m_y & (\gamma-1)m_y \\ \left(\frac{1}{2}(\gamma-1)q^2 - H\right)u_{\vec{m}} & H m_x - (\gamma-1) u u_{\vec{m}} & H m_y - (\gamma-1) v u_{\vec{m}} & \gamma u_{\vec{m}} \end{bmatrix}, \tag{6.22}$$

where $q^2 = u^2 + v^2$. In particular, if we let $\vec{m} = \vec{i}_x$ and $\boldsymbol{F} = \mathbf{F} \cdot \vec{i}_x$, and if we let $\vec{m} = \vec{i}_y$ and $\boldsymbol{G} = \mathbf{F} \cdot \vec{i}_y$, then

$$\frac{\partial \boldsymbol{U}}{\partial t} + \frac{\partial \boldsymbol{F}}{\partial x} + \frac{\partial \boldsymbol{G}}{\partial y} = 0, \tag{6.23}$$

where $\boldsymbol{F}$ and $\boldsymbol{G}$ are defined by

$$\boldsymbol{F} = \begin{bmatrix} \rho u \\ \rho u^2 + p \\ \rho u v \\ \rho u H \end{bmatrix}, \quad \boldsymbol{G} = \begin{bmatrix} \rho v \\ \rho u v \\ \rho v^2 + p \\ \rho v H \end{bmatrix}. \tag{6.24}$$

The eigenvalues $\Lambda_{\vec{m}} = \begin{bmatrix} \lambda^1_{\vec{m}} & \lambda^2_{\vec{m}} & \lambda^3_{\vec{m}} & \lambda^4_{\vec{m}} \end{bmatrix}$ (see Section 2.1.5) of $\mathbf{A}_{\vec{m}}$ are the roots of the equation

$$\det\left(\mathbf{A}_{\vec{m}} - \Lambda_{\vec{m}} \mathbf{I}\right) = 0. \tag{6.25}$$

They are

$$\lambda^i_{\vec{m}} = [\, u_{\vec{m}} - a \quad u_{\vec{m}} \quad u_{\vec{m}} + a \quad u_{\vec{m}} \,]. \tag{6.26}$$

The right eigenvectors, $\boldsymbol{R}_{\vec{m}} = \begin{bmatrix} \boldsymbol{r}^1_{\vec{m}} & \boldsymbol{r}^2_{\vec{m}} & \boldsymbol{r}^3_{\vec{m}} & \boldsymbol{r}^4_{\vec{m}} \end{bmatrix}$, are obtained by solving the equation

$$\mathbf{A}_{\vec{m}} \boldsymbol{r}^i_{\vec{m}} = \lambda^i_{\vec{m}} \boldsymbol{r}^i_{\vec{m}}. \tag{6.27}$$

They are the column vectors

$$\begin{bmatrix} \boldsymbol{r}^1_{\vec{m}} & \boldsymbol{r}^2_{\vec{m}} & \boldsymbol{r}^3_{\vec{m}} & \boldsymbol{r}^4_{\vec{m}} \end{bmatrix} = \begin{bmatrix} 1 & 1 & 1 & 0 \\ u - a m_x & u & u + a m_x & m_y \\ v - a m_y & v & v + a m_y & -m_x \\ H - a u_{\vec{m}} & \frac{1}{2} q^2 & H + a u_{\vec{m}} & u m_y - v m_x \end{bmatrix}. \tag{6.28}$$

The left eigenvectors satisfy the equation $\boldsymbol{l}^i_{\vec{m}} \mathbf{A}_{\vec{m}} = \lambda^i_{\vec{m}} \boldsymbol{l}^i_{\vec{m}}$. They can be found by taking the inverse of the right eigenvectors matrix. They are the row vectors given by

$$\begin{aligned} &\begin{bmatrix} \boldsymbol{l}^1_{\vec{m}} & \boldsymbol{l}^2_{\vec{m}} & \boldsymbol{l}^3_{\vec{m}} & \boldsymbol{l}^4_{\vec{m}} \end{bmatrix}^{\mathrm{T}} \\ &= \frac{1}{2a^2} \begin{bmatrix} \frac{1}{2}(\gamma - 1) q^2 + a u_{\vec{m}} & (1-\gamma) u - a m_x & (1-\gamma) v - a m_y & \gamma - 1 \\ 2a^2 - (\gamma - 1) q^2 & 2(\gamma - 1) u & 2(\gamma - 1) v & 2(1-\gamma) \\ \frac{1}{2}(\gamma - 1) q^2 - a u_{\vec{m}} & (1-\gamma) u + a m_x & (1-\gamma) v + a m_y & \gamma - 1 \\ 2a^2 (v m_x - u m_y) & 2a^2 m_y & -2a^2 m_x & 0 \end{bmatrix}. \end{aligned} \tag{6.29}$$

The quasilinear form in terms of conservative variables is

$$\frac{\partial \boldsymbol{U}}{\partial t} + \mathbf{A}\frac{\partial \boldsymbol{U}}{\partial x} + \mathbf{B}\frac{\partial \boldsymbol{U}}{\partial y} = 0. \tag{6.30}$$

Equation 6.30 follows from Equation 6.23 by applying the chain rule with the Jacobian matrices $\mathbf{A} = \partial \boldsymbol{F}/\partial \boldsymbol{U}$ and $\mathbf{B} = \partial \boldsymbol{G}/\partial \boldsymbol{U}$. The matrices $\mathbf{A}$ and $\mathbf{B}$ may be obtained from $\mathbf{A}_{\vec{m}}$ by letting $\vec{m} = \vec{i}_x$ and $\vec{m} = \vec{i}_y$, respectively:

$$\mathbf{A} = \begin{bmatrix} 0 & 1 & 0 & 0 \\ \frac{1}{2}(\gamma-1)q^2 - u^2 & (3-\gamma)u & (1-\gamma)v & \gamma-1 \\ -uv & v & u & 0 \\ u\left(\frac{1}{2}(\gamma-1)q^2 - H\right) & (1-\gamma)u^2 + H & (1-\gamma)uv & \gamma u \end{bmatrix}, \tag{6.31}$$

$$\mathbf{B} = \begin{bmatrix} 0 & 0 & 1 & 0 \\ -uv & v & u & 0 \\ \frac{1}{2}(\gamma-1)q^2 - v^2 & (1-\gamma)u & (3-\gamma)v & \gamma-1 \\ v\left(\frac{1}{2}(\gamma-1)q^2 - H\right) & (1-\gamma)uv & (1-\gamma)v^2 + H & \gamma u \end{bmatrix}. \tag{6.32}$$

The quasilinear form may also be written in terms of primitive variables:

$$\frac{\partial \boldsymbol{V}}{\partial t} + \tilde{\mathbf{A}}\frac{\partial \boldsymbol{V}}{\partial x} + \tilde{\mathbf{B}}\frac{\partial \boldsymbol{V}}{\partial y} = 0, \tag{6.33}$$

$$\boldsymbol{V} = \begin{bmatrix} \rho \\ u \\ v \\ p \end{bmatrix}, \ \tilde{\mathbf{A}} = \begin{bmatrix} u & \rho & 0 & 0 \\ 0 & u & 0 & 1/\rho \\ 0 & 0 & u & 0 \\ 0 & \rho a^2 & 0 & u \end{bmatrix}, \ \tilde{\mathbf{B}} = \begin{bmatrix} v & 0 & \rho & 0 \\ 0 & v & 0 & 0 \\ 0 & 0 & v & 1/\rho \\ 0 & 0 & \rho a^2 & v \end{bmatrix}, \tag{6.34}$$

where $\tilde{\mathbf{A}} = \boldsymbol{T}^{-1}\mathbf{A}\boldsymbol{T}$ and $\tilde{\mathbf{B}} = \boldsymbol{T}^{-1}\mathbf{B}\boldsymbol{T}$ follow from the transformation matrix $\boldsymbol{T} = \partial \boldsymbol{U}/\partial \boldsymbol{V}$ and its inverse:

$$\boldsymbol{T} = \begin{bmatrix} 1 & 0 & 0 & 0 \\ u & \rho & 0 & 0 \\ v & 0 & \rho & 0 \\ \frac{1}{2}q^2 & \rho u & \rho v & 1/(\gamma-1) \end{bmatrix}, \tag{6.35}$$

$$\boldsymbol{T}^{-1} = \begin{bmatrix} 1 & 0 & 0 & 0 \\ -u/\rho & 1/\rho & 0 & 0 \\ -v/\rho & 0 & 1/\rho & 0 \\ \frac{1}{2}(\gamma-1)q^2 & (1-\gamma)u & (1-\gamma)v & \gamma-1 \end{bmatrix}. \tag{6.36}$$

The right eigenvectors are given by $\tilde{\boldsymbol{R}} = \boldsymbol{T}^{-1}\boldsymbol{R}$, and the left eigenvectors are given by $\tilde{\boldsymbol{R}}^{-1} = \boldsymbol{R}^{-1}\boldsymbol{T}$.

In terms of symmetrizing variables, the Euler equations are

$$\frac{\partial \boldsymbol{Q}}{\partial t} + \breve{\mathbf{A}}\frac{\partial \boldsymbol{Q}}{\partial \xi} + \breve{\mathbf{B}}\frac{\partial \boldsymbol{Q}}{\partial \eta} = 0, \tag{6.37}$$

where $\breve{\mathbf{A}} = \boldsymbol{M}^{-1}\tilde{\mathbf{A}}\boldsymbol{M}$ and $\breve{\mathbf{B}} = \boldsymbol{M}^{-1}\tilde{\mathbf{B}}\boldsymbol{M}$, the matrix $\boldsymbol{M} = \partial \boldsymbol{V}/\partial \boldsymbol{Q}$ and its inverse are

$$\boldsymbol{M} = \begin{bmatrix} \rho/a & 0 & 0 & -1/a^2 \\ 0 & u/q & -v/q & 0 \\ 0 & v/q & u/q & 0 \\ \rho a & 0 & 0 & 1 \end{bmatrix}, \tag{6.38}$$

$$\boldsymbol{M}^{-1} = \begin{bmatrix} 0 & 0 & 0 & 1/\rho a \\ 0 & u/q & v/q & 0 \\ 0 & -v/q & u/q & 0 \\ -a^2 & 0 & 0 & 1 \end{bmatrix}, \tag{6.39}$$

and

$$\breve{\mathbf{A}} = \begin{bmatrix} \breve{u} & a & 0 & 0 \\ a & \breve{u} & 0 & 0 \\ 0 & 0 & \breve{u} & 0 \\ 0 & 0 & 0 & \breve{u} \end{bmatrix}, \quad \breve{\mathbf{B}} = \begin{bmatrix} 0 & 0 & a & 0 \\ 0 & 0 & 0 & 0 \\ a & 0 & 0 & 0 \\ 0 & 0 & 0 & 0 \end{bmatrix}. \tag{6.40}$$

The local streamwise coordinates are $\xi \sim ux/q + vy/q$ and $\eta \sim -vx/q + uy/q$, and $\breve{u} = q$ and $\breve{v} = 0$ are the velocity components along $(\xi, \eta)$, respectively.

### 6.4.2 Linearization

As for the scalar advection equation, we need to be able to linearize the equations such that conservation is preserved. We need to find linearized Jacobian matrices, $\hat{\mathbf{A}}$ and $\hat{\mathbf{B}}$, that satisfy

$$\Phi_{\mathrm{A}} = \oint_l \mathbf{F} \cdot \vec{n} \mathrm{d}l = -\int_A \nabla \cdot \mathbf{F} \mathrm{d}A = -A(\hat{\mathbf{A}}\hat{\boldsymbol{U}}_x + \hat{\mathbf{B}}\hat{\boldsymbol{U}}_y), \tag{6.41}$$

where the gradients $\hat{\boldsymbol{U}}_x$ and $\hat{\boldsymbol{U}}_y$ are consistent with the linearization. The linearization can be achieved by generalizing the Roe average (see Section 2.3.13) to two dimensions [56]. Introducing the parameter vector $\boldsymbol{Z} = (Z_1, Z_2, Z_3, Z_4)^{\mathrm{T}} = \sqrt{\rho}(1, u, v, H)^{\mathrm{T}}$, we have

$$\begin{aligned} \boldsymbol{U} &= \left(Z_1Z_1, Z_1Z_2, Z_1Z_3, Z_1Z_4/\gamma + \tfrac{1}{2}(\gamma-1)(Z_2^2 + Z_3^2)/\gamma\right)^{\mathrm{T}}, \\ \boldsymbol{F} &= \left(Z_1Z_2, Z_2^2 + (\gamma-1)(Z_1Z_4 - \tfrac{1}{2}(Z_2^2 + Z_3^2))/\gamma, Z_2Z_3, Z_3Z_4\right)^{\mathrm{T}}, \\ \boldsymbol{G} &= \left(Z_1Z_3, Z_2Z_3, Z_3^2 + (\gamma-1)(Z_1Z_4 - \tfrac{1}{2}(Z_2^2 + Z_3^2))/\gamma, Z_3Z_4\right)^{\mathrm{T}}. \end{aligned} \tag{6.42}$$

By inspection of (6.42), it follows that in terms of the components of $\boldsymbol{Z}$, the gradients $\boldsymbol{U}_{\boldsymbol{Z}}, \boldsymbol{F}_{\boldsymbol{Z}}$, and $\boldsymbol{G}_{\boldsymbol{Z}}$ are linear over each triangle. The gradients of $\boldsymbol{U}$, consistent with the linearization, are given by

$$\hat{\boldsymbol{U}}_x = \frac{1}{A}\iint_T \boldsymbol{U}_x \mathrm{d}\Omega = \frac{1}{A}\iint_T \frac{\partial \boldsymbol{U}}{\partial \boldsymbol{Z}}\frac{\partial \boldsymbol{Z}}{\partial x}\mathrm{d}\Omega = \frac{\partial \boldsymbol{U}}{\partial \boldsymbol{Z}}(\hat{\boldsymbol{Z}})\boldsymbol{Z}_x,$$

$$\hat{\boldsymbol{U}}_y = \frac{1}{A}\iint_T \boldsymbol{U}_y \mathrm{d}\Omega = \frac{1}{A}\iint_T \frac{\partial \boldsymbol{U}}{\partial \boldsymbol{Z}}\frac{\partial \boldsymbol{Z}}{\partial y}\mathrm{d}\Omega = \frac{\partial \boldsymbol{U}}{\partial \boldsymbol{Z}}(\hat{\boldsymbol{Z}})\boldsymbol{Z}_y,$$

$$\hat{\boldsymbol{Z}} = \tfrac{1}{3}(\boldsymbol{Z}_1 + \boldsymbol{Z}_2 + \boldsymbol{Z}_3),$$

where $\hat{\boldsymbol{Z}}$ is the average over the vertices of the triangle and the gradients $\boldsymbol{Z}_x$ and $\boldsymbol{Z}_y$ are constant, since the unknowns are linear over the triangle. The flow variables corresponding to $\hat{\boldsymbol{Z}}$ are

$$\begin{bmatrix} \hat{\rho} \\ \hat{u} \\ \hat{v} \\ \hat{H} \end{bmatrix} = \frac{1}{\frac{1}{3}\sum_{i=1}^{3}\sqrt{\rho_i}} \begin{bmatrix} \left(\frac{1}{3}\sum_{i=1}^{3}\sqrt{\rho_i}\right)^3 \\ \frac{1}{3}\sum_{i=1}^{3} u_i\sqrt{\rho_i} \\ \frac{1}{3}\sum_{i=1}^{3} v_i\sqrt{\rho_i} \\ \frac{1}{3}\sum_{i=1}^{3} H_i\sqrt{\rho_i} \end{bmatrix}.$$

Referring back to (6.41), the fluctuation is given by

$$\Phi_{\mathrm{A}} = -A\left(\frac{\partial \boldsymbol{F}}{\partial \boldsymbol{Z}}(\hat{\boldsymbol{Z}})\frac{\partial \boldsymbol{Z}}{\partial \boldsymbol{U}}(\hat{\boldsymbol{Z}})\hat{\boldsymbol{U}}_x + \frac{\partial \boldsymbol{G}}{\partial \boldsymbol{Z}}(\hat{\boldsymbol{Z}})\frac{\partial \boldsymbol{Z}}{\partial \boldsymbol{U}}(\hat{\boldsymbol{Z}})\hat{\boldsymbol{U}}_y\right).$$

Note that any of the forms in Section 6.4.1 may be used as long as the transformation matrices are evaluated at $\hat{\boldsymbol{Z}}$.

### 6.4.3 Fluctuation Distribution Matrix

For the scalar advection equation, the fluctuation $\phi_{\mathrm{A}}$ was distributed to the vertices of the triangle by means of the distribution coefficients $\alpha_i$, with the definition of the $\alpha_i$'s depending on the scheme used. For the Euler equations, the $\alpha_i$'s are replaced by distribution matrices $\boldsymbol{\alpha}_i$:

$$\oint_\Gamma \mathbf{F}\cdot\vec{n}\mathrm{d}l = \sum_i \mathbf{P}(\boldsymbol{\alpha}_i\Phi_A),$$

where $\mathbf{P} = \boldsymbol{TM}$ is the transformation matrix from $\partial \boldsymbol{Q}$ to $\boldsymbol{U}$. As in the scalar case, the $\boldsymbol{\alpha}_i$'s are constrained by

$$\sum_{i=1}^{3} \boldsymbol{\alpha}_i = \mathbf{I},$$

in order to satisfy conservation. The update takes the form

$$\boldsymbol{U}_i^{t+\Delta t} = \boldsymbol{U}_i^t + \frac{\Delta t}{\Omega_i} \sum_{\forall T_k} \mathbf{P}\left(\boldsymbol{\alpha}_i^k \Phi_A^k\right). \tag{6.43}$$

Equation 6.43 is the equivalent of Equation 6.17, however, for the integration of the Euler equations a different time integration might be used. The coefficients $\kappa_i$, defined by (6.16), are replaced by the matrix $\mathrm{K}_i = \frac{1}{2}\mathbf{A}_{\vec{n}} \cdot \vec{n}_i$, which can be expressed by

$$\mathrm{K}_i = \boldsymbol{R}_{\vec{n}_i} \Lambda_{\vec{n}_i} \boldsymbol{L}_{\vec{n}_i}.$$

The generalization of Equation 6.15 is given by

$$\Phi_{\mathrm{A}} = \oint_l \mathbf{F} \cdot \vec{n} \mathrm{d}l = -\sum_i \mathrm{K}_i \boldsymbol{U}_i.$$

Define $\lambda_j^{\pm} = \frac{1}{2}\left(\lambda_{\vec{n}_i}^j \pm \left|\lambda_{\vec{n}_i}^j\right|\right)$, $\Lambda_{\vec{n}_i}^{\pm} = \mathrm{diag}(\lambda_1^{\pm}, \lambda_2^{\pm}, \ldots)$, and $\mathrm{K}_i^{\pm} = \boldsymbol{R}_{\vec{n}_i} \Lambda_{\vec{n}_i}^{\pm} \boldsymbol{L}_{\vec{n}_i}$. Then, for the Euler equations the distribution matrix for the LDA scheme [18] is given by

$$\boldsymbol{\alpha}_i^{\mathrm{LDA}} = \left(\sum_{j=1}^{3} \mathrm{K}_j^+\right)^{-1} \mathrm{K}_i^+;$$

for the LW scheme [18] it is given by

$$\boldsymbol{\alpha}_i^{\mathrm{LW}} = \tfrac{1}{3}\mathbf{I} + \tfrac{2}{3}\left(\sum_{j=1}^{3} \left|\mathrm{K}_j^+\right|\right)^{-1} \mathrm{K}_i;$$

and for the linear N scheme [49] it is given by

$$\Phi_i^{\mathrm{N}} = -\mathrm{K}_i^+(U_i - U_-),$$

where the inflow state, $\boldsymbol{U}_-$, is computed by solving the following linear system:

$$\left(\sum_j \mathrm{K}_j^+\right)\boldsymbol{U}_- = \sum_j \left(\mathrm{K}_j^+ \boldsymbol{U}_i\right).$$

## 6.5 Floating Shock-Fitting Implementation

Figure 6.6 shows a small region of the computational plane with a shock front cutting the plane. The tessellation of the computational plane is done with a constrained Delaunay triangulation. The unknowns are defined at the vertices of the triangles, which are shown in the figure by the full circles. The shock is shown as the empty circles, which we will call *shock-nodes*, connected by a dashed line. A dual set of unknowns, corresponding to the values on the low (downstream) and high (upstream) pressure side of the shock, are defined at the shock-nodes. As shown in the figure, the left side of the shock corresponds to the high pressure side (downstream). The triangular mesh is treated as a background mesh over which the shock front moves (*floats*). At time level $t_0$ the solution is known at all mesh and shock-nodes. In addition, the shock speed is known. The first step consists of advancing the shock front from its position at $t_0$ to its position at $t_0 + \Delta t$. This process is discussed in Section 6.5.6 below. For now, let us assume that the shock front has been advanced to its position at $t_0 + \Delta t$. In order to continue advancing the solution to $t_0 + \Delta t$, the following steps are taken, as discussed in [163].

### 6.5.1 Cell Removal in Neighborhood of Shock Front

The background mesh in the immediate neighborhood of the shock front is not suitable for the integration of the governing equations. The first step, therefore, is to redefine the mesh near the shock front. We begin by removing all background cells that are crossed by the shock front and all nodes, which we will call *phantom* nodes,

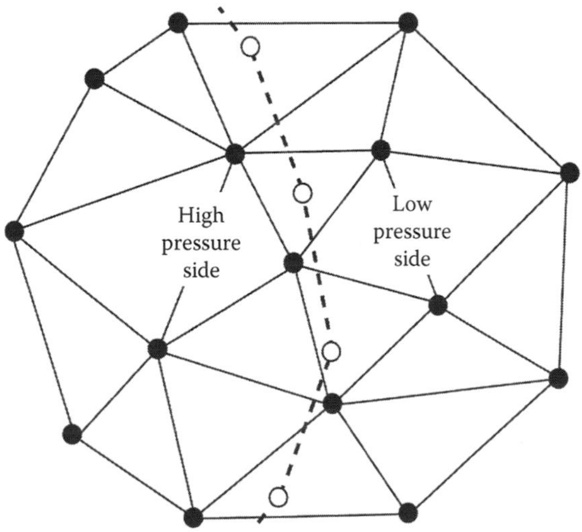


**FIGURE 6.6:** A small region of the tessellated plane with a shock (dashed line) cutting the plane.

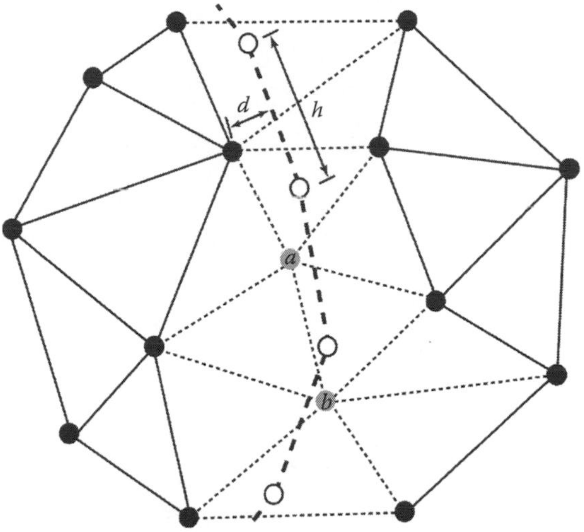


**FIGURE 6.7:** Background cells (thin dashed lines) and nodes $(a, b)$ that will be removed to rearrange mesh near shock front.

whose relative distance to the shock front is less than some user-specified tolerance, *tol*. The relative distance of a node point to the shock front is defined as the distance of the node to the nearest shock front edge, $d$, divided by the shock front edge length, $h$, as shown in Figure 6.7. All nodes for which $d/h \leq tol$ are deleted from the background mesh. All cells connected to the deleted nodes are also removed. The two nodes meeting the tolerance are labeled $a$ and $b$ in Figure 6.7. The cells to be removed are shown as thin dashed lines in the figure. After the cells are removed, a hole is left in the background mesh, as shown in Figure 6.8.

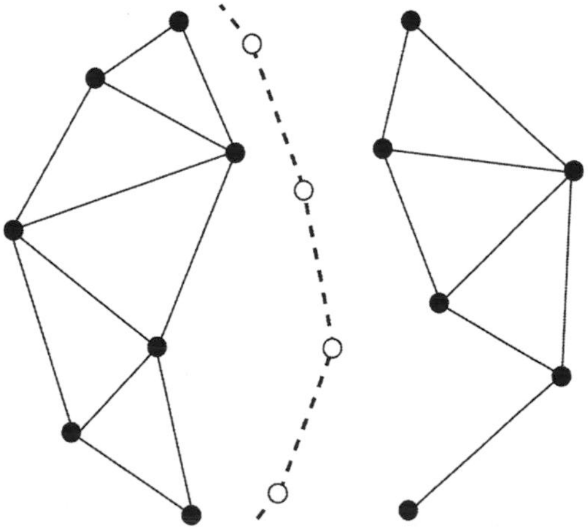

**FIGURE 6.8:** Hole left in background mesh after cell and nodes are removed.

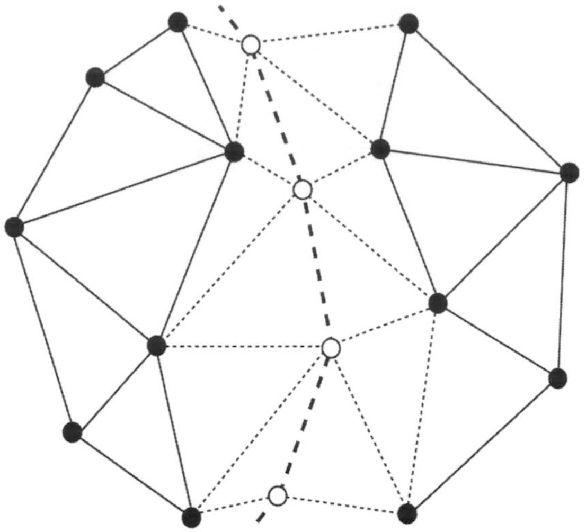

**FIGURE 6.9:** Resulting background mesh triangulation in the neighborhood of shock front. New mesh is constrained by the shock front and background mesh hole boundaries.

### 6.5.2 Local Re-Meshing near a Shock Front

We now need to re-mesh the hole left from the cell removal step. This is accomplished by using a constrained Delaunay triangulation [203] with the shock front and the boundaries of the hole specified as constraining boundaries for the triangulation, Figure 6.9. During this step, no new nodes are added to the mesh, i.e., only the nodes on the boundary of the hole and the nodes on the shock front are triangulated. The resulting mesh will be used in the calculation of the shock front and the neighboring points, and it must be of good quality. To insure a good quality mesh, we require that (1) the mesh nodes defining the hole boundary are not too close to the shock front; and (2) that the length of the shock edges, $h$, be approximately equal to the length of the triangle edges. The first condition is satisfied by requiring that $d/h \leq tol$, with $tol$ ranging between 0.2 and 0.5. The second requires readjusting the length of the shock front edges as the computation progresses, and is discussed below in Section 6.5.8.

For a time accurate computation, the grid motion created by the local re-meshing near the shock violates the space (or geometric) conservation law [81]. The goal should be to achieve the same accuracy on a deforming mesh as in a fixed mesh. This problem can be resolved by using an Arbitrary Lagragian Eulerian formulation (ALE), but this is not currently implemented in the code of Bonfiglioli and Paciorri, but could be easily implemented [19]. Of course, this issue does not affect the steady state results.

### 6.5.3 Computation of Shock Tangent and Normal Vectors

In order to satisfy the Rankine–Hugoniot jumps along the shock front, we need to define the unit tangent and normal vectors to the shock at each shock-node. Consider

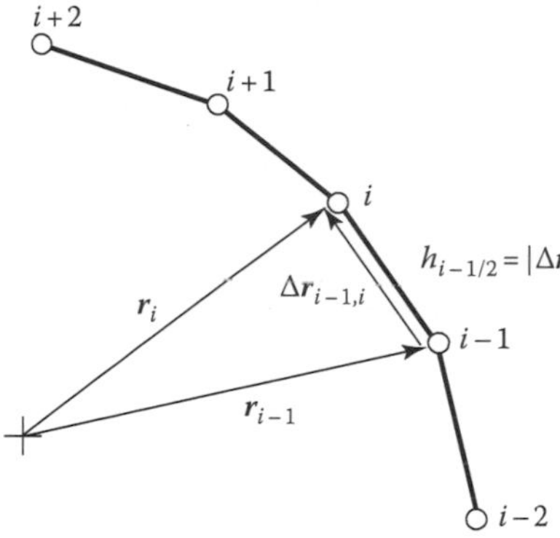


**FIGURE 6.10:** Details of shock front geometry used in tangent vector calculation at a shock-node.

Figure 6.10 depicting the shock front at $t_0 + \Delta t$ and in the neighborhood of shock-node $i$. Let $\boldsymbol{r}_i$ be the position vector to shock-node $i$. The vector $\Delta \boldsymbol{r}_{i,i-1} = \boldsymbol{r}_i - \boldsymbol{r}_{i-1}$ is a vector from shock-node $i-1$ to shock-node $i$, with length $h_{i-1/2}$, as shown in the figure. To compute the unit tangent and normal at $i$, we first need to determine if node points $i+1$ and $i-1$ are in the domain of dependence of node $i$. Three different cases may occur:

Case 1. Both shock-nodes, $i+1$ and $i-1$, are in the domain of dependence of node $i$.

Case 2. Only shock-node $i-1$ is in the domain of dependence of node $i$.

Case 3. Only shock-node $i+1$ is in the domain of dependence of node $i$.

We will discuss how to make this determination later.

Let us assume that we have established that Case 1 is valid. In this case, the unit tangent vector at $i$ is given by:

$$\boldsymbol{\tau} = \frac{\boldsymbol{v}_\tau}{|\boldsymbol{v}_\tau|},$$

where

$$\boldsymbol{v}_\tau = \Delta \boldsymbol{r}_{i,i+1} h_{i-1/2} + \Delta \boldsymbol{r}_{i-1,i} h_{i+1/2}. \tag{6.44}$$

Equation 6.44, used in [163], is second-order accurate if the shock-nodes are equally spaced, (see Problem 6.2 and Problem 6.3 for a more accurate tangent formula). The unit normal vector $\boldsymbol{n}$ is obtained by requiring that $\boldsymbol{n} \cdot \boldsymbol{\tau} = 0$ and that it points from high pressure (downstream) to low pressure (upstream):

$$\boldsymbol{n} = \pm \begin{bmatrix} 0 & 1 \\ -1 & 0 \end{bmatrix} \boldsymbol{\tau}. \tag{6.45}$$

Suppose Case 2 is valid. Since node $i+2$ is not in the domain of dependence, (6.44) is replaced by

$$\boldsymbol{v}_\tau = \Delta \boldsymbol{r}_{i-1,i}(h_{i-1/2} + h_{i-3/2}) - \Delta \boldsymbol{r}_{i-2,i} h_{i-1/2}, \tag{6.46}$$

which is also second-order accurate if the shock-nodes are equally spaced.

For Case 3, $\boldsymbol{v}_\tau$ is given by the mirror of (6.46):

$$\boldsymbol{v}_\tau = -\left(\Delta \boldsymbol{r}_{i+1,i}\left(h_{i+1/2} + h_{i+3/2}\right) - \Delta \boldsymbol{r}_{i+2,i} h_{i+1/2}\right). \tag{6.47}$$

To determine if shock-node $i$ is within the domain of dependence of point $i+1$ in a time interval $\Delta t$, we draw the velocity vector on the high pressure side of node $i+1$, multiply by $\Delta t$, and from the end point of this vector draw a circle with radius $a^t_{h,i+1}\Delta t$, where $a^t_{h,i+1}$ is the speed of sound on the high pressure side of node $i+1$, see the left panel of Figure 6.11. If node $i$ falls inside this circle, then it is influenced by node $i+1$. A simple (approximate) test for this condition is to require that

$$\left(a^t_{h,i+1} - \operatorname{sgn}\left[\boldsymbol{u}^t_{h,i+1} \cdot \tilde{\boldsymbol{\tau}}\right]\left|\boldsymbol{u}^t_{h,i+1}\right|\right)\Delta t > h_{i+1/2},$$

where the unit tangent to the shock, $\tilde{\boldsymbol{\tau}}$, is taken halfway between nodes $i$ and $i+1$, as indicated in Figure 6.11, and is defined by $\tilde{\boldsymbol{\tau}} = \Delta \boldsymbol{r}_{i,i+1}/\left|\Delta \boldsymbol{r}_{i,i+1}\right|$. A similar test is applied to node point $i-1$.

## 6.5.4 Solution Update

The computational mesh defined in Section 6.5.2 is used with the unstructured grid solver to obtain a solution at $t_0 + \Delta t$. Note that the shock front now divides the computational domain (much like a solid boundary) and mesh points on the high

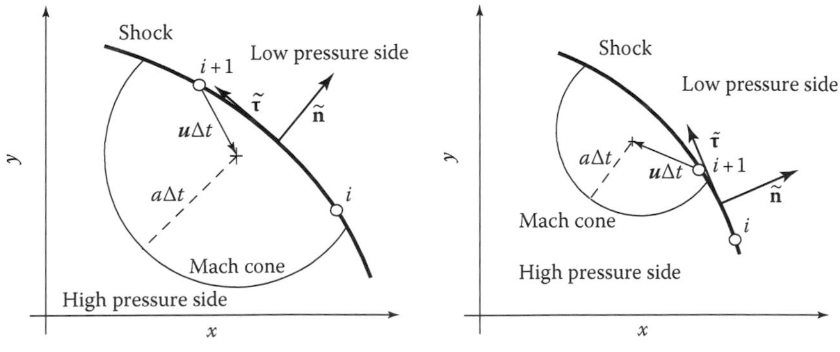


**FIGURE 6.11:** Illustration of how the domain of dependence is taken into account in the computation of the tangent and normal vectors to the shock front. The left panel shows that shock-node $i$ is influenced by $i+1$, but this is not the case on the right panel.

pressure (downstream) side of the shock are computed using only information from the high pressure side; similarly, mesh points on the low pressure (upstream) side of the shock are computed using only information from the low pressure side. For shock-nodes on the low pressure side, this is consistent with the domain of dependence, since the particle path and the Mach cone are directed toward the shock front. On the high pressure side, we have a problem. Here, we only have one equation (along the bicharacteristic reaching the shock from the high pressure side on a plane normal to the shock front) available to us, see Figure 6.12. Therefore, the updates that are made to the shock-nodes on the high pressure side are incorrect, except for one single piece of information, the Riemann variable:

$$Q_{h,i}^{t_0+\Delta t} = \frac{2}{\gamma - 1} a_{h,i}^{t_0+\Delta t} + \boldsymbol{u}_{h,i}^{t_0+\Delta t} \cdot \boldsymbol{n}, \tag{6.48}$$

where

- $\boldsymbol{n}$ is the unit normal to the shock pointing toward the low pressure side at node $i$,
- $a$ is the speed of sound, and
- $\boldsymbol{u}$ is the velocity vector.

Note that the Riemann variable $Q$ is reconstructed from the computed values of $a$ and $\boldsymbol{u}$ at $i$. The correct calculation of this variable is made possible by the upwind nature of the unstructured solver.

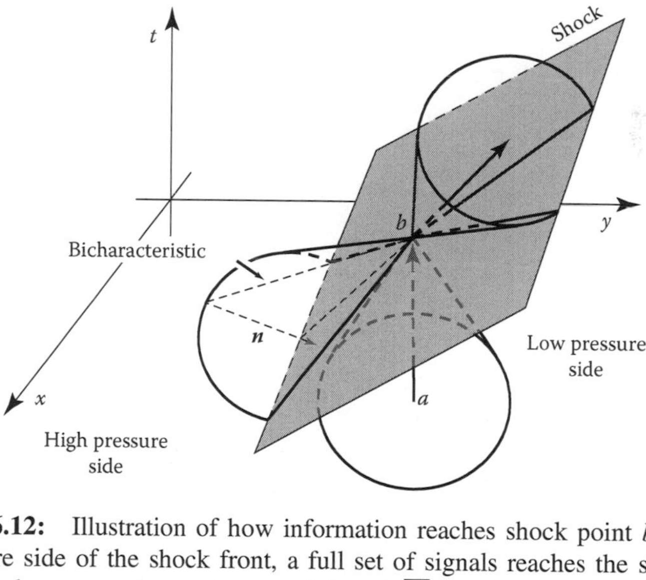


**FIGURE 6.12:** Illustration of how information reaches shock point $b$. From the low pressure side of the shock front, a full set of signals reaches the shock. They travel along the cone surface and the particle path $\overline{\text{ab}}$. On the high pressure side, only one piece of information reaches the shock. This is carried along the bicharacteristic line formed by the interception of the Mach cone reaching $b$ from the high pressure side and the plane normal to the shock front.

Once the solution update is completed, we have correct values on the low pressure side of the shock front, but need to recompute the values on the high pressure side and find the shock speed $w$. This is accomplished by solving the Rankine–Hugoniot jumps on a plane normal to the shock front together with the left-hand side of (6.48) which we accept as correct. Let subscripts l and h refer to values on the low and high pressure side of the shock front, respectively. The upstream normal component of the Mach number, relative to the shock front, at node $i$, at $t_0 + \Delta t$ is

$$M_{\mathrm{l,rel}} = \frac{\boldsymbol{u}_l \cdot \boldsymbol{n} - w}{a_l}, \tag{6.49}$$

where to simplify the notation we are dropping the $i$ subscript and $t_0 + \Delta t$ superscript. In terms of $M_{\mathrm{l,rel}}$ and the low pressure side values we have

$$\begin{aligned}
p_{\mathrm{h}} &= \frac{p_{\mathrm{l}}\left(2\gamma M_{\mathrm{l,rel}}^2 - (\gamma-1)\right)}{\gamma+1}, \\
\rho_{\mathrm{h}} &= \frac{\rho_{\mathrm{l}}(\gamma+1)M_{\mathrm{l,rel}}^2}{(\gamma-1)M_{\mathrm{l,rel}}^2+2}, \\
\frac{\rho_{\mathrm{h}}\left((\boldsymbol{u}_{\mathrm{h}} - \boldsymbol{u}_{\mathrm{l}})\cdot\boldsymbol{n} + a_{\mathrm{l}}M_{\mathrm{l,rel}}\right)^2}{\gamma p_{\mathrm{h}}} &= \frac{(\gamma-1)M_{\mathrm{l,rel}}^2+2}{2\gamma M_{\mathrm{l,rel}}^2-(\gamma-1)}, \\
\frac{2\sqrt{\gamma p_{\mathrm{h}}/\rho_{\mathrm{h}}}}{\gamma-1} + \boldsymbol{u}_{\mathrm{h}}\cdot\boldsymbol{n} &= Q_{\mathrm{h}}, \\
\boldsymbol{u}_{\mathrm{h}}\cdot\boldsymbol{\tau} &= \boldsymbol{u}_{\mathrm{l}}\cdot\boldsymbol{\tau}.
\end{aligned} \tag{6.50}$$

Equation 6.50 is a set of five equations for the five unknowns $(p_{\mathrm{h}}, \rho_{\mathrm{h}}, \boldsymbol{u}_{\mathrm{h}}\cdot\boldsymbol{n}, \boldsymbol{u}_{\mathrm{h}}\cdot\boldsymbol{\tau}, M_{\mathrm{l,rel}})$. Note that for consistency, all quantities, including $\boldsymbol{n}, \boldsymbol{\tau}$, and $w$ in (6.50) are evaluated at $t_0 + \Delta t$. The first four equations can be combined into a single equation for $M_{\mathrm{l,rel}}$:

$$\begin{aligned}
&\frac{1}{(\gamma-1)}\sqrt{\left(2\gamma M_{\mathrm{l,rel}}^2 - (\gamma-1)\right)\left((\gamma-1)M_{\mathrm{l,re}}^2+2\right)\Big/M_{\mathrm{l,rel}}^2} \\
&\quad + \left(M_{\mathrm{l,rel}}^2 - 1\right)\Big/M_{\mathrm{l,rel}} = \frac{(\gamma+1)}{2a_{\mathrm{l}}}(Q_{\mathrm{h}} - \boldsymbol{u}_{\mathrm{l}}\cdot\boldsymbol{n}),
\end{aligned} \tag{6.51}$$

which is the same equation we found in Section 5.1.6.3, Equation 5.29. The left-hand side of (6.51) varies almost linearly with $M_{\mathrm{l,rel}}$, thus (6.51) can be solved easily by Newton iteration using the value of $M_{\mathrm{l,rel}}$ at $t_0$ as a first guess. The shock speed follows from (6.49) once $M_{\mathrm{l,rel}}$ is known.

### 6.5.5 Interpolation at Phantom Nodes

In Section 6.5.1, all nodes for which $d/h \leq tol$ were deleted. These phantom nodes belong to the background mesh and will be needed in subsequent time steps,

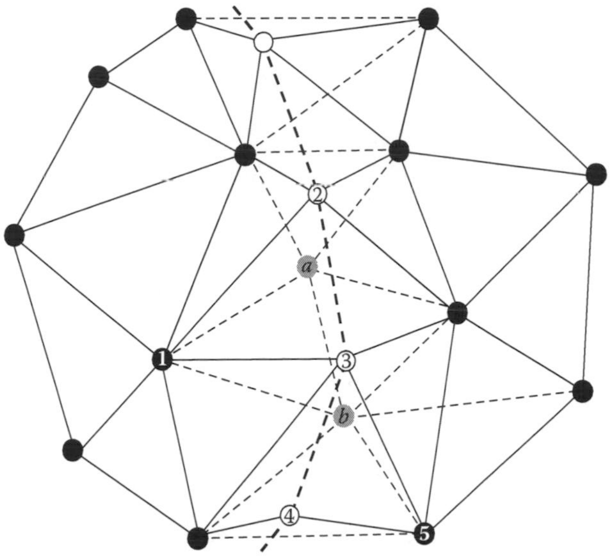


**FIGURE 6.13:** The values at the nodes that were removed, labeled $a$ and $b$ in the figure, during the cell removal step, Section 6.5.1, are found by interpolation using the surrounding nodes.

if the shock front moves sufficiently far away from its current position such that $d/h > tol$ at these nodes. Therefore, we need to update these nodes. We can either perform a search for these nodes or look them up on a list created when the nodes were removed. In either case, the values at the nodes are found by linear interpolation from surrounding nodes using (6.5). For example, with reference to Figure 6.13, the values at the phantom node labeled $a$ are evaluated by interpolating the values at the nodes labeled 1, 2, and 3 in the figure.

### 6.5.6 Shock Front Location Update

The shock calculation began with the update of the shock position, see Section 6.5, which we now describe as the beginning of the next step. The new shock front location is found from

$$\boldsymbol{r}_i^{t+2\Delta t} = \boldsymbol{r}_i^{t+\Delta t} + w_i^{t+\Delta t}\boldsymbol{n}_i^{t+\Delta t}\Delta t. \tag{6.52}$$

Equation 6.52 advances the shock front from its position at $t_0 + \Delta t$ to $t_0 + 2\Delta t$, setting the stage for the beginning of the next time step, see Figure 6.14. The update done by (6.52) is first-order accurate. This is sufficient if we are only interested in describing steady state solutions. If we want to describe unsteady flows to second-order accuracy, we will need some way to evaluate $w_t$. In Sections 4.3 and 5.1.6.2, we found an equation for $w_t$ by combining the characteristic reaching the shock from the high pressure side with the Rankine–Hugoniot jumps. This method has not been implemented within the unstructured grid frame work. Perhaps simpler to implement

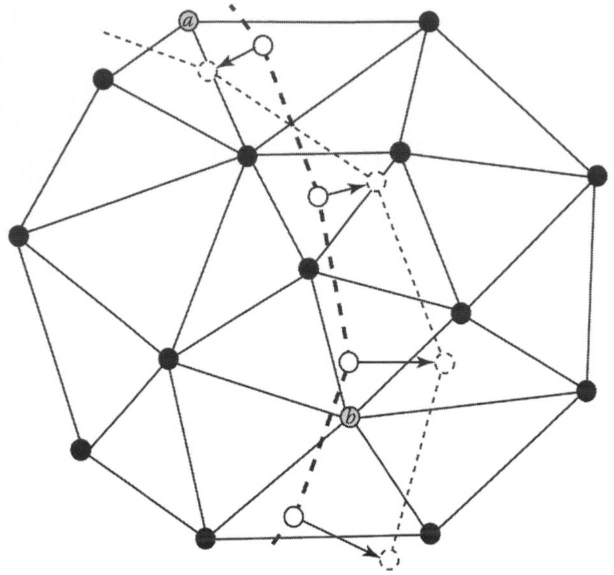


**FIGURE 6.14:** Shock front location update. The shock front position at $t_0 + \Delta t$ is shown as the heavy dashed line; the new position at $t_0 + 2\Delta t$ is shown as the thin dashed line.

in the unstructured grid frame work would be the approach described in Section 5.1.6.3. The method follows from (6.51). Let the left hand side of (6.51) be $g(M_{1,\mathrm{rel}})$, then, differentiating (6.51) with respect to $t$, we find

$$w_t = \left(\frac{\gamma+1}{2g'} + 1\right)\left[\frac{\partial \boldsymbol{u}_1}{\partial t} \cdot \boldsymbol{n} + \boldsymbol{u}_1 \cdot \frac{\partial \boldsymbol{n}}{\partial t}\right] - M_{1,\mathrm{rel}} \frac{\partial a_1}{\partial t} - \frac{\gamma+1}{2g'} \frac{\partial Q_{\mathrm{h}}}{\partial t}, \tag{6.53}$$

where $g' = \partial g / \partial M_{1,\mathrm{rel}}$. The difference between this approach and that in Sections 4.3 and 5.1.6.2 is that there we eliminated all time derivatives of terms coming from the high pressure side, but retained space derivatives of those terms. In the present method, we retain the term $\partial Q_{\mathrm{h}} / \partial t$, which has the space derivatives concealed in it. The remaining task is to obtain accurate discrete approximations to the time derivatives on the right-hand side of (6.53). The derivatives of the flow variables are available from the unstructured flow solver, and for $\partial \boldsymbol{n} / \partial t$, see Problem 6.6.

### 6.5.7 Interpolation at Shock-Crossed Nodes

As the shock front sweeps through the mesh, it will cross over mesh nodes of the background mesh. The situation is illustrated in Figure 6.15. The shock front is shown as the heavy dashed line. It corresponds to the new shock front position of Figure 6.14. Note from Figure 6.14 that the shock front has swept over the node labeled $b$. This node has values corresponding to the low pressure side. In order for the unstructured solver to do a proper computation, the values at $b$ must be corrected. We can make the

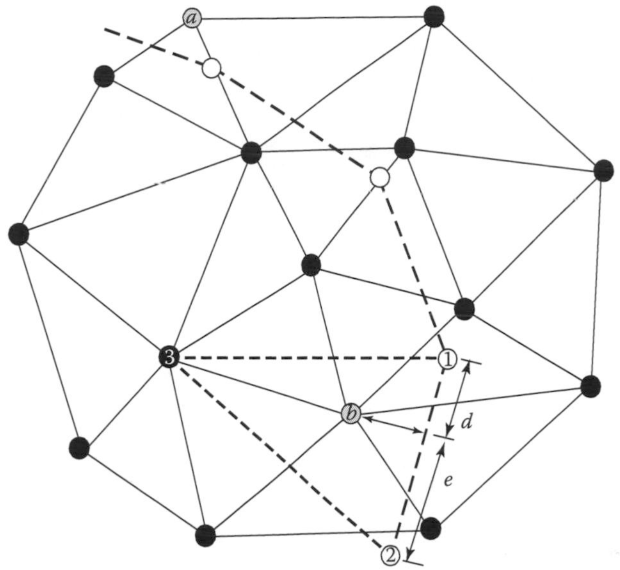


**FIGURE 6.15:** Update for background grid nodes crossed by shock front. In the illustration, the node labeled *b* was crossed by the shock front. The update could be done by either drawing the ⊥ from *b* to the shock edge 1–2 (to define the interpolating weights *d* and *e*), or by linearly interpolating from nodes 1, 2, and 3.

correction to node *b* in two different ways. One way is to draw the perpendicular from *b* to the shock edge that swept over it, edge 1–2 in Figure 6.15. This will define the distances *d* and *e* which are used as the weights to interpolate the values at node *b*. The other approach requires a search find node 3. With this approach, the values at node *b* are linearly interpolated using the values at nodes 1, 2, and 3.

### 6.5.8 Shock-Node Redistribution

As the flow field evolves, the size of the shock front edges could shorten or lengthen. If this is not corrected, the size of the shock front edges can become considerably different from the size of the edges of the background mesh leading to an unbalance in the discrete approximations. To avoid this problem, the shock front nodes need to be redistributed when some user-prescribed threshold is exceeded. For example, if a shock front edge becomes greater than or less than some percent of the average background mesh edge size, then the shock front nodes are redistributed, which might require the addition of new shock nodes, to meet the threshold.

### 6.5.9 Shock Detection

Thus far, unstructured floating shock-fitting has been applied only to problems where the shock fronts are defined as part of the initial conditions. That is, cases where the shock front develops as part of the flow field evolution have not been treated. To handle these problems, two things are needed. One, of course, is a shock detection algorithm. The other is the bookkeeping necessary to track new shocks and also delete shocks if they become too weak. The data infrastructure associated with unstructured grids should make this bookkeeping relatively easy and will not be discussed here. Instead we will outline the elements needed for a shock detection scheme.

The scheme we propose is the extension to the unstructured grid frame work of the scheme used in Section 4.3 and implemented in `DETECT_SHOCK.M` of the code `Q1D.M`. As a general rule, it is a good idea to start with a scheme that is known to work and make the necessary changes to accommodate a different environment than to start from scratch. The shock detection scheme can be broken down into the following steps:

1. Evaluate the pressure gradient, $\nabla p$, on each cell. This is a simple operation requiring the evaluation of the coefficients $c_1$ and $c_2$ of (6.5).

2. Create a list of cells where $|\nabla p| > tol_{\nabla p}$, where $tol_{\nabla p}$ is a user-specified tolerance. We will apply two other tests, so this test doesn't have to be too tight.

3. For each cell on the list evaluate the unit normal: $\boldsymbol{n} = -\nabla p/|\nabla p|$.

4. Interpolate four values of $\boldsymbol{u} \cdot \boldsymbol{n}$, bracketing the cell in question, onto a line parallel to $\boldsymbol{n}$ and running through the centroid of the cell. The values should be equally spaced with the spacing equal to the average cell edge size, which we will call $\bar{l}$. Label the values 1 through 4, as in Figure 6.16.

5. Use these $\boldsymbol{u} \cdot \boldsymbol{n}$ values to construct a third-order polynomial fit as described in Section 2.9 and find the determinant (2.154). Two tests are now made. For the first test we evaluate the Riemann variables: $Q_i = 2a_i/(\gamma - 1) - \boldsymbol{u}_i \cdot \boldsymbol{n}$, $R_i = \dfrac{2a_i}{\gamma - 1} + \boldsymbol{u}_i \cdot \boldsymbol{n}$, at the interpolation points 2 and 3, see Figure 6.16. Of course this requires interpolating the speed of sound at these points. Evaluate $\Delta Q = Q_2 - Q_3$ and $\Delta R = R_2 - R_3$. If $\Delta R > -0.21/\mathrm{d}x$ and $\Delta Q < 0.21/\mathrm{d}x$ are true no shock node is inserted and we move on to Step 3 above. The quantity $\mathrm{d}x$ is defined as $\mathrm{d}x = \min(0.02, \max(0.01, \bar{l}))$. These tolerances are a result of many tests with structured grid solvers. They would need to be fine-tuned for the unstructured grid solver. For the second test, we require that the determinant (2.154) be greater than $-0.1$. If this test fails no shock node is added, and we go to Step 3.

6. If the second test is met, a new shock node is added.

7. However, we cannot compute an isolated shock node. Therefore, a new *shock front* should not be introduced until a minimum of three neighboring shock nodes are detected, see Problem 6.7.

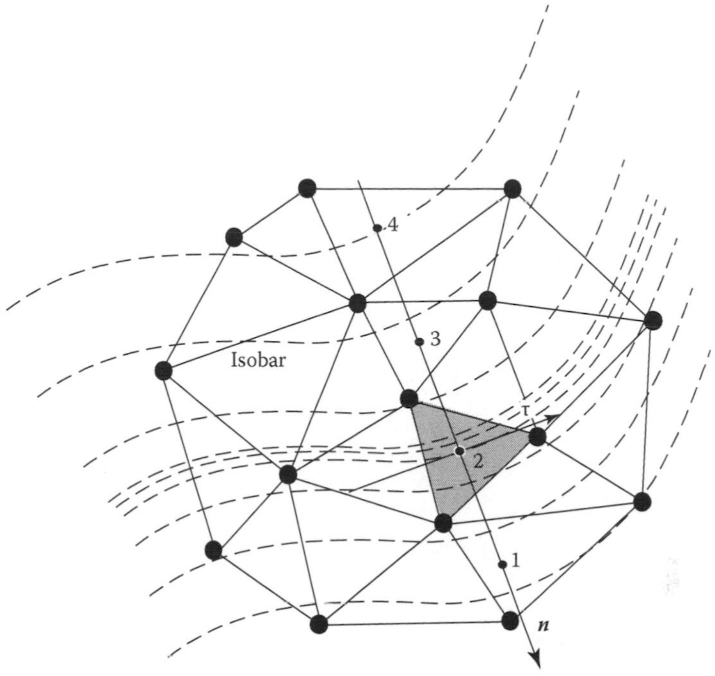


**FIGURE 6.16:** Illustration of shock formation detection. The dashed lines correspond to isobars drawn at equally spaced increments. The figure shows a high pressure gradient within the solid triangle. The normal $\boldsymbol{n}$ is drawn pointing toward the low pressure side.

### 6.5.10 Shock Interactions

Shock interactions in multi-dimensions are resolved in a plane normal to the shock front at the point of contact. In this plane, the interaction is not different from the one-dimensional shock interactions discussed in Section 4.3.1. Thus, to resolve a shock interaction we start by finding the unit normal to the shock front at the point of contact, projecting the velocity vector to this plane, resolving the interaction as in a one-dimensional problem, and finally adding the tangential velocity vector which is not affected by the interaction.

An example of a two-dimensional interaction is illustrated in Figure 6.17. It shows a planar shock front interacting with a circular disk. The *incident* shock Mach number is $M_s$ and the gas in front of the shock is at rest. In the $(x,t)$-plane the incident shock front makes an angle $\theta$ with the abscissa, upper panel of Figure 6.17, related to the shock speed, $w$, by $w = -1/\tan\theta$. The shock front makes contact with the disk at point $b$ at $t = t_1$, middle panel of Figure 6.17. The plane normal to the shock front at the point of contact is the $y = 0$ plane. The interaction with the disk requires a *reflected* shock front to turn the flow parallel to the disk, as illustrated in the figure. As the incident shock continues moving left it makes contact with the disk at a point

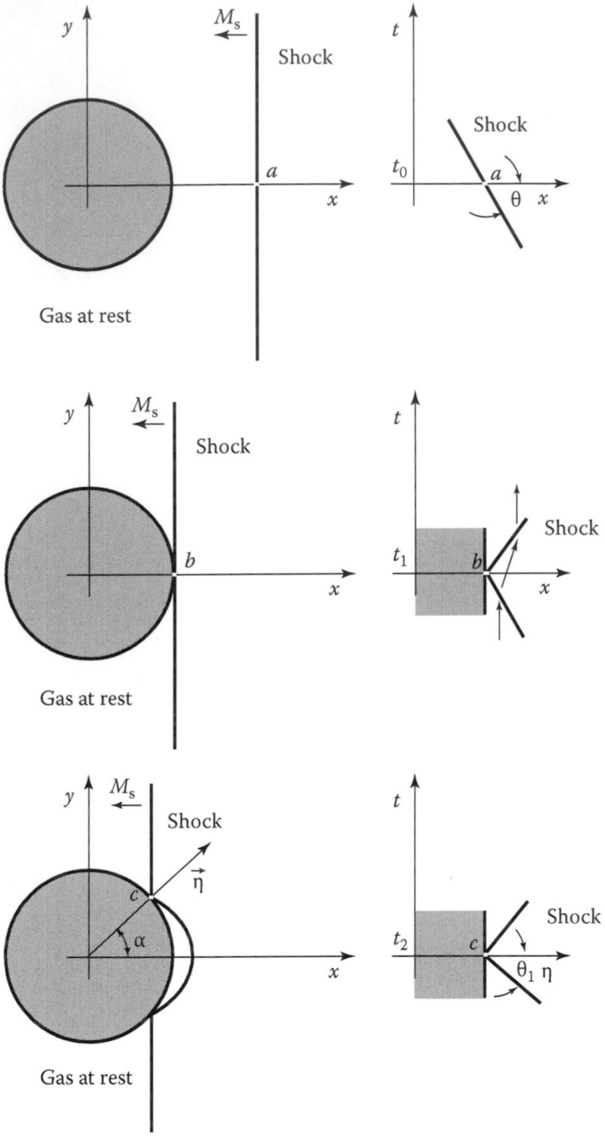


**FIGURE 6.17:** Sketch of planar shock front colliding with a disk.

$c$ located α-degrees from the abscissa, as shown in the lower panel of Figure 6.17 for $t = t_2$. The plane normal to the point of contact is the plane $y - \tan(\alpha)x = 0$. The unit normal at the point of contact is $\vec{\eta} = \cos\alpha\vec{i}_x + \sin\alpha\vec{i}_y$. The interaction between the shock front and the disk is resolved in the $(\eta, t)$ plane as shown in the right drawing of the lower panel of Figure 6.17. In this plane the shock front makes an angle $\theta_1$ with the abscissa. The unit normal to the incident shock front is $\vec{n}_s = \sin\theta\vec{i}_x + \cos\theta\vec{i}_t$, where $\vec{i}_x$ and $\vec{i}_t$ are the unit tangents to the $x$ and $t$ coordinates, respectively. The unit normal to the $\alpha =$ constant plane is $\vec{n}_\alpha = \sin\alpha\vec{i}_x - \cos\alpha\vec{i}_y$,

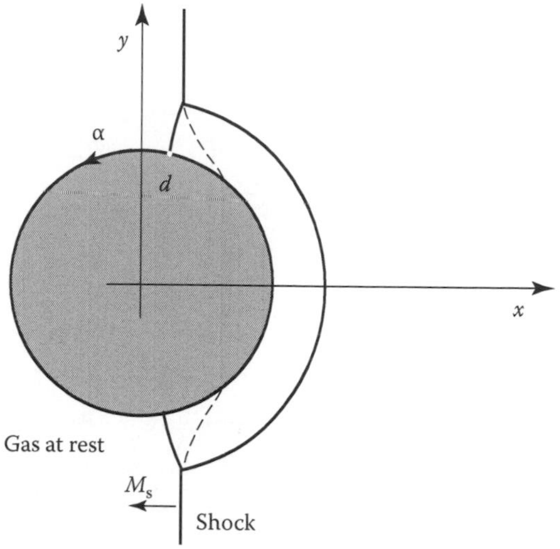


**FIGURE 6.18:** Sketch of Mach reflection for the diffraction of a planar shock front from a disk.

where $\vec{i}_y$ is the unit tangent to the $y$-coordinate. In terms of θ and α, the angle $\theta_1$ is given by $\sin\theta_1 = |(\vec{n}_s \times \vec{n}_\alpha) \times \vec{\eta}| = \sin\theta\cos\alpha$. The projection of the incident shock Mach number to the $\alpha$ − constant plane is $M_{s1} = M_s \cos\alpha$. As the incident shock front continues moving left, a point is reached beyond which a regular shock reflection can no longer turn the flow parallel to the disk in the $(\eta, t)$. At this point the regular shock reflection is replaced by a Mach reflection, as shown in Figure 6.18. Now, a Mach stem connects to the incident shock front at a *triple point* where we also have the reflected shock and a contact surface that separates the flow passing through the reflected shock from that passing through the Mach stem. The interaction at point $d$ of Figure 6.18 is now between the foot of the Mach stem and the disk and the boundary condition is to insure that the flow remains parallel to the disk after going through the Mach stem. A number of criteria have been proposed to define the transition point between regular and Mach reflection, see [15] for a recent review of this topic. For this problem, experimental results (Figure 7.48 of [83]) indicated that in the $M_s$ range of [1.5, 4.0] transition between regular and Mach reflection occurs at about 50°, see Figure 6.19, and in the $M_s$ range of [1.86, 5.96] the triple point path follows a 33° trajectory (Figure 7.44 of [83]), as shown in Figure 6.20. It is surprising that at low Mach numbers transition takes place in a plane where the effective shock Mach number is less than one, see Figure 6.19.

The triple point represents a new type of shock interaction not occurring in one-dimensional flows. The solution of this problem is not very different from those discussed for one-dimensional shock interaction problems. At the triple point, all conditions are known in front and behind the incident shock. We need to find the inclination of the reflected shock, the contact surface, and the Mach stem. We also must find the velocity vector defining of the triple point. The jump conditions across the reflected shock, contact surface, and Mach stem, the bicharacteristic reaching the

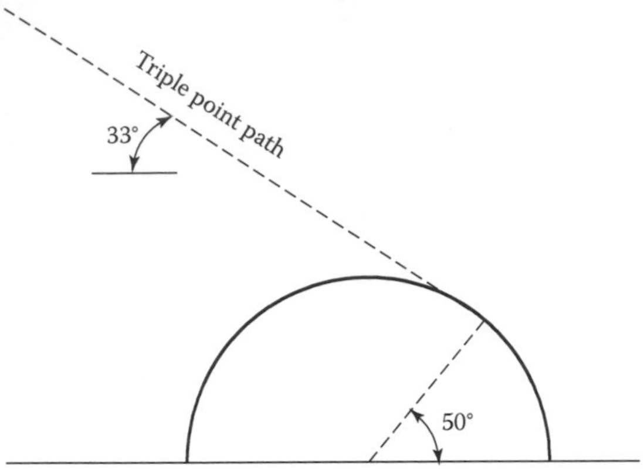


**FIGURE 6.19:** Solid line corresponds to locus of points for which the effective shock Mach number is sonic. The dashed line is for transition between regular and Mach reflection, from Figure 7.48 of [83]. The angle $\alpha_d$ corresponds to the transition angle.

Mach stem from the high pressure side, and the fact that the triple point lies on the incident shock provide the required number of equations to close the problem. Several problems of this type have been solved using shock-fitting techniques [135,239] and by an unstructured grid floating shock-fitting technique [162].

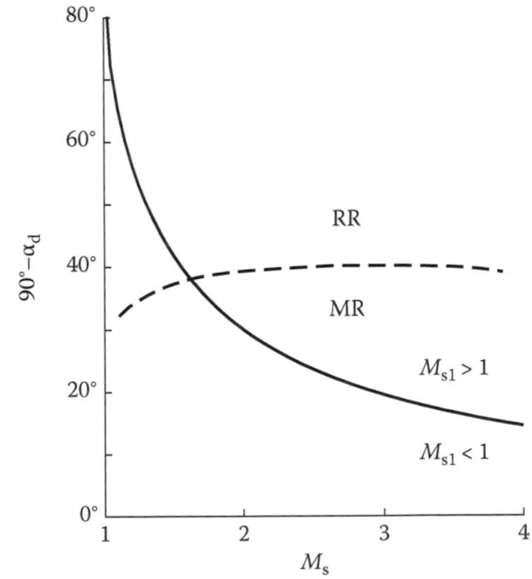


**FIGURE 6.20:** Experimental results reported in [83] indicate that regular reflection transitions to Mach reflection at about 50° and that the triple point path follows a 33° trajectory.

## 6.6 Unstructured Grids Shock-Fitting Results

### 6.6.1 Mach Reflection

The following results were obtained with the unstructured floating shock-fitting code developed by Bonfiglioli and Paciorri [163]. The first set is for Mach 2 flow over a 14° compression ramp, as illustrated in Figure 6.21. The incident shock generated by the ramp is reflected by a lower plate, but because the reflected shock is not able to turn the flow parallel to the plate an irregular reflection takes place. As shown in the illustration, the irregular reflection consists of a Mach disk, an incident shock, a reflected shock, and a contact surface all meeting at a triple point. The flow is supersonic everywhere except downstream of the Mach disk. The code of Bonfiglioli and Paciorri can treat shocks in a shock-capturing mode, a floating shock-fitting mode, or a mix mode where some shocks are captured while others are fitted. Figure 6.22 shows two sets of results. The upper panel corresponds to a fully shock captured solution. The upper right panel gives details of the captured solution near the Mach disk. The lower panel corresponds to a solution where the Mach disk and the reflected shock are fitted, but all other discontinuities are captured. The right lower panel gives the details near the Mach disk. The improvement in the quality of the results brought about by fitting these two shocks is quite obvious. The upper panel of Figure 6.23 is the same as that of Figure 6.22. The lower panel corresponds to a fully shock-fitted solution. Again, the improvement in the quality of the solution is remarkable. Actually, it is only remarkable because after some 40 years of shock-capturing dominance over computational fluid dynamics we are surprised by a solution that looks like it should. The only thing remarkable here is that we have accepted the results of the upper panel of Figure 6.23 for so long!

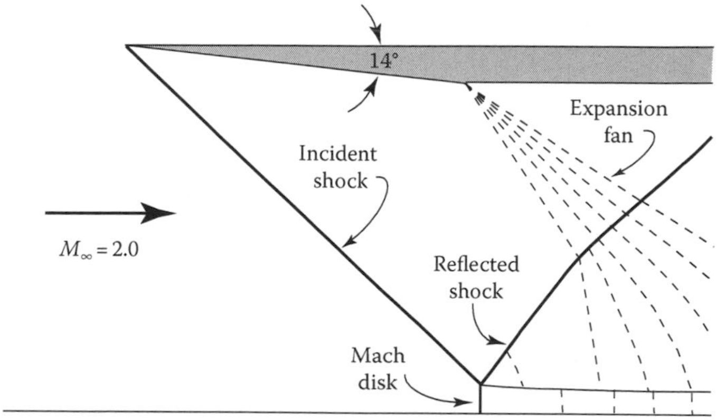


**FIGURE 6.21:** Sketch of main features for Mach reflection problem.

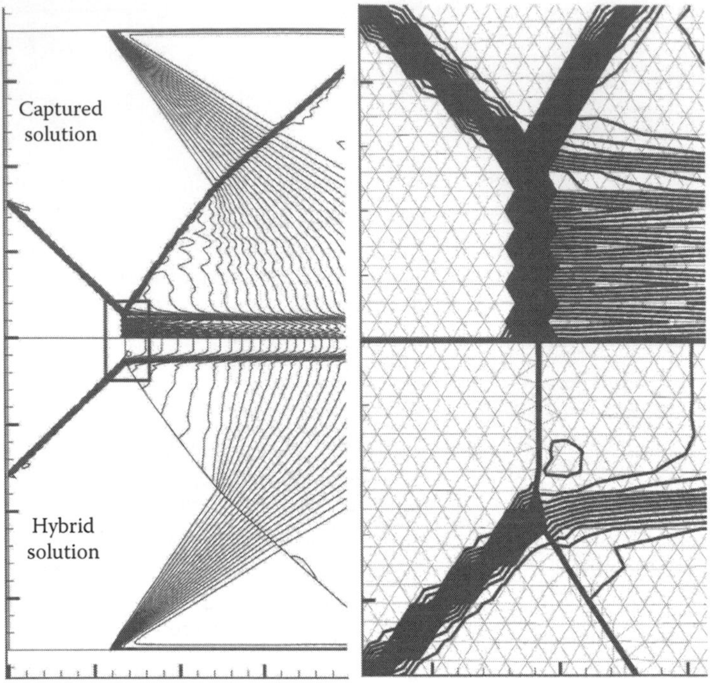


**FIGURE 6.22:** Upper panel corresponds to a fully shock-captured solution. Lower panel corresponds to a solution that only fits the Mach disk and reflected shock. Results presented courtesy of R. Paciorri. (From Paciorri, R. and Bonfiglioli, A., *Comp. Fluids*, 38, 715, 2009. With permission.)

## 6.6.2 Edney Type IV Shock Interaction

Edney [63] cataloged the interaction of an incident planar shock with the bow shock of a blunt body into six types depending on where the incident shock intercepts the bow shock. Type IV occurs when the incident shock intercepts the bow shock near its strongest (more normal to the incoming flow) location. At the point of interception, a triple point is created resulting in a transmitted shock and a contact surface, see Figure 6.24. The flow downstream of the transmitted shock is supersonic, unlike the flow behind the bow shock. The transmitted shock meets again with the bow shock at a second triple point where another shock and contact surface originate. The two contact surfaces form the boundaries of a supersonic jet that is imbedded in the subsonic flow created by the bow shock. The jet consists of a series of expansion and compression waves that channel the high speed flow into the surface of the blunt body. Just before the jet flow makes contact with the blunt body, a jet bow shock forms, further compressing and raising the temperature of the impinging flow. The effect of the hot jet hitting the surface of the blunt body can be catastrophic. Such was the experience on October 1967 of NASA's X-15–2 research airplane when flying at Mack 6.7. A Type IV shock interaction resulted in a structural failure of a pylon supporting a dummy scramjet engine. The plane survived, but never flew again.

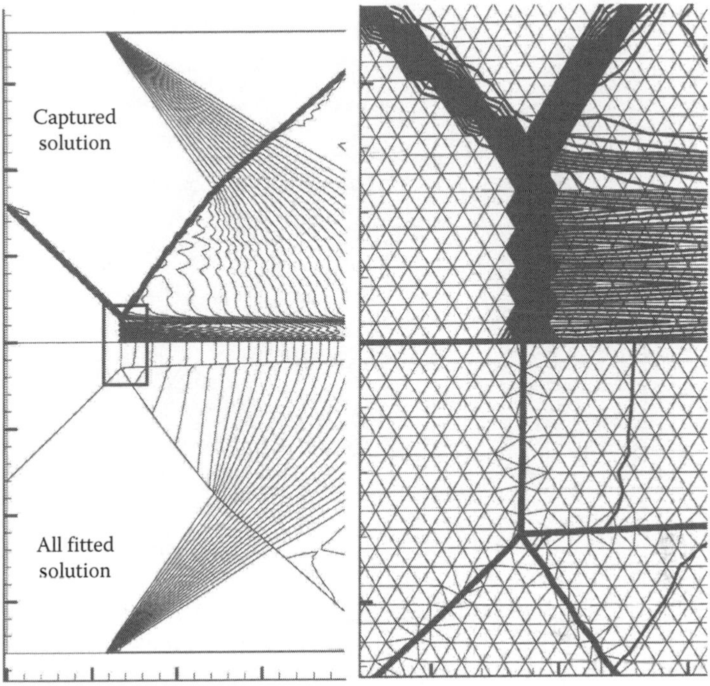


**FIGURE 6.23:** Upper panel corresponds to a fully shock-captured solution. Lower panel corresponds to a fully shock-fitted solution. Results presented courtesy of R. Paciorri. (From Paciorri, R. and Bonfiglioli, A., *Comp. Fluids*, 38, 715, 2009. With permission.)

Figures 6.25 through 6.28 show computed flow fields for a Type IV shock interaction created by an incident shock inclined at 10° from the free stream direction. The free stream Mach number is 10 and the blunt body is a circular

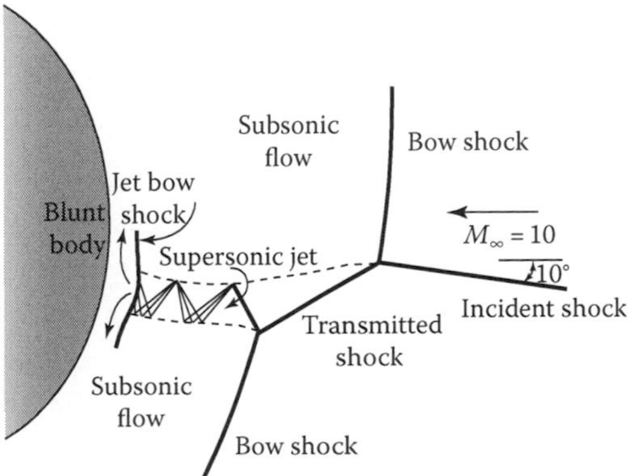


**FIGURE 6.24:** Sketch of main features of a Type IV shock interaction.

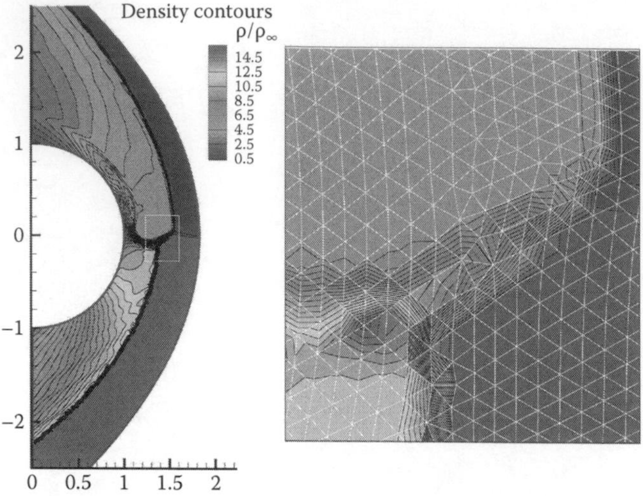


**FIGURE 6.25:** Computed Type IV shock interaction with full shock-capturing mode. Right panel shows details around shock interaction. Results presented courtesy of R. Paciorri. (From Paciorri, R. and Bonfiglioli, A., Numerical simulation of shock interactions with an unstructured shock-fitting technique, in *Sixth European Symposium on Aerothermodynamics for Space Vehicles*, Paris, France, 2008. With permission.)

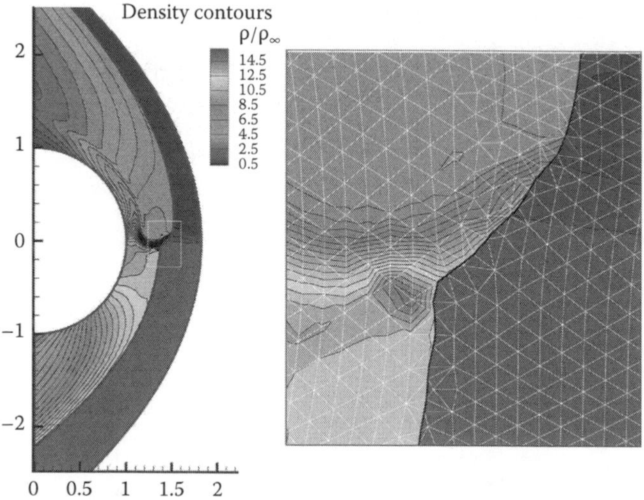


**FIGURE 6.26:** Computed Type IV shock interaction. The bow shock and the shock connecting the two triple points are fitted; all other discontinuities are captured. Right panel shows details around shock interaction. Results presented courtesy of R. Paciorri. (From Paciorri, R. and Bonfiglioli, A., Numerical simulation of shock interactions with an unstructured shock-fitting technique, in *Sixth European Symposium on Aerothermodynamics for Space Vehicles*, Paris, France, 2008. With permission.)

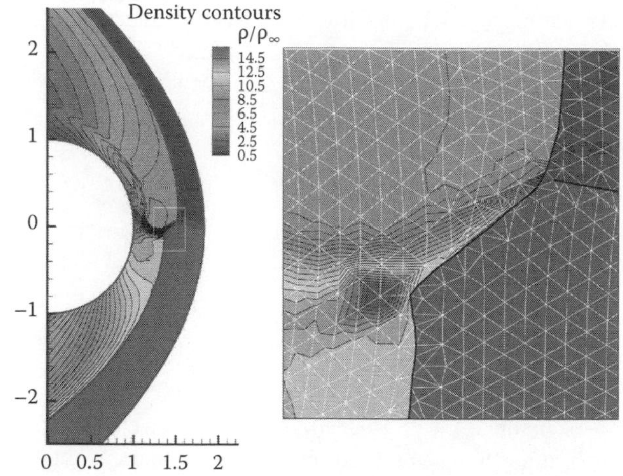


**FIGURE 6.27:** Computed Type IV shock interaction. The bow shock, incident shock and the shock connecting the two triple points are fitted. The jet is captured. Right panel shows details around shock interaction. Results presented courtesy of R. Paciorri. (From Paciorri, R. and Bonfiglioli, A., Numerical simulation of shock interactions with an unstructured shock-fitting technique, in *Sixth European Symposium on Aerothermodynamics for Space Vehicles*, Paris, France, 2008. With permission.)

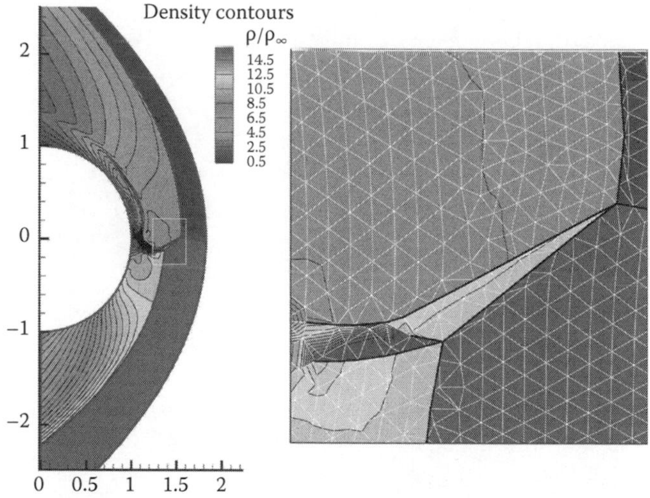


**FIGURE 6.28:** Computed Type IV shock interaction using fully shock-fitting mode. Right panel shows details around shock interaction. Results presented courtesy of R. Paciorri. (From Paciorri, R. and Bonfiglioli, A., Numerical simulation of shock interactions with an unstructured shock-fitting technique, in *Sixth European Symposium on Aerothermodynamics for Space Vehicles*, Paris, France, 2008. With permission.)

cylinder. The calculation used 10855 nodes, corresponding to 21127 triangles. The figures, beginning with a full shock-capturing mode in Figure 6.25 and ending in a full shock-fitting mode in Figure 6.28, show several levels of shock-capturing and shock-fitting. The full shock-capturing mode, Figure 6.25, at best, appears as a grotesque caricature of the actual flow field.

---

## Problems

**6.1** On the domain $0 \leq x \leq 1$, $0 \leq y \leq 1$, solve the advection equation (6.12) using the N scheme on a regular triangular mesh created from a Cartesian mesh by splitting the rectangular cells with diagonals all directed in the same direction. Let $\lambda_x = y$ and $\lambda_y = \frac{1}{2} - x$ and impose the following boundary conditions [55]:

$$\begin{aligned} u(0, y) &= 0, \\ u(x, 1) &= 0, \; x \geq 0.5, \\ u(x, 0) &= \begin{cases} 0, & x \leq 0.175, \\ 1, & 0.175 < x < 0.325, \\ 0, & 0.325 \geq x < 0.5. \end{cases} \end{aligned}$$

**6.2** Let the shape of a shock front be defined by $y = f(x)$. With the same notation as in Figure 6.10, show that the leading error term in the slope of (6.44), evaluated at $x_i$, is

$$\frac{f_{xx,i}}{4}\left((x_{i-1} - x_i) - (x_i - x_{i+1}) - (x_{i-1} - x_i)(x_i - x_{i+1})\frac{f_{x,i} f_{xx,i}}{\sqrt{1 + f_{x,i}^2}}\right).$$

A more general analysis can be done by Taylor expanding the position vector as a function of arc length. But the above analysis suffices for the leading error term.

**6.3** Following the same notation as in Problem 6.2, (a) show that the leading error term in the slope of the tangent vector defined by [4]:

$$\boldsymbol{v}_\tau = \Delta \boldsymbol{r}_{i,i+1} h_{i-1/2}^2 + \Delta \boldsymbol{r}_{i-1,i} h_{i+1/2}^2, \tag{6.54}$$

evaluated at $x_i$, is

$$\frac{f_{xx,i}^2 (x_{i-1} - x_i)(x_i - x_{i+1})}{2(1 + f_{x,i}^2)}\left(2 f_{x,i} + \frac{f_{xx,i}}{4}[(x_{i-1} - x_i) - (x_i - x_{i+1})]\right).$$

Equation 6.54 is the only linear combination of the vectors $\Delta \boldsymbol{r}_{i,i+1}$ and $\Delta \boldsymbol{r}_{i-1,i}$ that is second order accurate. (b) Prove that (6.54) is the tangent at $i$ to a circle passing through the nodes $i-1$, $i$, and $i+1$ of Figure 6.10. (c) Show that the one-sided approximation to the tangent vector:

$$\nu_\tau = \Delta \boldsymbol{r}_{i-1,i}\left(h^2_{i-1/2} + h^2_{i-3/2}\right) - \Delta \boldsymbol{r}_{i-2,i} h^2_{i-1/2}, \tag{6.55}$$

is second order accurate.

**6.4** The curvature vector at a point $i$ for a circle passing through the points $i-1$, $i$ and $i+1$ is given by

$$\vec{\kappa}_{c,i} = \frac{2}{h_{i+1/2} + h_{i-1/2}}\left(\frac{\Delta \boldsymbol{r}_{i,i+1}}{h_{i+1/2}} - \frac{\Delta \boldsymbol{r}_{i-1,i}}{h_{i-1/2}}\right), \tag{6.56}$$

and the curvature, the inverse of the radius of the circle, see Equation 6.4, is $\kappa_{c,i} = |\vec{\kappa}_{c,i}|$. If a curve is expressed using of polar coordinates $r, \theta$, then its curvature is $\kappa = (r^2 + 2r_\theta^2 - r r_{\theta\theta})/(r^2 + r_\theta^2)^{\frac{3}{2}}$. Consider the ellipse defined by

$$\left(\frac{r\cos\theta}{a}\right)^2 + \left(\frac{r\sin\theta}{b}\right)^2 = 1.$$

Let $a = 2$ and $b = 1$. In the quadrant $\theta = [0°, 90°]$ evaluate and compare the curvatures $\kappa_c$ and $\kappa$ by taking $\theta$ intervals $[10°, 5°, 2°]$.

If $s$ is the arc length along some curve, then tangent and normal vectors to the curve are related to the curvature by: $d\boldsymbol{\tau}/ds = \kappa \boldsymbol{n}$ and $d\boldsymbol{n}/ds = -\kappa\boldsymbol{\tau}$.

**6.5** (a) If the shock front is defined by $y = f(x,t)$, then the position vector to the shock front is $\boldsymbol{r} = x\vec{i}_x + f(x,t)\vec{i}_y$, $\boldsymbol{\tau} = -\partial \boldsymbol{r}/\partial x / |\partial \boldsymbol{r}/\partial x|$, and the shock front speed is $w = \boldsymbol{n}\cdot\partial\boldsymbol{r}/\partial t$. Show that

$$\partial \boldsymbol{n}/\partial t = \left(-w_x \boldsymbol{\tau}\cdot\vec{i}_x + w\partial\left(\boldsymbol{\tau}\cdot\vec{i}_x\right)/\partial x\right)\boldsymbol{\tau}. \tag{6.57}$$

(b) If the shock front is defined by $\boldsymbol{r} = x(s,t)\vec{i}_x + y(s,t)\vec{i}_y$, where $s$ is the arc length along the shock front, find the expression for $\partial\boldsymbol{n}/\partial t$.

**6.6** In Section 4.1.5, we found the exact solution to the simple wave region created by an accelerating piston (function `EXACT.M` in the `PISTON.M` code). The solution was found in the form $u = u(x,t)$. Express the solution as $u = u(x,y,t)$ by extending the piston in the $y$ direction. Then, on a regular triangular mesh created from a Cartesian mesh by splitting the rectangular cells with diagonals all directed in the same direction, use the exact solution (no unstructured flow solver is needed) to test the shock detection algorithm proposed in Section 6.5.9. Repeat this exercise by rotating the solution coordinates with respect to the mesh coordinates.

# *Appendix A: Codes Available in CD*

| Code | Chapter/Section | Comment |
|---|---|---|
| EXACTGAUSS.M | 2.6 | Study of shock formation with a Gaussian initial profile. |
| DETECTSHOCK.M | 2.9 | Study of shock detection using a cubic fit. |
| SHOCKCAP01.M | 2.3.16 | Integrates Burger's equation with shock capturing Runge–Kutta and Roe's approximate Riemann solver. |
| SHOCKFIT01.M | 2.5 | Uses shock-boundary fitting with Runge–Kutta and Roe's approximate Riemann solver to solve Burger's equation. Single region. |
| SHOCKFIT04.M | 2.7 | Uses shock-boundary fitting with Runge–Kutta and Roe's approximate Riemann solver, also non-conservative MacCormack scheme to solve Burger's equation. Uses two regions. |
| SHOCKFIT07.M | 2.9 | Floating shock-fitting. Solves Burger's equation. |
| SHOCKFIT12.M | 2.10 | Floating shock-fitting. Solves two non-conservative p.d.e.'s. |
| SHOCKFIT13.M | 2.10 | Floating shock-fitting. Solves two non-conservative p.d.e.'s, slightly different from SHOCKFIT12.M. |
| SHOCK_LAYER.M | 3.7 | One-dimensional, steady, Navier–Stokes shock layer. |
| PISTON.M | 4.3.2 | Piston driven flows, floating shock-fitting. |
| Q1D.M | 4.4.4 | Quasi-one-dimensional flows, floating shock-fitting. |

*(continued)*

**(continued)**

| Code | Chapter/Section | Comment |
|---|---|---|
| BLUNT.M | 5.1.9 | Blunt body code, 2D and axisymmetric, boundary shock-fitting. |
| CORNER_MU.M | 5.2.13 | External corner code, symmetric corner, boundary shock-fitting. |
| CONE.M | See "Problem" section in Chapter 5 | Potential solution compared to slender body theory. See Problem 5.6. |
| SUPERWING.M | 5.3.7 | Steady supersonic marching code for elliptical cross section wings. |

# References


[1] Adamson Jr., T. C., Messiter, A. F., and Liou, M. S. 1978. Large amplitude shock-wave motion in two-dimensional transonic channel flow. *AIAA J.* 16:1240–1247.

[2] Airy, G. B. 1841. Tides and waves. In *Encyclopedia Metropolitana*. Fellowes, London, U.K.

[3] Andronov, A. A., Leontovich, E. A., Gordon, J. J., and Maier, A. G. 1973. *Qualitative Theory of Second-Order Dynamic Systems*. Wiley, New York.

[4] Anoshkina, E. V., Belyaev, A. G., and Seidel, H.-P. 2002. Asymptotic analysis of three-point approximations of vertex normals and curvature. In *Vision, Modeling, and Visualization Proceedings*, Erlangen, Germany, pp. 211–216. IOS Press, Amsterdam, the Netherlands.

[5] Arbogast, L. F. A. 1791. Mémoire sur la nature des fonctions arbitraires qui entrent dans les intégrales des equations aux différences partielles. St. Petersburg, Russia.

[6] Bakker, P. G. 1983. On bifurcation phenomena in conical flow patterns. Delft University of Technology, Report LR-378. Delft, the Netherlands.

[7] Bakker, P. G. 1991. *Bifurcation in Flow Patterns*. Kluwer Academic Publishers, Dordrecht, the Netherlands.

[8] Bakker, P. G. and Bannink, 1974. Conical stagnation points in the supersonic flow around slender circular cones at incidence. Delft University of Technology, Report VTH-184. Delft, the Netherlands.

[9] Bakker, P. G. and Reyn, J. W. 1985. Conical flow near external axial corners as a bifurcation problem. *AIAA J.* 23:4–11.

[10] Bakker, P. G., Bannink, W. J., and Reyn, J. W. 1981. Potential flow near conical stagnation points. *J. Fluid Mech.* 105:239–260.

[11] Bannink, W. J. 1984. Investigation of the conical flowfield around external axial corners. *AIAA J.* 22:354–360.

[12] Barth, T. J. 1992. Aspects of unstructured grids and finite-element volume solvers for the Euler and Navier-Stokes equations. von Karman Inst., Lect. Ser., AGARD Publ. R-787.

[13] Batchelor, G. K. 1967. *An Introduction to Fluid Dynamics*. Cambridge University. Press, Cambridge, U.K.

[14] Belotserkovskii, O. M. 1958. On the numerical calculation of detached bow shock waves in hypersonic flow. *Prikl. Mat. Mekh.* 22:206–219. Transl. in *J. Appl. Math. Mech.* 22:279–296.

[15] Ben-Dor, G. 2007. *Shock Wave Reflection Phenomena*. Springer, Berlin.

[16] Biagioni, H. A. 1980. A nonlinear theory of generalized functions. *Lect. Notes Math.* 1421, 2nd ed. Springer-Verlag, New York.

[17] Bird, G. A. 1970. Direct simulation of the Boltzmann equation. *Phys. Fluids* 13:2672–2681.

[18] Bonfiglioli, A. 2000. Fluctuation splitting schemes for the compressible and incompressible Euler and Navier-Stokes equations. *Int. J. Comput. Fluid D.* 14:21–39.

[19] Bonfiglioli, A. and Paciorri, R. 2009. Private Communication.

[20] Boris, J. P. and Book, D. L. 1973. Flux-corrected transport, I: SHASTA, a fluid transport algorithm that works. *JCP* 11:38–69.

[21] Boyce, W. M. and DiPrima, R. C. 1965. *Elementary Differential Equations and Boundary Value Problems*. John Wiley & Sons, Inc., Hoboken, NJ, p. 20.

[22] Burgers, J. M. 1948. A mathematical model illustrating the theory of turbulence. *Adv. Appl. Mech.* 1:171–199.

[23] Burstein, S. Z. 1964. Numerical methods in multidimensional shocked flows. *AIAA J.* 2:2111–2117.

[24] Busemann, A. 1929. Drüke auf kegelförmige spitzen bei bewegung mit überschallgeschwindigkeit. *Z. Angew. Math. Mech.* 9:496–498.

[25] Butler, D. S. 1960. The numerical solution of hyperbolic systems of partial differential equations in three independent variables. *Proc. R. Soc. A* 255:232–252.

[26] Candler, G. V., Barnhardt M. D., Drayna T. W., Nompelis, I., et al. 2007. Unstructured grid approaches for accurate aeroheating simulations. AIAA Paper No. 2007–3959.

[27] Cauchy, A. 1827. Théorie de la propagation des ondes à la surface d'un fluide pesant d'une profondeur indéfinite. *Œuvres complètes D'Augustin Cauchy*. I^RE Série, Tome I, Académie des Sciences, Paris, France, pp. 5–318.

[28] Cauchy, A. 1899. *Œuvres complètes D'Augustin Cauchy*. I^RE Série, Tome XI, Académie des Sciences, Paris, France, p.178.

[29] Cauret, J. J., Colombeau, J. F., and Le Roux, A. Y. 1989. Discontinuous generalized solutions of nonlinear nonconservative hyperbolic equations. *J. Math. Anal. Appl.* 139:552–573.

[30] Chakravarthy, S. R., Anderson, D. A., and Salas, M. D. 1980. The split-coefficient method for hyperbolic systems of gas dynamic equations. AIAA Paper No. 1980-0268.

[31] Challis, J. 1848. On the velocity of sound. *Philos. Mag.* 32(III):494–499.

[32] Chapman, S. and Cowling, T. G. 1952. *The Mathematical Theory of Non-Uniform Gases*. Cambridge University Press, New York.

[33] Chew, L. P. 1987. Constrained Delaunay triangulation. In *Proceedings of the Third Annual Symposium on Computational Geometry*, Waterloo, ON, pp. 215–222.

[34] Chorin, A. J. and Marsden, J. E. 1992. *A Mathematical Introduction to Fluid Mechanics*, 3rd edn. Springer-Verlag, New York.

[35] Christoffel, E. B. 1877. Über die fortpflanzung von stössen durch elastische feste körper. *Annali di Matematica Pura e Applicata* 8:193–244.

[36] Cole, J. D. and Cook, L. P. 1986. *Transonic Aerodynamics*. North-Holland, Amsterdam, the Netherlands.

[37] Colombeau, J. F. 1985. *Elementary Introduction to New Generalized Functions. North-Holland Math. Studies 113*. North-Holland, Amsterdam, the Netherlands.

[38] Coulombel, J.-F., Benzoni-Gavage, S., and Serre, D. 2002. Note on a paper by Robinet, Gressier, Casalis & Moschetta. *J. Fluid Mech.* 469:401–405.

[39] Courant, R. 1943. Variational methods for the solution of problems of equilibrium and vibration. *Bull. Am. Math. Soc.* 49:1–43.

[40] Courant, R. and Friedrichs, K. O. 1944. Supersonic flow and shock waves, a manual on the mathematical theory of non-linear wave motion. *Appl. Math. Panel, Nat. Def. Res. Comm.*, AMP Report 38.2R.

[41] Courant, R. and Friedrichs, K. O. 1948. *Supersonic Flow and Shock Waves*. Interscience Publishers, Inc., New York.

[42] Courant, R. and Hilbert, D. 1966. *Methods of Mathematical Physics*. Vol. II, 3rd edn. Interscience Publishers, John Wiley & Sons, New York.

[43] Courant, R., Friedrichs, K. O., and Lewy, H. 1928. Über die partiellen differenzengleichungen der mathematischen physik. *Math. Ann.* 100:32–74.

[44] Courant, R., Friedrichs, K. O., and Lewy, H. 1967. On the partial difference equations of mathematical physics. *IBM J.* 11:215–234.

[45] Courant, R., Isaacson, E., and Rees, M. 1952. On the solution of nonlinear hyperbolic differential equations by finite differences. *Commun. Pure Appl. Math.* 5:243–255.

[46] Craik, A. D. D. 2004. The origins of water wave theory. *Annu. Rev. Fluid Mech.* 36:1–28.

[47] Craik, A. D. D. 2005. George Gabriel Stokes water wave theory. *Annu. Rev. Fluid Mech.* 37:23–42.

[48] Crocco, L. 1937. Eine neue stromfunktion für die erforschung der bewegung der gas emit rotation. *ZAMM* 17:1–7.

[49] Csík, Á., Ricchiuto, M., and Deconinck, H. 2002. A conservative formulation of the multidimensional upwind residual distribution schemes for general nonlinear conservation laws. *JCP* 179:286–312.

[50] d'Alembert, J. R. 1747. Recherche sur la courbe que forme une corde tendue mise en vibration. *Hist. Acad. Sci. Berlin* 3:214–219.

[51] d'Alembert, J. R. 1761. Recherches sur les vibrations des cordes sonores. *Opuscules Math.* 1:1–73.

[52] Darrigol, O. 2005. *Worlds of Flows: A History of Hydrodynamics from the Bernoullis to Prandtl*. Oxford University Press, New York.

[53] de Neef, T. and Hechtman, C. 1978. Numerical study of the flow due to a cylindrical implosion. *Comput. Fluids* 6:185–202.

[54] de Neef, T. and Moretti, G. 1980. Shock fitting for everybody. *Comput. Fluids* 8:327–334.

[55] Deconinck, H., Powell, K. G., Roe, P. L., and Struijs, R. 1991. Multi-dimensional schemes for scalar advection. AIAA Paper No. 1991-1532-CP.

[56] Deconinck, H., Roe, P. L., and Struijs, R. 1993. A multidimensional generalization of Roe's flux difference splitter for the Euler equations. *Comput. Fluids* 22:215–222.

[57] Delaunay, B. N. 1934. Sur la sphère vide. Izvestia *Akademia Nauk SSSR VII Seria, Otdelenie Matematicheskii I Estestvennyka Nauk* 7:793–800.

[58] Di Giacinto, M. and Valorani, M. 1989. Shock detection and discontinuity tracking for unsteady flows. *Comput. Fluids* 17:61–84.

[59] Diskin, B., et al. 2009. Comparison of node-centered, and cell-centered unstructured finite-volume discretizations, Part I: Viscous fluxes. AIAA Paper No. 2009-597.

[60] Dorodnitsyn, A. A. 1957. On a method of numerical solution of some nonlinear problems of aero-hydrodynamics. In *Proceedings of the Ninth*

*International Congress on Applied Mechanics I*, 485, University of Brussels, Brussels, Belgium.

[61] Duhem, P. 1891. *Hydrodynamique, Élasticité, Acoustique. Tome Premier.* A. Hermann, Paris.

[62] Earnshaw, S. 1860. On the mathematical theory of sound. *Phil. Trans. R. Soc. Lond.* 150:133–148. Reproduced in: [109].

[63] Edney, B. 1968. Anomalous heat transfer, and pressure distributions on blunt bodies at hypersonic speeds in the presence of an impinging shock. Aeronautical Research Institute of Sweden, FAA Report 115, Stockholm, Sweden.

[64] Emmons, H. W. 1944. The numerical solution of compressible fluid flow problems. NACA Tech. Note No. 932.

[65] Emmons, H. W. 1948. Flow of a compressible fluid past a symmetrical airfoil in a wind tunnel and in free air. NACA Tech. Note No. 1746.

[66] Engquist, B. and Osher, S. 1980. Stable and entropy satisfying approximations for transonic flow calculations. *Math. Comput.* 34:45–75.

[67] Euler, L. 1748. Sur la vibration des cordes. *Mém. Acad. Sci. Berlin* 4:69–85.

[68] Euler, L. 1757. Continuation des recherches sur la theorie du mouvement des fluides. *Mém. Acad. Sci. Berlin* 11:316–361.

[69] Euler, L. 1757. Principes généraux du movement des fluids. *Mém. Acad. Sci. Berlin* 11:274–315.

[70] Euler, L. 1761. Principia motus fluidorum. *Novi Commentarii Academiae Scientiarum Petropolitanae* 6:271–311. English translation: http://arxiv.org/PS_cache/arxiv/pdf/0804/0804.4802v1.pdf

[71] Euler, L. and translator: Blanton, J. D. 1988. *Introduction to Analysis of the Infinite, Book I.* Springer-Verlag, New York.

[72] Euler, L. and translator: Blanton, J. D. 2000. *Foundations of Differential Calculus*. Springer-Verlag, New York.

[73] Faddeeva, V. N. 1949. The method of lines applied to some boundary value problems. *Tr. Mat. Inst. Steklova, Acad. Nauk. SSSR* 28:73–103.

[74] Farassat, F. 1994. Introduction to generalized functions with applications in aerodynamics and aeroacoustics. NASA TP. 3428.

[75] Ferri, A., Vaglio-Laurin, R., and Ness, N. 1954. Mixed type conical flow without axial symmetry. Polytechnic Institute of Brooklyn, PIBAL Report No. 264.

[76] Folkerts, M. and Lindgren, U. eds. 1985. *Mathemata. Festschrift für Helmuth Geriecke*, Munich Steiner, pp. 493–510.

[77] Fox, R. 1971. *The Caloric Theory of Gases from Lavoisier to Regnault*. Clarendon Press, Oxford, U.K.

[78] Friedrichs, K. O. 1958. Asymptotic phenomena in mathematical physics. *Bull. Am. Math. Soc.* 61:485–504.

[79] Gabutti, B. 1983. On two upwind finite-difference schemes for hyperbolic equations in non-conservative form. *Comput. Fluids* 11:207–230.

[80] Gadd, G. E. 1960. The possibility of normal shock waves on a body with convex surfaces in inviscid transonic flow. *Z. Angew. Math. Phys.* 11:51–55.

[81] Geuzaine, P. and Farhat, C. 2003. Design and time-accurate analysis of ALE schemes for inviscid and viscous flow computations on moving meshes. AIAA Paper No. 2003–3694.

[82] Gilbarg, D. and Paolucci, D. 1953. The structure of shock waves in the continuum theory of fluids. *J. Ration. Mech. Anal.* 2:617–642.

[83] Glass, I. I. and Sislian, J. P. 1994. *Nonstationary Flows and Shock Waves*. Clarendon Press, Oxford, U.K.

[84] Glimm, J. and Lax, P. D. 1970. Decay of solutions of systems of nonlinear hyperbolic conservation laws. *Mem. Am. Math. Soc.* 101:204–214.

[85] Gnoffo, P. and White, J. A. 2004. Computational aerothermodynamic simulation issues on unstructured grids. AIAA Paper No. 2004–2371.

[86] Godunov, S. K. 1959. A finite-difference method for the numerical computation of discontinuous solutions of the equations of fluid dynamics. *Math. Sbornik* 47:271–306.

[87] Godunov, S. K. 1999. Reminiscences about difference schemes. *JCP* 153:6–25.

[88] Goldstein, S., ed. 1965. *Modern Developments in Fluid Dynamics*, Vol. I, Dover Publications, Inc., New York.

[89] Goodman, J. B. and LeVeque, R. J. 1985. On the accuracy of stable schemes for 2D scalar conservation laws. *Math. Comput.* 45:15–21.

[90] Grattan-Guinness, I. 2005. The École Polytechnique, 1794–1850: Differences over educational purpose, and teaching practice. *Am. Math. Mon.* 112:233–250.

[91] Griffith, W. C. 1981. Shock waves. *J. Fluid Mech.* 106:81–101.

[92] Gustafsson, B., Kreiss, H.-O., and Oliger, J. 1995. *Time Dependent Problems and Difference Methods*. John Wiley & Sons, Inc., New York.

[93] Hadamard, J. 1903. *Leçons Sur La Propagation Des Ondes Et Les Équations De L'hydrodynamique*. A. Hermann, Paris.

[94] Hadamard, J. 1964. *La théorie des équations aux dérivées partielles*. Editions Scientifiques, Pekin.

[95] Harten, A. 1983. High resolution schemes for hyperbolic conservation laws. *JCP* 49:357–393.

[96] Harten, A. and Osher, S. 1987. Uniformly high-order accurate nonoscillatory schemes. I. *SIAM J. Numer. Anal.* 24:279–309.

[97] Hawking, S. W. and Ellis, G. F. 1973. *The Large Scale Structure of Space-Time*. Cambridge University Press, Cambridge, U. K., p. 362.

[98] Hayes, W. D. 1957. The vorticity jump across a gas dynamic discontinuity. *J. Fluid Mech.* 2:595–600.

[99] Hayes, W. D. and Probstein, R. F. 1966. *Hypersonic Flow Theory*. 2nd ed. Academic Press, New York.

[100] Holt, M. 1949. Flow patterns, and the method of characteristics near a sonic line. *Q. J. Mech. Appl. Math.* 11(Pt. 2):246–256.

[101] Hughes, T. J. R. 1987. Recent progress in the development, and understanding of SUPG methods with special reference to the compressible Euler and Navier-Stokes equations. *Int. J. Numer. Meth. Fl.* 7:1261–1275.

[102] Hugoniot, P. H. 1887. Mémoire sur la propagation du movement dans les corps et plus spécialement dans les gaz parfaits, 1$^{e}$ Partie. *J. École Polytech. Paris* 57:3–97. Transl. in: [109].

[103] Hugoniot, P. H. 1889. Mémoire sur la propagation du movement dans les corps et plus spécialement dans les gaz parfaits, 2$^{e}$ Partie. *J. École Polytech. Paris* 58:1–125. Transl. in: [109].

[104] Hussaini, M. Y., van Leer, B., and Van Rosendale, J., eds. 1997. *Upwind and High-Resolution Schemes*. Springer-Verlag, Berlin.

[105] Iollo A. and Salas, M. D. 1999. On the propagation of small perturbations in two simple aeroelastic systems. *J. Sound Vib.* 222:152–162.

[106] Jameson, A. and Lax, P. D. 1986. Conditions for the construction of multi-point total variation diminishing difference schemes. *Appl. Numer. Math.* 2:335–345.

[107] Jaynes, E. T. 1980. The minimum entropy production principle. *Annu. Rev. Phys. Chem.* 31:579–601.

[108] Jerri, A. J. 1998. *The Gibbs Phenomenon in Fourier Analysis, Splines and Wavelet Approximations*. Kluwer Academic Publishers, Boston, MA.

[109] Johnson, J. N. and Chéret, eds. 1998. *Classic Papers in Shock Compression Science*. Springer, New York.

[110] Jones, D. J. 1969. Tables of inviscid supersonic flow about circular cones at incidence, $\gamma = 1.4$. AGARDograph 137.

[111] Jouguet, E. 1934. Résumé des théories sur la propagation des explosions. *La science aériene* 3:138–155.

[112] Jourdain, P. E. B. 1913. The origins of Cauchy's conception of a definite integral and of the continuity of a function. *Isis* 1:661–703.

[113] Kentzer, C. 1970. Discretization of boundary conditions on moving discontinuities. In *Proceedings of the Second International Conference on Numerical Methods for Fluid Dynamics*, Lafayette, IA, pp. 108–113.

[114] Kreiss, H.-O. and Lorenz, J. 1989. *Initial-Boundary Value Problems and the Navier-Stokes Equations*. Academic Press, Inc., Boston, MA.

[115] Krustal, M. D. 1974. In *Nonlinear Wave Motion*, ed. A. C. Newell, *Lect. Appl. Math.* 15:61–83, American Mathematical Society, Providence, RI.

[116] Kulikovskii, A. G., Pogorelov, N. V., and Semenov, A. Y. 2001. *Mathematical Aspects of Numerical Solution of Hyperbolic Systems*. Chapman & Hall/CRC, Boca Raton, FL.

[117] Lagrange, J. 1759. Recherches sur la nature et la propagation du son. *Miscellanca Taurinensia* 1:39–148.

[118] Lagrange, J. L. 1760. Nouvelles recherches sur la nature et la propagation du son. *Miscellanca Taurinensia* 2:151–332.

[119] Lamb, H. 1932. *Hydrodynamics*. 6th ed. Dover Publication, New York.

[120] Laplace, P.-S. 1816. Sur la vitesse du son dans l'air et dans l'eau. *Ann. Chim. Phys.* 3:238–241. Transl. in: Lindsay, R. B. ed. 1972. *Acoustics: Historical and Philosophical Development*, pp. 181–182. Dowden, Hutchinson, & Ross, Stroudsburg, PA.

[121] Lax, P. D. 1957. Hyperbolic systems of conservation laws, II. *Commun. Pure Appl. Math.* 10:537–566.

[122] Lax, P. D. and Wendroff, B. 1960. System of conservation laws. *Commun. Pure Appl. Math.* 13:217–237.

[123] LeFloch, P. G. 2002. *Hyperbolic Systems of Conservation Laws: The Theory of Classical and Nonclassical Shock Waves (Lectures in Mathematics)*. Birkhäuser, Basel, Switzerland.

[124] Lemmon, E. W. and Jacobsen, R. T. 2004. Viscosity and thermal conductivity equations for nitrogen, oxygen, argon and air. *Int. J Thermophys.* 25:21–69.

[125] Lieb, E. H. 1982. Comment on "Approach to equilibrium of a Boltzmann-equation solution." *Phys. Rev. Lett.* 48:1057.

[126] Lighthill, M. J. 1947. The shock strength in supersonic conical fields. *Philos. Mag. Ser. 7* 40:1202–1223.

[127] Lighthill, M. J. 1954. Mathematical methods in compressible flow theory. *Commun. Pure Appl. Math.* 7:1–10.

[128] Lighthill, M. J. 1956. Viscosity effects in sound waves of finite amplitude. In *Surveys in Mechanics*, eds. G. K. Batchelor and R. M. Davies, pp. 250–351. Cambridge University Press, New York.

[129] Lighthill, M. J. 1960. *An Introduction to Fourier Analysis and Generalized Functions*. Cambridge University Press, New York.

[130] Liu, T.-P. 1982. Nonlinear stability and instability of transonic flows through a nozzle. *Commun. Math. Phys.* 83:243–260.

[131] Lützen, J. 1982. *The Prehistory of the Theory of Distributions*. Springer-Verlag, New York.

[132] Lützen, J. 1983. Euler's vision of a general partial differential calculus for a generalized kind of function. *Math. Mag.* 56:299–306.

[133] Luzin, N. 1930. Function. *The Great Soviet Encyclopedia* 59:314–334. Transl. in: *Am. Math. Mon.*, (January) 1998.

[134] MacCormack, R. W. 1969. The effect of viscosity in hypervelocity impact cratering. AIAA Paper No. 1969-354.

[135] Marconi, F. 1983. Shock reflection transition in three-dimensional steady flow about interfering bodies. *AIAA J.* 21:707–713.

[136] Marconi, F. and Salas, M. D. 1973. Computation of three-dimensional flows about aircraft configurations. *Comput. Fluids* 1:185–195.

[137] Marconi, F., Salas, M. D., and Yaeger, L. 1976. Development of a computer code for calculating the steady super/hypersonic inviscid flow around real configurations, Vol. I—Computational technique. NASA CR-2675.

[138] Matano, J. 1982. Nonincrease of the lap-number of a solution for a one-dimensional semilinear parabolic equation. *J. Fac. Sci., Univ. Tokyo, Sec. I A, Math.* 29:401–441.

[139] Mavriplis, D. J. 1997. Unstructured grid techniques. *Annu. Rev. Fluid Mech.* 29:473–514.

[140] Mays, R. A. 1971. Inlet dynamics, and compressor surge. *J. Aicraft* 8:219–226.

[141] Messiter, A. F. and Adamson Jr., T. C. 1984. Forced oscillations of transonic channel and inlet flows with shock waves. *AIAA J.* 22:1590–1599.

[142] Monge, G. 1773. Mémoire sur la construction des fonctions arbitraries dans les intégrals des équations aux différences partielles. *Mémoires des mathématiques et de physique présentés a l'Académie...par divers sçavans...* **7**, 2$^e$:267–300.

[143] Morduchow, M. and Libby, P. A. 1949. On the complete solution of the one-dimensional equations of a viscous heat conducting compressible gas. *J. Aerosp. Sci.* 16:674–684.

[144] Morduchow, M. and Libby, P. A. 1965. On the distribution of entropy through a shock wave. *J. Mécanique* 4:191–213.

[145] Moretti, G. 1968. Inviscid blunt body shock layers: Two-dimensional symmetric, and axisymmetric flows. Polytechnic Institute of Brooklyn, PIBAL Report No. 68-15.

[146] Moretti, G. 1969. A critical analysis of numerical techniques: The piston-driven inviscid flow. Polytechnic Institute of Brooklyn, PIBAL Report No. 69-25.

[147] Moretti, G. 1973. Experiments in multi-dimensional floating shock-fitting. Polytechnic Institute of Brooklyn, PIBAL Report No. 73-18.

[148] Moretti, G. 1974. A pragmatical analysis of discretization procedures for initial- and boundary-value problems in gas dynamics and their influence on accuracy or Look ma, no wiggles! Polytechnic Institute of New York, POLY-AE/AM Report No. 74-15.

[149] Moretti, G. 1976. Conformal mappings for computations of steady three-dimensional, supersonic flows. In *Numerical/Laboratory Computer Methods in Fluid Mechanics*, eds. A. A. Pouring and V. I. Shah, pp. 13–28. ASME, New York.

[150] Moretti, G. 1979. The $\lambda$-scheme. *Comput. Fluids* 7:191–205.

[151] Moretti, G. 1982. Calculation of three-dimensional inviscid supersonic, steady flows. NASA CR-3573.

[152] Moretti, G. 1987. Computation of flows with shocks. *Ann. Rev. Fluid Mech.* 19:313–337.

[153] Moretti, G. 2000. Braggadocio, memoirs of America. Private communication.

[154] Moretti, G. and Abbett, M. 1966. A time-dependent computational method for blunt body flows. *AIAA J.* 4:2136–2141.

[155] Moretti, G. and DiPiano, M. T. 1983. An improved lambda-scheme for one-dimensional flows. NASA CR-3712.

[156] Morgan, K., Peraire, J., and Peiró, J. 1992. Unstructured grid methods for compressible flows. von Karman Inst, Lect. Ser., AGARD Publ. R-787.

[157] Oberguggenberger, M. 1992. *Multiplication of Distributions, and Applications to Partial Differential Equations*. Longman Scientific & Technical, London, U.K.

[158] Oppenheim, A. K., Urtiew, P. A., and Stern, R. A. 1959. Peculiarity of shock impingement on area convergence. *Phys. Fluids* 2:427–431.

[159] Osher, S. and Chakravarthy, S. 1984. Very high order accurate TVD schemes. ICASE Report No. 84-44,

[160] Oswatitsch, K. and Zierep, J. 1960. Das problem des senkrechten stosses an einer gekrümmten, Wand. *Z. Angew. Math. Mech.* Supplement:143–144.

[161] Oswatitsch, K. 1945. The drag as integral of the entropy flow. *Nachrichten der Akademie der Wissenschftenm in Göttingen, Mathemattisch-Physikalische* 1:88–90.

[162] Paciorri, R. and Bonfiglioli, A. 2008. Numerical simulation of shock interactions with an unstructured shock-fitting technique. In *Sixth European Symposium on Aerothermodynamics of Space Vehicles*, Paris, France.

[163] Paciorri, R. and Bonfiglioli, A. 2009. A shock-fitting technique for 2D unstructured grids. *Comput. Fluids* 38:715–726.

[164] Pandolfi, M. 1987. Upwind formulation for the Euler equations. von Karman Inst. Fld. Dyn., CFD Lecture Series 1987-04.

[165] Pandolfi, M. and D'Ambrosio, D. 2001. Numerical instabilities in upwind methods: Analysis and cures for the "Carbuncle" phenomenon. *JCP* 166:271–301.

[166] Penrose, R. 2005. *The Road to Reality: A Complete Guide to the Laws of the Universe*. Alfred A. Knopf, New York.

[167] Poisson, S. D. 1808. Mémoire sur la théorie du son. *J. École Polytech. Paris* 7:319–392. Transl. in: [109].

[168] Poisson, S. D. 1823. Sur la chaleur des gaz et des vapeurs. *Ann. Chim. Phys.* 23:337–353. Transl. in: Herapath, J. 1823. On the caloric of gases and vapours. *Philos. Mag.* 62:328–338.

[169] Polachek, H. and Seeger, R. J. 1951. On shock-wave phenomena: Refraction of shock waves at a gaseous interface. *Phys. Rev.* 84:922–930.

[170] Prandtl, L. 1906. Zur theorie des verdichtungsstosses. *Zeitschrift für der geramte Turbinen-Wesen* 3:241.

[171] Priestley, H. A. 1990. *Introduction to Complex Analysis*. Clarendon Press, Oxford.

[172] Rankine, W. J. M. 1851. On Laplace's theory of sound. *Philos. Mag.* 1(IV): 225–227.

[173] Rankine, W. J. M. 1870. On the thermodynamic theory of waves of finite longitudinal disturbances. *Philos. Trans. R. Soc. Lond.* 160:277–286. Reproduced in: [109].

[174] Rayleigh, L. 1910. Aerial plane waves of finite amplitude. *Proc. R. Soc. London.* A84:247–284. Reproduced in: [109].

[175] Reynolds, O. 1883. An experimental investigation of the circumstances which determine whether the motion of the water shall be direct or sinuous and of the law of resistance in parallel channels. *Philos. Trans. R. Soc.* 174:935–982.

[176] Riemann, B. 1860. Über die fortpflanzung ebener luftwellen von endlicher schwingungsweite. *Abhandlungen der Gesellschaft der Wissenschaften zu Göttingen, Mathematisch-physikalische Klasse* 8:43. Transl. in: [109].

[177] Ringleb, F. 1941. Exacte lösungen der differentialgleichunsen eineradiabatischen gasströmung. *ZAMM* 20:185–198.

[178] Rizzi, A. and Viviand, H. eds. 1981. Numerical methods for the computation of inviscid transonic flows with shock waves. Notes on Numerical Fluid Mechanics, 3, Friedr. Vieweg & Sohn, Braunschweig.

[179] Roache, P. J. 1998. *Verification and Validation in Computational Science and Engineering*. Hermosa Publishers, Albuquerque, NM.

[180] Roe, P. L. 1972. A result concerning the supersonic flow below a plane delta wing. ARC C.P. No. 1228.

[181] Roe, P. L. 1981. Approximate Riemann solvers, parameter vectors and difference schemes. *JCP* 43:357–372.

[182] Roe, P. L. 1990. "Optimum" upwind advection on a triangular mesh. ICASE Report No. 90-75.

[183] Roe, P. L. 1992. Beyond the Riemann problem, Part I. In *Algorithmic Trends in Computational Fluid Dynamics*, eds. M. Y. Hussaini, A. Kumar, and M. D. Salas, pp. 341–368. Springer-Verlag, New York.

[184] Rudinger, G. 1960. Passage of shock waves through ducts of variable cross section. *Phys. Fluids* 3:449–455.

[185] Runge, C. 1895. Ueber die numerische auflösung von differentialgleichungen. *Math. Ann.* 46:167–178.

[186] Rusanov, V. V. 1964. A three-dimensional supersonic gas flow past smooth blunt bodies. In *Proceedings of the 11th International Congress on Applied Mechanics*, Munich, Germany, pp. 774–778.

[187] Rusanov, V. V. 1976. A blunt body in a supersonic stream. *Annu. Rev. Fluid Mech.* 8:377–404.

[188] Salas, M. D. 1976. Shock fitting method for complicated two-dimensional supersonic flows. *AIAA J.* 14:583–588.

[189] Salas, M. D. 1980. Careful numerical study of flowfields about symmetric external conical corners. *AIAA J.* 18:646–651.

[190] Salas, M. D. 1980. Flow patterns near a conical sonic line. *AIAA J.* 18:227–234.

[191] Salas, M. D. 1982. Careful numerical study of flowfields about asymmetric external conical corners. *AIAA J.* 20:1661–1667.

[192] Salas, M. D. 1984. Local stability analysis for a planar shock wave. NASA TP 2387.

[193] Salas, M. D. 1991. Shock wave interaction with an abrupt area change. NASA TP 3113.

[194] Salas, M. D. 2007. The curious events leading to the theory of shock waves. *Shock Waves* 16:477–487.

[195] Salas, M. D. and Daywitt, J. 1979. Structure of the conical flowfield about external axial corners. *AIAA J.* 17:41–47.

[196] Salas, M. D. and Iollo, A. 1996. The entropy jump across an inviscid shock wave. *Theor. Comput. Fluid Dyn.* 8:365–375.

[197] Salas, M. D., Abarbanel, S., and Gottlieb, D. 1986. Multiple steady states for characteristic initial value problems. *Appl. Numer. Math.* 2:193–210.

[198] Schaaf, S. A. and Chambré, P. L. 1961. *Flow of Rarefied Gases.* Princeton University Press, Princeton, NJ.

[199] Schwartz, L. 1950, 1951. *Théorie des Distributions, Vols. I, II.* Hermann et Cie, Paris.

[200] Schwartz, L. 1954. Sur l'impossibilité de la multiplication des distributions. *C.R. Acad. Sci. Paris*, 239:847–848.

[201] Shapiro, A. H. 1954. *The Dynamics and Thermodynamics of Compressible Fluid Flow*, Vol. II. The Ronald Press Comp., New York.

[202] Sheehan, W., Kollerstrom, N., and Waff, C. B. 2004. The case of the pilfered planet. *Sci. Am.* (Dec). 291:92–99.

[203] Shewchuk, J. R. 1996. Triangle: Engineering a 2D quality mesh generator and Delaunay triangulator. *Lect. Notes Comput. Sci.* 1148:203–222.

[204] Shifrin, E. G. and Belotserkovskii, O. M. 1994. *Transonic Vortical Gas Flows*. John Wiley & Sons, New York.

[205] Sidilkover, D. 1989. Numerical solution to steady-state problems with discontinuities. PhD thesis, Weizmann Institute of Science, Rehovot, Israel.

[206] Sobolev, S. L. 1936. Méthode nouvelle á résoudre le probléme de Cauchy pour les équations linéaires hyperboliques normales. *Math. Sbornik* 43:39–72.

[207] Sonneveld, P. and van Leer, B. 1985. A minimax problem along the imaginary axis. *Nieuw Archief voor Wiskunde* 3:19–22.

[208] Spekreijse, S. P. 1988. Multigrid solution of the steady Euler equations. PhD thesis, Centre for Mathematics and Computer Science, Amsterdam, the Netherlands.

[209] Squire, L. C. 1979. Measured pressure distribution and shock shapes on a 'Butler' wing. Department of Engineering, Cambridge University, Report Aero/TR9.

[210] Steger, J. L. 1992. A viewpoint on discretization schemes for applied aerodynamic algorithms for complex configurations. In *Algorithmic Trends in Computational Fluid Dynamics*, eds. M. Y. Hussaini, A. Kumar, and M. D. Salas, pp. 3–11. Springer-Verlag, New York.

[211] Steger, J. L. and Warming, R. F. 1981. Flux vector splitting of the inviscid gasdynamic equations for application to finite-difference methods. *JCP* 40:263–293.

[212] Stokes, G. G. 1848. On a difficulty in the theory of sound. *Philos. Mag.*, 33(III):349–356. Original plus revisions reproduced in: [109].

[213] Stokes, G. G. 1864. On the discontinuity of arbitrary constants which appear in divergent developments. *Trans. Camb. Philos. Soc.* 10:106–128.

[214] Stokes, G. G. 1883. *Mathematical and Physical Papers*. Cambridge University Press, Cambridge, U.K.

[215] Struijs, R., Deconinck, H., and Roe, P. L. 1991. Fluctuation splitting schemes for the 2D Euler equations. *VKI LS 1991–01, Computational Fluid Dynamics*. von Karman Institute, Belgium.

[216] Sweby, P. K. 1984. High resolution schemes using flux limiters for hyperbolic conservation laws. *SIAM J. Numer. Anal.* 21:995–1011.

[217] Taton, R. 1950. Un texte inédit de Monge: Réflexions sur les équations aux différences partielles. *Osiris* 9:44–61.

[218] Taylor, G. I. and Maccoll, W. J. 1933. The air pressure on a cone moving at high speed. *Proc. R. Soc. Lond. A* 139:278–311.

[219] Theodorsen, T. and Garrick, I. E. 1933. General potential theory of arbitrary wing sections. NACA Report 452.

[220] Toro, E. F. 1997. *Riemann Solvers and Numerical Methods for Fluid Dynamics: A Practical Introduction*. Springer-Verlag, Berlin, Germany.

[221] Townsend, J. C. 1979. Surface pressure data on a series of analytic forebodies at mach numbers from 1.70 to 4.50 and combined angles of attack and sideslip. NASA TM 80062.

[222] Truesdell, C. 1953. Notes on the history of the general equations of hydrodynamic. *Am. Math. Mon.* 60:445–458.

[223] Truesdell, C. 1954. Rational fluid mechanics, 1687–1765, editor's introduction to Euleri Opera II 12. Zürich, Füssli, VII–CXXV.

[224] Truesdell, C. 1960. The rational mechanics of flexible or elastic bodies, 1638–1788, Euler, Opera Omnia, Series 2, 11, part 2:1–435.

[225] Truesdell, C. 1980. *The Tragicomical History of Thermodynamics, 1822–1854*. Springer-Verlag, New York.

[226] Truesdell, C. 1984. *An Idiot's Fugitive Essays on Science*, 2nd Part. Springer-Verlag, New York.

[227] van Albada, G. D., van Leer, B., and Roberts, W. W. 1982. A comparative study of computational methods in cosmic gas dynamics. *Astron. Astrophys.* 108:76–84.

[228] Van der Houwen, P. J. 1977. *Construction of Integration Formulas for Initial Value Problems*. North-Holland Publishing Comp., Amsterdam, the Netherlands.

[229] Van Dyke, M. D. and Gordon, H. D. 1959. Supersonic flow past a family of blunt axisymmetric bodies. NASA TR R-1.

[230] van Leer, B. 1973. Towards the ultimate conservative difference scheme. I. The quest for monotonicity. *Lect. Notes Phys.* 18:163–168.

[231] van Leer, B. 1984. On the relation between the upwind-differencing schemes of Godunov, Engquist-Osher and Roe. *J. Sci. Stat. Comput.* 5:1–20.

[232] van Leer, B. 1985. Upwind-difference method for aerodynamic problems governed by the Euler equations. Large-scale computations in fluid mechanics. *Lect. Appl. Math.* Prt. 2, 22:327–335. American Mathematical Society, Providence, RI.

[233] Vincenti, W. G. and Kruger, C. H. 1967. *Introduction to Physical Gas Dynamics*, 2nd printing. John Wiley & Sons, Inc., New York.

[234] von Kármán, T. 1923. Gastheoretische deuiung der Reynoldsschen kennzahl. *Z. Angew. Math. Mech.* 3:395–396.

[235] von Neumann, J. and Richtmyer, R. D. 1950. A method for the numerical calculation of hydrodynamic shocks. *J. Appl. Phys.* 21:232–237.

[236] Wesseling, P. 1991. *Principles of Computational Fluid Dynamics*. Springer-Verlag, Germany.

[237] Wiener, N. 1926. The operational calculus. *Math. Ann.* 95:557–584.

[238] Wilson, D. B. 1990. *The Correspondence between Sir George Gabriel Stokes and Sir William Thompson, Baron Kelvin of Largs, Vol. I, 1846–1869.* Cambridge University Press, Cambridge, U.K.

[239] Yamamoto, O., Anderson, D. A., and Salas, M. D. 1984. Numerical calculations of complex mach reflection. AIAA Paper No. 1984-1679.

[240] Yuschkevitsh, A. P. 1976. The concept of function up to the middle of the 19th century. *Arch. Hist. Exact Sci.* 16:37–85.

[241] Zannetti, L. and Colasurdo, G. 1981. Unsteady compressible flow: A computational method consistent with the physical phenomena. *AIAA J.* 19:852–856.

[242] Zannetti, L. and Moretti, G. 1981. Numerical experiments on leading edge flow field. AIAA Paper No. 1981-1011.

[243] Zemplén, G. 1905. Sur l'impossibilité des ondes de choc negatives dans les gaz. *CR Acad. Sci. Paris* 141:710–712.

[244] Zeytounian, R. Kh. 2002. *Theory and Applications of Nonviscous Fluid Flows*. Springer-Verlag, Berlin, Germany.

[245] Zhong, X-C. and Moretti, G. 1982. Comparison of different integration schemes based on the concept of characteristics as applied to the ablated blunt body problem. *Comput. Fluids* 10:277–294.

[246] Ziff, R. M., Merajver, S. D., and Stell, G. 1981. Approach to equilibrium of a Boltzmann-equation solution. *Phys. Rev. Lett.* 47:1493–1496.

# Index

## F

## G



## I



## J



## K



## L

## O



## P

## Q



## R



## S

## V



## W

Printed and bound by CPI Group (UK) Ltd, Croydon, CR0 4YY
21/10/2024
01777083-0015